NEW TECHNIQUES IN INTERVENTIONAL MUSCULOSKELETAL RADIOLOGY

NEW TECHNIQUES IN INTERVENTIONAL MUSCULOSKELETAL RADIOLOGY

Edited by

Mark E. Schweitzer

New York University School of Medicine

New York, New York, USA

Jean-Denis Laredo

Lariboisière Hôpital

Denis-Diderot Paris VII University

Paris, France

informa
healthcare

New York London

Informa Healthcare USA, Inc.
52 Vanderbilt Avenue
New York, NY 10017

Printed and bound by CPI Group (UK) Ltd, Croydon, CR0 4YY

Transferred to Digital Print 2012

International Standard Book Number-10: 0-8247-5445-X (Hardcover)
International Standard Book Number-13: 978-0-8247-5445-7 (Hardcover)

Library of Congress Cataloging-in-Publication Data

New techniques in interventional musculoskeletal radiology / edited by Mark E. Schweitzer, Jean-Denis Laredo.
p. ; cm.
Includes bibliographical references and index.
ISBN-13: 978-0-8247-5445-7 (hb : alk. paper)
ISBN-10: 0-8247-5445-X (hb : alk. paper) 1. Musculoskeletal system--Imaging. 2. Musculoskeletal system--Radiography. I. Schweitzer, Mark E., MD. II. Laredo, Jean-Denis, 1951-
[DNLM: 1. Musculoskeletal Diseases--radiography. 2. Musculoskeletal Diseases--therapy. 3. Musculoskeletal System--radiography. 4. Radiography, Interventional--methods. WE 140 N532 2007]

RC925.7.N49 2007
616.7'07572--dc22 2007021386

Visit the Informa Web site at
www.informa.com

and the Informa Healthcare Web site at
www.informahealthcare.com

Preface

While thinking about this book, several thoughts came to mind. The first concerns the importance of procedures in musculoskeletal radiology. Not superficially related to this, but nonetheless important in understanding the relevance of procedures, is politics. Allow me to explain.

Several authorities have commented that radiology has many more subspecialty societies than almost any other specialty. They have suggested that this dilutes the political interests of radiology and ultimately may be the seed of its demise and the dissemination of the field into the various clinical specialties that use its services.

On the other hand, the vibrancy of these subspecialties confirms the complexity of radiology and the importance of radiology to the clinical disciplines each of us interacts with every day. This interaction is what enhances the vigor of radiology and, for the authors of this book, the vigor of musculoskeletal radiology. Being in tune to your customers, while understanding imaging technology and three dimensional anatomy, allows us to be intellectually agile.

This intellectual agility of musculoskeletal radiology has led to the rapid evolution of this field within the last decade to decade and a half. We, the editors, consider ourselves fortunate to be on the cusp of this froth and to see the rapid changes that have occurred in musculoskeletal radiology over a relatively short period of time. We have rapidly transitioned from a subspecialty that is predominantly filmed to one that is more dependent, and in some ways codependent, on advanced imaging.

Interestingly, in this rapid evolution there has been no single disruptive technology. All that we've done is build on what we have done in the past and our understanding of what new technologies will allow us to do. This brings us to interventional procedures. There is a confluence here: three dimensional imaging and advances in percutaneous and minimally invasive tools combined with a better understanding of the natural history of disease and medical interventions the growth of MR and ultrasound has allowed us to obtain.

As a subspecialty we have progressed in less than two decades from a field where we were mostly limited to arthrography for our musculoskeletal procedures to today, where interventional radiologic techniques and procedures form the largest component of our daily work. This is the focus of this book.

This book is an attempt, hopefully successfully, to put together many experts in different specific procedures within radiology and to allow them to describe, in detail, the when, the where, the why, and the how of these procedures. We have done this in a detailed way for a sophisticated audience. We have also tried, in the beginning of the book, to add some background information about procedures that is not readily available: the histologic evaluation of tumors, the pharmacokinetics of local anesthetics, as well as several others. Each chapter following this is devoted to a specific procedure, which is covered in detail. We have selected not only the procedures that are most commonly used, but also procedures that are less commonly used, as those procedures may be more frequently used in the future.

Each of the chapter authors has done an outstanding job in bringing together this material. The politics of whether the growth of subspecialties within radiology is good or not aside, procedures are, to a large degree, where musculoskeletal radiology is heading. Because of the dependence upon imaging both for localization as well as assessment of therapeutic outcomes, we are confident that this material will remain within the confines of radiology, and it is our hope and belief that, in the future, they will be an even more important part of the subspecialty of musculoskeletal radiology.

We are reminded of a quote by a friend of ours—"I did not become a radiologist to be a doctor"—which to some would be the way bone radiology was practiced 20 years ago. Now the specialty has turned 180 degrees towards patient care. Leaving the humor of that quote aside, this is what many of us now do. We hope that our readers enjoy this book and are able to treat patients better because of it. That would make the labor that all these chapter authors have undertaken so well all the more worthwhile.

Mark E. Schweitzer
Jean-Denis Laredo

Contents

Contributors

Gerhard Adam Department of Diagnostic and Interventional Radiology, University Hospital Hamburg Eppendorf, Hamburg, Germany

Ronald S. Adler Division of Ultrasound and Body Imaging, Hospital for Special Surgery, Department of Radiology and Imaging, Weill Medical College of Cornell University, New York, New York, U.S.A.

Francesco Aparisi University Hospital La Fe and Nueve de Octubre Valencia, Valencia, Spain

Nathalie Azoulay Department of Bone and Joint Radiology, Lariboisière Hôpital, Assistance Publique–Hopîtaux de Paris, Paris, France

Laurence Bellaïche Department of Bone and Joint Radiology, Lariboisière Hôpital, Assistance Publique–Hopîtaux de Paris, Paris, France

Arno Bücker Department of Diagnostic Radiology, University of Technology, Aachen, Germany

Xavier Buy Department of Radiology B, University Hospital of Strasbourg, Strasbourg, France

John A. Carrino Department of Radiology and Radiological Science, Johns Hopkins University School of Medicine, Baltimore, Maryland, U.S.A.

Luc Ceugnart Service de Radiologie, Centre Oscar Lambret, Lille, France

Pierre Champsaur Department of Radiology, La Timone Hospital, Marseille, France

René Chapot Department of Neuroradiology, Krupp Krankenhaus, Essen, Germany

Hélène Chiavassa Department of Radiology, Purpan University Hospital, Place du Docteur Baylac, Toulouse, France

James J. Choi Department of Radiology, University of Wisconsin, Madison, Wisconsin, U.S.A.

Anne Cotten Service de Radiologie Ostéoarticulaire, Hôpital R. Salengro, Lille, France

Kirkland W. Davis Department of Radiology, University of Wisconsin, Madison, Wisconsin, U.S.A.

Xavier Demondion Service de Radiologie Ostéoarticulaire, Hôpital R. Salengro, Lille, France

Damian E. Dupuy The Warren Alpert Medical School of Brown University and Rhode Island Hospital, Providence, Rhode Island, U.S.A.

Robert G. Dussault Musculoskeletal Radiology, Laënnec Radiology, Montreal, and Reso-Concorde MRI, Laval, Quebec, Canada

Odile Enjolras Department of Neuroradiology, Lariboisière Hôpital, Assitance Publique–Hôpitaux de Paris, Paris, France

Erik Estivalezes Department of Radiology, Purpan University Hospital, Place du Docteur Baylac, Toulouse, France

Denise Galy Department of Radiology, Purpan University Hospital, Place du Docteur Baylac, Toulouse, France

Afshin Gangi Department of Radiology B, University Hospital of Strasbourg, Strasbourg, France

Bruno Grignon Guilloz Department of Radiology, University Hospital of Nancy, Nancy, France

Rolf W. Günther Department of Diagnostic Radiology, University of Technology, Aachen, Germany

Stephane Guth Department of Radiology B, University Hospital of Strasbourg, Strasbourg, France

Bassam Hamzé Department of Bone and Joint Radiology, Lariboisière Hôpital, Assistance Publique–Hopîtaux de Paris, Paris, France

Jean Pierre Imbert Department of Radiology B, University Hospital of Strasbourg, Strasbourg, France

Phoebe A. Kaplan Musculoskeletal Radiology, Laënnec Radiology, Montreal, and Reso-Concorde MRI, Laval, Quebec, Canada

David Karasick Department of Radiology, Thomas Jefferson University Hospital, Philadelphia, Pennsylvania, U.S.A.

Philip Lander Department of Radiology, University of Alabama at Birmingham, Birmingham, Alabama, U.S.A.

Jean-Denis Laredo Department of Bone and Joint Radiology, Lariboisière Hôpital, Assistance Publique–Hopîtaux de Paris, Paris, France

William B. Morrison Department of Radiology, Thomas Jefferson University Hospital, Philadelphia, Pennsylvania, U.S.A.

Jörg Neuerburg Department of Diagnostic Radiology, University of Technology, Aachen, Germany

Brian J. O'Hara Department of Pathology, Anatomy, and Cell Biology, Thomas Jefferson University Hospital, Philadelphia, Pennsylvania, U.S.A.

Antoni J. Parellada Diagnostic Imaging, Inc., Philadelphia, Pennsylvania, U.S.A.

Caroline Parlier-Cuau Department of Bone and Joint Radiology, Lariboisière Hôpital, Assistance Publique–Hopîtaux de Paris, Paris, France

Jean-Jacques Railhac Department of Radiology, Purpan University Hospital, Place du Docteur Baylac, Toulouse, France

Daniel I. Rosenthal Division of Musculoskeletal Radiology, Department of Radiology, Massachusetts General Hospital, Boston, Massachusetts, U.S.A.

Leon Rybak Department of Radiology, New York University/Hospital for Joint Diseases, New York, New York, U.S.A.

Nicolas Sans Department of Radiology, Purpan University Hospital, Place du Docteur Baylac, Toulouse, France

Mark E. Schweitzer New York University School of Medicine, New York, New York, U.S.A.

Jeff S. Silber North Shore/LIJ Health System, New Hyde Park, New York, U.S.A.

Pascal Swider Department of Radiology, Purpan University Hospital, Place du Docteur Baylac, Toulouse, France

Anthony C. Toppins Section of Musculoskeletal Radiology, Baylor University Medical Center, Dallas, Texas, U.S.A.

Martin Torriani Division of Musculoskeletal Radiology, Department of Radiology, Massachusetts General Hospital, Boston, Massachusetts, U.S.A.

Alexander R. Vaccaro Department of Orthopedics-Spine Division, Rothman Institute, Thomas Jefferson University, Philadelphia, Pennsylvania, U.S.A.

Michel Wassef Department of Pathology, Lariboisière Hôpital, Assistance Publique–Hopîtaux de Paris, Paris, France

Dominik Weishaupt Institute of Diagnostic Radiology, University Hospital, Zurich, Switzerland

Lisa Wong Department of Radiology B, University Hospital of Strasbourg, Strasbourg, France

Marc Wybier Department of Bone and Joint Radiology, Lariboisière Hôpital, Assistance Publique–Hopîtaux de Paris, Paris, France

1 Conscious Sedation

Mark E. Schweitzer
New York University School of Medicine, New York, New York, U.S.A.

INTRODUCTION

Adequate sedation for percutaneous musculoskeletal procedures requires a combination of patient information and reassurance, local anesthesia, and conscious sedation (1–4).

LOCAL ANESTHESIA

Local anesthetics are divided into ester- and amide-type agents. Ester agents include procaine, chloroprocaine, cocaine, and tetracaine. Amide-type agents include lidocaine, bupivacaine, mepivacaine, and etidocaine. Ester agents are hydrolyzed by plasma pseudocholinesterase. Amides undergo transformation through aromatic hydroxylation in the liver (3).

The mechanism of action of local anesthetics is by diffusion of the base form of the agent across the nerve sheath and membrane. Binding of the agent to the nerve receptor site results in inhibition of Na^{2+} transfer. The fastest-acting local anesthetics include chloroprocaine, mepivacaine, lidocaine, and etidocaine. Etidocaine is the strongest, mepivacaine is the weakest, and lidocaine is intermediate in potency. Bupivacaine is an amide agent similar to lidocaine, but lasts up to three times longer and is up to four times more potent (4).

Allergies to the local anesthetics are rare and are usually specific to ester-type agents. These allergic patients can usually be given amide-type agents because cross-reactivity between local anesthetics is rare. A true history of allergy will include the development of hives, wheezing, cardiac arrest, or shock. Most patients who claim a local allergy often have really had an adverse reaction. Symptoms of palpitations and nervousness may be a response to an additive such as Paraben or epinephrine. Patients may be describing the sequelae of inadvertent intervascular injections (4).

Lidocaine is the most commonly used amide-type agent. The maximum subcutaneous dosage of lidocaine is 7 mg/kg or less than 500 mg in a healthy adult. Two percent lidocaine contains 20 mg/mL; therefore, the maximum dose in the 70-kg adult is 25 mL of 2%, or 50 mL of 1% lidocaine. As a general rule, 2% lidocaine is preferred except when the clinical question is infection, where 1% is utilized. The lidocaine effects usually last for approximately 0.5 to 1 hour, but may be prolonged in elderly patients or those with liver failure (5).

Local anesthetics work best for soft tissue analgesia, but do not work well for intramedullary processes. The pain in this latter situation is caused by focal, local increase in intramedullary pressure. This occurs because when a needle enters the medullary cavity, it displaces local tissues. To ease this situation, a large amount of local anesthetic is given to the periosteum by scraping the local needle against the surface of the bone, as well as intravenous (IV) analgesia and patient reassurance.

Local anesthesia should be given prior to making the dermatomy. Before creating a skin wheal, anesthesia should be given a little deeper because the creation of the wheal is in itself somewhat painful. To test the adequacy of the skin anesthesia, we probe with the tip of the scalpel, touching the skin's surface without causing bleeding.

When using local anesthetics in areas of wounds, particularly devitalized areas, the lower local tissue pH decreases the lidocaine efficacy. This should be taken into consideration when using local anesthesia for percutaneous biopsies of suspected osteomyelitis in devitalized patients (6).

One should also pay careful attention to the indication for biopsy and the potential bacteriostatic or bactericidal effect of local anesthetics. This is particularly true when the indication for a biopsy is potential infection. Although the dose causing a bactericidal effect is quite large, and even the doses required for a bacteriostatic effect are fairly large, when doing a procedure for possible infection, IV sedation should be maximized and the use of local anesthesia should be minimized. That is also why 1% lidocaine is preferred when biopsying potential infections (7).

Epinephrine may be added to local anesthetics to cause local tissue vasoconstriction and consequently limit the uptake of the local anesthetic. Epinephrine should not be used in a patient who has a history of unstable angina pectoris and cardiac arrhythmia, or when performing a peripheral procedure on the fingers or toes.

CONSCIOUS SEDATION

The definition of conscious sedation is a state of depressed mentation achieved with a combination of amnesic, anxiolytic, and analgesic medications (8). Patients should retain the ability to protect their airway and respond appropriately to physical and verbal stimuli. Conscious sedation should result in mood alterations, amnesia, enhanced patient cooperation, and elevated pain thresholds. Patients should, however, maintain stable vital signs and have a rapid recovery (9,10).

Conversely, deep sedation is the controlled state of depressed consciousness with partial or complete loss of protective reflexes. Specific manifestations of deep sedation include the loss of ability to respond voluntarily to verbal or physical commands. This unconsciousness and unresponsiveness are not objectives of conscious sedation. Deep sedation may predispose a patient to respiratory depression, decreased response to fourth and a hypoxic stimulus, and cardiovascular depression (11,12).

Before giving sedation, it is important to understand four effects of IV anesthesia medications. The first is pain or analgesia relief (13). The second is amnesia, with loss of memory of the procedure. The third is sedation and relaxation (1). Part and parcel of sedation is an anxiolytic effect (14,15). Fourth and potentially fifth, there is hypotonia with loss of muscle tone and cardiorespiratory depression. The goal of conscious sedation is to maximize the first four effects, without significantly depressing cardiopulmonary function (16,17).

True conscious sedation is best achieved by a combination of narcotics, usually opiates and sedatives/hypnotics. Opiates provide analgesia by blocking the nerve response to painful stimuli, and anesthesia by causing drowsiness. Because of these effects, the shortest-acting opiates are ideal agents for percutaneous musculoskeletal procedures (18,19).

Opiates

Opiates are a class of drugs derived from the papaver somniferum and are used primarily for analgesia. Natural opiates contain up to 20 different alkaloids. Opiates can be natural, modified, or completely synthetic (19).

Patients with renal failure may have a prolonged ventilatory depression effect due to morphine and other opiates, particularly those excreted by the liver. As an overall rule, there are limited cardiovascular effects of opioids on occasion, but they may cause a dose-dependent bradycardia. Note should be made of the synergistic effect of benzodiazepams and opiates. There have been over 80 deaths directly attributed to these two medications in the setting of conscious sedation.

The shorter half-life opiates include morphine (about two hours), sufentanil (two to three hours), and meperidine (three to four hours). Only morphine and meperidine can be given orally. Although the agents described above have similar half-lives, the onset of action for fentanyl is much shorter than for morphine, with effects seen as early as 30 seconds after IV administration. Although fentanyl has a longer total elimination half-life, the duration of action is shorter than morphine. Fentanyl's increased total elimination time is related to its fairly high lipid solubility.

Fentanyl has a duration of action of 0.5 to 1 hour and is more potent than morphine. It also has less negative cardiovascular and gastrointestinal effects than morphine. One hundred milligrams of fentanyl is approximately equivalent to 10 mg of morphine or 75 mg of demerol.

Fentanyl is administered intravenously and we begin with a dose of approximately 12.5 to 25 mg. Similar to other opiate-type anesthetics, fentanyl is metabolized by the liver and should be used with care in the elderly or in patients with hepatic deficiency. Fentanyl has some respiratory depressive effects and should be used with care in those patients with borderline pulmonary status. Minor respiratory compromise can be treated by supplemental oxygen delivery. This is intended to keep blood oxygenation 95% or more. One should not continue to use this medication if it causes the blood oxygenation level to go to less than the low 90s even with oxygen supplementation.

Fentanyl is reversed with naloxone intravenously at a dose of 0.4 to 0.8 mg, with a response usually seen within five minutes. If there is no response within these five minutes, a supplemental dose of 2.4 mg can be given. Naloxone is an excellent reversal agent for fentanyl because this has a short half-life, similar to that of fentanyl.

Care should also be exercised in patients who have had prior long-term opiate use either as recreational pharmaceuticals or for pain relief. The possibility of a respiratory depressive effect in these patients at doses that are not adequate for anesthesia or analgesia is common. For these habituated patients, it is essential that the sedation be administered by an anesthesiologist, to carefully titrate medications and monitor vital functions. Also, in these patients, one should increase the dose of local anesthetic to compensate for the relative analgesic ineffectiveness of the opiates.

In a situation where there is renal or hepatic dysfunction, remifentanil may be utilized because unlike other opiates, it is broken down by nonspecific esterases in the blood rather than broken down in the liver or excreted by the kidney. This medication is 20 to 40 times more potent than fentanyl with an onset of action within one to three minutes and duration of action for only 5 to 10 minutes.

Benzodiazepams

Benzodiazepams are usually the best form of sedative/hypoxic agents with predominant anxiolytic and relaxation effects and no direct analgesic effects. These agents cause amnesia with loss of memory of the procedure and help to relieve patient anxiety (20,21).

Benzodiazepams cause anxiolysis, sedation, amnesia, suppression of seizure activity, and, if the dose is high enough, unconsciousness and respiratory depression. At its lowest concentrations, benzodiazepams produce only anxiolysis. High concentrations are needed for a sedative as well as an amnesic effect. The duration of retrograde amnesia is dose related. Higher doses of benzodiazepams produce unconsciousness.

Benzodiazepams affect the postsynaptic membranes of the central nervous system. At the GABA receptor complex, benzodiazepams enhance the binding of GABA to its receptor.

The three most often used benzodiazepams are diazepam, midazolam, and lorazepam. The duration of action ranges from up to 37 hours for diazepam, 10 to 20 hours for lorazepam, and one to four hours for midazolam. The ratio of potency of these is 1:5:3, respectively (22).

Midazolam (Versid) is the most commonly used benzodiazepam. Similar to amide-type local agents and opiates, Versid is metabolized by the liver. Versid is given intravenously at doses of 1 to 3 mg, under careful titration. Similar to opiates, side effects include respiratory depression, but occasional paradoxical stimulation can be seen. Most benzodiazepams can be reversed utilizing flumazenil. The recommended dose for reversal is 0.2 mg IV over 15 seconds. If no response is seen in less than a minute, a second dose of 0.2 mg can be injected, and this can be repeated up to four times at 60-second intervals (23).

Benzodiazepams, when combined with opiates, have a synergistic effect on respiratory depression. Benzodiazepams, however, have only a minimal effect on blood pressure, cardiac output, and systemic vascular resistance.

Other Medications

Occasionally promethazine hydrochloride (Phenergan) is given in conjunction with analgesics to potentiate the opiate effects and to combat the nausea that is associated with these medications.

Prophylactic antibiotics are given only rarely in specific situations, mainly for diskography.

PERIPROCEDURAL ISSUES

During the procedure, a running IV line should be available. IV fluids should be administered in all conscious sedation patients in order to maintain fluid and electrolyte balance (24). Isotonic solutions (isotonic saline or lactated Ringer's) are the preferred crystalloid IV maintenance fluid during the procedure.

Emergency medications including epinephrine and diphenhydramine should be available. Atropine should also be available in every procedure room.

Nothing by Mouth

A patient is given nothing by mouth (NPO) because of the risk of aspiration of gastric contents that may occur with as little as 25 mL. content in the stomach. Although all sedation has a risk of aspiration, avoidance of deep sedation can decrease this risk, or in specific circumstances, H2 blockers can be administered, which further decreases the risk of aspiration. Even though a patient is NPO, we allow clear liquids up to two or three hours before a scheduled procedure. Also, on the morning of the procedure, patients may take their typical oral medications but with only 2 or 3 oz. of water (25–27).

Ventilation

Oxygen by nasal cannulae will increase the blood oxygen by 3% to 4% for each liter delivered. Nasal cannulae rates greater than 4 L/min may lead to irritation, bleeding, and drying of mucous membranes. A simple oxygen mask allows delivery of oxygen at 40% to 60%. A nonbreathing mask allows the use of up to 100% oxygen when a tight seal is made. In a patient who has increased pulmonary secretions, increased humidified oxygen is indicate; however, humidified action requires high flow rates because of the high viscosity.

Patient Monitoring

Pulse oximetry is a noninvasive method for measuring blood oxygenation (27). It utilizes two wavelengths of light: red at 660 nm and infrared at 140 nm. One should be aware of the oxyhemoglobin disassociation curve when evaluating pulse oximetry readings. This curve has a steep drop-off below 90%, and therefore patients with a pulse oximetry less than 90% are significantly desaturated and possibly getting inadequate oxygen to their tissues. If additional oxygen is given, it will take approximately 20 seconds for the pulse oximetry readings to increase (28).

Special Circumstances

Specific considerations in older patients are based on the increased sensitivity to pharmacological agents in elderly patients, because the greater the proportion of body fat, the lesser the skeletal muscle mass, and the lesser the volume of intracellular fluid. There are negative cardiovascular, pulmonary, and neurological effects of aging seen even in patients who have normal physical examinations and/or laboratory results (29–31). Modifications should also be made in pediatric patients. These, however, are quite complex and should be made by a pediatric anesthesiologist (11,16,32,33).

Cardiac Monitoring

Cardiovascular monitoring is provided by intermittent blood pressure readings, continuous pulse readings, and a continuous electrical demonstration of the cardiac cycle (17). The electrocardiogram reflects the electrical activity of the heart in a graphic form, with the electrodes picking up electrode signals generated by the heart's conducting system (34). Normally electrodes are placed on four limbs and in six areas of the chest, however, standard bipolar leads used in procedural monitoring include one lead at the right arm, one at the left arm, and one at the left leg, or occasionally only one lead. The placement of these leads causes large alterations in the EKG readout and experience is necessary to understand these effects on the formal reading.

POST-PROCEDURE PAIN RELIEF

In the vast majority of cases, post-procedure pain relief is not necessary. In the small number of cases in which it is necessary, over-the-counter analgesics are suggested (35). Aspirin is to be avoided after percutaneous biopsies because of its potential anticoagulant effect. In the situations where there is inadequate pain relief with over-the-counter analgesics, consultation with the patient's primary physician is suggested. When that physician is not available, Tylenol with codeine (one tablet with 30–60 mg of codeline is suggested) every four hours up to three doses.

REFERENCES

1. Kempe AR, Gelazis R. Patient anxiety levels. An ambulatory surgery study. AORN J 1985; 41:390–391.
2. Balboa WT. For a better prepared patient. AANA J 1977; 45:621–624.
3. De Negri P, Ivani G, Tirri T, Favullo L, Nardelli A. New drugs, new techniques, new indications in pediatric regional anesthesia. Minerva Anestesiol 2002; 68:420–427.
4. Chen AH. Toxicity and allergy to local anesthesia. J Calif Dent Assoc 1998; 26:683–692.
5. Chan SK, Karmakar MK, Chui PT. Local anesthesia outside the operating room. Hong Kong Med J 2002; 8:106–113.
6. Apfelbaum JL, Walawaner CA, Grasela TH, et al. Eliminating intensive post-operative care in same day surgery patients using short-acting anesthetics. Anesthesiology 2002; 97:66–74.
7. Schweitzer ME, Deely DM, Beavis K, Gannon F. Does the use of lidocaine affect the culture of percutaneous bone biopsy specimens obtained to diagnose osteomyelitis? An in vitro and in vivo study. AJR 1995; 164:1201–1204.
8. Merrick PA, Ramsby GR. Conscious sedation for imaging and interventional studies. Radiol Manage 1994; 16:35–40.
9. Luginbuhl M, Remonda L. Interventional neuroradiology. Recent developments and anaesthesiologic aspects. Minerva Anestesiol 1999; 65:445–454.
10. Lang EV, Chen F, Fick LJ, Berbaum KS. Determinants of intravenous conscious sedation for arteriography. J Vasc Interv Radiol 1998; 9:407–412.
11. Malviya S, Voepel-Lewis T, Prochaska G, Tait AR. Prolonged recovery and delayed side effects of sedation for diagnostic imaging studies in children. Pediatrics 2000; 105:42.
12. Lundgren S. Sedation as an alternative to general anaesthesia. Acta Anaesthesiol Scand Suppl 1988; 88:21–23.
13. Hildebrandt NB, Host E, Mikkelsen AL. Pain experience during transvaginal aspiration of immature ococytes—conscious sedation is well accepted. Acta Obstet Gyn Scan 2001; 80:1043–1045.
14. Lang EV, Joyce JS, Spiegel D, Hamilton D, Lee KK. Self-hypnotic relaxation during interventional radiological procedures: effects on pain perception and intravenous drug use. Int J Clin Exp Hypn 1996; 44:106–119.
15. Medina LS, Racadio JM, Schwid HA. Computers in radiology. The sedation, analgesia, and contrast media computerized simulator: a new approach to train and evaluate radiologists' responses to critical incidents. Pediatr Radiol 2000; 30:299–305.
16. Leelataweedwud P, Vann WF. Adverse events and outcomes of conscious sedation for pediatric patients—study of an oral sedation regimen. J Am Dent Assoc 2001; 132: 1531–1539.
17. Jenner CA, Wilson JA. Continuous patient monitoring. Anaesthesia 2001; 56:591–592.
18. Breivik H. Opioids in cancer and chronic non-cancer pain therapy—indications and controversies. Acta Anaesth Scand 2001; 45:1059–1066.
19. Carter JG, Gergis SD. Anaesthesia for outpatient surgery. Middle East J Anaesthesiol 1980; 5: 539–544.
20. Watson DS. The use of the benzodiazepine antagonist flumazenil. AORN J 1993; 57: 497–502.
21. Mason KP, Zurakowski D, Karian VE, Connor L, Fontaine PJ, Burrows PE. Sedatives used in pediatric imaging: comparison of IV pentobarbital with IV pentobarbital with midazolam added. AJR 2001; 177:427–430.
22. Kinoshita K, Sakurai A, Mera K, et al. Midazolam for continuous sedation in Japanese critical care patients: Phase II study. J Int Med Res 2001; 29:342–348.
23. Deng XM, Xiao WJ, Luo MP, et al. The use of midazolam and small-dose ketamine for sedation and analgesia during local anesthesia. Anesth Analg 2001; 93:1174–1177.
24. Pandit UA, Pandit SK. Fasting before and after ambulatory surgery. J Perianesth Nurs 1997; 12:181–187.
25. Kreimeier U. Pathophysiology of fluid imbalance. Crit Care 2000; 4:3–7.
26. Kramer FM. Patient perceptions of the importance of maintaining preoperative NPO status. AANA J 2000; 68:321–328.

27. Maziarski FT, Simonson D. NPO status prior to surgery: two difference approaches. CRNA 1994; 5:59–62.
28. Bohnhorst B, Peter CS, Poets CF. Pulse oximeters' reliability in detecting hypoxemia and bradycardia: comparison between a conventional and two new generation oximeters. Crit Care Med 28:1565–1568.
29. Perry JF Jr. Surgery in the aged. Postgrad Med 1967; 41:414–417.
30. Janis KM. Outpatient anesthesia in a geriatric setting. Int Anesthesiol Clin 1982; 20: 89–97.
31. Crawford FJ. The elderly patient. AORN J 1985; 41:356–359.
32. Nguyen MT. Greenberg SB, Fitzhugh KR, et al. Pediatric imaging: sedation with an injection formulation modified for rectal administration. Radiology 2001; 221:760–762.
33. Chumpa A, Kaplan RL, Burns MM, et al. Nalmefene for elective reversal of procedural sedation in children. Am J Emerg Med 2001; 19:545–548.
34. Blike GT, Surgenor SD, Whalen K, et al. Specific elements of a new hemodynamics display improves the performance of anesthesiologists. J Clin Monitor Comp 2000; 16: 485–491.
35. Hylka SC. Discharge instructions for ambulatory surgery patients. Perioper Nurs Q 1986; 2:61–64.

2 Biochemistry of Local Anesthetics and Steroids Used for Selective Injections

Dominik Weishaupt
Institute of Diagnostic Radiology, University Hospital, Zurich, Switzerland

INTRODUCTION

Percutaneous selective injections including selective joint and soft-tissue injections as well as selective nerve root blocks are commonly performed by radiologists. Percutaneous selective injections are considered standard techniques in the diagnosis and treatment of various pain sources originating from the musculoskeletal system. Selective injections may be of great help in identifying the pain generator and can provide pain relief, which may maximize treatment and improve functional outcome. The rationale of these procedures is to deliver local anesthetics and/or steroids precisely within the joint cavity or to the immediate vicinity of either the nerve or the inflamed tissue. For this purpose, a needle is placed next to the presumed affected tissue target. Needle guidance may be achieved by fluoroscopy, computed tomography, or magnetic resonance imaging.

In daily practice, selective injections are most frequently performed in the management of pain affecting the lumbosacral spine. The spectrum of selective lumbosacral injections includes selective nerve root block, facet joint injection or facet joint block, and injections of the sacroiliac joint (1–7). Less frequently, patients are referred for selective injections of the cervical and thoracic spine as well as for injection into other joint cavities or injections into other nonarticular structures of the appendicular skeletal system.

This chapter is focused on biochemical and pharmacological aspects of local anesthetics and steroids that are used for selective injections. Special attention is drawn to physiological effects of these drugs to enable a rational use of these substances in clinical practice.

LOCAL ANESTHETICS

Local anesthetics are drugs that block nerve conduction when applied locally to nerve tissue in appropriate concentrations (8). They act on any part of the nervous system and on every type of nerve fiber. Thus, a local anesthetic in contact with the nerve trunk can cause both sensory and motor paralysis in the area innervated. The necessary practical advantage of the compounds that are labeled as local anesthetics is that their action is reversible, i.e., their use is followed by complete recovery of nerve function with no evidence of structural damage to nerve fiber cells (8,9).

The synthesis of procaine is considered a milestone in the development of local anesthetics. Procaine is a synthetic substitute for cocaine, and it was used as the first local anesthetic in patients. Since the synthesis of procaine in 1905 by Einhorn, several local anesthetics have been developed. The goals of these efforts were reduction of local irritation and tissue damage, minimization of systemic toxicity, shorter onset of action, and longer duration of action. Currently, lidocaine, mepivacaine, bupivacaine, and ropivacaine are most commonly used for selective infiltrations in the musculoskeletal system (1–7).

Chemistry

Figure 1 shows the chemical structure of frequently used local anesthetics for selective injections. All these drugs consist of a lipophilic group (aromatic ring) connected by an amide intermediate chain to a ionizable group (usually a tertiary amine) (8,9). The lipophilic domain determines the anesthetic potential and toxicity of the drug. Because local anesthetics can penetrate only as an uncharged (non-ionized) fraction into cells, the ionization potential of the amine group correlates

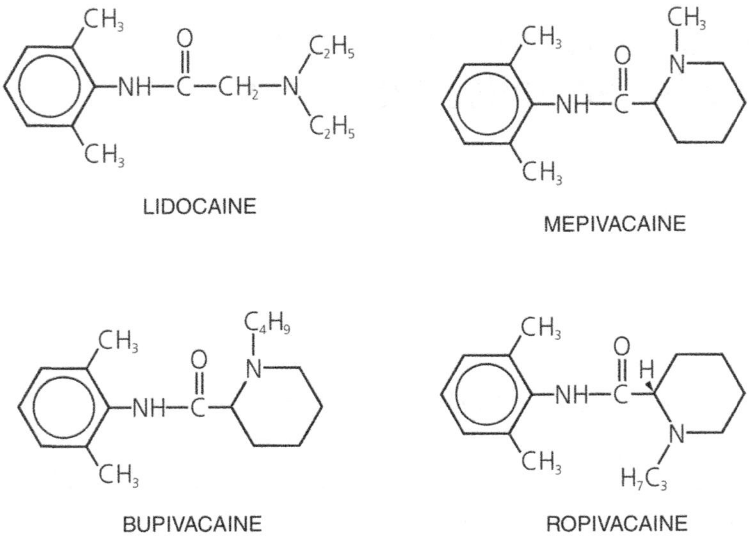

FIGURE 1 Chemical structure of selected local anesthetics used for selective injections.

with the speed of onset of the drug. The amine link determines another pharmacological property of these agents. The amine links are less susceptible to hydrolysis by plasma esterase than ester links (as in procaine), which result in a longer duration of action (8,9).

The clinical characteristics of the most commonly used local anesthetics for selective injections are displayed in Table 1. For selective injections, these local anesthetics are usually injected without epinephrine. The drugs vary with respect to potency, speed of onset, diffusibility, duration of action, and relative toxicity. Basic requirements for selective blocks for diagnostic purposes include a rapid speed of onset and short duration of action. For therapeutic use, long duration is more desirable. Robivacaine requires special attention for selective injections in musculoskeletal radiology. Robivacaine is a potent long-duration local anesthetic similar to bupivacaine. However, robivacaine is characterized by a less pronounced effect on motor nerve fibers.

Mechanism of Action

Local anesthetics prevent the generation and the conduction of a nerve impulse. Their primary site of action is the cell membrane where local anesthetics interact with one or more specific sites within Na^+ channels (10). Local anesthetics block conduction by decreasing or preventing the large transient increase in the permeability of excitable membranes to Na^+ that is normally produced by a slight depolarization of the membrane. This action of local anesthetics is due to their direct interaction with voltage-gated Na^+ channels. When progressively increasing the concentrations of a local anesthetic, the threshold for electrical excitability gradually increases, the rate of rise of the action potential progressively declines, impulse connection slows down, and finally, the ability to generate an action potential is abolished.

TABLE 1 Clinical Characteristics of Selected Local Anesthetics Used for Selective Injections

	Lidocaine	Mepivacaine	Bupivacaine	Ropivacaine
Characteristics	Low potency, short duration	Intermediate potency and duration	Potent, long duration	Potent, long duration
Latency (speed of onset)	Fast	Moderate	Fast	Fast
Penetrance (diffusibility)	Marked	Moderate	Moderate	Moderate
Duration	Moderate	Moderate	Long	Long

In addition to Na^+ channels, local anesthetics also can bind to other membrane proteins. In particular, the drugs can block K^+ channels. However, because the interaction of local anesthetics with K^+ channels requires higher concentrations of the drug, conduction blockade is not accompanied by any large or consistent change in the resting membrane potential due to the blocking of K^+ channels.

The sensitivity of nerve fibers to local anesthetics is variable, depending on the type of nerve fiber. In general, autonomic fibers, small unmyelinated C fibers (mediating pain sensations), and small myelinated Aδ fibers (mediating pain and temperature sensations) are blocked before the larger myelinated Aγ, Aβ, and Aα fibers (carrying postural, touch, pressure, and motor information) (11). The differential of block exhibited by fibers of varying sizes is of practical importance and explains why local anesthetics affect the sensory functions of most nerves in predictable order. Following the local administration of these drugs, the sensation of pain usually is the first modality to disappear; it is followed in turn by the sensations of cold, warmth, touch, deep pressure, and finally, by motor function, although there exists broad variation among individuals. Local anesthetics preferentially block small fibers because the distance over which such fibers can passively propagate an electrical impulse is shorter. Another important reason for the preferential blockade of sensory fibers follows directly from the state-dependent mechanism of action of local anesthetics. Blockade using these drugs is more marked at higher frequencies of depolarization and with longer depolarization. Sensory fibers, especially pain fibers, have a high firing rate and a relatively long action potential duration. Motor fibers fire at a slower rate and have a shorter action potential duration.

The effect of local anesthetics is dependent on the extracellular pH value. In an inflamed tissue, the local anesthetics may be less effective due to the lower extracellular pH. The decrease in extracellular pH results in a very low fraction of non-ionized local anesthetic available for diffusion into the cell (8).

Adverse Reactions

When local anesthetics for selective injections are administered in a correct fashion, adverse effects are infrequent. Very rarely, individuals are hypersensitive to local anesthetics. The reaction may manifest itself as allergic dermatitis or as a typical asthma attack (8,12).

When applied at excessively high concentrations, all local anesthetics can be toxic to the nerve tissue. However, this is not a real problem with the doses that are usually administered for selective injections.

In case of inadvertent intravascular injection, local anesthetics may have effects on both the central nervous system (CNS) and the cardiovascular system (8,9). Following absorption, local anesthetics may cause stimulation of the CNS resulting in restlessness and tremor that may proceed to clonic convulsions. Central stimulation is followed by depression; death is usually caused by respiratory failure. Following systemic administration, local anesthetics act on the cardiovascular system by decreasing the electric excitability, conduction rate, and force of contraction of the myocardium. Hypotension and asystolie may result.

STEROIDS

Shortly after the introduction of cortisone and hydrocortisone for systemic use in the management of rheumatoid arthritis in the late 1940s, Thorn tested the effect of hydrocortisone by injecting 10 mg of hydrocortisone into the knee joint of a patient with rheumatoid arthritis. After the injection, not only was there an improvement of the knee locally, but there was also a general improvement in the patient's condition; it was concluded that the improvement resulted from systemic absorption of the intra-articularly injected material.

This early experience showing the clinical benefit of local steroid administration has triggered the interest of many clinicians in taking advantage of the therapeutic effect of selective steroid injection for the management of inflammatory conditions in degenerative diseases as well as for treatment of true inflammatory conditions in the musculoskeletal system. Selective injection of steroids (often administered simultaneously with local anesthetics) is currently extensively used as an adjunct treatment for radicular pain, pain originating from the facet joints, sacroiliac joints, and injections of joints of the appendicular skeletal system. Steroids can provide a mid- to long-term pain relief if the cause of the pain is a reversible inflammatory process (13). Steroids have an anti-inflammatory effect (14–16) and, after approximately 24 hours, a direct analgesic effect (17).

Chemistry

The basic structure of steroids consists of a 21-carbon steroid molecule (Fig. 2). The major biologically active glucocorticoid is cortisol (hydrocortisone). Steroids with an 11-ketone group (cortisone and prednisone) are biologically inactive until reduced to a 11-hydroxial compound (hydrocortisone and prednisolone, respectively) (18). Therefore, preparations such as cortisone and prednisone, which require bioactivation in the liver, can only be used systemically. Hence, cortisone and prednisone are not useful for selective injections. For selective injections, biologically active hydroxicorticoid preparations such as methylprednisolone, triamcinolone, or other steriods must be used (Table 2) (3,5–7,19).

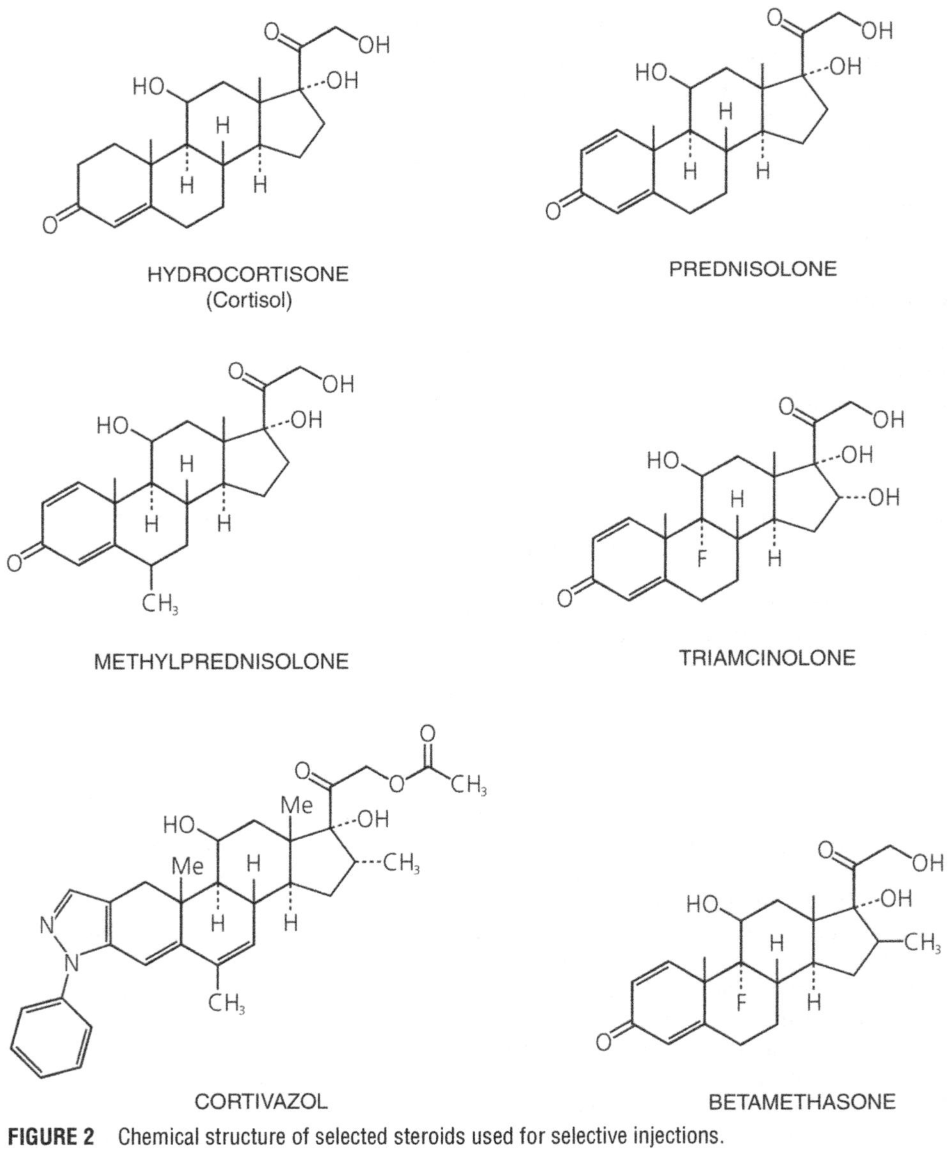

FIGURE 2 Chemical structure of selected steroids used for selective injections.

TABLE 2 Comparative Potencies and Duration of Action of Selected Steroids Used for Selective Injections

Drug	Anti-inflammatory potency (cortisol=1)	Mineralocorticoid potency (cortisol=1)	Duration of action	Nonproprietary name (trade name)[a]
Hydrocortisone	1	1	S	Hydrocortisone sodium phosphate (hydrocortone phosphate), hydrocortisone sodium succinate (Solucortef, A-Hydrocort) (7)
Methylpredni-solone	5	0.5	I	Methylprednisolone acetate (Depo-Medrol) (19,20)
Triamcinolone	5	0	I	Triamcinolone diacetate (Kenacort) (5)
Cortivazol			I	Cortivazol (Altim) (3,21)
Betamethasone	25	0	L	Betamethasone sodium phosphate (Celestone Phosphate), Betamethasone sodium phosphate and acetate (Celestone Soluspan) (2,6)

[a]Refers to reference.
Abbreviations: S, short (i.e., 8–12 hour biological half-life); I, intermediate (i.e., 12–36 hour biological half-life); L, long (i.e., 63–72 hour biological half-life).

Steroids are grouped according to their relative potency of Na^+ retention, effects on carbohydrate metabolism (i.e., hepatic deposition of glycogen and gluconeogenesis), and anti-inflammatory effects. In general, the potencies of steroids, as judged by their ability to sustain life in the adrenalectomized animal, closely parallel those determined for Na^+ retention. Potencies based on effects on glucose metabolism closely parallel those for anti-inflammatory effects. The effects on Na^+ retention and carbohydrate/anti-inflammatory effects are not closely related. Based on these differential potencies, the steroids are classically divided into mineralocorticoids and glucocorticoids.

For selective injections, steroid preparations of various potencies, as compared to the effect of prednisone, are available (Table 2). The crystalloid constitution of all these drugs is important for local action. Alteration in the molecular structure has yielded preparations with useful differences in potency, mineralocorticoid activity, and pharmacokinetic profiles. Although some investigators believe that the more potent preparations are also more efficacious, the clinical evidence for this is only moderate, and comparative studies of different preparations are not reported. Many clinicians have settled on a preparation that seems to have, in their opinion, good benefit and few side effects. Some prefer, for example, a preparation of a short-acting solution and long-acting suspension of betamethasone (Celestone Soluspane®). They believe that the solution works rapidly and prevents the possibility of postinjection flare. Betamethasone has the longest biological half-life and, therefore, the longest duration in suppressing inflammation.

Mixing the steroid suspension with a local anesthetic avoids the injection of high concentrations, which may produce soft-tissue atrophy or pericapsular calcification. However, one has to be concerned about the compatibility of the two preparations, in particular, when the steroid is mixed with local anesthetics containing preservatives. This may result in flocculation of the steroid.

Mechanism of Action

The anti-inflammatory and immunosuppressive actions of steroids are complex, and, to date, not fully understood. The effects of steroids are mediated through the binding of these steroids to cytoplasmatic glucocorticoid receptors in target cells. Glucocorticoid receptors are expressed on almost every type of human cell, including lymphocytes, monocytes, and neutrophils (22). Steroids have several effects on the pathogenesis of inflammation. In general, steroids suppress cellular immunity more than humoral immunity. Steroids inhibit the production of multiple cell factors (i.e., lipocortin, prostaglandin, cytokine, and adhesion molecules) that are critical in generating the inflammatory response (23). As a result, there is decreased release of vasoactive and chemoattractive factors, diminished secretion of lipolytic and proteolytic enzymes, and decreased extravasation of leukocytes to areas of inflammation. Steroids also exert profound effects on specific host immune responses, at least partly through their effects on cytokine production. Factors that are inhibited include interferon gamma, granulocyte/monocyte colony-stimulating factor, interleukines, and tumor necrosis factor α.

The anti-inflammatory properties of steroids may explain the efficacy of selective injections in pain relief among patients with radicular pain. Some authors have proposed that radicular pain is a result of "chemical radiculitis" induced by leakage of breakdown products such as phospholipase A2 from a degenerating nucleus pulposus (14). Other causes of nerve root irritation include mechanical compression by disk herniation or indirect effects such as vascular compression (15,16). The local application of steroids is considered to relieve reversible inflammatory changes or processes, such as vascular congestion related to mechanical obstruction (24). In addition, in experimental animal study, it was reported that the effect of an epidural steroid injection relates to the inhibition of phospholipase A2 (25).

Adverse Reactions

In general, few adverse reactions occur following local selective steroid injections (26,27). Infection, the most serious complication, is extremely rare. High concentrations of local steroids may result in atrophy of the overlaying skin or periarticular calcifications around the site of injection. When injected into the epidural space, corticosteroids cause no meningeal or neural damage. Inadvertent intrathecal injection of high doses of corticosteroids may result in arachnoiditis (28,29). However, based on an animal study, Latham et al. (30) concluded that small volumes of betamethasone injected intrathecally in humans are unlikely to cause arachnoiditis. A variable fraction of the steroid escapes from the site of the selective injection into the systemic circulation. Elevation of blood sugar, flushing, hormonal suppression, and inhibition of ovulation have been reported. Systemic effects of locally administered steroids are also infrequent.

REFERENCES

1. Chevrot A, Drape JL, Gedofroy D, et al. Image guided spinal steroid injections. Semin Musculoskel Radiol 1997; 1:221–240.
2. Dussault RA, Kaplan PA, Anderson MW. Fluoroscopy-guided sacroiliac joint injections. Radiology 2000; 214:273–277.
3. Gangi A, Dietermann JL, Mortazavi R, Pfleger D, Kauff C, Roy C. CT-guided interventional procedures for pain management in the lumbosacral spine. Radiographics 1998; 18:621–633.
4. North RB, Kidd DH, Zahurak M, Piantadosi S. Specificity of diagnostic nerve root blocks: a prospective, randomized study of sciatica due to lumbosacral spine disease. Pain 1996; 65:77–85.
5. Pfirrmann CWA, Oberholzer PA, Zanetti M, et al. Selective nerve root blocks for the treatment of sciatica: evaluation of injection site and effectiveness—a study with patients and cadavers. Radiology 2001; 221:704–711.
6. Riew D, Yin Y, Gilula L, et al. The effect of nerve-root injections on the need for operative treatment of lumbar radicular pain. J Bone Joint Surg 2000; 82-A:1589–1593.
7. Zennaro H, Dousset V, Viaud B, Allard M, Dehais J, Sénégas Caillé J. Periganglionic foraminal steroid injections performed under CT control. Am J Neuroradiol 1998; 19:349–352.
8. Catterall W, Mackie K. Local anesthetics. In: Catterall W, Mackie K, Hardman JG, et al., eds. The Pharmacological Basis of Therapeutics. New York: McGraw-Hill, 1996:331–347.
9. Tucker GT. Pharmacokinetics of local anesthetics. Br J Anaesth 1986; 58:717–731.
10. Butterworth JF, Strichartz GR. The molecular mechanisms by which local anesthetics produce impulse blockade. Anesthesiology 1990; 72:711–734.
11. Raymond SA, Gissen AJ. Mechanism of differential nerve block. In: Raymond SA, Gissen AJ, Strichartz GR, eds. Local Anesthetics. Handbook of Experimental Pharmacology. Berlin: Springer, 1987:95–164.
12. Covino BG. Pharmacology of local anesthetic agents. Br J Anaesth 1986; 58:701–716.
13. Derby R, Bodgduk N, Kine G. Precision percutaneous blocking procedures for localizing spinal pain. Part 2. The neuroaxial compartment. Pain Digest 1993; 3:175–188.
14. Franson RC, Saal JS, Sall JA. Human disc phospholipase A2 is inflammatory. Spine 1992; 17:S129–S132.

15. Nygaard OP, Mellgren SI, Osterud B. The inflammatory properties of contained and noncontained lumbar disc herniation. Spine 1997; 22:2484–2488.
16. Olmarker K, Blomquist J, Stromberg J, Nannmark U, Thomsen P, Rydevik B. Inflammatogenic properties of nucleus pulposus. Spine 1995; 20:665–669.
17. Johannsson A, Hao J, Sjolund B. Local corticosteroid application blocks transmission in the normal nocireceptive C-fibers. Acta Anaesthesiol Scan 1990; 34:335–338.
18. Schimmer BP, Parker KL. Adrenocorticotropic hormone: adrenoscortical steroids and their synthetic analogs; inhibitors of the synthesis and actions of adrenocortical hormones. In: Schimmer BP, Parker KL, Hardman JG, et al., eds. The Pharmocological Basis of Therapeutics. New York: McGraw-Hill, 1996:1459–1485.
19. Carette S, Leclaire R, Marcoux S, et al. Epidural corticosteroid injections for sciatica due to herniated nucleus pulposus. N Engl J Med 1997; 336:1634–1640.
20. Hove B, Gyldensted C. Cervical analgesic facet joint arthrography. Neuroradiology 1990; 32:456–459.
21. Parlier-Cuau CP, Wybier M, Nizard R, Champsaur P, Le Hir P, Laredo JD. Symptomatic lumbar facet joint synovial cyst: clinical assessment of facet joint steroid injection after 1 and 6 months and long-term follow-up in 30 patients. Radiology 1999; 210: 509–513.
22. Barnes PM, Adcock I. Antiinflammatory actions of steroids: molecular mechanism. Trends Pharmacol Sci 1993; 14:436–441.
23. Fauci AS, Dale DC, Balow JE. Glucocorticoid therapy: mechanisms of action and clinical considerations. Ann Intern Med 1976; 84:304–315.
24. Olmarker K, Byrod G, Cornefjord M, Nordborg C, Rydevik B. Effects of methylprednisolone on nucleus pulposus-induced nerve root injury. Spine 1994; 19:1803–1808.
25. Lee HM, Weinstein HM, Meller ST, Hayashi N, Spratt KF, Gebhard GF. The role of steroids and their effects on phospholipase A2: an animal model of radiculopathy. Spine 1998; 23:1191–1196.
26. Gatter RA. Arthrocentesis technique and intrasynovial therapy. In: Gatter RA, McCarty DJ, Koopman WJ, eds. Arthritis and Allied Conditions. Philadelphia: Lea & Febiger, 1993:711–720.
27. Nelson DA. Dangers from methylprednisolone acetate therapy by intraspinal injection. Arch Neurol 1988; 45:804–806.
28. Cicala RS, Turner R, Moran E, Henley R, Wong R, Evans J. Methylprednisolone acetate does not cause inflammatory changes in the epidural space. Anesthesiology 1990; 72:556–558.
29. Delaney TJ, Rowlingson JC, Carron H, Butler A. Epidural steroid effects on nerves and meninges. Anesth Analg 1980; 58:610–614.
30. Latham JM, Fraser RD, Moore RJ, Blumbergs PC, Bodgduk N. The pathologic effects of intrathecal betamethasone. Spine 1997; 22:1558–1562.

3 Preoperative Evaluation, Interpretation, and Laboratory Data

Antoni J. Parellada
Diagnostic Imaging, Inc., Philadelphia, Pennsylvania, U.S.A.

Mark E. Schweitzer
New York University School of Medicine, New York, New York, U.S.A.

PREOPERATIVE EVALUATION

Choice of Imaging Modality for Guidance

Although the basic technique for imaging-guided bone and soft-tissue biopsy has not undergone substantial changes, the proliferation of imaging modalities has added new emphasis on preoperative planning.

Often, patients are referred for a biopsy with imaging studies obtained at different institutions, with reports generated by radiologists not directly connected with the receiving hospital. Therefore, before even attempting to schedule the patient, all images and reports should be collected, and a decision made as to the appropriateness of the indication for biopsy. Sometimes, additional imaging can reveal a more accessible lesion than the one presented for biopsy, minimizing risks and procedure time.

End plate Modic changes can be dramatic (Fig. 1), and after the intravenous administration of contrast material, can exhibit florid enhancement, leading to a heightened concern for osteomyelitis. Although there is no single rule to exclude spine infection based on imaging techniques, there is a defined pattern of false-positive interpretations, including patients with no fever or changes in their white blood cell counts, erythrocyte sedimentation rate (ESR), or other markers of infection. A more critical analysis may forgo the need for a biopsy if the adjacent end plates are spared and there is no fluid-like signal in the disc. Careful evaluation of the facet joints may reveal marked degenerative changes, and flexion and extension views may be indicated in this context to diagnose underlying instability. Sometimes, scrutiny of gradient echo images, including scannograms, will exclude the possibility of infection if a vacuum phenomenon is identified. In addition, marked involvement of the facet joints implies neuropathic spine.

Once the review of available imaging studies is complete, and the lesion has been identified and deemed amenable to percutaneous biopsy, several technical decisions must be made, starting with the imaging modality that will be utilized. Operator preference is a major determinant factor, as well as scheduling flexibility and availability of different modalities within the department of radiology.

Fluoroscopy offers some well-defined advantages, particularly if a C-arm or biplane is available (Fig. 2). Patients can be positioned comfortably, and the approach to the lesion selected without restrictions related to the limited craniocaudal tilting of a computed tomography (CT)-scanner gantry. Real-time guidance is only matched by CT-fluoroscopy at the expense of a substantial increase in the amount of radiation delivered to the patient and operator. The working space between the patient and the image intensifier, a unique advantage, allows for comfort and ease during the procedure.

Some limitations are intrinsic to the projectional nature of fluoroscopy and its lack of low-contrast resolution. Hence, many lesions are inadequately visualized on conventional radiographs, making fluoroscopy inadequate as a guiding modality. Similarly, lesions close to a neurovascular bundle may pose an increased technical risk. Fluoroscopy also fails to provide accurate measurements of the distance between the tip of the needle and the lesion, and stroke margins cannot be calculated when using spring-loaded guns. Therefore, fluoroscopic guidance is most suitable for use with fine-needle aspiration and coring needles.

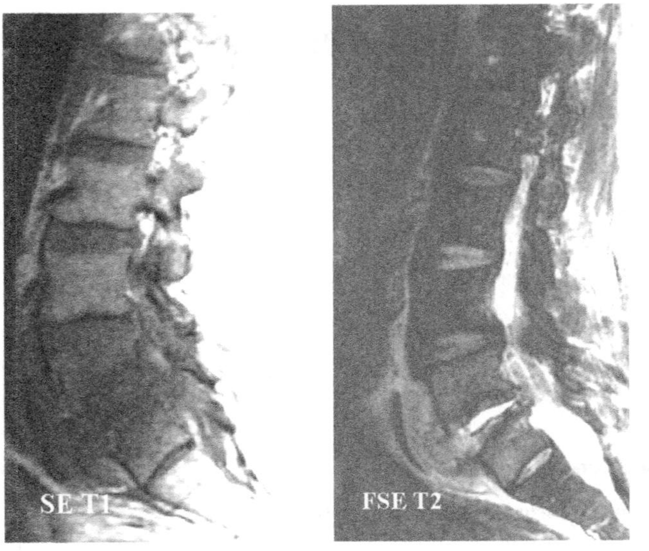

FIGURE 1 End plate changes can assume an extremely aggressive appearance on magnetic resonance. In this case, the biopsy showed reactive bone without acute inflammatory signs. The marked changes at L4–L5 were likely due to instability.

Ultimately, fluoroscopy is more operator dependant than CT scan, and requires familiarity with osseous landmarks. Systematic use of fluoroscopy for spine procedures such as diskography, vertebroplasty, or selective nerve root injections has a natural extension in its application to percutaneous spine biopsies.

CT offers improved low-contrast resolution, allowing increased lesion detectability. Likewise, vital structures adjacent to the lesion can be avoided. Therefore, CT scan is an excellent modality in the cervical spine because of its intrinsic anatomical difficulty. In the thoracic spine, CT scan will help minimize the chances of creating a pneumothorax. Small osseous lesions inconspicuous to conventional radiography and soft-tissue masses are prime indications for CT guidance (Fig. 3). The needle tip can be clearly identified due to the conspicuous streak artifact cast distally. This allows accurate estimates of the stroke margin, enabling the use of spring-loaded guns for soft tissue and destructive osseous lesions.

The advent of CT flouroscopy has reduced the length of procedures. However, CT continues to be limited by its initial conception as a cross-sectional technique. In this regard, it offers limited flexibility for angled approaches.

Ultrasound guidance presents an alternative approach to lesions with a soft-tissue component. Naturally, cystic lesions are exquisitely depicted with ultrasound. Ultrasound

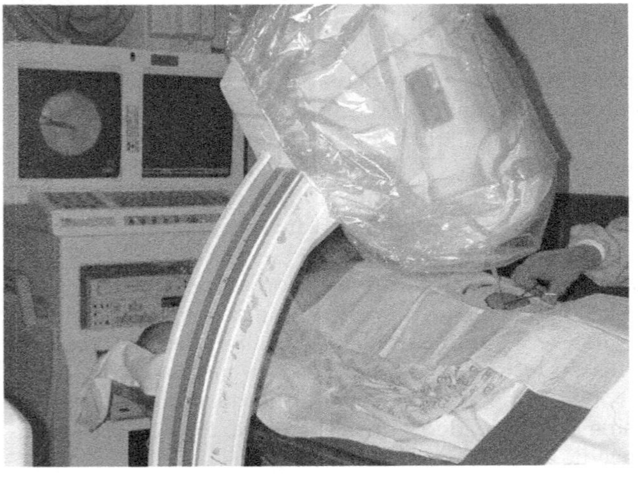

FIGURE 2 A C-arm provides flexibility in the approach to the lesion, and enables the operator to verify needle placement in multiple projections.

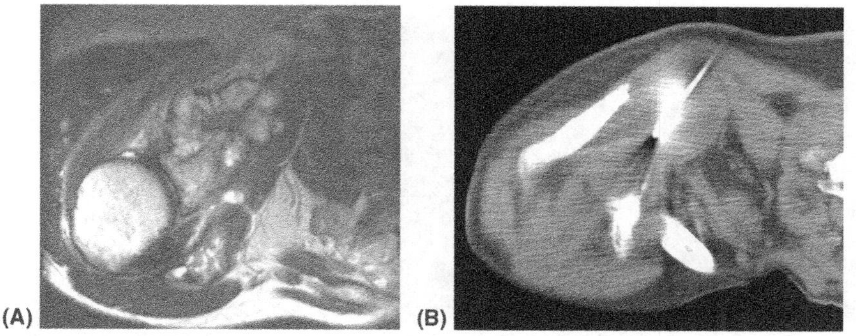

FIGURE 3 Prostate carcinoma metastatic to the scapula. Although primarily a bone lesion, there is a large soft-tissue mass displacing the infraspinatus (**A**), amenable to a soft-tissue biopsy with a spring-loaded gun (**B**).

flexibility in needle placement is virtually unlimited, and its real-time feedback as to the needle position is unmatched by any other modality. It is almost a platitude to mention its additional advantage of avoiding ionizing radiation. Naturally, it is not a suitable technique to target strictly intraosseous lesions, and as with any ultrasound procedure, operator expertise is fundamental to the success.

The introduction of nonferromagnetic magnetic resonance (MR)-compatible biopsy needles has opened the door to the use of MR as a guiding technique for needle placement, with great potential based on the superb soft-tissue contrast afforded by MR, which enables visualization of lesions otherwise inconspicuous or poorly defined under alternative imaging methods. Open and closed systems have been used for this purpose albeit this technique is of less use in the musculoskeletal system than the brain (1,2).

Preoperative Assessment of the Lesion

There are three different types of situations in musculoskeletal interventional radiology: soft-tissue lesions, mixed density osseous lesions, and intraosseous lytic lesions. Each one of these situations requires a slightly different approach and technique. Logan et al. devised a structured algorithm to workup lesions pertaining to each separate category (3). In their proposed scheme, there is an arbitrary cutoff diameter of 3 cm, after which the authors start using automated cutting needles. We believe with Jelinek et al. (4) that the size of the lesion should not be a decisive factor, provided the stroke margin is appropriate.

As in any other part of the anatomy, fine-needle aspiration biopsies may provide enough cellular material to diagnose a lesion. Malignancy can be demonstrated in many cases without need for histologic specimens. However, a more thorough evaluation of the lesion is obtained by simultaneously sampling histologic core specimens. Biopsy cores preserve the architecture of the tissue, raising their diagnostic yield. In addition, in the musculoskeletal system, purely sclerotic lesions make poor candidates for fine-needle sampling, and require a trephine bone biopsy needle.

Moulton et al. showed that the systematic addition of histologic analysis to fine-needle aspiration samples increases the accuracy of percutaneous biopsies by providing a more specific diagnosis. Cutting needles provide more consistent specimens, lessening the result dependency on the ability of the operator and the availability of an experienced cytopathologist. Cutting needles have the greatest impact on the diagnosis of benign lesions (5). Schweitzer et al. further reflect on the appropriateness of using automated cutting needles in lytic lesions of the musculoskeletal system. The study is based on 25 biopsies, with 24 final diagnoses obtained with the use of a gun, despite obtaining only two cores (6).

Soft-tissue lesions can be reliably sampled with a fine-needle aspiration [van Sonnenberg or Franseen (Cook, Bloomington, Indiana, U.S.A.)], or a spring-loaded core-biopsy device (Fig. 4A) (reusable gun or disposable single-use devices such as Achieve (Allegiance, Mc Gaw Park, Illinois, U.S.A.) (3,7). Spring-loaded core-biopsy needles have a predetermined stroke related to the size of the collecting area and the bevel in front of it. Ideally, the collecting area

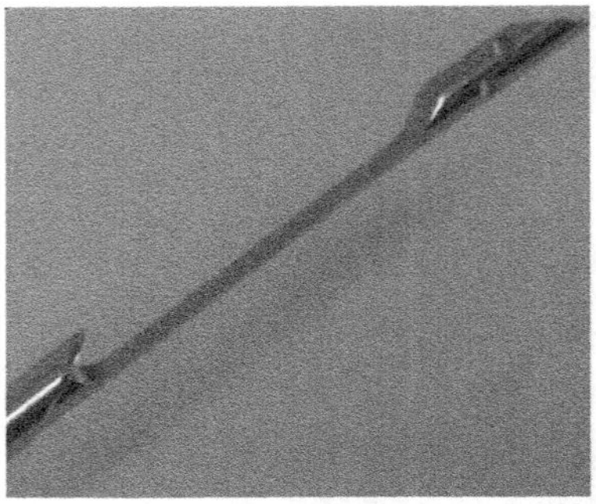

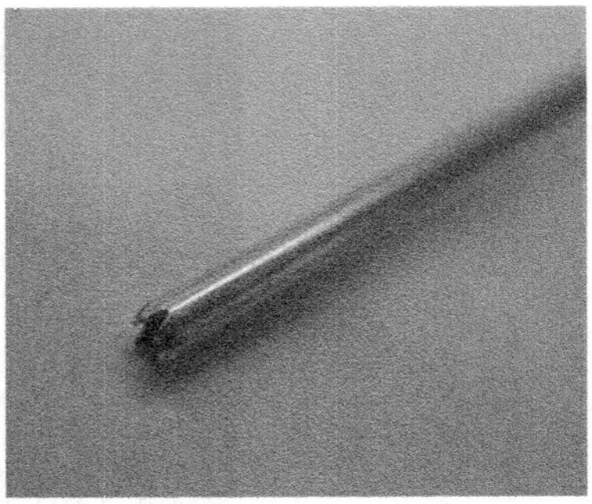

FIGURE 4 (**A**) Core-biopsy cutting needles are most suitable for sampling soft-tissue lesions. (**B**) Bone biopsy needles do not have a capturing collecting area, and when attempting to sample soft-tissue lesions, the yield is often limited to serosanguineous material.

should be deployed entirely within the lesion (Fig. 5). Therefore, the stroke margin should be larger than the needle stroke. Although needle strokes are variable, the most widely used needles have a throw of approximately 2.5 cm. When the presence of vital structures behind the lesion limit the stroke margin, adequate cores can be obtained with a needle design consisting of a stylet tightly fitted into the cutting cannula, providing a suction effect upon progressive withdrawal of the stylet. The operator manually exerts in-and-out excursions of the needle at each progressive interval withdrawal of the stylet. Eventually a core is suctioned within the cutting cannula [Vacu-Cut (Bard, Covington, Georgia, U.S.A.)] (Fig. 6). Alternatively, needles which retract rather than advance can be utilized, although these types of needles yield overall poorer specimens.

For limb-salvage surgery to be an option in cases of soft-tissue sarcomas, it is essential that the compartmental anatomy be respected during the percutaneous procedure. Violation of fascial planes at the time of the biopsy may alter the eventual surgical approach by limiting therapy to amputation. To minimize the chances of inadvertently violating compartments, Anderson et al. summarize several guidelines: always choose the shortest possible route to the lesion; use the same path that the surgeon will select for resection if the lesion is proven malignant; and avoid transgressing fascial planes or passing close to a neurovascular bundle (8). Also important, large soft-tissue lesions tend to necrose and produce cystic cavities, which are prone to lower the yield of the biopsy. Consequently, it is important to obtain a contrast-enhanced MR or CT before performing the biopsy to select a solid-appearing part of the mass.

Open-ended type needles with a serrated edge (trephine needles), on the other hand, afford dependable purchase on sclerotic bone lesions, avoiding the need for a capturing mechanism. Not only sclerotic neoplastic lesions, but also end plate biopsies in patients with

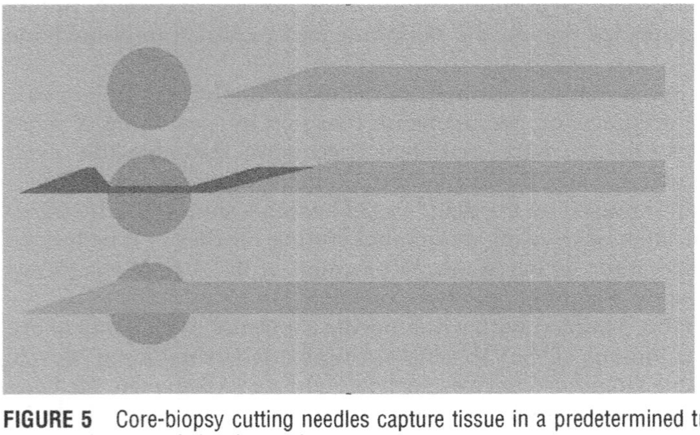

FIGURE 5 Core-biopsy cutting needles capture tissue in a predetermined trough in the inner stylet, over which the outer cutting cannula is advanced.

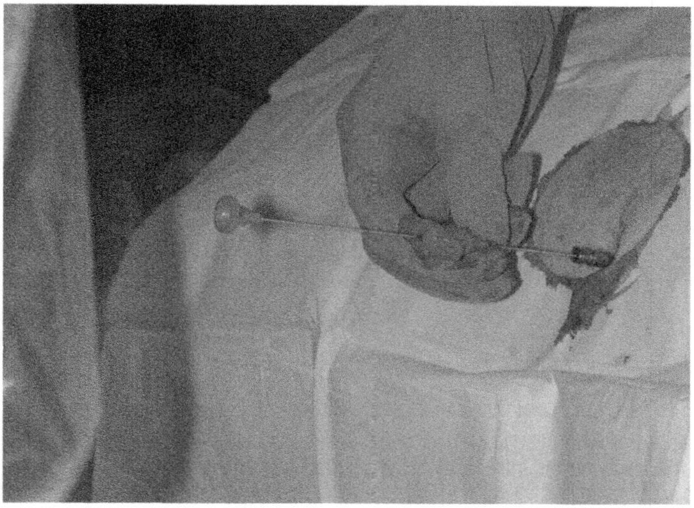

FIGURE 6 Vacuum-assisted core-biopsy needle (Vacu-Cut®, Bard, Covington, Georgia, U.S.A.).

suspected vertebral osteomyelitis, are good indications for the use of this type of needle. Although most possess a diamond-shaped or beveled tip, there are also models with a hollow introducer that allows advancement over a fine needle. Traditionally, bone biopsies are performed with Ackerman (Cook, Bloomington, Indiana, U.S.A.) (Fig. 43 and Fig. 7) or Jamshidi-type needles (MDTech Bone Marrow Biopsy Needle, Gainesville, Florida, U.S.A.). This latter type precludes the use of a coaxial system, practically limiting the biopsy to one single pass. Threaded-tip needles provide a more controlled and gradual introduction, usually without as much physical demand on the operator [Ostycut (Bard, Covington, Georgia, U.S.A.)]. However, the use of reusable handgrips facilitates the introduction of Ackerman-type needles.

A subset of lesions is basically composed of soft tissue, yet they lie entirely within the medullary cavity of the bone, sheltered by an intact cortex. These lesions are better accessed in two separate steps, (Fig. 8) as described by White et al. in a coaxial approach with a trephine needle used to penetrate the cortex, and the final tissue sampling performed with a biopsy gun (9). Notably, in their description, a Van Sonnenberg needle was used to guide the Ackerman to the cortex, and to negotiate the 18-g introducer for the biopsy gun beyond the created hole in the cortex. However, this Seldinger exchange over the Van Sonnenberg can be obviated with the use of eccentric drilling needles (10).

(A) 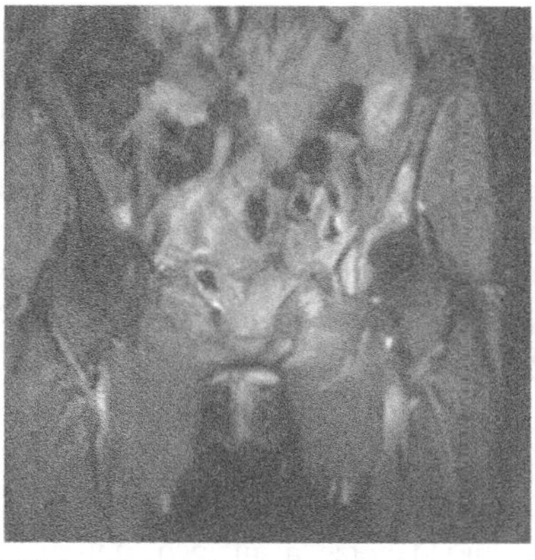 (B) 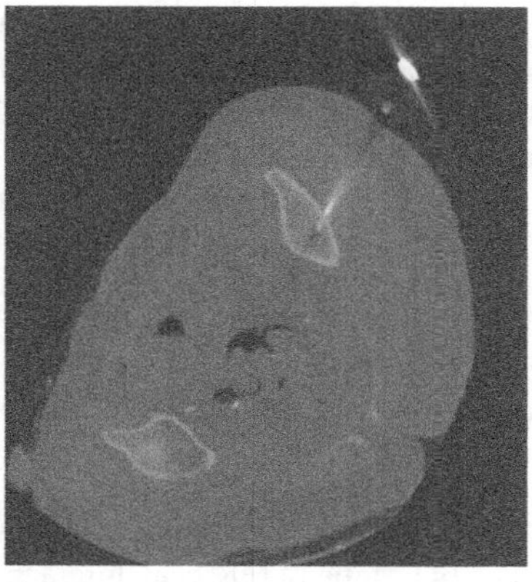

FIGURE 7 Metastatic urethral carcinoma to the left iliac bone producing minimal changes in the underlying bone (**A**), sampled with an Ackerman needle (**B**).

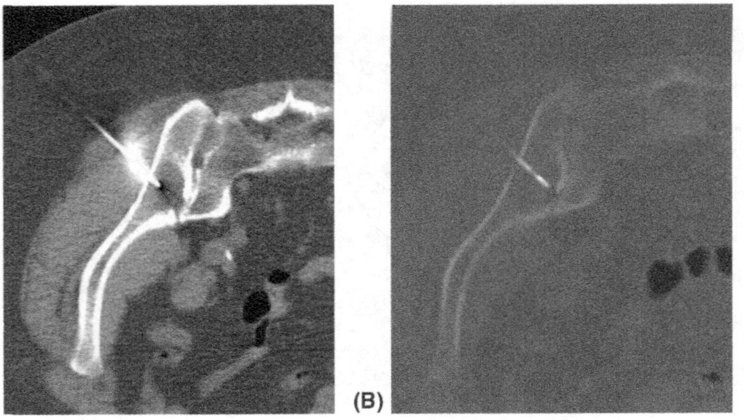

(A) (B)

FIGURE 8 A metastatic focus on hypernephroma with overlying intact cortex is accessed in two stages. Initially, the cortex is transgressed with an eccentric drill, allowing the introducer to be secured past the cortical bone (**A**). After removing the eccentric drill, several passes are obtained with a spring-loaded needle, shown with the collecting chamber open in (**B**).

Preoperative Evaluation of the Patient

We have already emphasized the importance of reviewing any available imaging studies to assess the indication for the biopsy, as well as to select an appropriate guiding technique and plan the route of access and the final target. Equally important, the patient must be screened for systemic conditions and medications that may interfere with the performance of the biopsy. Often, this preoperative contact can be established over the telephone. Any pertinent clinical information relating to past and present diseases is recorded. Patients are questioned about any history of allergic reactions. A list of current medications is also obtained, with special emphasis on anticoagulants and aspirin. If the patient is on aspirin, the biopsy is postponed for six days. In urgent cases, a period of three days may be sufficient to prevent a problem with bleeding. Alternatively, the use of fine needles, or simply the assumption of risk, may be adequate when there is a true medical urgency, and appropriate consent has been obtained from the patient. These situations tend to be limited to patients with suspected spine infection, or when there is an impending threat to the spinal cord, and emergency radiotherapy is considered without prior histologic confirmation of a malignant lesion. Coumadin can always be circumvented by administering fresh frozen plasma before the procedure. Heparin has a short half-life (30–60 minutes for unfractioned heparin), and simply waiting 10 half-lives will negate its effects, which usually allows for the discontinuation of heparin and the performance of the biopsy on the same day. We obtain a platelet count, prothrombin time and partial thromboplastin time, and INR before undertaking a deep-bone biopsy (11). Care should be utilized if the patient is on Lovenox as the anticoagulation may be greater than the usual laboratory tests suggest.

We perform the majority of our biopsies under conscious sedation with midazolam and fentanyl. These medications tend to depress respiration, and therefore, preoperative screening for asthma, sleep apnea, or other respiratory problems becomes important. Also, they may affect the patient's ability to remain supine throughout the procedure. Conscious sedation requires arrangements to be made for someone to drive the patient home after the procedure.

In our department, this initial contact is carried out by the musculoskeletal fellows, and a log book is kept with all the data gathered for each patient, including contact telephone numbers and addresses of the patient and referring physician, imaging studies available, and pertinent clinical and laboratory information obtained.

REFERENCES

1. Silverman SG, Collick BD, Figueira MR, et al. Interactive MR-guided biopsy in an open-configuration MR imaging system. Radiology 1995; 197(1):175–181.
2. Salomonowitz E. MR imaging-guided biopsy and therapeutic intervention in a closed-configuration magnet: single-center series of 361 punctures. Am J Roentgenol 2001; 177(1):159–163.

3. Logan PM, Connell DG, O'Connell JX, Munk PL, Janzen DL. Image-guided percutaneous biopsy of musculoskeletal tumors: an algorithm for selection of specific biopsy techniques. Am J Roentgenol 1996; 166(1):137–141.
4. Jelinek J, Buick M, Shmookler B. Image-guided percutaneous biopsies of musculo-skeletal lesions. Am J Roentgenol 1996; 167(2):532–533.
5. Moulton JS, Moore PT. Coaxial percutaneous biopsy technique with automated biopsy devices: value in improving accuracy and negative predictive value. Radiology 1993; 186(2):515–522.
6. Schweitzer ME, Deely DM. Percutaneous biopsy of osteolytic lesions: use of a biopsy gun. Radiology 1993; 189(2):615–616.
7. Kattapuram SV, Rosenthal DI. Percutaneous biopsy of skeletal lesions. Am J Roentgenol 1991; 157(5):935–942.
8. Anderson MW, Temple HT, Dussault RG, Kaplan PA. Compartmental anatomy: relevance to staging and biopsy of musculoskeletal tumors. Am J Roentgenol 1999; 173(6):1663–1671.
9. White LM, Schweitzer ME, Deely DM. Coaxial percutaneous needle biopsy of osteolytic lesions with intact cortical bone. Am J Roentgenol 1996; 166(1): 143–144.
10. Ahlstrom KH, Astrom KG. CT-guided bone biopsy performed by means of a coaxial biopsy system with an eccentric drill. Radiology 1993; 188(2):549–552.
11. Silverman SG, Mueller PR, Pfister RC. Hemostatic evaluation before abdominal interventions: an overview and proposal. Am J Roentgenol 1990; 154(2):233–238.

4 Percutaneous Biopsy of Musculoskeletal Lesions

Jean-Denis Laredo and Bassam Hamzé
Department of Bone and Joint Radiology, Lariboisière Hôpital, Assistance Publique–Hopîtaux de Paris, Paris, France

Pierre Champsaur
Department of Radiology, La Timone Hospital, Marseille, France

INTRODUCTION

A histopathological proof is often needed in musculoskeletal lesions to establish a definitive diagnosis, especially in cases of suspicion of tumor or infection. In such cases, percutaneous biopsy (PB) under radiologic control has become a routine procedure. The percutaneous approach is less invasive than surgery. Radiologic control allows needle guidance during lesion approach and also accurate needle placement.

EQUIPMENT AND INSTRUMENTATION

Radiological Guidance

Permanent radiologic control of the needle position is needed during PB procedures (1). This can be accomplished by using either fluoroscopic, computed tomography (CT) scan or ultrasonographic guidance, and in selective cases magnetic resonance imaging (MRI), depending on the biopsy site. A biopsy of most superficial lesions can usually be achieved using single-plane fluoroscopy. In most cases, PB of deep musculoskeletal lesions, such as vertebral biopsy, may be achieved with single-plane fluoroscopy (for radiologists experienced with this technique), a C-arm, biplane fluoroscopy or CT scan. CT scan is the method of choice for biopsy of soft-tissue components and small deep lesions close to vital structures, such as vertebral lesions involving the neural arch. Biopsy under ultrasound guidance is discussed in Chapters 5 and 6.

Biopsy Needles

Bone Trephine Needles

We will only review disposable bone trephine needles (BTN) available. There are two main types of BTN, which differ by their type of placement against bone through the soft tissue. BTN may be placed using a coaxial Seldinger technique or directly. Needles that are placed using a Seldinger technique are useful for deep bone biopsy, especially vertebral biopsy through a posterolateral route (Table 1). Seldinger technique allows coaxial placement of an external biopsy cannula and BTN in the same track as the thin anesthetic needle using a blunt guidewire. This reduces the risk of wounding the anatomical structures with the large BTN.

BTN placed directly are simpler to use, easier to handle, and more convenient for superficial bone biopsies or bone biopsy in anatomical areas where there is no vital structure such as bone biopsy of pelvic bones (Table 2).

Most BTN include an external cannula that remains in place, whereas the trephine needle is withdrawn with the sample. This allows multiple sampling without need for needle repositioning.

Some bone needles have a special cutting tip designed to penetrate hard cortical or sclerotic bone (Table 3). Manual cutting needles and semiautomatic or automatic biopsy guns with either a side notch or an end notch are used to sample osteolytic or soft-tissue lesions (Table 4). These needles may be used alone or introduced through an introducer for coaxial use (for example Tru-Guide Bard, 7–18cm, 12–20 G).

TABLE 1 Disposable Bone Trephine Needles for Direct Approach

Needle name	Manufacturer	Length (cm)	Diameter (outer needle) (G)
Osteosite M_1 (side bevel)	COOK, Bloomington, U.S.A.	10, 15	11
Osteosite M_2 (diamond bevel)	COOK, Bloomington, U.S.A.	10, 15	11
Manan	Manan, Northbrook, U.S.A.		
Lar 1000	Laurane Medical, Saint Arnoult, France	10, 15	11
Jamshidi	Baxter Healthcare, Valencia, U.S.A.		

Source: From Refs. 22–33.

TABLE 2 Disposable Bone Trephine Needles Placed with a Seldinger Technique

Needle name	Manufacturer	Length (cm)	Diameter (outer needle) (G)
Geremia	COOK, Bloomington, U.S.A.	15	16
Elson	COOK, Bloomington, U.S.A.	6.5, 14	12
Cook guided bone biopsy set	COOK, Brisbane, Australia	20	14
KyphX bone biopsy set	Kyphon Inc., Sunnyvale, U.S.A., Kyphon Europe, Zaventem, Belgium	18.5	10
Lar 2000	Laurane Medical, Saint Arnoult, France Cardinal Health, Maurepas, France	10, 15	11

Source: From Refs. 22–33.

TABLE 3 Disposable Bone Trephine Needles for Cortical and Sclerotic Bone

Needle name	Manufacturer	Length (cm)	Diameter (outer needle) (G)
Bonopty	Bard Inc., Covington, U.S.A.	12–16	15
Ostycut	Bard Inc., Covington, U.S.A.	5–15	14–17
Lar 4000	Laurane Medical, Saint Arnoult, France; Cardinal Health, Maurepas, France	10, 15	11

Source: From Refs. 22–33.

TABLE 4 Disposable Needles for Biopsy of Osteolyse or Soft-Tissue Lesions

Needle name	Manufacturer	Length (cm)	Diameter (outer needle) (G)
MAX CORE (side notch)	Bard Inc., Covington, U.S.A.	10–25	14–20
QUICK-CORE (side-notch)	COOK Inc., Bloomington, U.S.A.	6–20	14–20
ASAP-18 (side notch)	Medi-Tech, Boston Scientific, Watertown, U.S.A.		
Manan (side notch)	Manan Medical, Northbrook, U.S.A.		
Achieve (side notch)	Cardinal Health, U.S.A.	6–20	14–20
Sure-Cut (end notch)	Nycomed, Princeton, U.S.A.	7–23	15–22
Biopince (end notch)	Amedic, Sweden		18
Coaxial Temno (end notch)	Cardinal Health, U.S.A.	11–20	14–20

Source: From Refs. 22–33.

PREOPERATIVE ASSESSMENT (SEE ALSO CHAPTER 2)

Each decision is made in consultation with the referring physician. Prior to PB, the patient's file should be evaluated to determine whether noninvasive procedures might yield the desired diagnosis. Patient hemostasis must be checked in the days before the procedure. The most accessible skeletal lesion and the proper anatomic approach are chosen on the basis of all available imaging procedures. Bone scintigraphy and MRI may identify additional lesions more accessible to PB than the initial detected abnormality (2). In all cases, radiographs in two projections, CT and MRI of the lesion are required prior to biopsy. In our experience, however, CT scan is indispensable in the selection of an appropriate biopsy site, because biopsy of an osteolytic area of soft-tissue tumor usually provides better results than sampling of sclerotic bone. Areas of sclerotic bone are often less easily detected with an MRI. In musculoskeletal malignant tumors, gadolinium-enhanced MRI scan also helps delineate areas of osteolytic bone or soft-tissue mass, which show the most enhancement and are more likely to be high grade. On the contrary, cystic areas, which often correspond to necrotic tissue, should be avoided (3). Contrast-enhanced CT or MRI may be useful when a highly vascular lesion is suspected: frank hypervascularization may call for needle aspiration rather than trephine biopsy. If a primary bone tumor is thought to be a possible diagnosis, the site of skin incision and lesion approach should be decided in consultation with the orthopedic surgeon so as to not compromise surgical treatment, and the biopsy track is marked with carbon to be able to resect it at time of surgery.

Immediate Preparation

Skeletal biopsies are usually performed under local anesthesia or conscious sedation (see Chapter 1 on Conscious Sedation), with the exception of children, restless patients, especially drug-addicted patients, and patients with lesions that are expected to be very painful at biopsy, who are placed under general anesthesia or heavy sedation. In our experience, some infectious lesions such as postoperative disc space infection and osteomyelitis of the shaft of long bones are often very painful and may require deep analgesia or even general anesthesia. Hospitalization for at least 24 hours is required following a vertebral biopsy. In most other locations, the biopsy can be done on an outpatient basis. Percutaneous bone biopsies are performed under strict aseptic conditions, including the use of sterile drapes, gloves, and gowns.

TECHNIQUE OF BONE TREPHINE BIOPSY UNDER FLUOROSCOPIC GUIDANCE (4)

Lumbar Spine

Posterolateral Approach

The lumbar spine is approached through a posterolateral route either right- or left-sided depending on the location of the lesion (Fig. 1). The needle is inserted at 7–10 cm (close to 7 cm

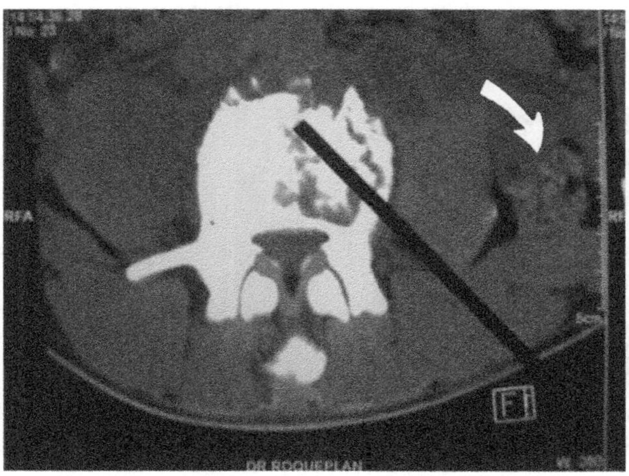

FIGURE 1 Posterolateral approach for lumbar spine percutaneous biopsy. Point of skin puncture is at 7 to 10 cm from the midline. The lumbar spine is approached at an angle of 40° to 60° with the sagittal plane. In this case, the left colon (*curved arrow*) is very close. This points out the usefulness of computed tomography scan prior to biopsy.

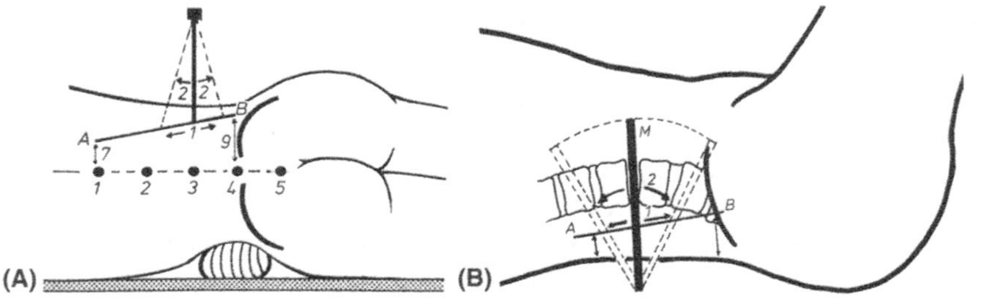

FIGURE 2 Approach to the lumbar spine. Lateral (**A**) and upper (**B**) views of the patient placed on their side. The point of skin puncture is at 7 to 9 cm from the midline (line *A–B*). Both level of needle insertion (*1*) and angle of cephalad or caudal approach (*2*) are determined using a metallic ruler (*M*) under fluoroscopic control.

for the upper lumbar spine and close to 9–10 cm for the lower lumbar spine) from the midline figured by the spinous processes, depending on the biopsy level and patient's build (Fig. 2). Angle of approach is at 40° to 60° with the sagittal plane (Fig. 1).

The patient is placed on his side on the X-ray table and must be made comfortable and stable. A radiolucent block is placed beneath the flank in order to correct a lateral deviation of the spine. Perfect lateral positioning of the patient is crucial to a correct approach and is carefully checked by fluoroscopic control at each step of the procedure. The gantry is tilted to profile the disc space. The exact level of skin puncture and the angle of cephalad or caudad approach are determined using a metallic ruler placed on the patient's side in order to simulate the approaching needle on the lateral fluoroscopic view. In disc biopsies, the approach must be as parallel as possible to the vertebral end plates (Figs. 3 and 4). Both disc and vertebral end plates must be sampled in cases where there is a suspicion of disc-space infection. On the other hand, vertebral bodies are preferably biopsied with a cephalad or caudad angulation approach to sample the entire vertebral body (Figs. 3 and 5). T12 and L1 must be approached through an ascending route (Fig. 6). Once the point of skin puncture and the optimal approach have been determined, the skin is prepared, and the superficial planes are anesthetized. The needle is inserted and advanced under fluoroscopic control toward the lumbar spine at an angle of 40° to 60° with the sagittal plane (depending on patient build and lesion site) and in a cephalad or caudad direction (as determined above). The nature of any obstacle encountered can be determined from the lateral view fluoroscopic screen (transverse processes at the level of the pedicles or facet joints at the level of the disc space or lower half of the vertebral body). In this event, the needle is withdrawn 2–3 cm and then advanced in a more sagittal direction. If needed, the patient may be turned to a prone-oblique position to profile the needle with the X-ray beam. This view allows clear assessment of the needle tip position in regard to the spine. In most cases, contact with the spine should be obtained when the needle reaches the posterior third of the vertebral body or disc space. When proper placement of the needle is verified on both anteroposterior and lateral views, the periosteum is carefully anesthetized, and the trephine needle is inserted in the track of the anesthesia needle.

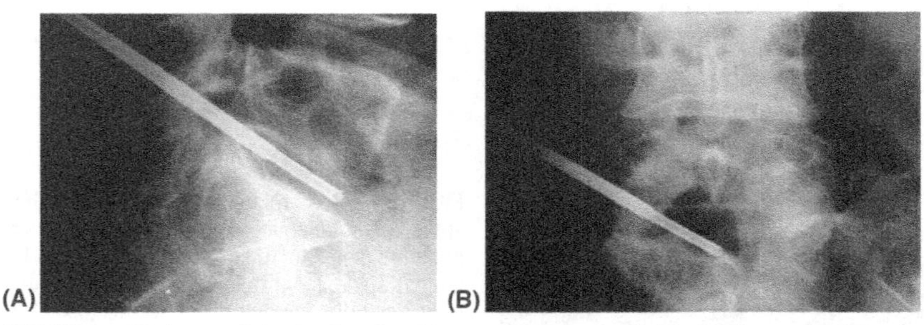

FIGURE 3 (**A**) Approach for lumbar disc biopsy should be parallel to the disc space (**B**) approach for lumbar vertebral body biopsy is oblique in order to avoid transverse processes and to sample the entire vertebral body.

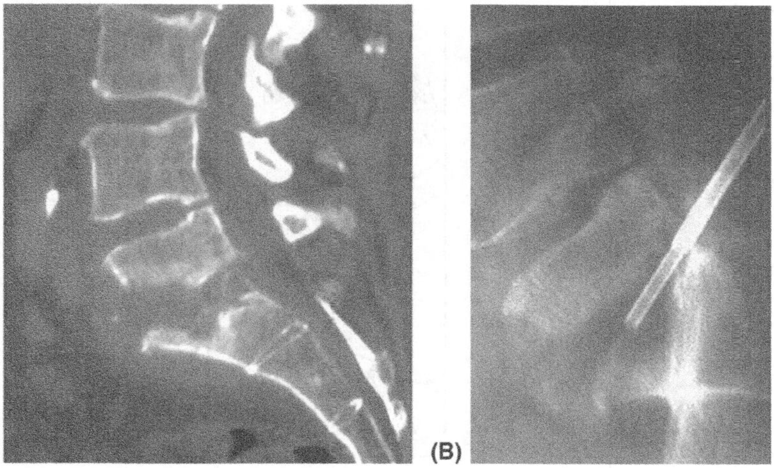

FIGURE 4 Spinal infection: postoperative contamination at L5–S1 interspace. (**A**) Computed tomography scan demonstrates a prevertebral abscess. (**B**) The appropriate level of puncture and caudal inclination of the approach are determined in order to orientate the biopsy needle parrallel to the disk space.

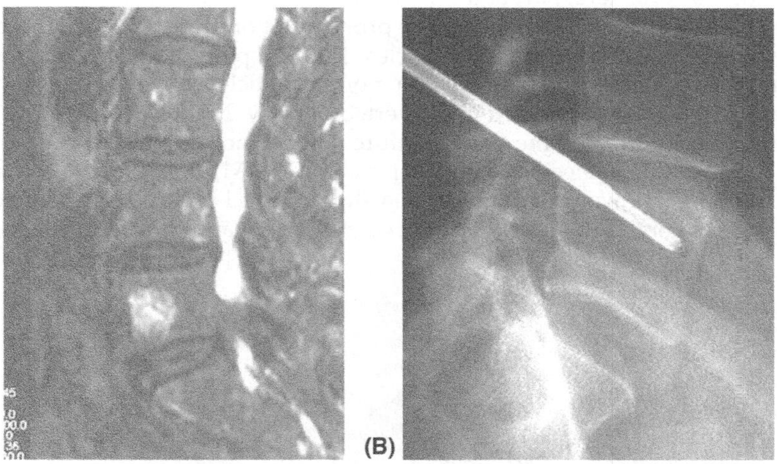

FIGURE 5 (**A**) Metastatic nodule of L5 with high-signal intensity on T2-weighted magnetic resonance image (**B**) oblique descending approach.

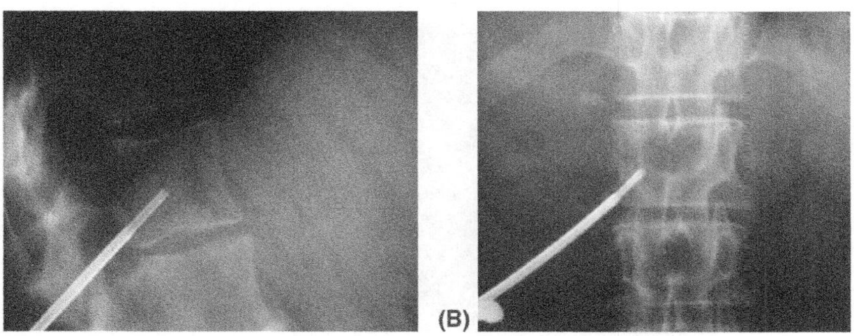

FIGURE 6 Biopsy of L1 using an ascending route in a patient with bone metastasis.

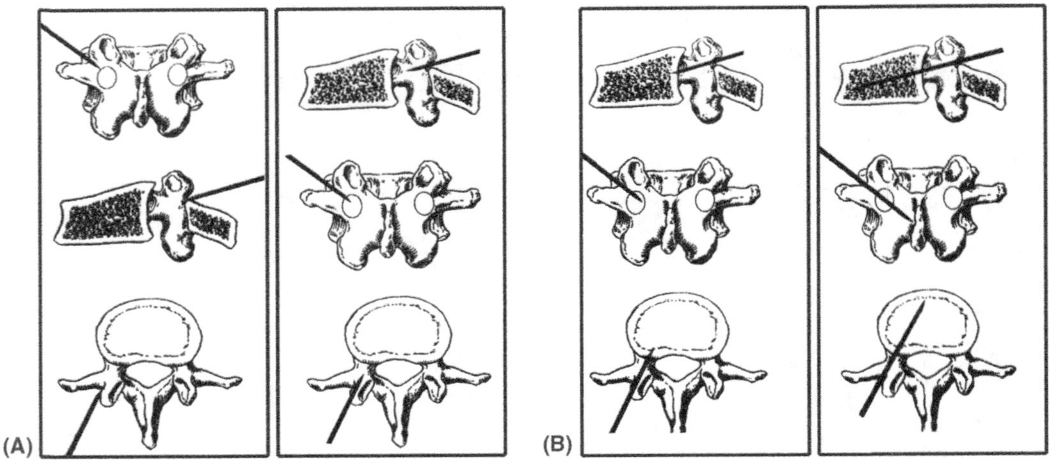

FIGURE 7 The four steps of the transpedicular approach at the lumbar level. (**A**) Bone contact and penetration of the vertebral pedicle. (**B**) Penetration of the vertebral body and final position. *Source*: Reproduced with the kind permission of Kyphon Inc.

Transpedicular Approach

Lesions of the pedicles and posterolateral part of the vertebral body are more easily reached via a transpedicular route (Fig. 7). The patient is placed in prone position under posteroanterior fluoroscopy. The X-ray beam is tilted to profile the pedicles. The skin puncture point is usually situated 5 mm above and lateral to the superolateral corner of the pedicle. However, this will be adapted to precise location of the lesion to reach in the vertebral body. The needle track will be more oblique (more toward the midline) if the skin-puncture point is more lateral (up to 15 mm lateral to the pedicle). After local anesthesia, a direct approach BTN (M1 or M2 Cook, Manan, LAR 1000) is inserted posteriorly toward the pedicle in a slight medial and caudad direction. Bone is abutted at the superolateral outer cortex of the pedicle (Figs. 7 and 8). Then the needle

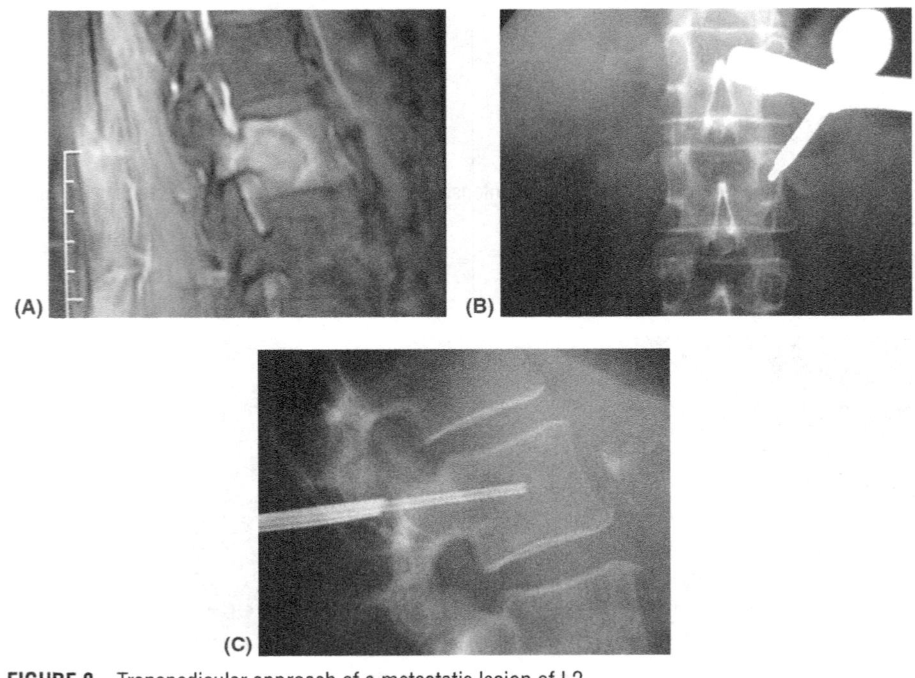

FIGURE 8 Transpedicular approach of a metastatic lesion of L2.

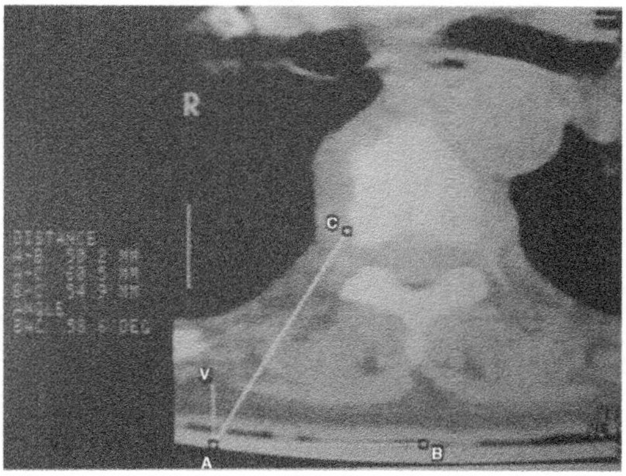

FIGURE 9 Posterolatera approach to the thoracic spine drawn on a computed tomography picture. Note that puncture site (*A*) is 5 cm from the midline and the angle of approach (*BAC*) is close to 55° from the coronal plane and needle path between pleura and spinal canal.

crosses the pedicle toward its medial cortex. In any case, the needle tip should not cross the medial cortex of the pedicle on the anteroposterior view before it reaches the posterior cortex of the vertebral body on a lateral view. The transpedicular approach may be more difficult at the upper lumbar level where the pedicles may be very thin.

Thoracic Spine

Posterolateral Approach in the Oblique Position

Following the technique previously reported (1,5,6), the thoracic vertebras from T2 to T12 can be reached, under single-plane fluoroscopic, a C-arm or CT-scan guidance, via an oblique, postero-lateral, intercostal approach at an angle of 35° from the patient's sagittal plane (Fig. 9). A coaxial trephine needle is used (Table 2). Preoperative radiographs and CT or MRI help to determine the side of the approach. The presence of a paravertebral abscess or mass facilitates the puncture. Depending on the target of the biopsy (vertebral body or disc space) and the spinal level, the needle should be oriented in a cephalad or caudad direction (Figs. 10 and 11). Using single-plane fluoroscopic guidance, the patient is placed in a 35° prone-oblique position (Fig. 12). With C-arm guidance, the C-arm can be titled 35° obliquely, which prevents the patient from turning. The

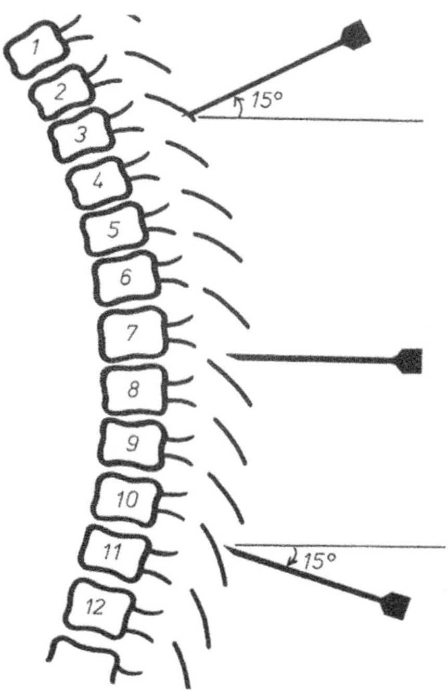

FIGURE 10 Changes in angle of approach that must be made for various vertebral levels.

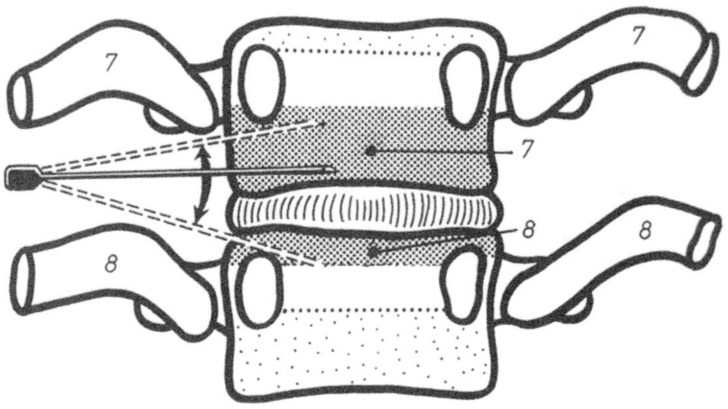

FIGURE 11 Shaded and striped areas shows portions of disc and vertebras that can be reached through an intervertebral approach. As an example, the T-7 and T-8 vertebras and the intervening intervertebral disc are shown.

puncture point is located at the level of the lesion, 4.5 cm from the midline (5,6). A radiopaque mark is placed at this point, and a radiograph is made in this position. The vertebral body or disc lesion can be reached through a "puncture area" whose upper and lower borders are the heads of the ribs, posterior border is the anterior margin of the zygoapophyseal joint, and anterior border is the vertebral body or pleura (Fig. 13). The spinal canal and the pleural space can thus be avoided during the procedure. A 20-G needle is used to anesthetize the superficial planes, and the thin needle or guide of the trephine set is then introduced through a small skin incision. The needle is advanced under fluoroscopic guidance and is vertically oriented to the puncture area, following the direction of the gantry (Fig. 14). At the same time, the needle is used to anesthetize the deeper planes. The vertebral body or disc is usually reached at a depth of 6 cm (5,6). After checking the accurate placement of the needle tip, the thin needle is replaced by the trephine needle. Once the trephine needle abuts the lesion, the patient can be moved to a strict prone position. Two radiographs (a posteroanterior view and a cross-table lateral view) are obtained to check for the correct positioning of the trephine needle. Biopsies are then performed using the cutting cannula, and additional radiographs are obtained (Fig. 5). Pain is usually mild or absent. If there is severe pain, a technical error must be suspected.

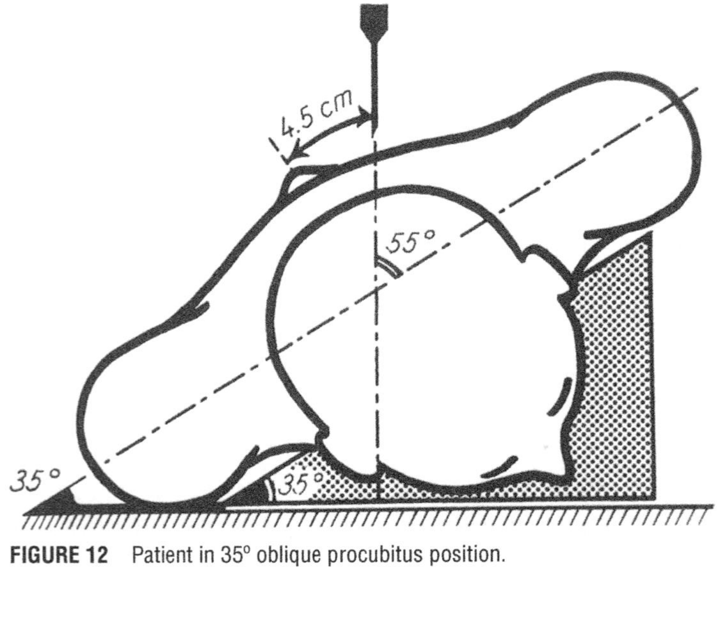

FIGURE 12 Patient in 35° oblique procubitus position.

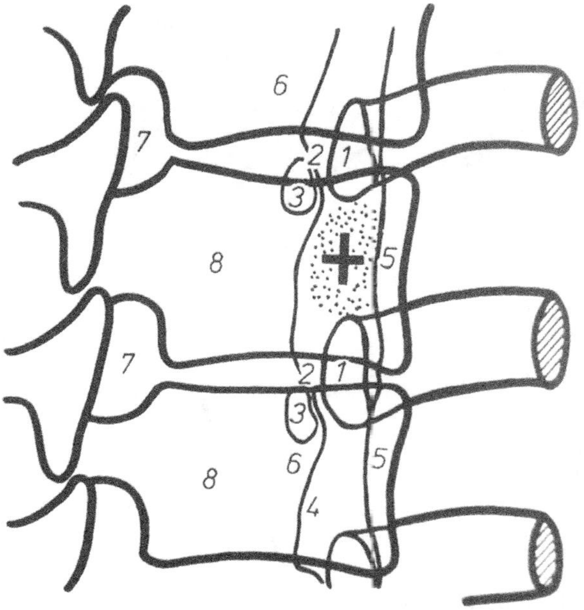

FIGURE 13 Diagram of anatomical relations in 35° oblique procubitus position. (*1*) Rib head, (*2*) costovertebral joint, (*3*) transverse process, (*4*) external edge of the articular process, (*5*) line of pleural reflection, (*6*) vertebral body, (*7*) contralateral lamina, (*8*) spinal canal. The puncture area is indicated by the *dotted* area.

Transpedicular and Posterolateral Approach in the Prone Position

The transpedicular and posterolateral approaches can be achieved in the prone position on a C-arm unit. The transpedicular approach is done with the same landmarks as described above at the lumbar level, using a direct approach BTN (Fig. 15). It is well suited for lesions involving the upper part of the thoracic vertebral bodies. In the case of a lesion of the lower part of the vertebral body or the disc space, a posterolateral approach can be achieved with similar landmarks as in the transpedicular route. The needle should cross the intervertebral foramen, imagining an additional pedicle between the vertical lines passing through the medial and lateral cortices of the pedicles of the superior and inferior vertebras used as guiding landmarks (Fig. 6). The needle tip should contact bone on a vertical line passing through the lateral cortex of the pedicles and penetrate the vertebral body on the lateral view before crossing the vertical line passing through the medial cortex of the pedicles on the anteriopcsterior view (Fig. 15). The

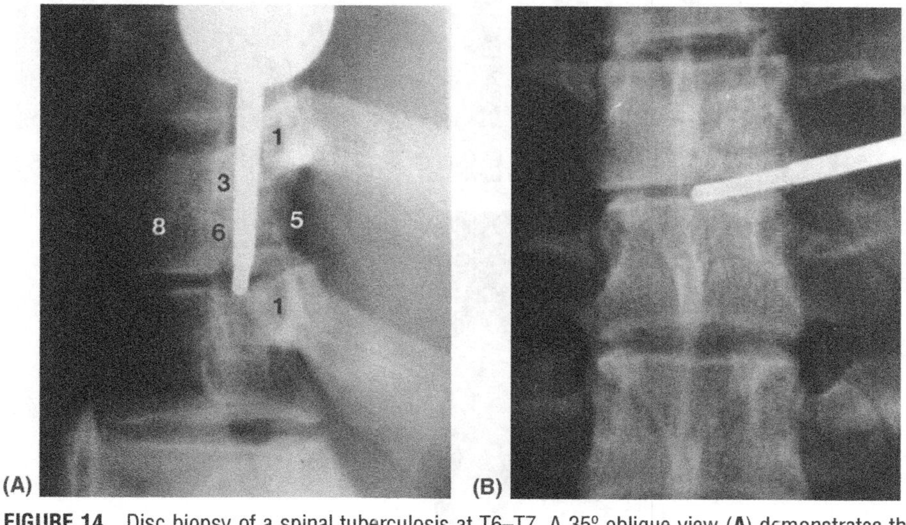

FIGURE 14 Disc biopsy of a spinal tuberculosis at T6–T7. A 35° oblique view (**A**) demonstrates the same anatomic relations as those in Figure 13. (**B**) Anteroposterior view.

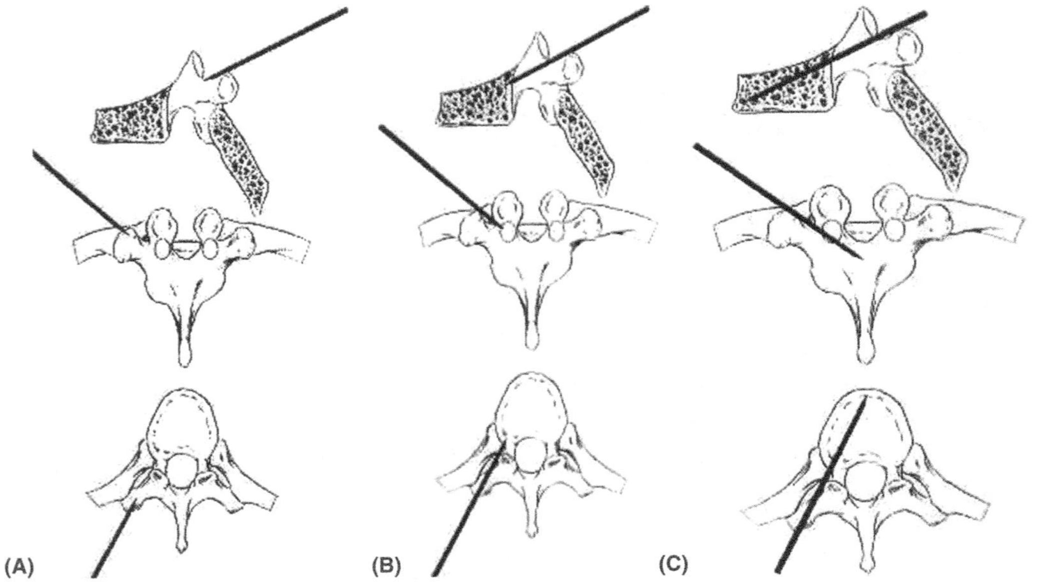

FIGURE 15 Transpedicular approach at the thoracic level. (**A**) Point of bone contact, (**B**) needle projections when entering the vertebral body, and (**C**) final placement. *Source*: Reproduced with the kind permission of Kyphon Inc.

disc space may also be approached with the patient in the prone position with the technique of the "imaginary pedicle" (Fig. 16).

Cervical Spine

The approach to cervical vertebras from C2 to C7 is anterolateral, between the larynx and carotid arteries or lateral, behind the large vessels, at the posterior border of the sterno-cleido-mastoid muscle (Figs. 17, 18 and 19) (1). Lesions involving predominantly the neural arch may be

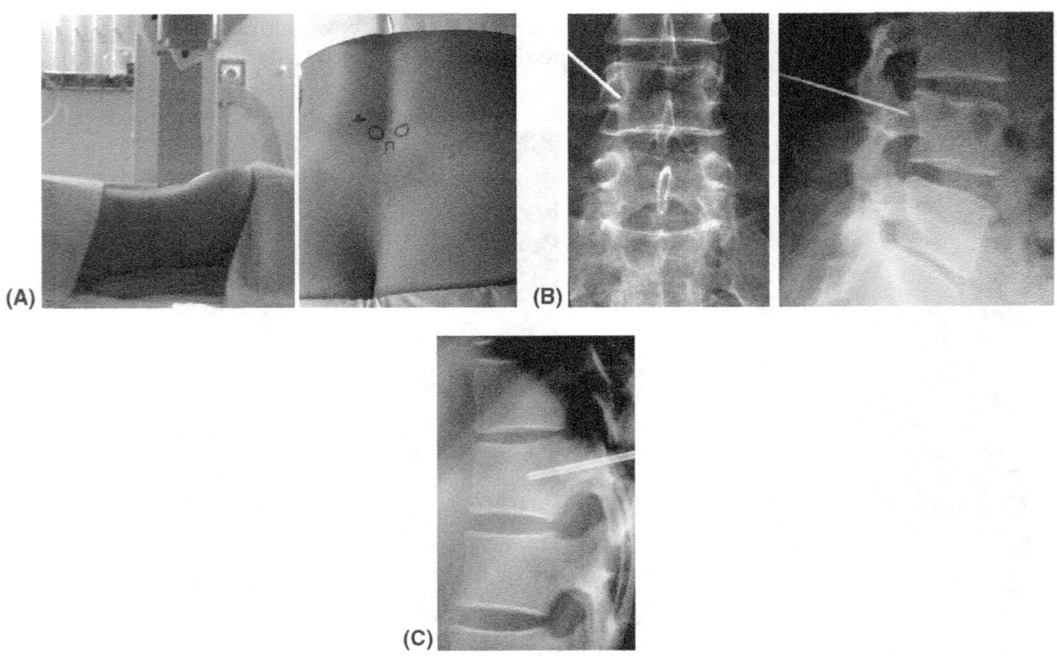

FIGURE 16 Biopsy of a thoracic lesion in prone position through an "imaginary" pedicle.

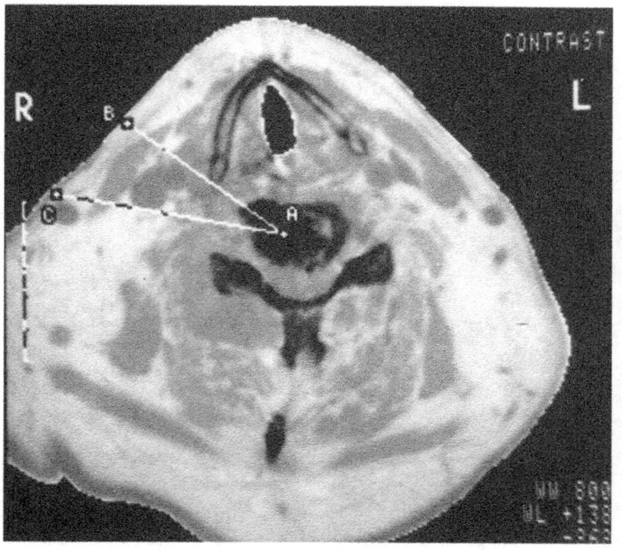

FIGURE 17 Approach to the cervical spine is anterolateral (*B*) before carotid artery or lateral (*C*) behind the large vessels.

sampled through a direct posterior approach under CT-scan control. C1 biopsy is very critical and can only be achieved through an anterior transoral approach under biplane fluoroscopic guidance and general anesthesia.

Sacroiliac Joint

The sacroiliac joint is approached, with the patient in prone position, via a direct posterior route under CT control (Fig. 20), which allows sampling of the joint space and both iliac and sacral subchondral bone (4). However, if only the anterior aspect of the joint is involved, it may be necessary to use a posterolateral route through the gluteal muscles and iliac bone under fluoroscopic or CT guidance (Fig. 21) (7,8). The inferior portion of the sacroiliac joint, 1cm above the inferior margin of the joint, is usually the best site for biopsy with both approaches. At this level, the thickness of the iliac bone, which must be crossed prior to penetrating the sacroiliac joint, is much less than at a higher level and the joint space is simply anteroposteriorly oriented.

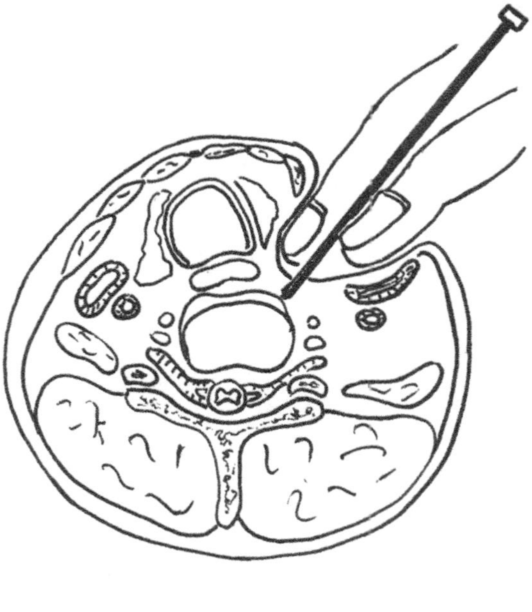

FIGURE 18 Anterolateral approach to the cervical spine between tracheal tree and main vessels, which is displaced posteriorly with the fingers.

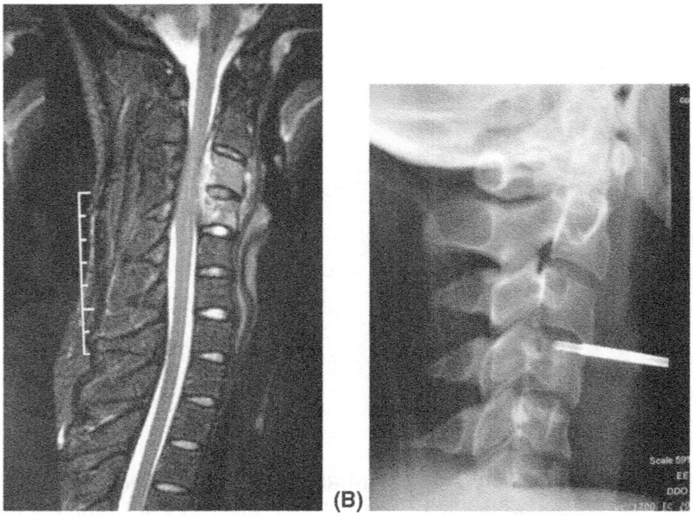

FIGURE 19 (A) Fat-sat T2-weighted magnetic resonance image of a spinal infection at C3–C4. (B) Percutaneous biopsy through an anterolateral route.

In the transiliac approach, the needle is inserted at 9 to 11 cm (usually 10 cm) from the midline, at an angle of 30° to 40° with the sagittal plane and advanced to the iliac bone, which is reached immediately laterally to the external margin of the sacroiliac joint, if fluoroscopic guidance is used. During the procedure, an arthrogram is performed to check that the joint space has been sampled.

Peripheral Bones

Most peripheral bones can be biopsied (1). A CT scan study and good knowledge of anatomy are useful in selecting the appropriate approach to the lesion. In lesions involving long bones, the needle is usually introduced perpendicular to the shaft to prevent it from sliding off the round cortex (Fig. 22). Use of BTN for cortical bone (Table 3) makes trephination of the cortex relatively easy and allows bone marrow aspiration. Once the cortex is trephinated, a soft tissue-cutting needle (Table 4) may be introduced through the outer cannula of the first needle to sample the medullar cavity or central soft lesion. Penetration of the medullar cavity of long bones is often very painful (especially in the case of osteomyelitis) and may require general anesthesia.

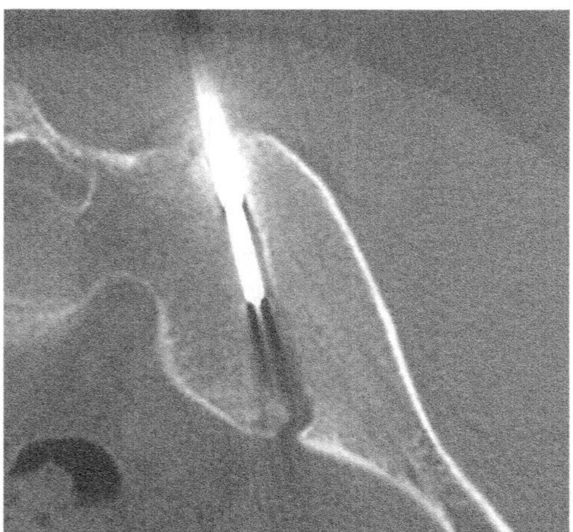

FIGURE 20 Biopsy of the sacroiliac joint with a posterior direct approach under computed tomography guidance.

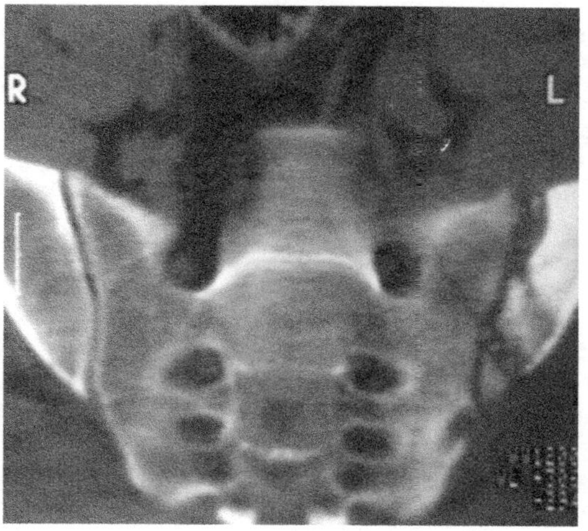

FIGURE 21 Oblique transiliac approach to the sacroiliac joint. Postbiopsy computed tomography scan showing the biopsy tracks.

Flat Bones and Vertebral Neural Arch

In the case of biopsy of osteolytic lesions of the rib and when a vital structure is close, CT rather than fluoroscopic guidance should be used. For flat bones such as the scapula, ribs, sternum, iliac bone, and even the cranial vault, a tangential approach is preferred to sample the maximum amount of bone possible and to avoid damage to the underlying structures (Figs. 23 and 24). In lesions of the neural arch of the vertebras, the tangential approach also protects from the risk of damage to the spinal cord or thecal sac (Fig. 25).

Sclerotic Lesions

Preoperative enhanced CT scan and MRI are especially useful in such situations to help in locating lytic-enhancing areas to biopsy within bone sclerosis (Fig. 26).

A BTN needle able to penetrate the cortical bone should be used (Table 3). With other trephine needles, advancement may become very difficult after a few millimeters (1). Excessive pressure on the trephine should be avoided by withdrawing the needle, removing the plug of cortical bone, and replacing the needle in the same hole to continue the biopsy. This process is repeated if needed and is less damaging to the biopsy samples. A motor-driven or manual hand-drill can also be used.

SPECIMEN HANDLING

Numerous samplings have to be performed in different sites of the lesion (1). Most authors believe that the larger the sample of tissue obtained by the biopsy, the easier the histologic

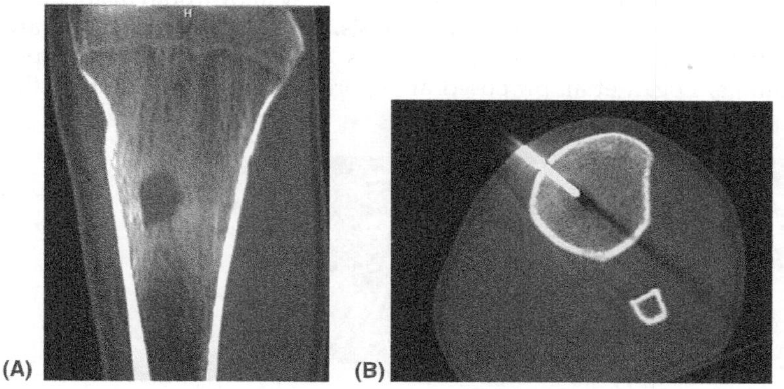

FIGURE 22 Brodie tibial abscess. (**A**) Coronal computed tomography (CT)-reformation. (**B**) Biopsy under CT-scan guidance.

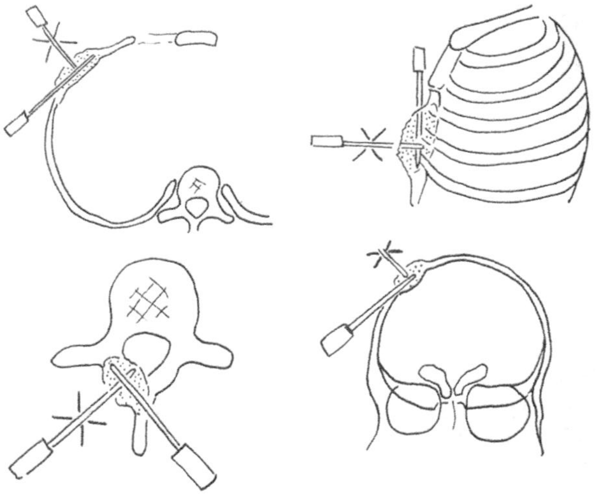

FIGURE 23 Examples of the tangential approach for biopsy of the ribs and sternum, vertebral neural arch, and cranial vault.

diagnosis (9,10). Material should be obtained for both histologic diagnosis and bacteriological studies. Core biopsies for histopathologic examination are fixed in 10% formalin, decalcified, mounted in paraffin block, sectioned transversally, and stained. For electron microscopy, the tissue must be preserved in buffered glutaraldehyde solution. Prints of the tissue core are taken or smears of the needle aspirate are prepared for cytologic examination each time a tumor is suspected. Blood aspirated from the lesion is fixed and analyzed as a bone specimen. It may contain neoplastic cells sometimes not found in the solid tissue. In infection, a lavage of the lesion with several millimeters of sterile saline solution is done if no pus is aspirated. At least one tissue fragment should also be cultured. Samples must be delivered immediately to the laboratories. Blood cultures are sometimes positive in the hours following the biopsy due to the discharge of microorganisms by the lesion opening and should be performed in the hours following the biopsy even if the patient has no fever, each time an infection is suspected. If any question arises as to the proper specimen handling, the microbiologist or the pathologist should be consulted prior to the procedure. Close cooperation with them is imperative. In some cases, when a fragile microorganism is suspected, the bacteriologist should be present in the biopsy room to ensure immediate seeding of specific culture milieus. They must be given all clinical information available. The exact origin of the samples must be indicated in a diagram of the lesion.

TREPHINE OR ASPIRATION BIOPSY

For most authors, trephine biopsy is more efficient than aspiration biopsy in establishing a pathologic diagnosis (9,10). Moreover, in infectious diseases, bacteriological examinations may remain negative despite the aspiration of pus, whereas trephine biopsy allows both histopathologic and bacteriologic evalution. Thus, in bone tuberculosis, histopathologic results are obtained much more rapidly than those of bacteriologic cultures, allowing rapid set up of specific treatment. In bone tumors, Logan et al. proposed an algorithm for the selection of the

(A) 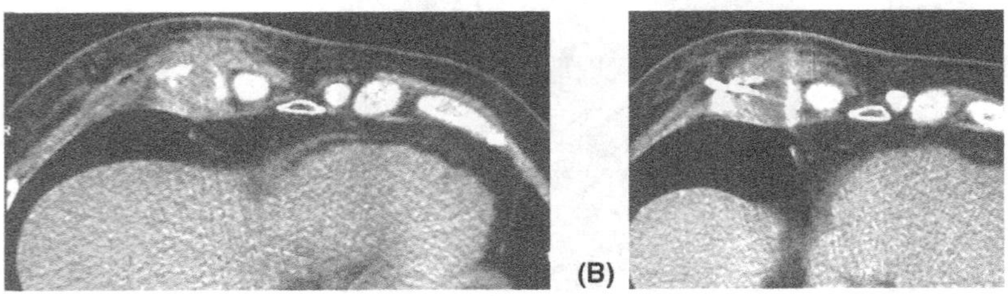 (B)

FIGURE 24 Biopsy of a lytic metastatic lesion of an anterior rib.

(A) (B)

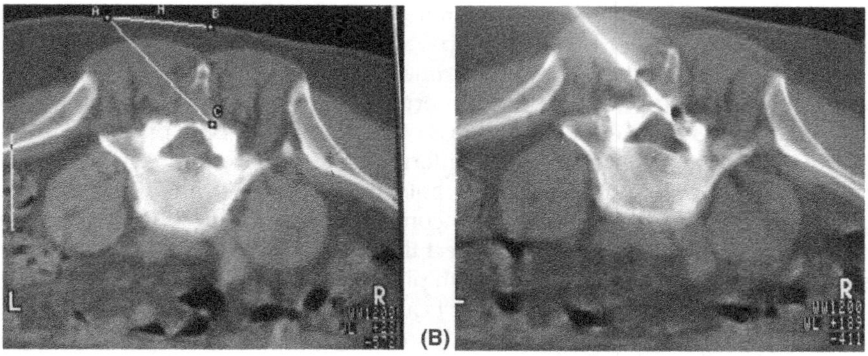

FIGURE 25 Percutaneous biopsy under computed tomography scan control of a myeloma located at the neural arch of L5. (**A**) Preoperative measurements; (**B**) biopsy with a soft tissue, side-notch, cutting needle.

biopsy technique. They used fine-needle aspiration to sample small (less than 3 cm) soft-tissue masses, a bone-cutting core biopsy needle to sample sclerotic bone tumors, and a combination technique to sample lytic bone tumors with an intact bony shell. With this algorithm, they achieved an overall diagnostic accuracy rate of 96% in musculoskeletal tumors (11). In a series of 138 patients from whom percutaneous skeletal aspiration and core biopsy were both obtained, Schweitzer et al. found that core biopsy had more procedural utility than aspiration in 28 patients, while conversely, aspiration was more useful than core biopsy in 18 (12). These authors concluded that aspiration and core skeletal biopsy have a complementary role and should be both routinely performed.

COMPLICATIONS

The overall complication rate of PB is low. In a series of 500 biopsies, Laredo et al. (1) reported only one paravertebral hematoma that resolved spontaneously. Murphy et al., in a large review of 9500 percutaneous skeletal biopsies reported in the English literature, identified approximately 22 complications (0.2%) considered important enough to mention. They reported nine pneumothoraxes, most of them treated conservatively. There were also more serious complications such as transient or permanent paraplegia or quadriplegia, and other neural injuries, such as those with resultant footdrop. Five spinal cord injuries were recorded, including one case of meningitis and two with resultant deaths. Therefore, serious neurologic injury occurred in 0.08% of procedures and death in 0.02% (10).

In primary tumors, open or closed biopsy may result in complications and also adversely affect the care of the patients (13). In the cooperative study reported by Mankin et al., a problem with the biopsy forced the surgeon to carry out a different and often more complex operation or to use adjunctive irradiation or chemotherapy in 19.3% of patients. This resulted in a change in

(A) (B)

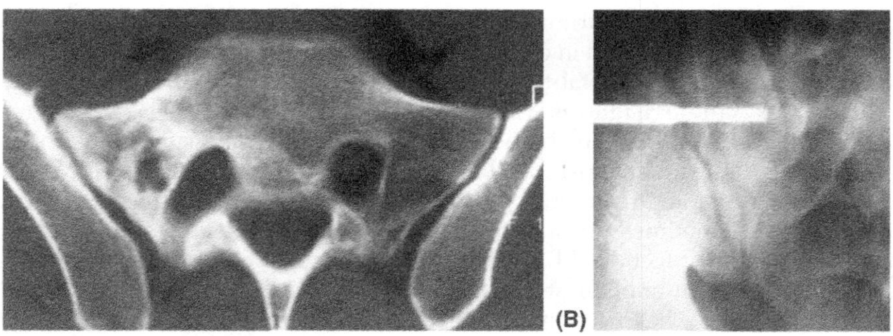

FIGURE 26 Tuberculous osteomyelitis of the sacral wing. (**A**) A preoperative computed tomography scan shows a central lytic area to biopsy. (**B**) Biopsy of this lytic area.

the outcome (disability, loss of function, local recurrence, or death) in 10.1% of patients. Errors and complications were much greater when the biopsy was done in a nonspecialized treatment center instead of in a referral institution (13). In the same study, closed needle biopsy was found to be less accurate than open biopsy (60% vs. 76% of correct diagnosis), but also much less likely to cause complications. Therefore, biopsy technique is of great importance in primary musculoskeletal tumors and should be achieved only in centers with a multidisciplinary team able to manage the pretherapeutic evaluation and the whole treatment of primary bone tumors. Puncture site and approach have to be selected in consultation with the orthopedic surgeon who will operate on the patient. He will have to resect the needle track during surgery to avoid tumor seeding. During the biopsy, an external sheath placed up to the tumor margin should be used to protect the path from the cutting needle. It is also wise to mark the biopsy track with carbon, which is amorphous and persists for several weeks, and to tattoo the skin entry point with india ink to allow easy location during surgery (11).

RESULTS

The overall accuracy of PB of bone lesions reported in the literature varies from 66% to 96% (1,10,11,14–16). However, the method of evaluation of skeletal PB accuracy varies greatly in the different series reported. Because PB is also useful in excluding a diagnosis, evaluation of accuracy must take into account both true-positive and true-negative results. However, true-negative and false-negative results are difficult to confirm (10). They require a second test or adequate follow-up (10). Nevertheless, the accuracy of PB of skeletal lesions is reliably evaluated in some series. Debnam and Staple reported that an accurate diagnosis was made or disease was excluded in 81% of patients or 74% of biopsy sites (17). Murphy et al. found a 94% overall accuracy in 160-closed biopsies (10).

The accuracy of trephine biopsy depends on many different factors: the nature of the lesion, its location and radiologic appearance, the accuracy of the preoperative radiologic assessment, careful selection of the biopsy site and choice of adequate instruments, approach and method of radiologic guidance, biopsy technique with emphasis on specimen adequacy, especially sampling of multiple cores in different locations, and expertise of the pathologist and the bacteriologist and close cooperation between them and the radiologist.

PB has a high diagnostic accuracy in bone metastases ranging from 79% (16) to 96% (1), and PB should be preferred to surgical open biopsy when a bone metastasis is suspected (Fig. 27). In some cases, such as bone metastases from hepatocarcinoma or thyroid carcinoma, the histopathologic examination may indicate the precise nature of the primary neoplasm. Most authors accept that some primary neoplasms of bone such as myeloma, plasmacytoma, lymphoma, and Ewing's tumors can be accurately diagnosed by PB. However, the value of PB in the diagnosis of other primary bone neoplasms such as bone sarcomas is still under debate. Most authors found that needle biopsy provides lower diagnostic accuracy than open biopsy in primary musculoskeletal tumors (13,18,19). Mankin et al. obtained a 60% rate of accuracy with needle biopsy as compared with 76% of incisional biopsies in primary musculoskeletal tumors (13). The diagnostic accuracy rate of PB in primary bone tumors was 79% in a referral cancer center (18). In another study, the diagnostic accuracy of PB was only 78% in primary soft-tissue tumors (19). However, PB also causes less complications than open surgery in primary musculoskeletal tumors (13). In each individual case, however, the decision must be taken in consultation with the orthopedic surgeon.

Diagnostic accuracy of PB in skeletal tuberculosis is very high. In a series of 21 cases of vertebral tuberculosis, an accurate diagnosis was obtained in 20 cases (95.2%) (1). This high rate of accuracy may be due to several factors: both histologic and bacteriologic examination may lead to a correct diagnosis; bone tuberculosis produces a large amount of pus (Fig. 28); and diagnosis is not impaired by antibiotic treatment prescribed without specific diagnosis. However, pathologic examination is sometimes nonspecific, and may, in some cases, mimic those of a nontuberculous pyogenic bone infection (14). Morre et al. obtained an 80% accuracy rate in 117-pyogenic bone infections (16). In another study, the causative microorganism was isolated in only 15 (55.5%) of 27 pyogenic disc infections (1). Histological features suggestive of pyogenic infection were found in half the cases with negative bacteriologic examination (1). This relatively

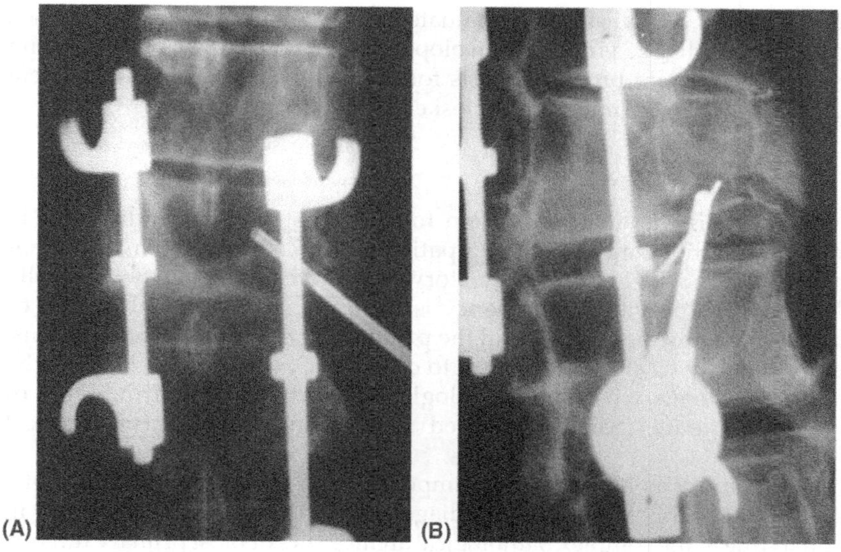

FIGURE 27 Anteroposterior (A) and oblique views (B) taken during the percutaneous biopsy of a lytic metastasis of T11, which was carried out despite posterior metallic fixation. Previous laminectomy failed to demonstrate the metastatic nature of the vertebral collapse.

low accuracy rate of PB in pyogenic disc infection as compared to spinal tuberculosis has several explanations: definitive diagnosis is only possible when the bacteriologic examination is positive, patients often received antibiotics prior to biopsy, and spontaneous resolution with elimination of the microbial agent is possible.

Finally, percutaneous image–guided biopsy is a cost-effective procedure for most musculoskeletal lesions (11,20). In a study by Fraser-Hill et al., the estimated cost of CT-guided biopsy was $442 compared with $1658 for open surgical biopsy (20). In a series including 83% of

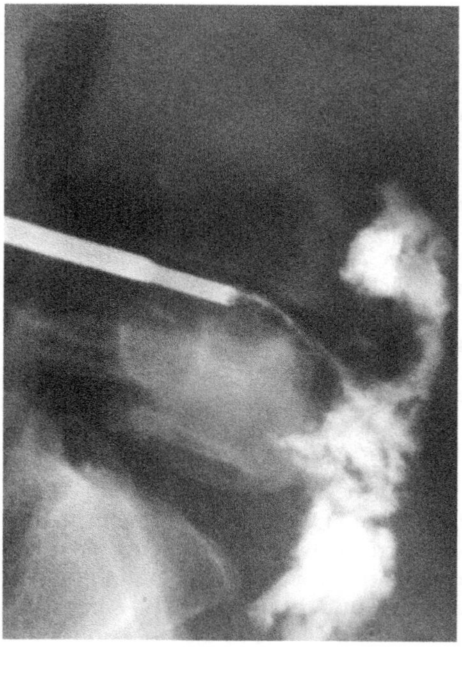

FIGURE 28 Disc biopsy in spinal tuberculosis at L4–L5 level with opacification of a prevertebral abscess.

primary musculoskeletal tumors, Skrzynski et al., evaluated the hospital charges for the closed biopsy to be $1106, compared with $7234 for the open biopsy (19). According to Springfield and Rosenberg, it is likely that these data provide a basis for the recommendation of outpatient needle biopsies for properly selected primary musculoskeletal tumors (21).

INDICATIONS

Suspicion of bone metastases is the main indication for PB. Most frequent indications in bone metastases are suspected bone metastasis in a patient with no known primary malignancy, single lesion in patients with a previous history of primary malignancy especially when the relationship with the previous malignancy is unclear due to unusual clinical or radiologic manifestations, and new bone lesions in the presence of multiple primary lesions, apparently stabilized previously treated metastases, to determine activity, to determine the nature of bone changes in patients who have radiologic abnormalities after irradiation of the skeleton, and to differentiate radiation-induced necrosis from metastasis or local tumor extension.

As discussed above, biopsy is a step of first importance in primary musculoskeletal tumors (13). Needle biopsy probably yields a lower diagnostic accuracy than open biopsy in musculoskeletal sarcomas due to the complex pathological architecture of most primary tumors. However, needle biopsy also causes less complications than open biopsy (13). Especially, the risk of extensive spread of tumor cells is probably lower after closed biopsy than after open biopsy (15). In any case, if a closed biopsy is perform, the puncture site and approach will be carefully selected together with the surgeon and the needle track will be tatooed with carbon to be excised "en block" with the tumor.

PB is indicated in skeletal infection when the nature of the causative microorganism has not been ascertained clinically. PB is frequently indicated to differentiate sacroiliac infection from unilateral sacroiliitis due to Reiter's syndrome and psoriatic arthritis.

PB of bone may be indicated for diagnostic confirmation in some disorders that may also simulate bone neoplasms, such as fibrous dysplasia, eosinophilic granuloma, and Paget's disease. PB is indicated in vertebral collapse when the clinical and radiologic findings are not sufficient to differentiate malignant from benign causes, mainly osteoporosis.

CONTRAINDICATIONS

There is no absolute contraindication to PB of bone and soft tissue. The risk of a particular biopsy is always measured against the risk from alternative diagnosis methods or against the risk of no specific diagnosis being obtained. Biopsy is not performed if the result would not affect treatment or management. PB should not be performed in patients with platelet counts below 50,000 per mm^3 or abnormal hemostasis unless they are adjusted prior to biopsy. Hemorrhage may be significant during PB of highly vascularized lesions such as giant cell tumor, aneurysmal bone cysts, and bone metastases from kidney and thyroid carcinomas. These lesions may exhibit a rather characteristic roentgenographic appearance, with a bubbly lytic pattern with cortical expansion and bony septation. Skeletal metastases from kidney and thyroid can be biopsied only if bleeding is not excessive during aspiration with thin needles. In case of important venous bleeding at biopsy, gel foam can be injected through the trephine needle to obtain hemostasis.

CONCLUSION

PB of the skeletal system is not a recent technique. Its effectiveness, however, has dramatically improved during recent years owing to a more careful selection of patients, refinements in preoperative radiologic assessment, improvement of biopsy instruments, and diversification of the modalities of radiologic guidance. Optimal results are obtained by close cooperation among physicians, orthopedic surgeons, radiologists, pathologists, and bacteriologists. Almost all musculoskeletal locations are approachable. Open biopsy should be restricted to contraindications and failures of PB.

REFERENCES

1. Laredo JD, Bard M, Cywiner-Golenzer C, Chretien J. Percutaneous biopsy of musculoskeletal lesions. In: Bard M, Laredo JD, eds. Interventional Radiology of Bone and Joint. Wien, New-York: Springer-Verlag, 1988:3–50.
2. Resnick D. Needle biopsy of bone. In: Resnick D, Niwayama G, eds. Diagnosis of Bone and Joint Disorders. Philadelphia: WB Saunders, 1995:506–519.
3. Jelinek J, Buick M, Shmookler B. Image-guided percutaneous biopsies of musculoskeletal lesions. AJR 1996; 167:532.
4. Hendrix RW, Lin JP, Kane WJ. Simplified aspiration of injection technique for the sacroiliac joint. J Bone Joint Surg 1982; 64-A:1249–1252.
5. Laredo JD, Bard M, Leblanc G, Folinais D, Cywiner-Golenzer C. Technique et résultats de la ponction-biopsie transcutanée radioguidée du rachis dorsal. Rev Rhum 1985; 52:283–287.
6. Laredo JD, Bard M. Thoracic spine: percutaneous trephine biopsy. Radiology 1986; 160:485–489.
7. Chevrot A, Godefroy D, Horreard P, Pallardy G. Biopsie osseuse profonde au trocart sous radioscopie télévisée dans les arthrites sacro-iliaques. Rev Rhum 1981; 48:95–99.
8. Vinceneux P, Lasserre PP, Grossin M. Technique de ponction-biopsie percutanée au trocart de l'articulation sacro-iliaque pour le diagnostic bactériologique et histologique des sacro-iliites. Rev Rhum 1981; 48:93–98.
9. Fyfe JS, Henry AP, Mulholland RC. Closed vertebral biopsy. J Bone Joint Surg 1983; 65-B:140–145.
10. Murphy WA, Destouet JM, Gilula LA. Percutaneous skeletal biopsy: a procedure for radiologists. Results, review and recommendations. Radiology 1981; 139:545–549.
11. Logan PM, Connell DG, O'Connell JX, Munk PL, Janzen DL. Image-guided percutaneous biopsy of musculoskeletal tumors: an algorithm for selection of specific biopsy techniques. AJR 1996; 166:137–141.
12. Schweitzer ME, Gannon FH, Deely DM, O'Hara BJ, Juneja V. Percutaneous skeletal aspiration and core biopsy: complementary techniques. AJR 1996; 166:415–418.
13. Mankin HJ, Mankin CJ, Simon MA. The hazards of the biopsy, revisited. For the members of the Musculoskeletal Tumor Society. J Bone Joint Surg 1996; 78-A:656–663.
14. Chevrot A, Godefroy D, Horreard P, Pallardy G. Biopsie osseuse profonde disco-vertébrale au trocart sous contrôle de la radioscopie télévisée. Ann Med Intern (Paris) 1980; 131: 448–451.
15. Deeley TJ. The drill biopsy of bone lesions. Clin Radiol 1972; 23:536–540.
16. Morre TM, Myers MH, Patzakis MJ, Terry R, Harvey JP. Closed biopsy of musculoskeletal lesions. J Bone Joint Surg 1979; 61-A:375–380.
17. Debnam JW, Staple TW. Trephine bone biopsy by radiologists. Results of 73 procedures. Radiology 1975; 116:607–609.
18. Ayala AG, Zornosa J. Primary bone tumors: percutaneous needle biopsy. Radiology 1983; 149:675–679.
19. Skrzynski MC, Biermann JS, Montag A, Simon MA. Diagnostic accuracy and charge-savings of outpatient core needle biopsy compared with open biopsy of musculoskeletal tumors. J Bone Joint Surg 1996; 78-A:644–649.
20. Fraser-Hill MA, Renfrew DL, Hilsenrath PE. Percutaneous needle biopsy of musculoskeletal lesions. 2. Cost-effectiveness. AJR 1992; 158:813–818.
21. Springfield DS, Rosenberg A. Biopsy: complicated and risky. Editorial. J Bone Joint Surg 1996; 78-A: 639–643.
22. Ackerman W. Vertebral trephine biopsy. Ann Surg 1956; 143:373–385.
23. Aström KG, Ahlström KH, Magnusson A. CT-guided transsternal core biopsy of anterior mediastinal masses. Radiology 1996; 199:564–567.
24. Ahlström KH, Aström KGO. CT-guided bone biopsy performed by means of a coaxial biopsy system with an eccentric drill. Radiology 1993; 188:549–552.
25. Cohen MA, Zornoza J, Finkelstein JB. Percutaneous needle biopsy of long-bone facilitated by the use of a hand drill. Radiology 1979; 139:750–751.
26. Craig FS. Vertebral body biopsy. J Bone Joint Surg 1956; 38-A:93–102.
27. Kreula J. Effect of sampling technique on specimen size in fine needle aspiration biopsy. Invest Radiol 1990; 25:1294–1299.
28. Kreula J, Virkkunen P, Bondestam S. Effect of suction on specimen size in fine-needle aspiration biopsy. Acta Radiol 1990; 25:1175–1181.
29. Laredo JD, Bard M, Patrux C. Intérêt d'un nouveau matériel pour biopsie osseuse vertébrale percutanée. J Radiol 1984; 65:297–300.
30. Levernieux J, Huber-Levernieux C, de Seze D. La ponction-biopsie du corps vertébral. Sem Hop Paris 1972; 48:1029–1037.
31. Morrisson R, Deeley TJ. Drill biopsy: a technique using a high-speed drill. J Faculty Radiol 1955: 287–289.
32. Schweitzer ME, Deely DM. Percutaneous biopsy of osteolytic lesions: use of a biopsy gun. Radiology 1993; 189:615–616.
33. White LM, Schweitzer ME, Deely DM. Coaxial percutaneous needle biopsy of osteolytic lesions with intact cortical bone. AJR 1996; 166:143–144.

5 Percutaneous Biopsy of the Synovial Membrane

Jean-Denis Laredo and Bassam Hamzé
Department of Bone and Joint Radiology, Lariboisière Hôpital, Assistance Publique–Hopîtaux de Paris, Paris, France

Pierre Champsaur
Department of Radiology, La Timone Hospital, Marseille, France

INTRODUCTION

In selected cases, examination of the synovium may provide precise diagnostic clues or useful information about the nature of arthritis of unclear etiology. Biopsy of the synovium can be performed through open surgery or a percutaneous approach, or as part of an arthroscopy procedure during which the biopsy area can be visualized.

Percutaneous biopsy of the synovium (PBS) permits removal of specimens from several regions of the joint lining while causing a minimum of trauma. However, because laboratory examination of the synovium is useful for diagnostic evaluation in a limited range of diseases, the precise indications for PBS must be well known.

PBS was first introduced by Forestier in 1932 (1). In 1951, Polley and Bickel reported a large series of PBS of the knee performed with a specific instrument with a diameter of 5 mm (2). In 1963, Parker and Pearson described a small-caliber synovial biopsy needle that permitted the removal of synovial membrane from the knee and other superficial joints such as the ankle, shoulder, elbow, and wrist, while causing a minimum of trauma (3). In 1979, Aignan reported a technique for PBS of the hip under fluoroscopic control, utilizing a large bone trocar and a Watanabe forceps originally designed for knee arthroscopy (4). In 1988, Laredo and Bard and then Beaulé et al. described a new technique of PBS of large joints using a manual side notch needle with placement of the needle parallel to the anterior aspect of the joint (5,6).

TECHNIQUES

PBS under fluoroscopic guidance is performed with disposable side notch-cutting needles (SNCN) (5,6). The basic principle of this technique consists in bringing the cutting side notch of the needle parallel to the surface of the joint. Therefore, the whole specimen notch of the needle can be placed within the joint or joint recess, allowing better and safer samplings of the synovium. Automatic or semiautomatic cutting needles with a large cutting side notch (10–20 mm, depending on the joint sampled) are very convenient (see Chapter 4 on percutaneous bone biopsy, Table 4). Use of an introductor (such as Achieve®, Allegiance through which the cutting needle is introduced) facilitates samplings of different parts of the joint cavity: the introductor allows orientation of the cutting needle tip in different areas of the joint. The needle is advanced with the side notch closed until the specimen notch is within the tissue to be biopsied. The notch is then opened and quickly closed. All the large joints (hip, ankle, shoulder, elbow, and wrist) may be sampled with this technique (5,6). PBS of the knee does not require radiological control and will not be described here. The technique of PBS of the hip will be first detailed. Specific considerations for the other joints will be then mentioned.

Technique for Percutaneous Biopsy of the Synovium of the Hip Joint

A decision is made after a complete radiological evaluation of joint disease. The degree and location of synovial proliferation are systematically evaluated prior to PBS by means of magnetic resonance imaging (MRI) with intravenous administration of gadolinium or arthrography combined with computed tomography (CT). Gadolinium-enhancement MRI is especially

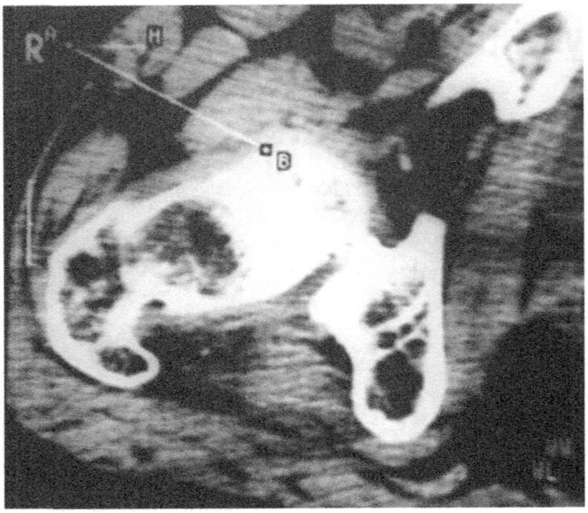

FIGURE 1 Simulation of needle approach to the hip joint on a transverse image. Computed tomography R: right; H: horizontal line; A: point where the needle tip first reaches the bone; A–B: simulation of needle approach.

valuable in localizing synovial hypertrophy and in determining the biopsy site. The patient's hemostasis must be checked in the days preceding the procedure. A sedative drug (e.g., 100 mg hydroxyzine dichlorhydrate) is administered orally one hour prior to the examination, and a perfusion of antalgic drugs (e.g., Propacetamol Chlorhydrate) is given during the procedure.

PBS is performed under local anesthesia on an outpatient basis. However, because the patient received sedative drugs before the procedure, the patient has to be escorted back home. PBS is achieved under single plane fluoroscopy guidance. An SNCN allows for multiple sampling in the inferior and medial joint recess, which is a frequent site of synovial proliferation and the preferred site for tissue sampling (Fig. 1). The patient is placed supine on the X-ray table with the leg stabilized in internal rotation by sandbags. The hip joint is approached through an anterolateral route (Figs. 1 and 2). The point of skin puncture is determined with the aid of palpation and fluoroscopy. It is located 3 to 4 cm medial to the anterior aspect of the greater trochanter (Fig. 3A). The appropriate level of puncture is determined under fluoroscopic control by placing a metallic ruler over the patient's hip in order to simulate the approaching needle on the fluoroscopic screen. The ruler is appropriately tilted so that it projects over the junction of the femoral head and neck as simulated in Figure 2. After skin preparation, the joint is draped and superficial and deeper planes are anesthetized with 1% lidocaine. The joint is first approached with a 20-gauge needle. If present, synovial fluid is aspirated for bacteriologic and cytologic examination. If no fluid is aspirated, sterile saline solution is injected and then reaspirated to be sent for bacteriological analysis. Distension of the joint with contrast media,

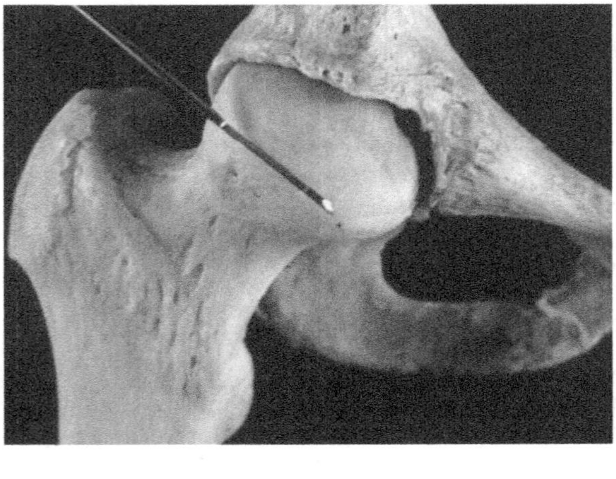

FIGURE 2 Approach for percutaneous biopsy of the synovium of the hip.

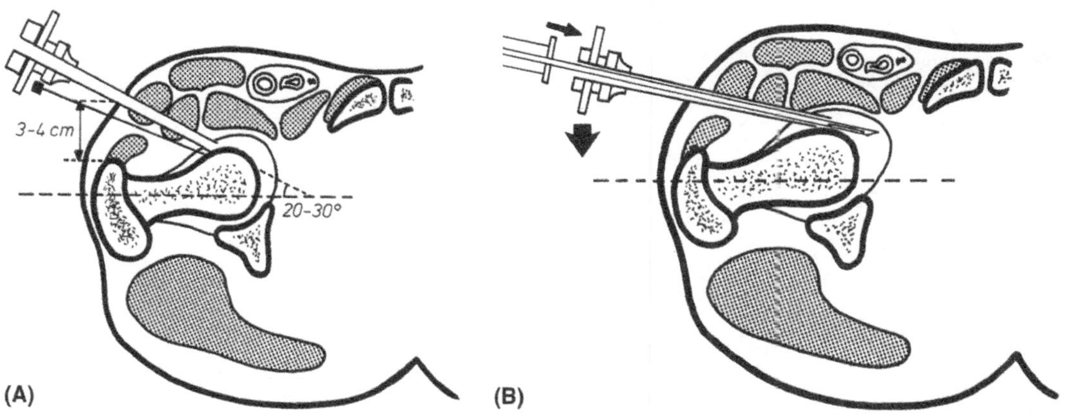

FIGURE 3 Needle approach for percutaneous biopsy of the synovium of the hip. The introducer is first advanced to the lateral aspect of the junction of the femoral head and neck (**A**). The lateral notch-cutting needle is then introduced into the joint through the trephine *(small arrow)* (**B**). Gentle alteration in trephine inclination *(large arrow)* facilitates advancement of the lateral notch-cutting needle.

1% lidocaine, and saline solution makes easier the penetration of the biopsy instruments into the joint space (7). Because repeated filling of the joint cavity during the procedure also facilitates the biopsy of multiple samplings, a lockable three-way stopcock is placed over the 20-gauge needle. The introductor and then the SNCN are inserted by means of a skin stab and advanced toward the lateral aspect of the junction of the femoral head and neck. The needle is simultaneously directed downward at an angle of 20° to 30° to the horizontal plane (Fig. 3A). When the needle contacts the bone at the correct point, it is withdrawn 2 cm and advanced again 2 cm in a direction more horizontal toward the anterior aspect of the joint (Fig. 3B). A gentle alteration in the position and inclination of the needle is usually needed to allow the SNCN to slide along the anterior aspect of the bone. Once a correct placement of the SNCN has been achieved (the cutting window must be tangential to the surface of the joint), the biopsy is performed. The SNCN is removed with each specimen while the introductor remains in place. Return of fluid through the lumen of the SNCN or through the introductor confirms a correct biopsy site. The same process is repeated in adjacent areas of the joint space using slight alterations in inclination and position of the biopsy needle. Repeated fillings of the joint with contrast media or saline solution helps in obtaining multiple specimens. Radiographs are performed during the procedure to document the exact site of tissue sampling (Figs. 4 and 5). After the

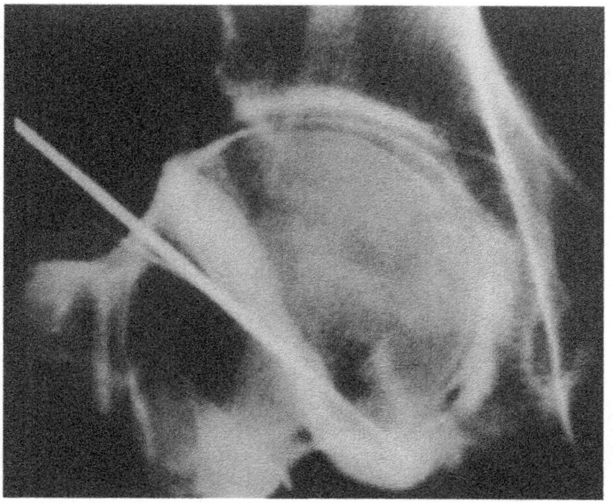

FIGURE 4 Percutaneous biopsy of the synovium in hip infection due to *Escherichia coli*.

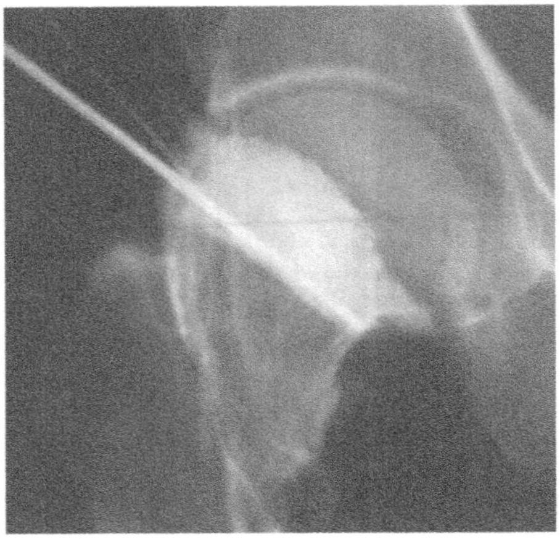

FIGURE 5 Percutaneous biopsy of the synovium of the hip in a case of early inflammatory arthritis.

procedure, patients are advised to rest for 24 hours. The need for rapid consultation in case of fever, abnormal increase in pain, swelling, or any signs and symptoms that may indicate complications is carefully explained.

Specimen biopsy must be carefully examined during the procedure. Synovial tissue has a pale pink color that can be distinguished from muscle and from yellow-white fibrinous exudate or necrotic material (8). To minimize the sampling error, many specimens are taken from various parts of joint lining (8). Several specimens are fixed in neutral formalin. When gout, calcium pyrophosphate dehydrate, or apatite crystal–related arthropathies are suspected, absolute alcohol is also used (9). Some other specimens and synovial fluid are taken for bacteriologic examination.

Technique for Percutaneous Biopsy of the Synovium of the Ankle Joint

The patient is placed supine on the table with the legs extended and stabilized in slight internal rotation with sandbags. The biopsy is performed in the anterior recess of the ankle joint. The SNCN is inserted anterior to the lateral malleolus at the level of the tibiotalar joint space and advanced horizontally toward the medial aspect of the joint (Figs. 6 and 7).

Technique for Percutaneous Biopsy of the Synovium of the Shoulder Joint

The anterior aspect of the glenohumeral joint is biopsied using a descending approach through the rotator cuff interval. The patient is positioned supine under the fluoroscope with the hand

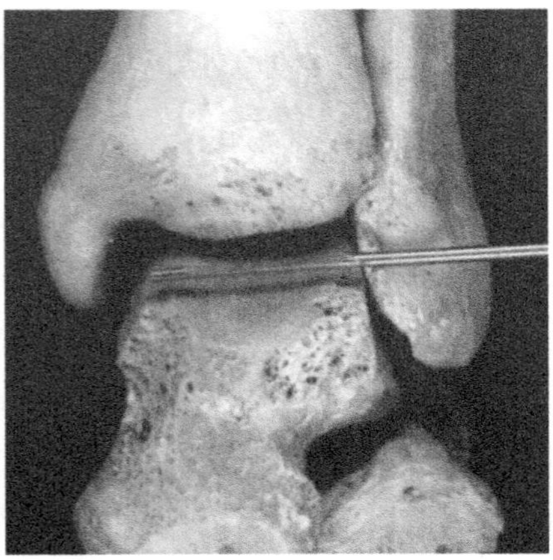

FIGURE 6 Approach for percutaneous biopsy of the synovium of the ankle.

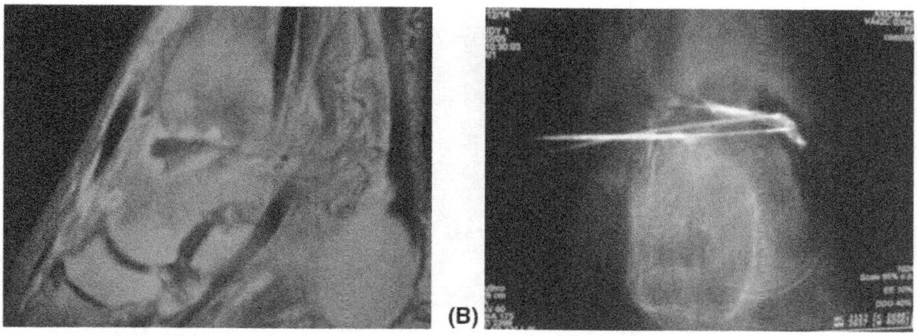

FIGURE 7 Percutaneous biopsy of the synovium (PBS) of the ankle through an anterolateral approach in septic arthritis due to *Staphylococcus aureus.* (**A**) Gadolinium-enhanced T1-weighted sagittal magnetic resonance image prior to biopsy. (**B**) Arthrography during PBS.

in external rotation anchored with a sandbag to protect the long biceps tendon. The SNCN is inserted at a point equidistant from the acromioclavicular joint and the coracoid process, and 1 cm lateral to the vertical line passing through the glenohumeral interspace. The needle is advanced downward with a 20° to 30° posterior and slightly medial inclination. The needle must reach the bone and then slide over the anterior and medial aspects of the humeral head (Figs. 8 and 9).

Technique for Percutaneous Biopsy of the Synovium of the Elbow Joint

The biopsy is performed at the posterolateral aspect of the elbow, between the olecranon process medially and the lateral epicondyle anteriorly. The patient is placed supine with the elbow semipronated and flexed approximately 60°. The needle is inserted 1 cm lateral and 1 cm proximal to the upper extremity of the olecranon process and advanced downward with slight (20–30°) anterior inclination to cross the posterolateral aspect of the joint (Fig. 10).

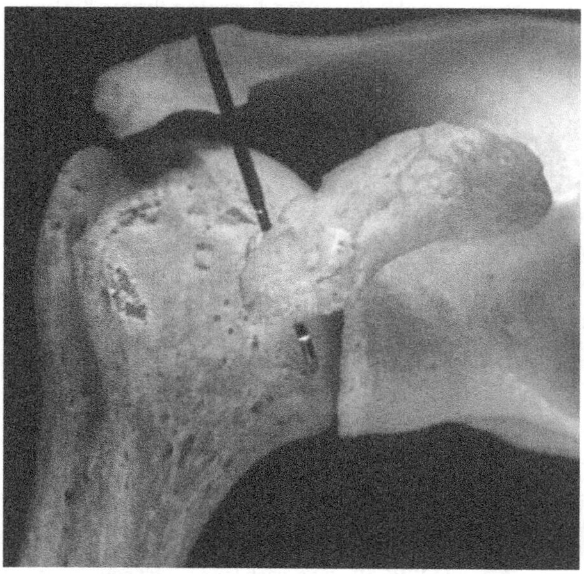

FIGURE 8 Approach for percutaneous biopsy of the synovium of the shoulder.

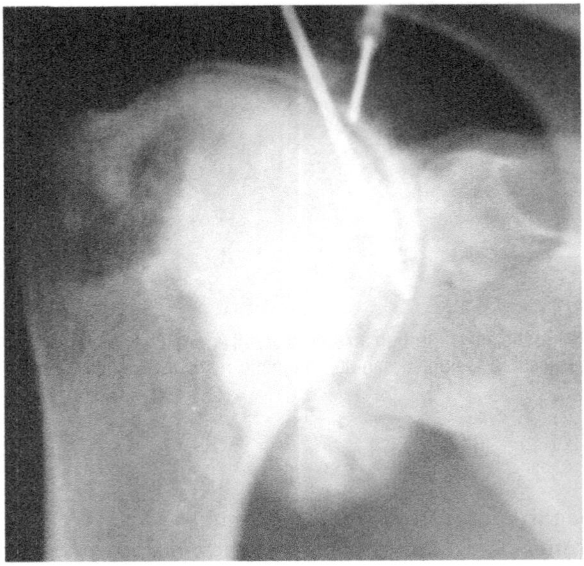

FIGURE 9 Percutaneous biopsy of the synovium of the shoulder.

Technique for Percutaneous Biopsy of the Synovium of the Wrist Joint

The posterior aspect of the radiocarpal joint is biopsied through a horizontal approach. The patient is placed supine on the X-ray table with his arm under the fluoroscope and the hand pronated. The needle is inserted at the posteromedial aspect of the wrist between extensor carpi ulnaris and extensor tendons and then advanced horizontally toward the radiocarpal joint (Figs. 11 and 12).

COMPLICATIONS

Complications of PBS are rare. Some patients experience mild pain and tenderness in the days following the procedure. Joint effusion and hemarthrosis (7) are very rare. Use of a strict aseptic technique makes a joint infection unlikely. However, this risk is much higher in patients with immunodeficiency, especially in those with chronic renal failure undergoing hemodialysis. With these patients, special attention must be paid to the indication of PBS.

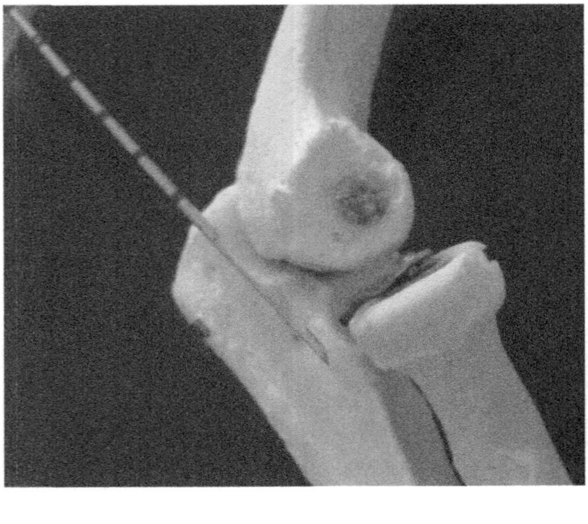

FIGURE 10 Approach for percutaneous biopsy of the synovium of the elbow.

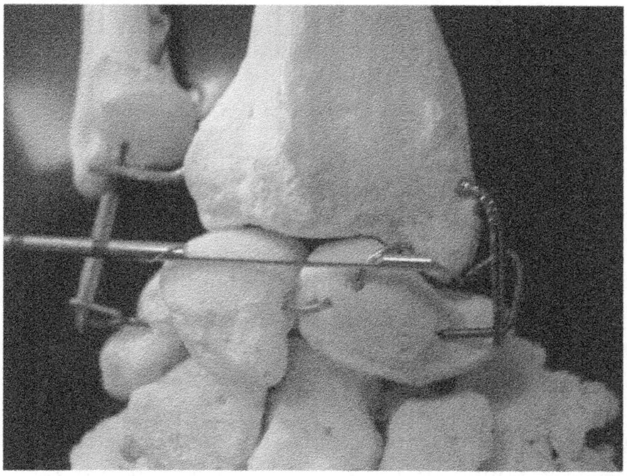

FIGURE 11 Approach for percutaneous biopsy of the synovium of the wrist—retroulnar approach to the posterior aspect of the joint.

RESULTS

The reported accuracy of PBS in obtaining adequate synovial specimens has varied according to the specific joint being evaluated. In most series of PBS of the knee, the accuracy rate is high, with adequate specimens obtained in 86.2% (8) to 96.2% (10) of cases Reports concerning PBS of other joints are rare (5–8,10). The absence of synovial tissue upon laboratory examination, despite apparently appropriate technique and samples, is more frequent in PBS in areas other than the knee, especially in biopsies of the hip and shoulder. The articular cavity of these joints is relatively small, and access to the synovium is more difficult than in the knee joint (7). Moon et al. obtained synovial tissue in 15 of 22 (68.2%) PBSs of the wrist, elbow, and ankle using the Franklin-Silverman needle (7). Beaulé et al. (5) obtained synovial membrane in 81% of PBSs using Tru-Cut needles (Travenol Labs, Deerfield, Illinois, U.S.A.) and, more precisely, 88% for the hip and 79% for other joints (shoulder, elbow, wrist, and ankle). The success rate for obtaining samples of the synovial membrane also varies with the pathologic process. In the hip joint, synovial tissue is obtained by means of PBS in only 48% of cases of degenerative diseases and, by contrast, in 83% of cases of rheumatic diseases or septic arthritis. The significant hypertrophy of the synovial membrane in rheumatic diseases and septic arthritis explains the high success rate for obtaining synovial membrane in these diseases. By contrast, in osteoarthritis of the hip, a true synovial lining is often lacking at histopathological examination, which only shows capsular tissue and small cartilage pieces despite an adequate technique authenticated by the presence of fluid reflux during the procedure. The atrophy of the synovial membrane, which is common in degenerative joint diseases, may be one explanation for the relatively low success rate for obtaining synovial membrane (48%) in such cases (5).

(A) (B)

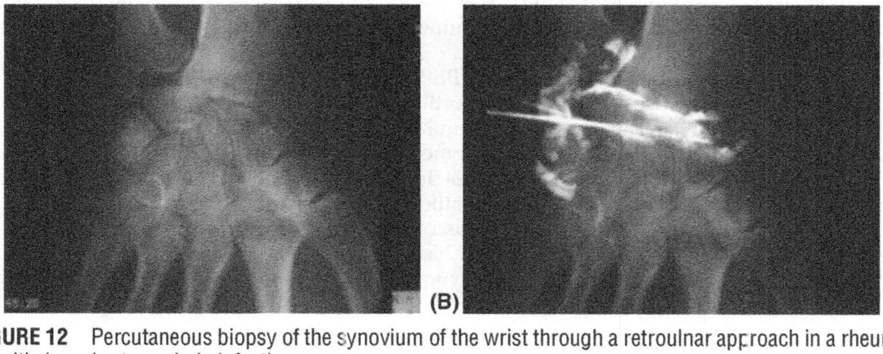

FIGURE 12 Percutaneous biopsy of the synovium of the wrist through a retroulnar approach in a rheumatismal monoarthritis in order to exclude infection.

PBS results have been found to be of major diagnostic value in 38% (8) to 53% (11) of diseases undiagnosed prior to the biopsy. Other studies have described a higher percentage, up to 70.7% (7), of biopsies helpful to the diagnosis. However, all reported series are not comparable and, as noted by Schumacher and Kulka, some of them include many patients with a previously established diagnosis (8).

INDICATIONS

Results of laboratory examination of the synovium may provide a definitive diagnosis in a limited range of articular diseases (9). Of these conditions, suspicion of septic arthritis, either pyogenic or tuberculous, is by far the most important indication for PBS. Erosive osteoarthritis (especially rapidly destructive osteoarthritis), which may mimic septic arthritis, is another relatively frequent indication of PBS.

The nonspecificity of many other synovial reactions including those of rheumatoid arthritis (12,13), seronegative spondylarthropathies, and connective tissue diseases has already been emphasized by several authors (8,14–19). Even the most typical picture of rheumatoid arthritis synovitis cannot lead to a definitive diagnosis, and may be seen in other rheumatic diseases including seronegative spondylarthropathies and systemic lupus erythematosus (9). In these conditions, synovial biopsy only contributes to a diagnosis, which depends on the addition of several criteria. Finally, PBS in joints other than the knee is rarely indicated in these diseases, unless intercurrent secondary joint infection is suspected.

REFERENCES

1. Forestier J. Instrumentation pour medical. CR Soc Biol (Paris) 1932; 110:186.
2. Polley HF, Bickel WH. Punch biopsy of synovial membrane. Ann Rheum Dis 1951; 10:277–287.
3. Parker RH, Pearson CM. A simplified synovial biopsy needle. Arthritis Rheum 1963; 6:172–175.
4. Aignan M. La ponction biopsie synoviale de la hanche. Med Hyg 1979; 37:115.
5. Beaulé V, Laredo JD, Cywiner C, Bard M, Tubiana JM. Synovial membrane: percutaneous biopsy. Radiology 1990; 177:581–585.
6. Laredo JD, Bard M. Percutaneous biopsy of the synovial membrane. In: Laredo J-D, Bard M, eds. Interventional Radiology in Bone and Joints. Wien, New York: Springer-Verlag, 1988:51–64.
7. Moon MS, Kim T, Kim JM, Lee HS, Atin YP. Synovial biopsy by Franklin-Silverman needle. Clin Orthrop Rel Res 1980; 150:224–228.
8. Schumacher HR, Kulka JP. Needle biopsy of the synovial membrane: experience with the Parker-Pearson technique. N Engl J Med 1972; 286:416–419.
9. Goldenberg DL, Cohen A. Synovial membrane histopathology in the differential diagnosis of rheumatoid arthritis, gout, pseudogout, systemic lupus erythematosus, infectious arthritis and degenerative joint disease. Medicine 1978; 57:239–249.
10. Guzman L, Arinoviche R. Intraarticular fracture of synovial biopsy needle. Arthritis Rheum 1978; 21:742.
11. Saaibi DL, Schumacher HR. Percutaneous needle biopsy and synovial histology. Bailliere's Clin Rheumatol 1996; 10:535–554.
12. Schumacher HR, Bautista BB, Krauser RE. Histologic appearance of the synovium in early rheumatoid arthritis. Sem Arthr Rheum 1994; 23:224–233.
13. Kirkham BW, Navarro FJ, Corkill MM, Panayi GS. In vivo analysis of disease modifying drug therapy activity in rheumatoid arthritis by sequential immunohistological analysis of synovial membrane interleukin 1B. J Rheum 1994; 21:1615–1619.
14. Sherman MS. Non-specificity of synovial reactions. Bull Hosp Joint Dit 1951; 12:110–125.
15. Sokoloff. Biopsy in rheumatic diseases. Med Clin North Am 1961; 45:1171–1180.
16. Beutler AM, Wittum-Hudson JA, Nanagara R. Intracellular location of inapparently infecting chlamydia in synovial tissue from patients with Reiter's syndrome. Immunol Res 1994; 13:163–171.
17. Bubul R, Rothfuss S, Clayburne G. Studies on synovial fluid and synovium in psoriatic arthritis. Effects of medications and implications for diagnosis and pathogenesis. Arthritis Rheum 1995; 38:S380.
18. Chang C, Schumacher HR. Light and electron microscopic observations on the synovitis of synovitis of ankylosing spondylitis. Sem Arthr Rheum 1992; 22:54–65.
19. Gallagher PJ, Blake DR, Lever JV. Audit of closed synovial biopsy in the diagnosis of inflammatory joint disease. Scand J Rheum 1985; 14:307–314.

6 Histology and Cytology of Biopsies of Tumors and Histology and Microbiology for Evaluation of Infection

Brian J. O'Hara
Department of Pathology, Anatomy, and Cell Biology, Thomas Jefferson University Hospital, Philadelphia, Pennsylvania, U.S.A.

GENERAL CONSIDERATIONS

Biopsy of a skeletal lesion, with subsequent histologic examination, is often performed as part of the diagnostic procedure by the traditional triangle of surgeon, radiologist, and surgical pathologist. Needle core biopsy can be used to confirm the clinical and radiological impression or to procure adequate tissue necessary for ancillary studies. This has increasingly translated into the use of percutaneous needle core biopsy (1,2). In essence, the surgical pathology laboratory has the task of extrapolating maximal information from, at times, minimal tissue samples. This information may be used to plan subsequent excisions or the addition of adjuvant therapy, or may result in no further action. The mainstay of pathologic diagnosis remains histologic examination of tissues, utilizing the hematoxylin–eosin stain. Additional special stains, microbiologic cultures, immunohistochemical studies, electron microscopy, flow cytometry, and cytogenetic and molecular diagnostic techniques are also employed when required (3–6). The following chapter briefly covers the histologic and cytologic features of osteomyelitis and selected bone tumors.

PREPARATION OF TISSUE

Due to the presence of mineralized matrix within resection and biopsy specimens, routine processing techniques, utilized in soft-tissue specimens, cannot be performed. Bony tissue received within the laboratory is customarily fixed in 10% formalin. If a hematologic, lymphoproliferative process or other small round blue cell tumor is clinically suspected, the specimen should be sent fresh (without fixative). Tissue can then be harvested for flow cytometric analysis, molecular diagnostics, or cytogenetic studies. Additionally, microbiologic cultures can be performed. A tissue imprint (smear) can provide preliminary information as to adequacy of sample or aid in the diagnosis. A frozen section may be attempted if tissue is deemed adequate and without significant osseous matrix. After formalin fixation, a decalcification step must be performed prior to further processing. This may involve a commercially available acid-based solution (Decal) or an ethylenediaminetetraacetic acid-containing reagent if dealing with small biopsies. A delay in diagnosis may occur if the specimen is large or involves dense cortical bone, thus prolonging the decalcification step. Most needle core biopsies require less than 24 hours. Another alternative is to use a methyl methacrylate–based embedding technique in which the decalcification step is unnecessary. This option allows for excellent cytologic and histologic preservation of tissue. However, it is both time consuming and requires specialized equipment, which may be too expensive for most surgical pathology laboratories.

INFECTIONS (OSTEOMYELITIS)

Infections involving bones may be a consequence of hematogenous spread or by direct inoculation following trauma or surgical intervention (secondary osteomyelitis). Hematogenous bacterial osteomyelitis results from a source elsewhere in the body and has decreased in incidence over the past 25 years. The typical case involves the metaphyseal portion of long bones in children aged 15 years or below. *Staphylococcus aureus* is the most common organism cultured; however,

Hemophilus influenza may be present in children aged less than three years (7). Other bacteria such as Group B streptococcus and coliforms may be seen in neonates. *Staphylococcus epidermidis* is another pathogen sometimes cultured and may follow surgical procedures or invasive techniques such as an intravenous catheter placement. Vertebral body involvement, especially in older patients with genitourinary tract infection (gram-negative rods), can occur via Batson's venous plexus. Infections involving the small bones of the hands and feet usually manifest in elderly diabetics (8).

Acute Hematogenous Osteomyelitis

The metaphyseal region in the long bones of children and adolescents is frequently affected due to the presence of unique capillary loops. Blood flow slows in these areas, allowing bacteria to escape more easily through the fenestrated capillaries, adhere to the surface of trabecular bone, and multiply. Depending on the virulence of the organism, the ensuing proliferation results in edema and an increased pressure on adjacent blood vessels, which further compromises the vascular supply, resulting in bone necrosis. *S. aureus* is the most common organism cultured in acute hematogenous osteomyelitis (9). In addition to possessing receptors to bone surface proteins, *S. aureus* can secrete a mucopolysaccharide coating, enhancing its ability to adhere to trabecular bone, while resisting penetration by antibiotics and phagocytosis by tissue macrophages (10). A biopsy of the involved area would initially show edema within the tissues, increased acute inflammatory cells (neutrophils), and perhaps necrotic bone. Marrow fibrosis and reactive bone formation may not yet be present. As the infection progresses, bone destruction occurs because cytokines released by inflammatory cells stimulate osteoclasts to resorb bone. Access to both Haversian and Volkmann canals allows the infection to spread to the periosteum, resulting in periosteal elevation and abscess formation. Local cortical necrosis (sequestrum) develops because periosteal-based cortical blood vessels are compromised. The elevated periosteum produces reactive new bone, which envelops the area (involucrum) (11).

Secondary Osteomyelitis

Direct inoculation of the bone may result from traumatic events, adjacent soft-tissue infection or may be iatrogenic in nature. Post-traumatic osteomyelitis has become the most important type of inflammatory bone process (12–14). Accurate diagnosis can be complicated by vague symptomatology or masked by concurrent disease processes. A biopsy coupled with microbiologic culture will confirm the clinical suspicion in most cases.

Subacute Osteomyelitis

Subacute osteomyelitis follows an initial infection. The patient's immune system may be able to contain but not totally obliterate the infective focus. Although acute inflammatory cells are still present within the edematous granulation-type tissue, lymphocytes and plasma cells are also noted. Symptoms can be vague, such as low-grade fever or ill-defined pain, or nonexistent. Imaging studies will show a circumscribed lytic area surrounded by reactive bone (Brodie's abscess).

Chronic Osteomyelitis

Failure to obliterate acute osteomyelitis (hematogenous or secondary) leads to chronic osteomyelitis, which can result in significant disability. Long-term treatment with antibiotics and possible surgical debridement of the active focus may be necessary. Most cases are associated with open fractures (secondary osteomyelitis); however, occasional cases are attributed to hematogenous routes of infection. *S. epidermidis, S. aureus, Pseudomonas aeruginosa, Serratia marcescens,* and *Escherichia coli* are commonly isolated in patients with chronic osteomyelitis (15). Symptoms tend to be intermittent and may persist for years. Complications include development of sequestra, draining sinus tracts, and rare cases of squamous cell carcinoma (16).

Chronic Recurrent Multifocal Osteomyelitis

This entity represents a rare variant of osteomyelitis. Children and young adults, predominantly women, are usually affected. Clinical symptoms include malaise with local swelling and pain in the affected bone. Multifocal lesions may be present, which resolve spontaneously over a span of years. Etiology is unknown; however, some authors have reported a high incidence of preceding throat infections, whereas others postulate an indolent organism such as mycoplasma. The results of bacterial, fungal, and viral cultures have been negative (17,18).

Granulomatous Osteomyelitis

Granulomatous inflammation of the bone can be seen with either tuberculosis (TB) or fungus. Sarcoid has been reported less commonly. After a decline in the 1940s, a recent increase in TB cases has been reported. The reasons for this increase may be the result of an overall increase in the number of immunocompromised patients, as well as travel to or immigration from a region in which TB is endemic, or the emergence of resistant strains of TB. Fungal involvement of the bone is usually seen against the background of an immunocompromised patient or after prolonged use of corticosteroids.

Histology and Cytology of Osteomyelitis

Inflammation and reactive bone encountered in bone biopsy material are not necessarily pathognomonic of osteomyelitis. A number of conditions may elicit an inflammatory response. These include fracture, infarct, arthritides, and neoplasm. Also, the ability to categorize with certainty whether a particular case represents acute, subacute, or chronic osteomyelitis by microscopic examination is limited due to the considerable histological overlap. This limitation may be further magnified by the small amount of tissue sample in needle core biopsies. Most authors will agree that, in general, acute suppurative osteomyelitis will contain acute inflammatory cells as the prominent feature with an increase in osteoclastic activity (Fig. 1). In the subacute category, there is quantitatively less acute inflammation, while a mixture of plasma cells and lymphocytes are present. Granulation tissue is often identified. Chronic osteomyelitis will contain lymphocytes, plasma cells, and marrow fibrosis (Fig. 2). Necrotic bone tends to be more common in chronic osteomyelitis. The histology of chronic recurrent multifocal osteomyelitis can vary from collections of acute inflammatory cells to a mixture of lymphocytes, plasma cells, and even peculiar granulomas. Granulomatous inflammation may be present in either fungus or TB, with caseation favoring the latter. Cultures along with special stains, such as a modified acid-fast, Periodic Acid Schiff (PAS), and silver stains, can differentiate between the two. Immunocompromised patients may not show typical caseating granuloma formation, and a background population of histiocytes may be the only histologic evidence. Acid-fast stains will highlight intracellular bacilli, which can be numerous in the case of Mycobacterium avium intracellulare.

Differential Diagnosis

The histologic differential diagnosis of osteomyelitis is usually not problematic when dealing with the acute (suppurative) form. Chronic osteomyelitis, with its prevalence of lymphocytes,

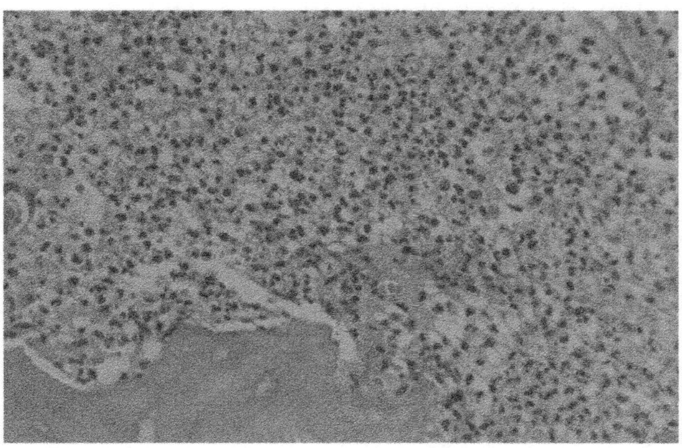

FIGURE 1 Acute osteomyelitis. Acute inflammatory cells (neutrophils) within marrow space adjacent to nonviable trabecular bone (hematoxylin and eosin, original magnification × 40).

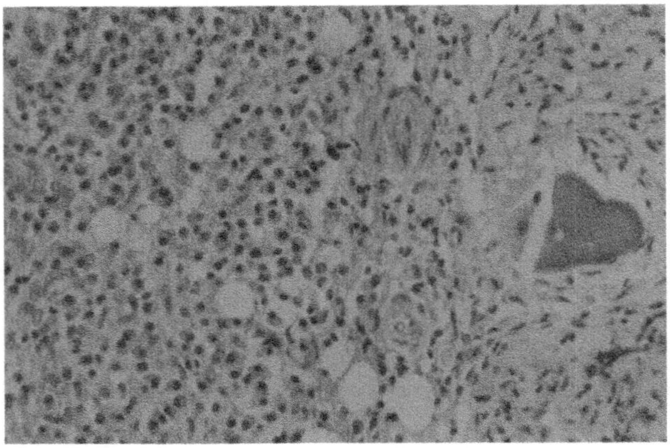

FIGURE 2 Chronic osteomyelitis. Prominent plasma cell focus with background lymphocytes. Focal fibrosis and nonviable bone fragment (hematoxylin and eosin, original magnification × 40).

can simulate lymphoma of bone. If a prominent plasma cell component is biopsied, the possibility of multiple myeloma/plasmacytoma may be entertained. A diagnosis of Langerhans cell histiocytosis (eosinophilic granuloma) may be considered if a biopsy involves a heterogeneous mixture of plasma cells, lymphocytes, macrophages, scattered neutrophils, and eosinophils. Immunohistochemical stains (T and B-cell markers, and/or immunoglobulin class) are utilized to demonstrate clonality in the case of suspected lymphoma or plasma cell dyscrasia. Langerhans cell histiocytosis is characterized by expression of the Langerhans cell phenotype (S-100+, CD 1a+). Another small round blue cell tumor to consider is Ewing's sarcoma/peripheral neuroectodermal tumor (PNET). Immunohistochemical stains and cytogenetic studies may aid in the diagnosis by demonstrating the MIC2 gene product (CD99) and reciprocal translocation of chromosome 11 and 22, which are associated with Ewing's sarcoma/PNET (19).

BONE-FORMING TUMORS

Osseous tumors can be found in both the axial and appendicular skeleton. The majority are benign, with primary malignant bone tumors accounting for 0.1% to 0.2% of all new cancers per year, exclusive of multiple myeloma and other hematopoietic neoplasms. Tumors in adolescents and young adults account for the preponderance of cases. The bony matrix can vary from dense compact lamellar bone present in an osteoma to the wispy osteoid sometimes found in osteogenic osteosarcoma (20).

Osteoma and Bone Islands

These sclerotic, slow-growing lesions are considered hamartomatous by some authors. They can be divided as per sites of involvement. Thus sinonasal osteomas are seen affecting the paranasal sinuses, facial bones, and orbit. Juxtacortical osteomas affect the surfaces of long bones. "Ivory exostoses" are found in the mandibular and calvarial bones, whereas enostoses are seen within the medullary cavity.

Histology and Cytology

The above lesions usually contain dense compact lamellar bone (Fig. 3). Sino-orbital osteomas sometimes contain areas that show prominent osteoblastic rimming as well as osteoclastic activity, and are histologically similar to osteoblastomas. These features have given reason for some authors to consider sino-orbital osteomas as benign bone-forming tumors and not hamartomas (21).

Differential Diagnosis

This is usually not a problem when dealing with small lesions and most are not biopsied unless there are cosmetic considerations or symptomatology. A large slowly growing bone island or

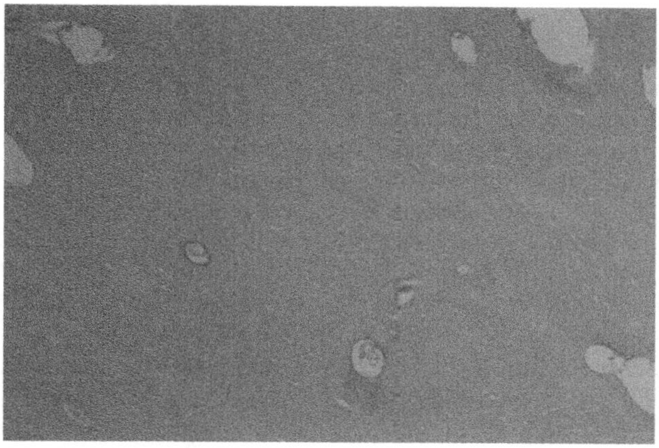

FIGURE 3 Osteoma. Dense cortical lamellar bone is present in this example from the skull (hematoxylin and eosin, original magnification × 4).

surface osteoma of a long bone may cause suspicion of the sclerotic variant of osteosarcoma or parosteal osteosarcoma, respectively. However, the spindle cell stroma typical of both osteosarcoma subtypes will be lacking on histologic examination.

Osteoid Osteoma and Osteoblastoma

The osteoid osteoma is a benign, circumscribed, frequently cortical-based lesion that typically involves the long bones in teenagers and young adults. The richly vascular nidus generally measures less than 1 cm and contains numerous unmyelinated nerve fibers, which were thought to be a prime reason for the pain often associated with this entity (22,23). Some authors now believe that the elaboration of high levels of prostaglandin E2 and prostacyclin are not only responsible for the pain but also for the reactive sclerosis and nonspecific inflammatory changes that are invariably present (24). Although histologically similar, osteoblastoma differs from osteoid osteoma in that it is larger (usually >2 cm), and more often involves the vertebral column. The symptom of nocturnal pain that can be relieved by aspirin is usually encountered in osteoid osteoma (25).

Histology and Cytology

The nidus of an osteoid osteoma consists of interlacing thin bony trabeculae against the background of a loose vascularized connective tissue. There is prominent osteo-blastic rimming of the

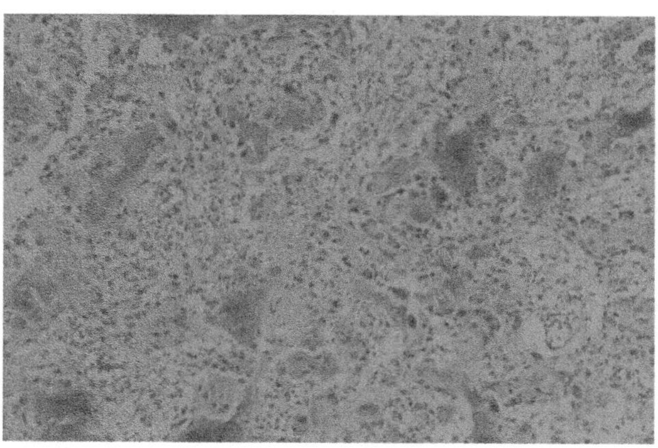

FIGURE 4 Osteoid osteoma: nidus of osteoid osteoma containing numerous irregular trabeculae of woven bone, adjacent active osteoblasts, scattered osteoclast-like giant cells, and fibro-vascular marrow (hematoxylin and eosin, original magnification × 10).

trabeculae, which can display different levels of mineralization. Mitoses can be seen, but atypical mitotic figures should not be present. Multinucleated osteoclast-like giant cells are also noted (Fig. 4). The nidus can be very cellular and may show a zonal architecture with the more mature, less cellular elements located centrally and surrounded by a sclerotic rim. Aggressive osteoblastoma bears morphologic similarity to both osteoid osteoma and osteoblastoma. Several authors have documented the presence of "epithelioid osteoblasts." These osteoblasts are larger, round, possess eosinophilic cytoplasm, and contain an eccentrically located nucleus (26–28).

Differential Diagnosis

The histologic differential diagnosis of an osteoid osteoma and osteoblastoma includes osteomyelitis, especially if the nonspecific inflammatory component is misinterpreted. Additionally, needle biopsy of the cellular region may contain increased osteoblastic activity and occasional mitoses, which can be misinterpreted as osteosarcoma or Ewing's sarcoma. The lack of atypical mitotic figures or significant pleomorphism should help avoid this pitfall. The gross identification of the nidus may be assisted by the injection of methylene blue via computed tomography needle localization (29). Another method is administration of tetracycline prior to biopsy. The curettage material will fluoresce when exposed to an ultraviolet lamp if the nidus is present (30). Well-differentiated "osteoblastoma-like" osteosarcoma must be considered when dealing with an aggressive osteoblastoma.

Osteosarcoma

This malignant tumor of bone accounts for approximately 20% of all primary bone sarcomas. There is a bimodal age distribution, with the first peak occurring in the second decade and a smaller peak seen in patients above the age of 50. Males predominate and there is a tendency to involve the appendicular skeleton, with the majority of cases being reported in the metaphyseal region about the knee. The morphologic classification encompasses various subtypes; however, the common determinant rests on the identification of osteoid or woven bone within the tumor. A four tier histologic grading system has traditionally been utilized. However, there were problems with subjectivity and reproducibility. This has given way to the current trend of dividing conventional osteosarcoma into high- and low-grade categories based on the degree of cellular pleomorphism. The quality and amount of osteoid production can vary greatly between histologic variants (27). There is an association with the Rb gene, a tumor-suppressor gene originally identified in retinoblastoma. Patients who inherit mutations in the Rb gene have an increased risk of developing osteosarcoma (up to 500-fold) (31,32).

Histology and Cytology

There is considerable histologic and cytologic variability within conventional osteosarcomas. The predominant matrix pattern has been used to divide osteosarcoma into its variants, with osteoblastic osteosarcoma being more common than the chondroblastic and fibroblastic histologic subtypes. Mixtures of the patterns can and do occur. Tumor growth patterns are usually intramedullary but may be located on the bone surface.

Conventional Intramedullary Osteosarcoma

The usual presentation is that of a poorly differentiated sarcoma with osteoblastic differentiation. The oval and round tumor cells contain significant pleomorphism with a brisk mitotic rate as well as atypical mitoses. The cytoplasm can be densely eosinophilic with an eccentric hyperchromatic nucleus. The tumor cells are usually larger than the osteoblasts that they sometimes resemble (Fig. 5). The osteoid production may consist of broad islands to wispy trabecules and can be densely mineralized or undergo little mineralization. When thin and delicate interlacing osteoid trabeculae become mineralized, a "filigree" histologic pattern can emerge (33).

Chondroblastic Osteosarcoma

Some osteosarcomas consist of prominent chondroblastic foci that are cytologically medium-to-high grade. These tumors may need to be extensively sampled with a meticulous hunt performed for osteoid production by the neoplastic cells. Examination at the periphery of or in between cartilaginous lobules will often reveal the diagnostic regions.

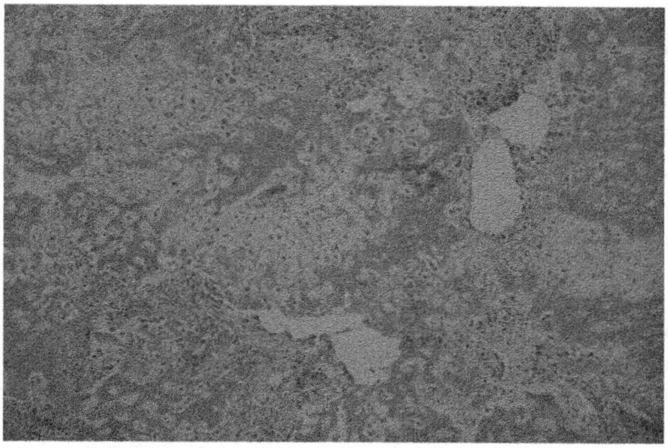

FIGURE 5 Osteoblastic osteosarcoma. Abundant osteoid matrix surrounds sarcomatous tumor cells (hematoxylin and eosin, original magnification × 10).

Fibroblastic Osteosarcoma

The spindle cell stroma of this histologic variant can contain a deceptively bland cytology with a "herringbone" pattern, reminiscent of a fibrosarcoma. Another presentation is that of a high-grade spindle cell sarcoma mimicking a malignant fibrous histiocytoma (MFH) (34). In either case, there may be only focal osteoid production, which will require careful microscopic examination.

Giant Cell-Rich Osteosarcoma

In this variant, large numbers of benign appearing stromal giant cells are present, which can mask the sarcomatous component. Once again, a thorough search needs to be performed to identify osteoid production and definitive stromal pleomorphism in order to arrive at the correct diagnosis.

Small-Cell Osteosarcoma

Sheets of undifferentiated small round cells with minimal cytoplasm characterize this histologic entity. Three subtypes are recognized based on cell size. The first is the Ewing's sarcoma-like pattern. The tumor cells are round to oval with coarse clumped chromatin and mild cellular pleomorphism (Fig. 6). The second type is the lymphoma-like pattern, where cells are slightly larger. The chromatin is vesicular and prominent nucleoli are present. The third pattern contains crowded spindle-shaped cells with scant cytoplasm. Mixtures of these patterns may occur. Glycogen may be present within the cytoplasm similar to Ewing's sarcoma (35,36).

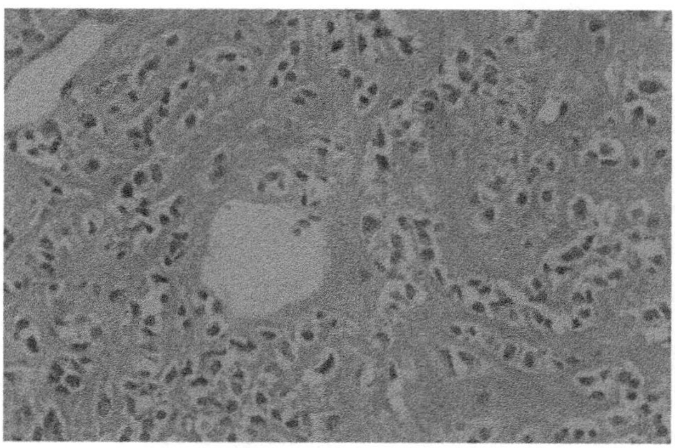

FIGURE 6 Small cell osteosarcoma. Small round tumor cells are embedded within osteoid (hematoxylin and eosin, original magnification × 40).

Telangiectatic Osteosarcoma

This lytic variant is made up of cystically dilated vascular spaces. The septa between the spaces contain conventional osteosarcoma cells. Two histologic variants are described. The first is the hemorrhagic and necrotic type. The malignant cells are often scattered and sometimes widely separated by the blood and necrosis. Osteoid tends to be minimal but can be absent in up to 20% of the biopsy cases. The second type is morphologically similar to aneurysmal bone cyst if examined on low magnification. The malignant cells contain hyperchromatic nuclei and are usually present in the cyst walls. Atypical mitotic figures are also noted. Osteoid, when present, is wispy and lace-like (37).

Intraosseous Well-Differentiated Osteosarcoma

The hallmark of this type is the cytologically bland fibrous and osseous components that make up this entity. The fibroblasts have an "activated" appearance and mild atypia may be present; however, obvious pleomorphism is not present. Mitoses are present and number approximately 1 to 2 per 10 high-powered microscopic field. The osteoid production can be conspicuous, mature appearing, and hard to separate from surrounding normal bone. A minority of cases will contain only scant osteoid production, while a subset can resemble the histologic appearance of fibrous dysplasia (38,39).

Parosteal Osteosarcoma

Morphologically similar to its well-differentiated intraosseous counterpart, this variant is located in the juxtacortical region of long bones and tends to grow out from the cortex like a bony protrusion. Most occur in the posterior aspect of the distal femur. The woven bone trabeculae are often long, narrow, and separated by fibroblastic tissue with only minimal cytologic atypia (Fig. 7). Islands of osteoid production may be ill defined. Only scattered mitotic figures are identified and usually no atypical forms are present. Foci of high-grade osteosarcoma are sometimes found in these lesions, and, when present, are termed "dedifferentiated parosteal osteosarcoma" (40,41).

Periosteal Osteosarcoma

This surface osteosarcoma grows around the bone and encases it. It is more prevalent in the diaphyseal portion of the long bone as opposed to the metaphyseal region as in the other osteosarcoma variants. Its histologic appearance is that of cellular cartilage lobules with cytologic atypia, separated by fibrous stromal bands of tissue (Fig. 8). The neoplastic osteoid necessary for diagnosis can usually be found within the fibrous stroma (42,43).

High-Grade Surface Osteosarcoma

Histologic examination of this rare variant will demonstrate a morphology very similar to conventional osteosarcoma. However, it arises from the surface of bone and will contain cortical

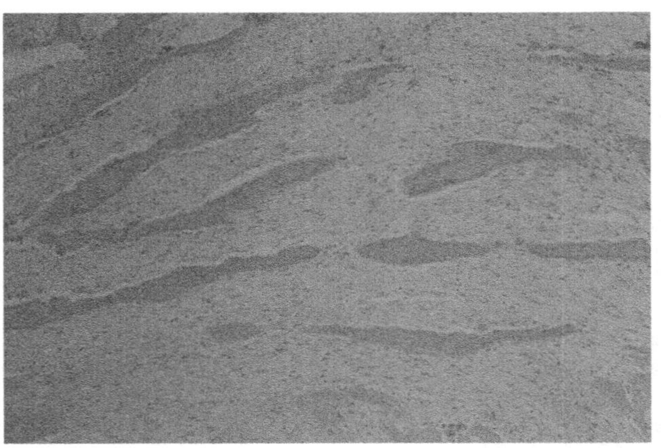

FIGURE 7 Parosteal osteosarcoma. A bland fibroblastic stroma and elongated bony trabeculae are present. Mitoses are rare and cellular pleomorphism is minimal (hematoxylin and eosin, original magnification $\times$ 4).

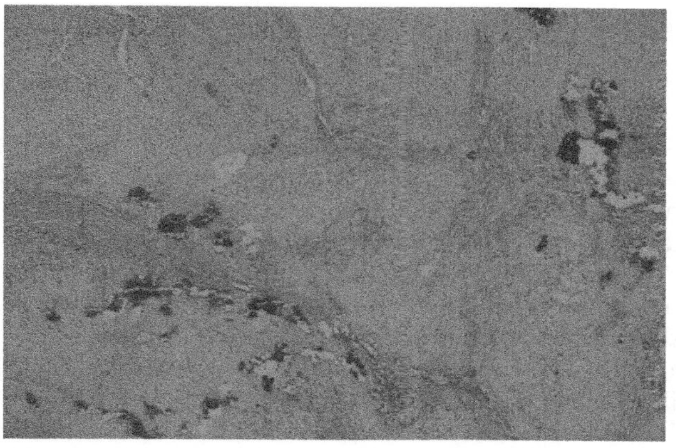

FIGURE 8 Periosteal osteosarcoma. A prominent lobulated cartilaginous component with calcification is noted. Osteoid production can be focal and found in the spindle cell septa between lobules (hematoxylin and eosin, original magnification × 4).

invasion in most cases. The tumor cells are pleomorphic with hyperchromatic nuclei. Mitotic figures as well as osteoid production are present (44).

Differential Diagnosis

The diagnosis of osteosarcoma is usually uncomplicated when typical histology is complimented by clinical and radiographic information. The histologic variants when examined without benefit of the aforementioned information may give rise to other diagnostic possibilities. Osteoblastoma and fracture callus may bear some resemblance to conventional intramedullary osteosarcoma, if dealing with minimal tissue samples. The findings of cellular anaplasia, atypical mitotic figures, and an infiltrative border will help rule out osteoblastoma, whereas non-neoplastic cartilage and reactive bony trabeculae lined by osteoblasts are typically seen in a fracture callus. Chondroblastic osteosarcoma can be distinguished from chondrosarcoma by detecting definitive osteoid formation. Distinguishing between giant cell tumor of bone and giant cell osteosarcoma can be difficult. The finding of pleomorphism in the mononuclear cells and atypical mitotic figures help argue against giant cell tumor of bone. Additionally, giant cell tumors tend to occur in the epiphysis of skeletally mature individuals, while the preponderance of osteosarcomas are in the metaphyseal region of the skeletally immature individuals. Another histologic puzzle is the finding of a small round blue cell tumor, especially if osteoid is not readily identified. The separation of small-cell osteosarcoma will rely on the findings of ancillary studies. This may include immunohistochemistry, flow cytometry, and molecular analysis to rule out lymphoma, Ewing's sarcoma, Langerhans cell histiocytosis, and osteomyelitis. Intraosseous and well-differentiated osteosarcoma may be confused with benign fibro-osseous lesions such as fibrous dysplasia. A careful search for osteoid and cellular atypia is necessary to make the distinction. Telangiectatic osteosarcoma can mimic aneurysmal bone cyst. The histologic identification of osteoid formation, atypical mitoses, and cellular pleomorphism in the walls of the vascular spaces will call attention to the true malignant nature of the neoplasm.

CARTILAGE-PRODUCING TUMORS

Cartilaginous lesions can be hamartomatous, benign, or malignant. Osteochondroma is an example of a hamartomatous exostosis, which can be solitary or part of a hereditary syndrome (multiple hereditary exostoses). A common benign intramedullary lesion is the enchondroma, which frequently affects the small bones of the hands and feet. It may also be part of a syndrome such as Ollier's disease or Maffucci's syndrome. Its surface counterpart is the periosteal chondroma. Chondroblastoma and chondromyxoid fibroma are two benign cartilaginous lesions in which the cartilage cells are immature and associated with an extracellular matrix. Chondrosarcomas are malignant cartilage tumors, which rank as the second most common bone tumor following osteosarcoma. These tumors usually are found in older patients, with the pelvis being the most common site.

Osteochondroma

The solitary, sporadic osteochondroma is a cartilage-capped outgrowth or hamartomatous projection of bone that maintains its connection with the medullary canal. They are thought to arise from islands of epiphyseal plate cartilage left behind during active skeletal growth. The cartilaginous island grows and ossifies similar to the adjacent bone, stopping at skeletal maturity. Most are asymptomatic and true prevalence is unknown. Resection may follow cosmetic concerns or be secondary to pain caused by compressive symptomatology. A thin fibrous periosteum covers the hyaline cartilage cap (45). The cartilage cells are evenly distributed and usually normal in appearance. Some mild or even moderate nuclear atypia can be seen. Where the cartilage cap meets the underlying bone, chondrocytes line up into columns similar to the epiphyseal plate (Fig. 9). The underlying marrow is usually fatty; however, hematopoietic cells may be present.

Enchondroma

These islands of mature hyaline-type cartilage are found within the medullary canal of the metaphysis or metadiaphysis. They are thought to result from displaced nests of cartilage from the growth plate during development and consequently are found only in bones that undergo enchondral ossification. Particularly common sites are the hands and feet. They comprise lobules of hyaline cartilage, which are usually surrounded by a thin band of reactive bone into which they blend. Chondrocytes are typically small with round nuclei, a dark homogeneous chromatin pattern, and surrounded by abundant chondroid matrix (Fig. 10). The nuclei lie within a single lacuna. Cellularity may vary and occasional binucleated forms can be present. Occasionally, necrosis and myxoid changes may be seen (28). The differential diagnosis mainly rests with chondrosarcoma, especially if the lesion is cellular. In this instance, a low-grade chondrosarcoma may be virtually impossible to separate from enchondroma by histologic means. The finding of an "aggressive" radiographic presentation and pain in the absence of fracture may be the only clues to separate the two entities.

Periosteal Chondroma

This is an uncommon mature cartilage lesion, which is located underneath the periosteum and does not involve the medullary cavity. This hyaline cartilage tumor is lobulated with variable cellularity of chondrocytes. The lobules may be separated by fibrous tissue or mature lamellar

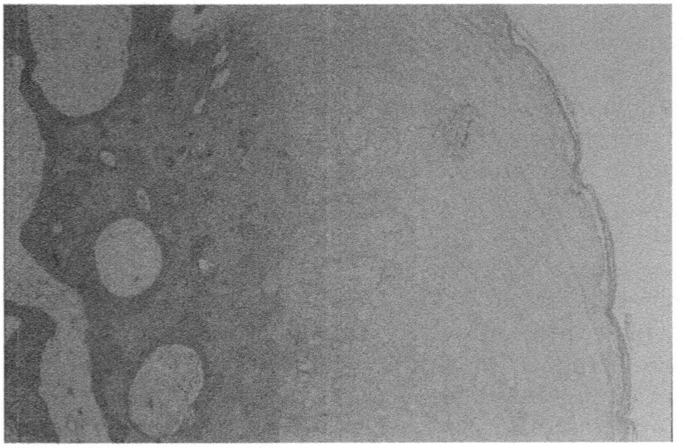

FIGURE 9 Osteochondroma. A moderately cellular hyaline cartilage cap is covered by thin fibrous perichondrium. Cancellous bone with endochondral ossification is present underlying cap with adjacent fatty bone marrow (hematoxylin and eosin, original magnification $\times$ 4).

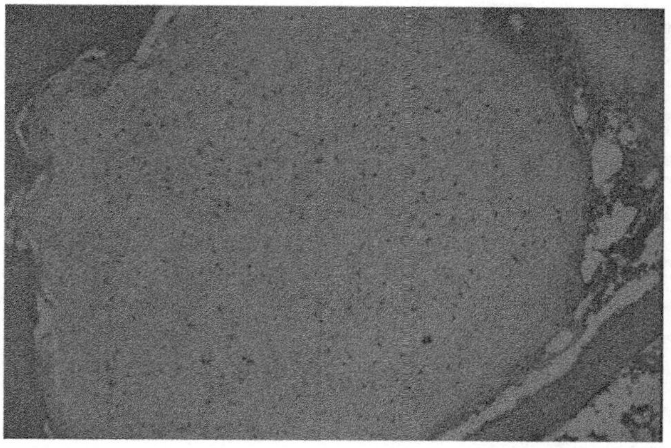

FIGURE 10 Enchondroma. Bland, mildly cellular hyaline cartilage with evenly distributed cells. This circumscribed lesion is bounded by lamellar bone (hematoxylin and eosin, original magnification × 10).

bone. In addition to hypercellular areas, chondrocytes may be binucleated, enlarged, or hyperchromatic. Myxoid change within the matrix can occur. The differential diagnosis would again focus on a possible chondrosarcoma in this location. In general, there is more cytologic atypia seen throughout the lesion compared to periosteal chondroma. Individual cases may show some overlap (46,47).

Chondroblastoma

In contrast to the aforementioned mature hyaline cartilage tumors, this benign lesion consists of round-to-spindled cells resembling immature chondrocytes. The great preponderance involves the epiphyseal region of the distal femur, proximal tibia, and proximal humerus. Both cytology and histology may vary. Most tumor cells will show a sharp cytoplasmic border. The cytoplasm may be lightly eosinophilic to clear, and the round-to-oval nuclei may contain grooves, clefts, or invaginations. Cellular atypia can be identified on occasion. Mitoses are present, and may number as high as 4 per 10 high-power microscopic fields; however, atypical mitoses should be absent. Spindle cells are identified as either separate collections or scattered among the chondroblastic cells. Chondroid matrix may be present and is usually eosinophilic rather than the typical basophilic matrix of hyaline cartilage. Calcifications are either within the matrix or surround individual cells in a "chicken-wire" type configuration (Fig. 11). This is one

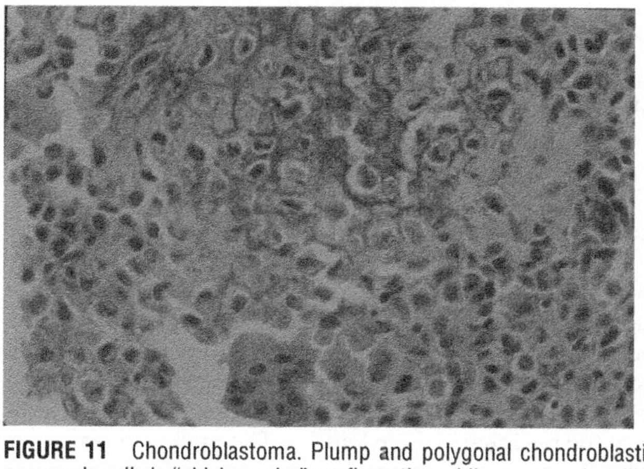

FIGURE 11 Chondroblastoma. Plump and polygonal chondroblastic cells with hyperchromatic nuclei. Calcification surrounds cells in "chicken-wire" configuration while an osteoclast-like giant cell is present superiorly (hematoxylin and eosin, original magnification × 40).

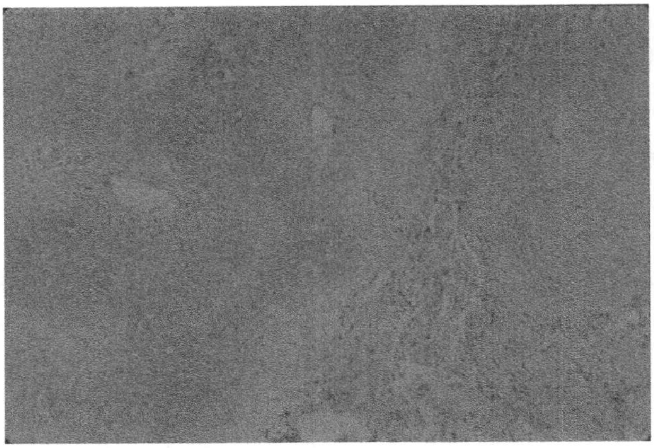

FIGURE 12 Chondromyxoid fibroma. Lobulated myxoid and eosinophilic cartilaginous matrix containing spindle and stellate cells. Blood vessels are present within the intervening fibrous septa (hematoxylin and eosin, original magnification × 10).

of the benign tumors, which can contain multinucleated cells and osteoclast-like giant cells (48). The differential diagnosis would include giant-cell tumor and clear-cell chondrosarcoma. Unlike giant-cell tumor, the background mononuclear cells of chondroblastoma resemble chondroblasts and not histiocytes. A chondroid matrix is noted and an S-100 immunohistochemical stain will be positive (49). Clear-cell chondrosarcoma typically contains sheets of tumor cells with clear or nearly clear cytoplasm, whereas this is usually a focal occurrence in chondroblastoma.

Chondromyxoid Fibroma

This circumscribed benign tumor is predominantly found in the metaphysis of long bones, and is composed of lobules of variable myxoid to chondroid stroma containing spindle and stellate cells. The fibrous component is small and consists of thin septa separating the lobules (Fig. 12). Osteoid, blood vessels, and multinucleated giant cells can be found within the septa. The cellular component can be fibroblastic or comprise cells containing eosinophilic cytoplasm and one or more cytoplasmic processes. The nuclei can vary from a hyperchromatic and clumped chromatin pattern to being vacuolated. Cellularity is usually increased at the periphery of the lobule along with the more spindled cells. Cellular pleomorphism has been reported in up to 30% of cases. Mitoses, however, are rare (50,51). The differential diagnosis, once again, includes chondrosarcoma. The circumscription from adjacent normal bone and lack of mitotic activity can help separate this tumor from a chondrosarcoma.

Chondrosarcoma

This malignant cartilage tumor represents the second most common malignant bone tumor following osteosarcoma. It tends to involve the axial skeleton and occurs in older patients when compared to osteosarcoma. Although there are various subtypes, conventional intramedullary chondrosarcoma represents about 90% of tumors that occur. Low-grade tumors are morphologically similar to enchondroma and often difficult-to-impossible to separate by microscopic evaluation alone. These slow-growing tumors are locally destructive and have little capacity for distant metastasis. High-grade lesions are very aggressive and can metastasize, sometimes early in the course of the disease. Chondrosarcomas may be primary or secondary and are divided into central, peripheral, or juxtacortical. A three-grade system is utilized in histologic evaluation. Grade-1 lesions contain chondrocytes with small dark nuclei, which are evenly dispersed in chondroid background stroma. Binucleated and mildly pleomorphic cells can be seen, but there is no increase in mitoses or cellularity. Grade-2 tumors have less matrix and increased cellularity that often manifests at the periphery of the lobules. The nuclei are enlarged with a more vesicular chromatin pattern. Background stroma is often myxoid and necrosis can be present (Fig. 13). Grade-3 tumors contain an even greater increase in both cellularity

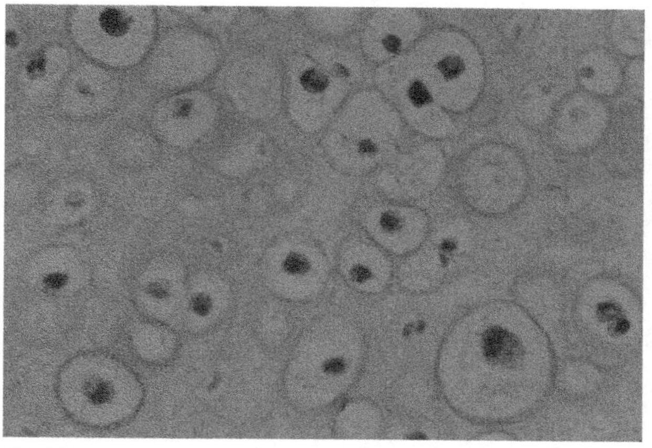

FIGURE 13 Chondrosarcoma. Increased cellularity and moderate cellular atypia are present in this grade-2 chondrosarcoma (hematoxylin and eosin, original magnification × 40).

and pleomorphism. Chondroid matrix may be totally absent and the background stroma is usually myxoid. Tumor cells are stellate or irregular, and the nuclei spindle-shaped and large. Some authors use the presence of mitoses as a criterion for qualification for grade-3 tumors, whereas others contend that significant pheomorphism and cellularity are sufficient criteria (52,27).

Dedifferentiated Chondrosarcoma

This is a high-grade sarcoma that arises against the background of a low-grade chondrosarcoma. The "dedifferentiated" component may be in the form of MFH, osteosarcoma, fibrosarcoma, rhabdomyosarcoma, or even an angiosarcoma. It most likely represents origin from a mesenchymal stem cell rather than from a true dedifferentiation of mature cartilage. The histologic picture is that of a grade-1 chondrosarcoma with a sudden transition to a poorly differentiated neoplasm (Fig. 14) (53,54).

Clear-Cell Chondrosarcoma

This uncommon histologic variant contains tumor cells with abundant clear cytoplasm and little matrix. Involvement of the epiphysis is common. Although the cells have abundant clear cytoplasm with sharply defined cell borders, the growth pattern is still lobular. Nuclei are large and may be binucleated. The limited matrix production is chondroid and calcifications may be present (55,56).

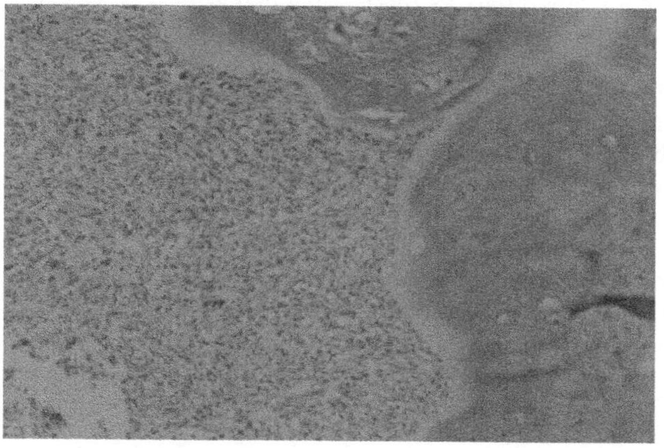

FIGURE 14 Dedifferentiated chondrosarcoma. There is an abrupt transition between the low-grade chondrosarcoma and the high-grade spindle cell component (hematoxylin and eosin, original magnification × 10).

FIGURE 15 Mesenchymal chondrosarcoma. There is a mixture of small round primitive tumor cells and intervening cartilaginous foci (hematoxylin and eosin, original magnification × 10).

Mesenchymal Chondrosarcoma

A predominance of small round blue cells is present in this rare tumor. Blood vessels can be irregular in configuration similar to the staghorn-like appearance in hemangiopericytoma. The small tumor cells are round, oval, or spindle shaped. Pleomorphism is mild, with clumped chromatin and small nucleoli (Fig. 15). Mitoses can be sparse or on occasion easily found. Islands of low-grade cartilage are present, but usually as a minor component and well demarcated (57,37).

Juxtacortical Chondrosarcoma

This chondrosarcoma is found on the surface of bones, and although it may erode the cortex, it does not involve the medullary canal. The histologic and cytologic features are similar to conventional chondrosarcoma, with most lesions being low grade.

Differential Diagnosis

A chondroblastic osteosarcoma enters the differential for conventional chondrosarcoma. The detection of osteoid formation is key to the histologic diagnosis and may be lacking in small tissue samples. Attention must be paid to the age of the patient as well as to the clinical and radiologic presentation to avoid this pitfall. Another problem area occurs with low-grade (grade-1) chondrosarcoma versus enchondroma. The histologic presentation may be identical and again will require clinical and radiologic correlation to arrive at the correct diagnosis. Dedifferentiated chondrosarcoma will typically show a sharp transition from a low-grade chondrosarcoma to a high-grade sarcoma. This transition may be missed in small samples and could lead to misinterpretation on microscopic evaluation. The high-grade component may be osteosarcoma, fibrosarcoma, or MFH. Clear-cell chondrosarcoma will be positive for cytoplasic glycogen by PAS stain. The tumor cells will also be positive for vimentin and S-100. Epithelial markers will be negative. The diagnosis of mesenchymal chondrosarcoma may be assisted by immunohistochemistry to rule out other small round blue cell tumors and to highlight foci of cartilage.

FIBROUS BONE TUMORS

These lesions are made up of proliferations of fibrous tissue, as in nonossifying fibroma (NOF), or contain an osseous component, as in fibrous dysplasia. Benign and malignant counterparts exist with varying cellularity and histologic presentations.

Histology and Cytology

Nonossifying Fibroma

This benign metaphyseal fibrohistiocytic process has also been termed fibrous cortical defect and metaphyseal fibrous defect and is usually seen in long tubular bones of skeletally immature

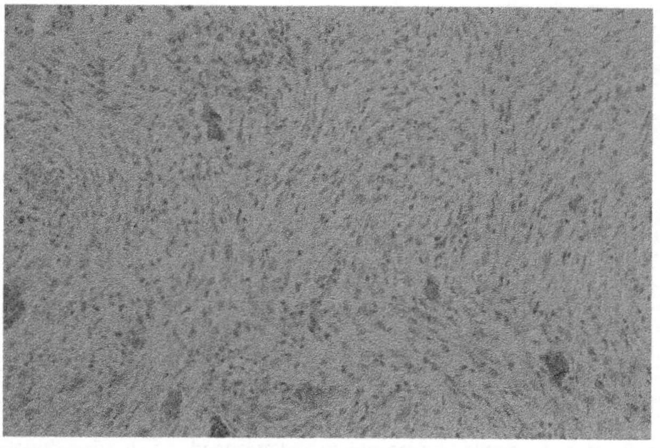

FIGURE 16 Nonossifying fibroma. The fibrous tissue shows a vague storiform pattern with interspersed multinucleated giant cells (hematoxylin and eosin, original magnification × 10).

patients. The spindle cell fibrous stroma is variably cellular with a whorled or storiform pattern (Fig. 16). Collections of foamy histocytes are scattered, as are multinucleated giant cells. Pleomorphism of the fibroblastic cells is minimal and mitotic figures are rare. Hemosiderin-containing macrophages may also be encountered. The presence of reactive new bone can be seen if there is an associated fracture (58,59).

Fibrous Dysplasia

This benign intramedullary process is made up of a proliferation of fibrous stroma containing irregular woven bone trabeculae without significant rimming by osteoblasts (Fig. 17). Most lesions involve only one site; however, polyostotic disease accounts for up to 20% of cases. This lesion can also be a component of a syndrome (i.e., McCune Albright) It is thought that fibrous dysplasia is caused by a somatic activating mutation of the Gs alpha subunit of protein G, resulting in an increased cAMP concentration and thus abnormalities of osteoblast differentiation (60). There is also an increased interleukin-6–induced osteoclastic bone resorption. The variably cellular fibrous stroma lacks significant cellular pleomorphism or mitotic activity. The woven trabecular bone may be branched, C-shaped, or cementicle-like in configuration (9,61).

Osteofibrous Dysplasia

This lesion presents in the first decade of life, predominantly affecting the anterior cortex of the diaphysis of the tibia and fibula. Similar to fibrous dysplasia, woven trabecular bone is present

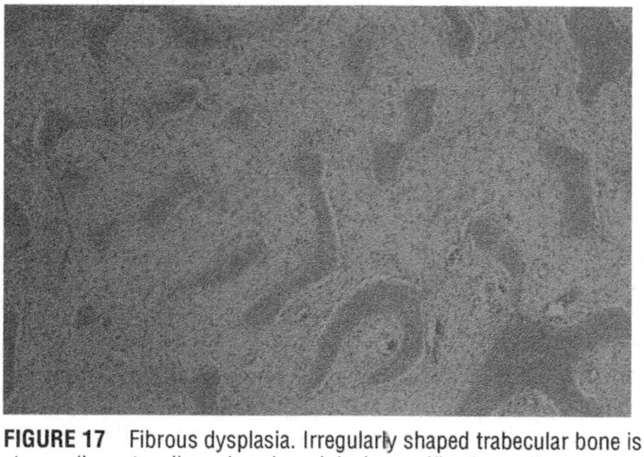

FIGURE 17 Fibrous dysplasia. Irregularly shaped trabecular bone is present with a moderately cellular, bland fibrous stroma (hematoxylin and eosin, original magnification × 4).

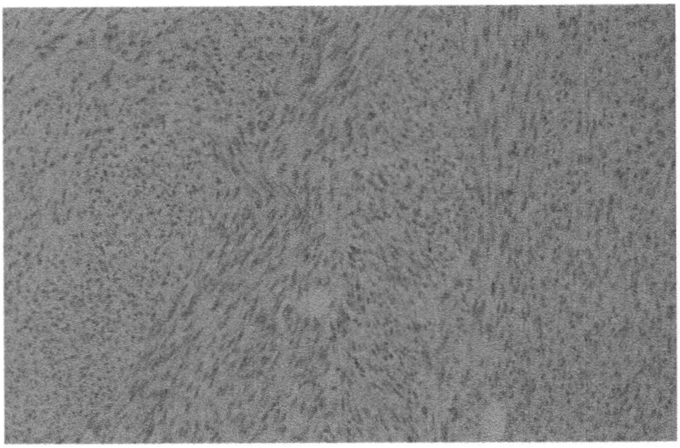

FIGURE 18 Fibrosarcoma. Cellular, low-grade fibrous tumor with mild pleomorphism and arranged in a "herring-bone" pattern (hematoxylin and eosin, original magnification × 10).

within a variably cellular, benign fibrous stroma. In contrast to fibrous dysplasia, there is prominent osteoblastic rimming of the trabeculae. A zonal architecture is present with more mature lamellar bone merging with normal cortical bone at the periphery (62,28). There is an ongoing speculation regarding the relationship of osteofibrous dysplasia with adamantinoma. Similar sites of involvement, reactivity with cytokeratin, and cytogenetic studies support this contention (63). This argument does not address the issue of prevalence of adamantinoma in adolescents and adults while osteofibrous dysplasia is seen in young children. Other authors believe that this simply represents a variant of fibrous dysplasia.

Benign Fibrous Histiocytoma

There is histologic similarity between this lesion and NOF. The benign fibrous stroma has a storiform pattern containing multinucleated giant cells and foamy macrophages, which may be numerous. Occasional mitoses are present but atypical forms and cytologic atypia are not seen. The diagnosis is made after exclusion of other tumors with similar histology (64).

Fibrosarcoma

This malignant spindle cell tumor shows fibroblastic differentiation without osteoid or chondroid matrix production. It parallels osteosarcoma as to sites of involvement, but the age range does not show the second decade peak as in osteosarcoma. Tumors can range from those that are cellular with little collagen and necrotic foci to those with heavy collagen deposition. There is a consistent arrangement of fibroblastic cells in a "herring-bone" pattern (Fig. 18). Histologically, a three grade system is employed and is based on number of mitoses as well as cellular pleomorphism. Grade-1 tumors are cytologically bland with up to four mitotic figures per 10 high-powered microscopic fields. Grade-2 lesions have larger cells with mild cellular pleomorphism, numerous mitoses, and less collagen. Grade-3 tumors show significant pleomorphism, mitoses, and atypical mitotic figures (65,37).

Malignant Fibrous Histiocytoma

This high-grade sarcoma can exhibit all the varied histologic subtypes similar to its soft-tissue counterpart: storiform-pleomorphic, giant-cell-rich, myxoid, and inflammatory. The most common is the pleomorphic/storiform pattern in which there is an admixture of fibroblast-like spindle cells and highly pleomorphic tumor cells in a storiform histologic architecture. The nuclei of the tumor cells can be large and hyperchromatic (Fig. 19). The cytoplasm may be densely eosinophilic or contain vacuolar changes. Atypical mitotic figures are usually present and multiple. Malignant giant tumor cells as well as benign osteoclastic giant cells may be present (66). Most authors divide the tumors into low and high grade.

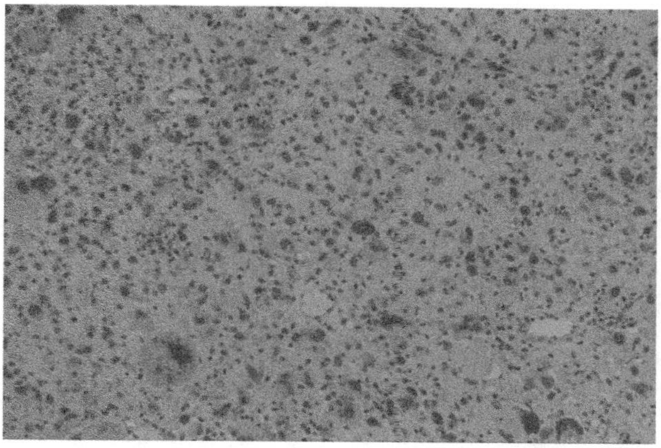

FIGURE 19 Malignant fibrous histiocytoma. Needle core biopsy containing pleomorphic spindle and histiocytoid giant cells (hematoxylin and eosin, original magnification × 10).

Differential Diagnosis

The similar histologic appearance of NOF, metaphyseal fibrous defect, and benign fibrous histiocytoma underscores the importance of clinical and radiographic correlation when making their distinction. Reactive new bone formation within the tumor may be confused with fibrous dysplasia; however, prominent osteoblastic rimming of the trabeculae and storiform pattern of the fibrous stroma should not be present in fibrous dysplasia. The presence of multinucleated giant cells in a cellular background may cause one to consider giant cell tumor of bone. The giant cells are usually not diffuse and the typical background mononuclear cells are not evident. Osteofibrous dysplasia and well-differentiated osteosarcoma may be considered when dealing with fibrous dysplasia. Osteofibrous dysplasia is almost exclusively seen in tibia and fibula and the bony trabeculae will usually possess prominent osteoblastic rimming. The fibrous stroma in well-differentiated osteosarcoma will contain larger nuclei and atypia as compared to that of fibrous dysplasia. Fibrosarcoma-like areas can be seen in osteosarcoma, MFH, and dedifferentiated chondrosarcoma. The absence of osteoid will rule out osteosarcoma, whereas MFH will contain more atypia as well as neoplastic giant cells. Additionally, fibrosarcoma tends to have a "herring-bone" pattern when compared to the storiform configuration often present in MFH. The abrupt transition from a low-grade chondrosarcoma to a spindle-cell sarcomatous component will help separate dedifferentiated chondrosarcoma from fibrosarcoma.

SMALL ROUND BLUE CELL TUMORS

This category of bone tumors may contain remarkable histologic and cytologic similarity and includes both benign and malignant lesions. It is in this area that the importance of ancillary studies is appreciated in separating the various entities.

Histology and Cytology

Langerhans Cell Histiocytosis

Eosinophilic granuloma, Hand-Schuller-Christian disease, and Letterer-Siwe disease comprise this category. The disease can be focal as in eosinophilic granuloma (80% of cases), or involve multiple skeletal and extraskeletal sites as in Letterer-Siwe disease. The histologic and cytologic findings are similar in all subtypes. The tumor is made up predominantly of two components, Langerhans cells and eosinophils in varying quantities. Other inflammatory cells may also be present. The Langerhan cell is a round-to-oval histiocyte-like cell containing an oval nucleus and clearly demarcated cytoplasm. A prominent nuclear groove, which runs along the long axis, is present in most of these cells (Fig. 20). Cellular atypia as well as mitoses is minimal. Multinucleated giant cells and ordinary macrophages can be present. Eosinophils may be so

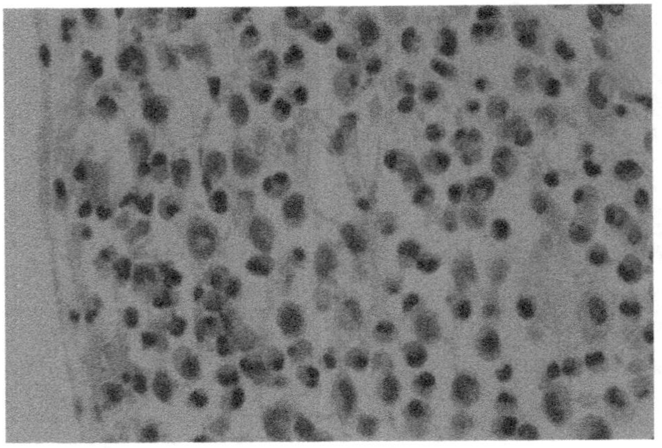

FIGURE 20 Langerhans cell histiocytosis. This high-power magnification contains eosinophils and Langerhans cells (hematoxylin and eosin, original magnification × 40).

prominent in the sample material that the term "eosinophilic abscess" has been used (67,9). Immunohistochemical stains will be positive for S-100 and CD1a (68).

Ewing's Sarcoma/PNET

This tumor family tends to involve the diaphysis of long bones but also affects the pelvis and ribs. It is made up of undifferentiated mesenchymal cells growing in dense solid sheets with minimal intervening stroma. The nucleus is round and central, and possesses little associated cytoplasm. The chromatin pattern is finely granular. Nucleoli are present and may be multiple but not prominent (Fig. 21). Sometimes a population of light and dark cells is seen with the dark cells representing apoptotic cells. Rosette formation is occasionally noted but if it is more than 20%, one should consider a diagnosis of PNET. Necrosis is a frequent feature. The tumor cells permeate the cortex and involve the haversian system and Volkman canals. Some tumors have slightly larger tumor cells, contain vacuoles or are spindle shaped, and have been termed "atypical" Ewing's sarcoma. Intracellular glycogen is found in most tumors and can be demonstrated by a PAS stain. Immunohistochemical stains will be positive for the MIC2 gene product (CD99) in approximately 90% of tumors; however, this marker is also positive in some pediatric lymphomas and rhabdomyosarcoma as well as mesenchymal chondrosarcoma. Thus a panel of stains tailored to rule out these other entities should be performed. If tissue is adequate, molecular studies can identify the Ewing's sarcoma–associated reciprocal translocation between chromosome 11 and 22. For a tumor to

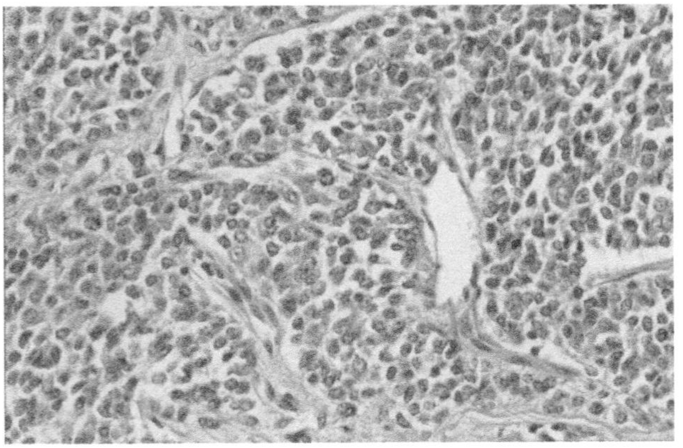

FIGURE 21 Ewing's sarcoma. Monomorphic population of small round-to-oval tumor cells with hyperchromatic nuclei and no matrix production (hematoxylin and eosin, original magnification × 40).

be classified as a PNET, there must be positivity for neural markers, with greater than 20% rosette formation (27,69).

Lymphoma of Bone

Primary non-Hodgkins lymphoma of bone is rare but similar in morphology to its nodal and soft-tissue counterpart. To be considered as a primary lymphoma of bone, a four- to six-month interval is required without development of extraskeletal disease. Most are diffuse large-cell lymphomas (large cell or mixed small- and large-cell types); however, other types have been described. Ancillary studies such as immunohistochemistry, flow cytometry, and molecular diagnostic studies are desirable to confirm diagnosis in most cases. Most of the lymphomas are of B-cell phenotype using the various lymphoid marker studies (28).

Multiple Myeloma/Plasma Cell Dyscrasia

This malignant proliferation of plasma cells is the most frequent neoplasm presenting as a skeletal lesion, with 90% diagnosed in patients over 40 years of age. The proliferations may be either plasmacytic or plasmablastic. The plasmacytic histologic variant contains round-to-oval cells with an eccentric nucleus, irregular chromatin distribution, and prominent nucleolus. The cytoplasm is dense and eosinophilic (Fig. 22). Some of the cells can be binucleated and larger and contain prominent cellular atypia. Some tumors will contain plasmablastic features such as bizarre immature cells, which are difficult to recognize as plasma cell lineage. Immunohistochemistry using CD38 and immunoglobulin monoclonal markers for kappa and lambda light chains are diagnostic. Additionally, serum protein electrophoresis with immuno-fixation will identify monoclonal immunoglobulin, usually of the IgG type (9,70).

Differential Diagnosis

The above tumors may contain similar morphologic features by routine microscopic examination. Clinical and radiologic correlation is essential in guiding the pathologist in his use of ancillary studies to arrive at the diagnosis.

GIANT CELL TUMOR OF BONE

This tumor occurs in the epiphyseal region of long bones in the skeletally mature, with 50% reported in the vicinity of the knee. The usual histologic appearance shows multinucleated giant cells uniformly distributed against the background of mononuclear cells, which are round

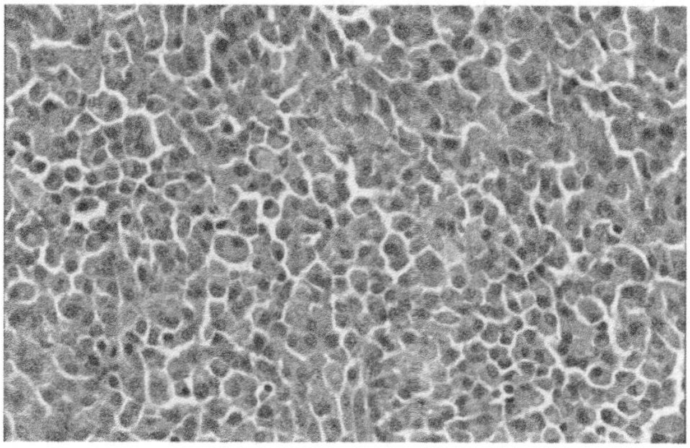

FIGURE 22 Plasmacytoma/multiple myeloma. Monomorphic population of plasma cells with occasional binucleation. The nuclear chromatin is irregularly distributed (hematoxylin and eosin, original magnification $\times$ 40).

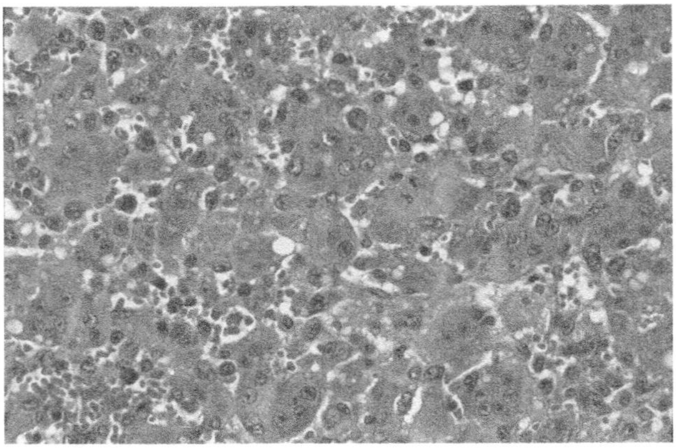

FIGURE 23 Giant cell tumor of bone. Evenly distributed multinucleated giant cells and associated mononuclear cells. The nuclei of both the giant cells and mononuclear cells are similar (hematoxylin and eosin, original magnification × 40).

to oval in shape. There is cytologic similarity between the nuclei of the mononuclear cells and multinucleated giant cells (Fig. 23). Mitotic activity can be present and may be brisk in the mononuclear cells, but atypical forms are absent. Cartilage is not present unless there is a fracture. Osteoid formation can be identified in about half of the lesions in the vicinity of the advancing tumor edge. Secondary aneurysmal bone cyst change may complicate and, at times, obscure the giant cell tumor. Histologic grading of giant cell tumors has not been helpful in predicting the recurrence, local aggressiveness, or metastasis of these lesions (71,20).

Differential Diagnosis

The differential diagnosis would include other giant cell-containing tumors, both malignant and benign. The "brown tumor" of hyperparathyroidism may show similar histology; however, the multinucleated giant cells are fewer in number and not as evenly distributed within a background stroma, which is spindled, and more collagenized. Serum parathyroid hormone levels will be increased. Likewise, fibrous dysplasia, chondroblastoma, NOF, benign fibrous histiocytoma, and giant cell-rich osteosarcoma can be ruled out by identification of other pertinent histologic components and clinicoradiologic correlation.

BONE CYSTS

Bone cysts can be composed of single or multiple septate compartments and lined by fibrous connective tissue. They may contain clear yellowish fluid, blood, mucinous material, or keratinaceous debris. Secondary cystic change may also occur in the background of other bony lesions such as fibrous dysplasia and giant cell tumor.

Unicameral Bone Cyst

This bone cyst is also termed "simple cyst" or "solitary bone cyst" and is thought to represent a reactive or developmental process rather than neoplastic. Other studies have suggested a reaction to trauma as an underlying etiology. Most cases involve the proximal humerus and femur. Most specimens are curettage fragments. Histologic examination is usually limited to the cyst lining consisting of thin fragments of unremarkable fibrous tissue. Occasionally there may be reactive new bone formation, granulation tissue, or scattered lymphocytes.

Aneurysmal Bone Cyst

These cysts are benign and not neoplastic. They can be rapidly expansile and locally destructive. It is thought that they are secondary to arteriovenous malformation or trauma, or arise against the background of a pre-existing lesion.

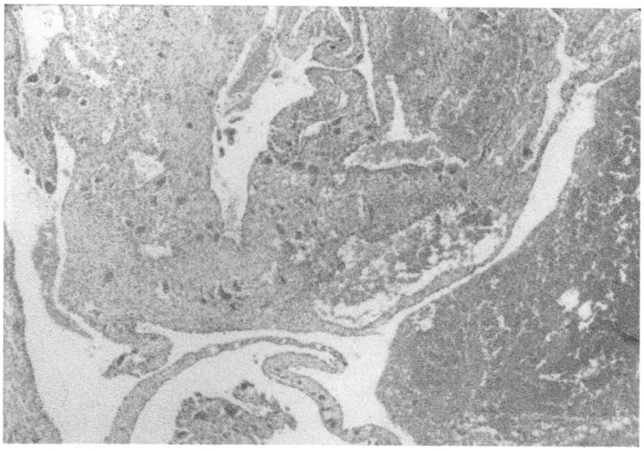

FIGURE 24 Aneurysmal bone cyst. Varying sized blood-filled spaces are separated by loose fibrous tissue. The blood-filled spaces usually lack a definitive lining. Multinucleated giant cells can be found within the septa (hematoxylin and eosin, original magnification × 4).

Histology and Cytology

Microscopic examination of unicameral bone cyst is usually relegated to the thin fibrous cyst lining, which may contain dilated vessels and scattered multinucleated giant cells. Occasionally there is reactive new bone formation, granulation tissue, and scattered lymphocytes. Fibrin deposition within the wall can become mineralized, resulting in the formation of cementum-like bodies. Aneurysmal bone cyst will contain multiple blood-filled spaces. The septa contain fibrous tissue with multinucleated giant cells, red blood cells, hemosiderin-containing macrophages, and small capillaries (Fig. 24). There is usually no discernable lining. The multinucleated giant cells and macrophages tend to be located just underneath the surface of the septa.

More solid areas have a granulation tissue quality. Reactive bone formation and long strands of osteoid with osteoblasts are commonly seen (37).

Differential Diagnosis

Because the predominant histologic feature of unicameral bone cyst is fibrous tissue, there can be confusion with the fibrous component of other solid tumors. This is especially true if the cystic nature of the lesion is not communicated to the surgical pathologist. The diagnosis of aneurysmal bone cyst rests on the ability to rule out an underlying bone tumor that has undergone secondary cystic change. The entire specimen should be processed whenever possible. Many benign lesions including giant cell tumor, chondroblastoma, osteoblastoma, and fibrous dysplasia may undergo secondary aneurysmal bone cyst change. Recognition of the histologic features of the other entities is therefore essential. The most important pitfall is differentiation from a telangiectatic osteosarcoma. A thorough search must be made for the presence of atypical mitoses or anaplastic sarcomatous tumor cells within the septal walls dividing the blood-filled lakes.

METASTATIC TUMORS

Most metastatic tumors are adenocarcinomas with both histologic and cytologic features dependent on their site of origin. The more common primary sites include breast, lung, prostate, kidney, and thyroid (Figs. 25 and 26). Less likely candidates in the adult include small cell carcinoma of the lung, transitional cell carcinoma, and gynecologic carcinomas. A variety of epithelial immunohistochemical markers are now available which can help identify the site of origin for many of these tumors. This is particularly helpful if the tumor is poorly differentiated, thus limiting classification by histologic examination alone. In the pediatric population, the possibility of a metastatic neuroblastoma or rhabdomyosarcoma needs to be considered and

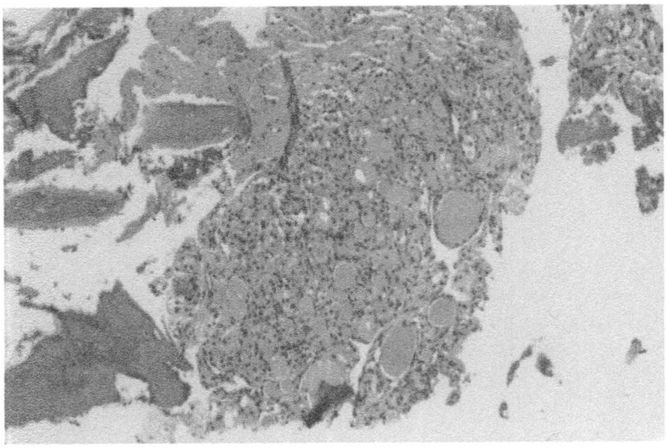

FIGURE 25 Metastatic follicular thyroid carcinoma. Metastatic well-differentiated follicular thyroid carcinoma is easily identified within this needle core biopsy of a vertebral mass (hematoxylin and eosin, original magnification × 10).

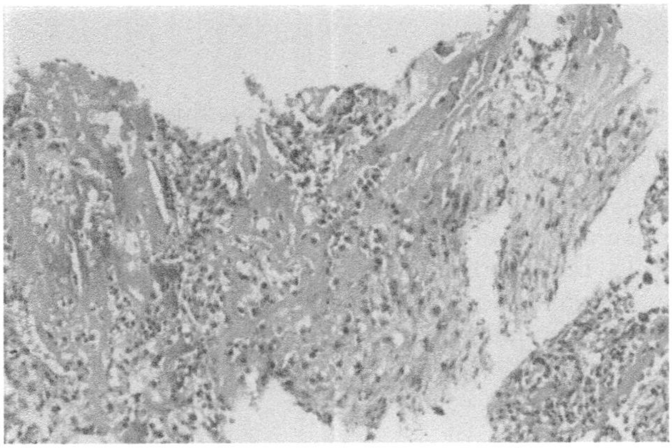

FIGURE 26 Metastatic renal cell carcinoma, clear cell type. This needle core biopsy demonstrates metastatic renal cell carcinoma. The clear cytoplasm of the tumor cells is typical of this neoplasm (hematoxylin and eosin, original magnification × 10).

added to the differential list of small round blue cell tumors (27). A detailed clinical history is beneficial not only for diagnosis but also invaluable when selecting a panel of relevant immunohistochemical stains when working up a particular case.

CONCLUSION

This chapter has reviewed the basic histologic and cytologic aspects of benign and neoplastic bone tumors and infection. Histologic overlap between various tumor subtypes occurs, which on occasion can cause a diagnostic dilemma. Advances in the area of molecular diagnostics, cytogenetics, and immunohistochemistry have made a tremendous impact on diagnostic accuracy, especially in the area of small round blue cell tumors. Even though these advances are significant, the importance of correlation with both radiographic and clinical information cannot be overstated. Therefore, the close working relationship between surgeon, radiologist, and surgical pathologist remains the basis for accurate and timely diagnosis.

REFERENCES

1. Schweitzer ME, Gannon FH, Deeley DM, O'Hara BJ, Junja V. Percutaneous skeletal aspiration and core biopsy: complimentary techniques. AJR 1996; 166:415–418.

2. Hodge JC. Percutaneous biopsy of the musculoskeletal system: a review of 77 cases. Can Assoc Radiol J 1999; 50(2):121–125.
3. Dickersin GR. Diagnostic Electron Microscopy–A Text/Atlas. New York: IGAKU-SHOIN Medical Publishers Inc., 1988:157–167.
4. Coughlan B, Feliz A, Ishida T, Czerniak B, Dorfman HD. p53 expression and DNA ploidy of cartilage lesions. Hum Pathol 1995; 26(6):620–624.
5. Stephenson A, Chatten J, Bertoni F, Miettinen M. CD99 (p30/p32 mic2) neuroectodermal sarcoma antigen as an immunohistochemical marker: review of more than 600 tumors and the literature experience. Appl Immunohistochem 1994; 2(4):231–240.
6. Luna LG. Histopathologic methods and color atlas of special stains and tissue artifacts. Downers Grove, Illinois: Johnson Printers, 1992.
7. Blyth MJ, Kincaid R, Craigen MA, Bennet GC. The changing epidemiology of acute and subacute haematogenous osteomyelitis in children. J Bone Joint Surg–British Volume 2001; 83(1):99–102.
8. Lipsky BA. Osteomyelitis of the foot in diabetic patients. Clin Infect Dis 1997; 25(6):1318–1326.
9. Bullough PG. Atlas of Orthopedic Pathology With Clinical and Radiologic Correlations. 2nd ed. New York: Gower Medical Publishing, 1992:4.1–4.24.
10. Ciampolini J, Harding KG. Pathophysiology of chronic bacterial osteomyelitis. Why do antibiotics fail so often? Postgrad Med J 2000; 76(898):479–483.
11. Song KM, Sloboda JF. Acute hematogenous osteomyelitis in children. J Am Acad Orthoped Surg 2001; 9(3):166–175.
12. Rubin E, Gorstein F, Rubin R, et al. Rubin's Pathology, 4th ed. Lippincott, Williams and Wilkins, 2005:1327–1331.
13. Bohm E, Josten C. What's new in exogenous osteomyelitis? Pathol Res & Pract 1992; 188(1–2): 254–258.
14. Tsukayama DT. Pathophysiology of posttraumatic osteomyelitis. Clin Orthopaed Relat Res 1999; (360):22–29.
15. Carek PJ, Dickerson LM, Sack JL. Diagnosis and management of osteomyelitis. Am Fam Physician 2001; 63(12):2413–2420.
16. McGrory JE, Pritchard DJ, Unni KK, Ilstrup D, Rowland CM. Malignant lesions arising in chronic osteomyelitis. Clin Orthopaed Relat Res 1999; (362):181–189.
17. Stanton RP, Lopez-Sosa FH, Doidge R. Chronic recurrent multifocal osteomyelitis. Orthopaed Rev 1993; 22(2):229–233.
18. Handrick W, Hormann D, Voppmann A, et al. Chronic recurrent multifocal osteomyelitis– report of eight patients. Pediat Surg Int 1998; 14(3)195–198.
19. Durbin M, Randall RL, James M, Sudilovsky D, Zoger S. Ewing's sarcoma masquerading as osteomyelitis. Clin Orthopaed Relat Res 1998; (357):176–185.
20. Dahlin DC, Unni KK. Bone Tumors: General Aspects and Data on 8,542 Cases. 4th ed. Springfield: Charles C. Thomas, 1986:84–87.
21. Greenspan A, Steiner G, Knutzen R. Bone island (enostosis): clinical significance and radiologic and pathologic correlations. Skeletal Radiol 1991; 20(2):85–90.
22. O'Connell JX, Nanthakumar SS, Nielsen GP, Rosenberg AE. Osteoid osteoma: the uniquely innervated bone tumor. Mod Pathol 1998; 11 (2):175–180.
23. Hasegawa T, Hirose T, Sakamoto R, Seki K, Ikata T, Hizawa K. Mechanism of pain in osteoid osteomas: an immunohistochemical study. Histopathology 1993; 22(5):487–491.
24. Wold LE, Pritchard DJ, Bergert J, Wilson DM. Prostaglandin synthesis by osteoid osteoma. Mod Pathol 1988; 1(2):129–131.
25. Greenspan A. Benign bone-forming lesions: osteoma, osteoid osteoma and osteoblastoma. Clinical, imaging, pathologic, and differential considerations. Skeletal Radiol 1993; 22(7):485–500.
26. Rocca CD, Huvos AG. Osteoblastoma: varied histological presentations with a benign course—an analysis of 55 cases. Am J Surg Pathol 1996; 20(7):841–850.
27. Dorfman HD, Czerniak B. Bone Tumors. Philadelphia: Mosby, 1998:85–125.
28. Fechner RE, Mills SE. Atlas of Tumor Pathology: Tumors of Bones and Joints. 3rd series. Washington, D.C.: Armed Forces Institute of Pathology, fascicle 8, 1993:25–72.
29. Magre GR, Menendez LR. Preoperative CT localization and marking of osteoid osteoma: description of a new technique. J Comput Assist Tomogr 1996; 20(4):526–529.
30. Ayala AG, Murray JA, Erling MA, Raymond AK. Osteoid osteoma: intraoperative tetracycline-fluorescence demonstration of the nidus. J Bone Joint Surg 1986; 68(A):747–751.
31. Thomas DM, Carty SA, Piscopo DM, Lee JS, Wang WF, Hinds PW. The retinoblastoma protein acts as a transcriptional coactivator required for osteogenic differentiation. Mol Cell 2001; 8(2): 303–316.
32. Sellers WR, Kaelin WG Jr. Role of the retinoblastoma gene in the pathogenesis of human cancer. J Clin Oncol 1997; 15(11):3297–3298.
33. Rosai J. Rosai and Ackerman's Surgical Pathology. Vol. 2. 9th ed. 2004:2150–2158.
34. Balance WA, Mendelsohn G, Carter JR, Abdul-Karim FW, Jacobs G, Makley J. Osteogenic sarcoma: malignant fibrous histiocytoma subtype. Cancer 1988; 62:763–771.

35. Ayala AG, RO JY, Raymond AK, et al. Small cell osteosarcoma. A clinicopathologic study of 27 cases. Cancer 1989; 64(10):2162–2173.
36. McCarthy EF, Frassica FJ. Pathology of bone and joint disorders with clinical and radiologic correlation. Philadelphia: W.B. Saunders, 1998:216–217.
37. Huvos AG, Rosen G, Bretsky SS, Butler A. Telangiectatic osteogenic osteosarcoma: a clinicopathologic study of 124 patients. Cancer 1982; 49:1679–1689.
38. Bertoni F, Bacchini P, Fabbri N, et al. Osteosarcoma. Low-grade intraosseous-type osteosarcoma resembling parosteal osteosarcoma, fibrous dysplasia, and desmoplastic fibroma. Cancer 1993; 71(2):338–345.
39. Kurt AM, Unni KK, McLeod RA, Pritchard DJ. Low-grade intraosseous osteosarcoma. Cancer 1990; 65(6):1418–1428.
40. Sheth DS, Yasko AW, Raymond AK, et al. Conventional and dedifferentiated parosteal osteosarcoma. Diagnosis, treatment, and outcome. Cancer 1996; 78(10):2136–2145.
41. Lindell MM, Shirkhoda A, Raymond AK, Murray JA, Harle TS. Parosteal osteosarcoma: radiologic-pathologic correlation with emphasis on CT. Am J Roentgenol 1987; 148(2):323–328.
42. Bertoni F, Boriani S, Laus M, Campanacci M. Periosteal chondrosarcoma and periosteal osteosarcoma. Two distinct entities. J Bone Joint Surg–British Volume 1982; 64(3):370–376.
43. Ritts GD, Pritchard DJ, Unni KK, Beaubout JW, Eckardt JJ. Periosteal osteosarcoma. Clin Orthopaed Relat Res 1987; (219):299–307.
44. Wold LE, Unni KK, Beabout JW, Pritchard DJ. High-grade surface osteosarcomas. Am J Surg Pathol 1984; 8:181–186.
45. Rubin E, Gorstein F, Rubin R, et al. Rubin's Pathology, 4th ed. Lippincott, Williams & Wilkins, 2005:1319.
46. Bauer TW, Dorfman HD, Latham JT. Periosteal chondroma. A clinicopathologic study of 23 cases. Am J Surg Pathol 1982; 6(7):631–637.
47. Boriani S, Bacchini P, Bertoni F, Campanacci M. Periosteal chondroma. A review of twenty cases. J Bone Joint Surg-American Volume 1983; 65(2):205–212.
48. Kurt AM, Unni KK, Sim FH, McLeod RA. Chondroblastoma of bone. Hum Pathol 1989; 20(10): 965–976.
49. Monda L, Wick MR. S-100 protein immunostaining in the differential diagnosis of chondroblastoma. Hum Pathol 1985; 16(3):287–293.
50. Zillmer DA, Dorfman HD. Chondromyxoid fibroma of bone: thirty-six cases with clinicopathologic correlation. Hum Pathol 1989; 20(10):952–964.
51. Bleiweiss IJ, Klein MJ. Chondromyxoid fibroma: report of six cases with immunohistochemical studies. Mod Pathol 1990; 3(6):664–666.
52. Evans HL, Ayala AG, Romsdahl MM. Prognostic factors in chondrosarcoma of bone: a clinicopathologic analysis with emphasis on histologic grading. Cancer 1977; 40(2):818–831.
53. Bertoni F, Present D, Bacchini P, et al. Dedifferentiated peripheral chondrosarcoma: a report of seven cases. Cancer 1989; 63:2054–2059.
54. Frassica FJ, Unni KK, Beabout JW, Sim FH. Dedifferentiated chondrosarcoma: a report of the clinicopathologic features and treatment of seventy-eight cases. J Bone Joint Surg-American Volume 1986; 68(8):1197–1205.
55. Present D, Bacchini P, Pignatti G, Picci P, Bertoni F, Campanacci M. Clear cell chondrosarcoma of bone. A report of 8 cases. Skeletal Radiol 1991; 20(3):187–191.
56. Unni KK, Dahlin DC, Beabout JW, Sim FH. Chondrosarcoma: clear-cell variant. A report of sixteen cases. J Bone Joint Surg–American Volume 1976; 58:676–683.
57. Nakashima Y, Unni KK, Shives TC, Swee RG, Dahlin DC. Mesenchymal chondrosarcoma of bone and soft tissue. A review of 111 cases. Cancer 1986; 57(12):2444–2453.
58. Moser RP Jr, Sweet DE, Haseman DB, Madewell JE. Multiple skeletal fibroxanthomas: radiologic-pathologic correlation of 72 cases. Skeletal Radiol 1987; 16:353–359.
59. Arata MA, Peterson HA, Dahlin DC. Pathologic fractures through non-ossifying fibromas: review of the Mayo Clinic experience. J Bone Joint Surg–American Volume 1981; 63(6):980–988.
60. Shenker A, Weinstein LS, Sweet DE, Spiegel AM. An activating Gs alpha mutation is present in fibrous dysplasia of bone in McCune-Albright syndrome. J Clin Endocrinol Metabol 1994; 79(3):750–755.
61. Voytek TM, RO JY, Edeiken J, Ayala AG. Fibrous dysplasia and cemento-ossifying fibroma. A histologic spectrum. Am J Surg Pathol 1995; 19(7):775–781.
62. Campanacci M, Laus M. Osteofibrous dysplasia of the tibia and fibula. J Bone Joint Surg–American Volume 1981; 63:367–375.
63. Bridge JA, Dembinski A, DeBoer J, Travis J, Neff JR. Clonal chromosomal abnormalities in osteofibrous dysplasia. Implications for histopathogenesis and its relationship with adamantinoma. Cancer 1994; 73(6):1746–1752.
64. Clarke BE, Xipell JM, Thomas DP. Benign fibrous histiocytoma of bone. Am J Surg Pathol 1985; 9(11):806–815.

65. Bertoni F, Capanna R, Calderoni P, Patrizia B, Campanacci M. Primary central (medullary) fibrosarcoma of bone. Sem Diagn Pathol 1984; 1(3):185–198.
66. Huvos AG, Heilweil M, Bretsky SS. The pathology of malignant fibrous histiocytoma of bone. A study of 130 patients. Am J Surg Pathol 1985; 9(12):853–871.
67. Nauert C, Zornosa J, Ayala AG, Harle TS. Eosinophilic granuloma of bone: diagnosis and management. Skeletal Radiol 1983; 10:227–235.
68. Emile JF, Wechsler J, Brousse N, et al. Langerhans cell histiocytosis: definitive diagnosis with the use of monoclonal antibody O10 on routine paraffin-embedded samples. Am J Surg Pathol 1995; 19(6):636–641.
69. Fellinger EJ, Gari-Chesa P, Glasser DB, Huvos AG, Rettig WJ. Comparison of cell surface antigen HBA71 (p30/32 mic2), neuron-specific enolase and vimentin in the immunohistochemical analysis of Ewing's sarcoma of bone. Am J Surg Pathol 1992; 16(8):746–755.
70. Strickler JG, Audeh MW, Copenhaver CM, Warnke RA. Immunophenotypic differences between plasmacytoma/multiple myeloma and immunoblastic lymphoma. Cancer 1988; 61(9):1782–1786.
71. Campanacci M, Baldini N, Boriani S, Sudanese A. Giant-cell tumor of bone. J Bone Joint Surg-American Volume 1987; 69(1):106–114.

7 Percutaneous Ultrasound-Guided Peritendinous Injections

Ronald S. Adler
Division of Ultrasound and Body Imaging, Hospital for Special Surgery, Department of Radiology and Imaging, Weill Medical College of Cornell University, New York, New York, U.S.A.

INTRODUCTION

The rapid development of ultrasound technology in recent years, particularly with regard to small-parts imaging, has paralleled the increasing number of clinical applications of this modality. One such application is the use of ultrasound to provide image guidance for therapeutic delivery of corticosteroid and local anesthetic. A number of publications have appeared, illustrating the efficacy of this method for the performance of musculoskeletal interventional procedures (1–7). The real-time nature of ultrasound allows continuous observation of needle position, thereby ensuring proper placement and providing continuous monitoring of the distribution of the therapeutic agent during administration of long-acting corticosteroid and anesthetic. The potential deleterious effects of not attaining proper needle placement are well documented (8–13).

The current generation of high-frequency small-parts transducers further allows excellent depiction of soft tissue detail. For this reason, needle placement in structures such as a nondistended tendon sheath or bursa can be performed. Ultrasound guidance has broad appeal in as much as it does not involve ionizing radiation; this feature is particularly advantageous in the pediatric population and during pregnancy.

Following a discussion of sonographic technique, ultrasound-guided peritendinous injections will be reviewed with particular attention to injection of tendon sheaths, bursae, and ganglion cysts. Emphasis will be directed toward the most commonly requested injections performed at our institution, which is an Orthopedic and Rheumatology specialty hospital. The most common clinical indication for ultrasound-guided injections generally relates to pain that does not respond to other conservative measures, regardless of the anatomic site. This may occur as the result of a chronic repetitive injury in the work environment, a sports-related injury, or from an underlying inflammatory disorder such as rheumatoid arthritis. We will concentrate on presenting illustrative examples of each of the most relevant studies performed in our practice, without dwelling on the specific clinical entities.

TECHNICAL CONSIDERATIONS

Diagnostic and subsequent interventional examinations are performed using either linear or curved phased array transducers, based on depth and local geometry. Needle selection is based on specific anatomic conditions (i.e., depth and size of the region of interest). We employ a freehand technique in which the basic principle is to ensure needle visualization as a specular reflector (Fig. 1). This relies on orienting the needle so that it is perpendicular (or nearly so) to the insonating beam. The needle then becomes a specular reflector, often having a strong ring-down artifact (7).

Patient positioning to ensure comfort and optimal visualization of the anatomy should first be assessed. It is important to keep in mind that tendons display inherent anisotropy (14). It is, therefore, necessary that the transducer is oriented to maximize tendon echogenicity in order to avoid false interpretation of the tendon as being complex fluid or synovium. An offset may be required at the skin entry point of the needle relative to the transducer to allow for the appropriate needle orientation. Deep structures, such as tendons about the hip, are often better imaged using a curved linear or sector transducer, operating at center frequencies of

(A) (B)
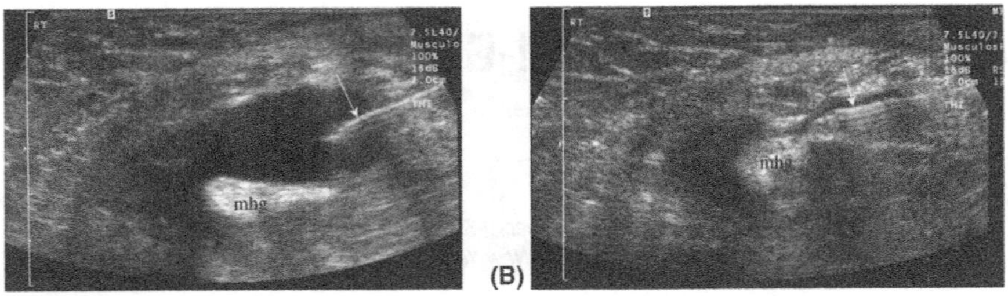

FIGURE 1 Ultrasound-guided aspiration of a Baker cyst. (**A**) Transverse sonogram over the posteromedial aspect of the knee shows a hypoechoic collection in close association with the tendon of the medial head of the gastrocnemius muscle (*mhg*). A needle (*arrow*) is present within the cyst, appearing as an echogenic reflector with a strong posterior reverberation artifact. (**B**) The needle is unchanged in position, and the cyst is observed to be decompressed about the needle tip.

approximately 3.5 to 7.5 MHz. Superficial, linearly oriented structures, such as in the wrist or ankle, are best approached using a linear array transducer with center frequencies greater than or equal to 10 MHz. These factors should be assessed prior to skin preparation.

In our experience, a short-axis approach affords the best opportunity to avoid intratendinous injections. The latter have been associated with collagen breakdown and potential tendon rupture (8,9,11,13). The curvature of the extremity of interest, being greater in short axis, often permits a shorter trajectory and, therefore, greater flexibility in repositioning the needle. This is particularly relevant when performing injections of superficial tendons in the wrist and ankle. Once the transducer is properly positioned, the degree of tendon sheath or bursal distension can be assessed, as well as the needle position relative to the tendon. When positioning the needle into a distended tendon sheath or bursa, the presence of surrounding fluid and/or synovium often provides a standoff to better visualize the needle tip. Alternatively, when injecting into a nondistended structure, a test injection with local anesthetic often permits improved visualization of the needle tip by introducing microbubbles, as well as providing some fluid distension of the sheath or bursa (Fig. 2) (1). The immiscible nature of the steroid–anesthetic mixture may likewise produce temporary contrast effect (Fig. 3).

In every case, the area in question is cleaned with iodine-based solution, and draped with a sterile drape. The transducer is immersed into iodine-based solution and surrounded by a sterile drape; a drape is also placed over portions of the ultra-sound unit. A sonologist or radiologist positions the transducer, while a radiologist positions the needle and performs the procedure. We use 1% lidocaine (Abbot Laboratories, North Chicago, Illinois, U.S.) for local anesthesia. Once the needle is in position, the procedure is undertaken, while imaging in real time. Depending on anatomic location, either a 1.5 in or spinal needle with a stylet is used to administer the anesthesia–corticosteroid mixture typically, [0.5 cc 1% lidocaine, 0.5 cc 0.5% bupivacaine (Sensorcaine–Astra Pharmaceuticals, Westborogh, Massachusetts, U.S.), and 1 cc (40 mg) triamcinolone (Kenalog–Apothecon, a Bristol Myers Squibb Company, Princeton, New Jersey, U.S.), or some equivalent long-acting agent.

INJECTION OF SUPERFICIAL TENDONS

Peritendinous injection of anesthetic and long-acting corticosteroid is an effective means to treat tenosynovitis, bursitis, and ganglion cysts in the hand, foot, and ankle (2,3,5–7,15–18). These structures, being superficially located, are well imaged on ultrasound. Ultrasound-guided injections have been shown to be an effective means to ensure correct localization of therapeutic agent (2,3,5–7).

Foot and Ankle

The most commonly requested peritendinous injections in the foot and ankle, in our experience, are in patients with chronic achillodynia, or those with medial or lateral ankle pain due to posterior tibial or peroneal tendinosis/tenosynovitis, respectively. The large majority of patients

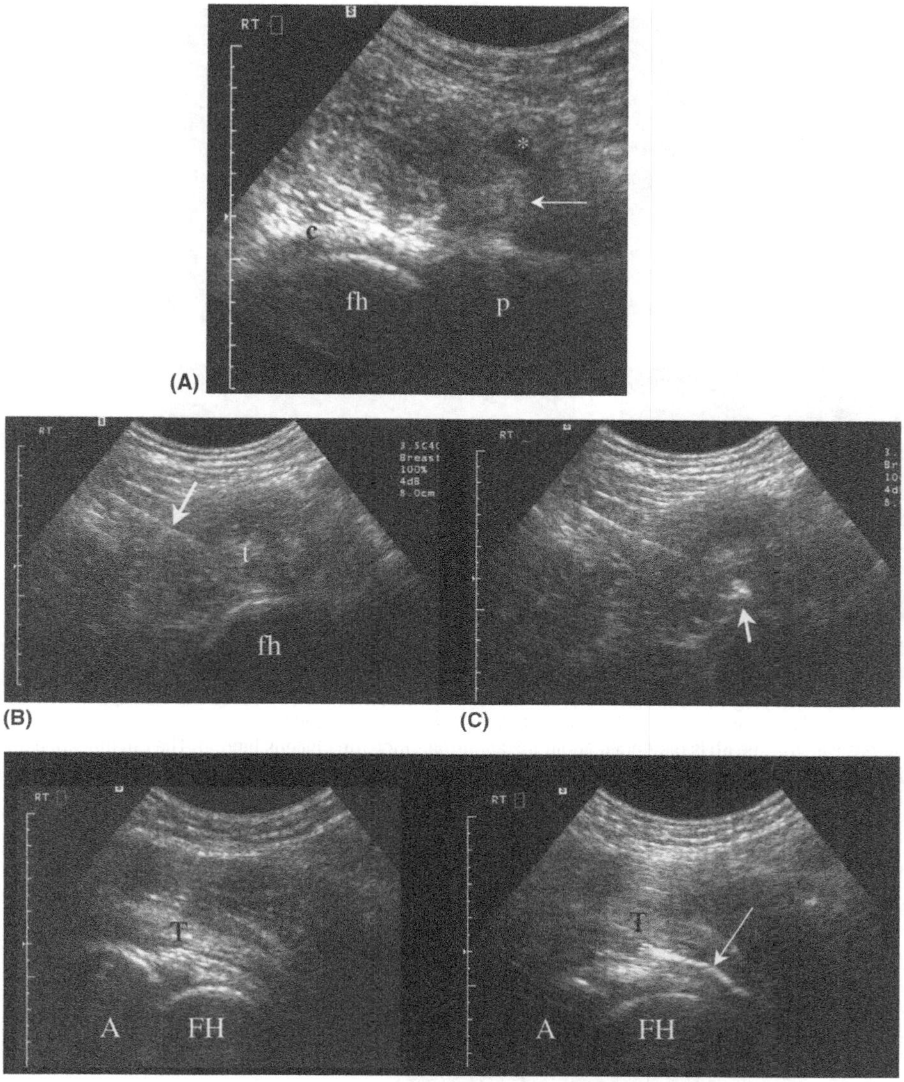

FIGURE 2 Iliopsoas tendon sheath injection. (**A**) Transverse sonogram obtained over the anterior capsule (*c*) of the hip in a patient with a "snapping iliopsoas tendon." A small bursal fluid collection (∗) is seen immediately superficial to the tendon. The iliopsoas tendon (*arrow*) is inhomogeneous-containing intrasubstance clefts. "fh" and "p" denote the femoral head and pubis, respectively. (**B**) Using a curved linear transducer and lateral approach, a 22-gauge spinal needle is advanced toward the deep surface of the iliopsoas tendon (t). (**C**) Improved visualization of the needle tip along the deep surface of the tendon is achieved by test injection of a small amount of local anesthetic. The presence of echogenic microbubbles along the deep surface of the tendon confirms needle position. (**D**) and (**E**) Longitudinal sonograms over the iliopsoas tendon before (**D**) and following (**E**) administration of anesthetic and corticosteroid. In (**E**), echogenic microbubbles (*arrow*) are seen to distribute along the deep surface of the tendon. "A" and "FH" denote acetabulum and femoral head, respectively. *Source*: From Ref. 7.

with achillodynia have pain referable to the enthesis, with associated retrocalcaneal bursitis and achilles tendinosis (2,16). When the pathology in this location is confirmed on gray-scale imaging, retrocalcaneal bursal injection may help alleviate local pain and inflammation (2). We scan the patient in a prone position with the ankle in mild dorsiflexion, using a 7.5 MHz or higher-frequency linear transducer. A short 1.5 in needle usually suffices in these situations, with placement of the needle using a lateral approach. The deep retrocalcaneal bursa is usually

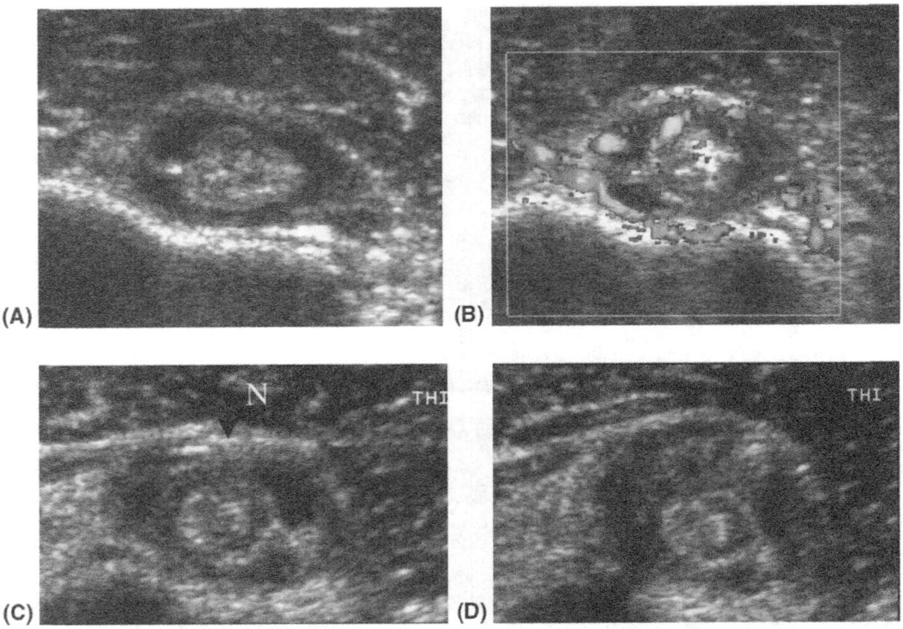

FIGURE 3 Bicipital tenosynovitis. Transverse gray scale (**A**) and power Doppler (**B**) sonograms of the long head of the biceps tendon shows it to be inhomogeneous and surrounded by soft tissue and/or fluid. Increased peritendinous blood flow seen on power Doppler imaging (**B**) is indicative of an associated tenosynovitis. (**C**) Using a lateral approach, the tip of a 22-gauge spinal needle (*N*) is seen within the superficial aspect of the distended tendon sheath. (**D**) Following injection and needle removal, the sheath is distended about the abnormal-appearing biceps tendon. The mixture appears echogenic, likely due to small scatterers produced within the steroid–anesthetic mixture.

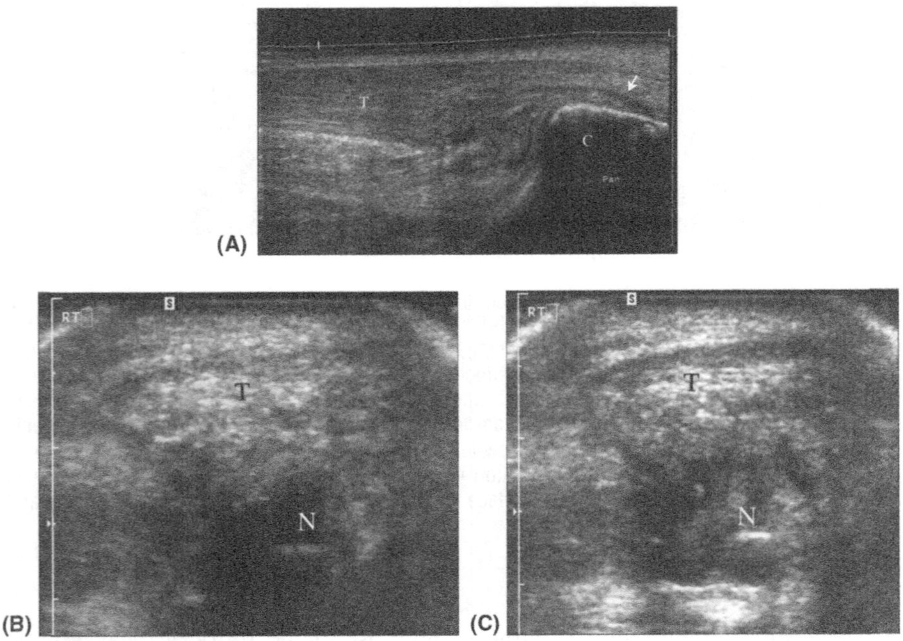

FIGURE 4 Patient with retrocalcaneal pain. (**A**) Extended field of view longitudinal sonogram obtained over the Achilles tendon (*T*), showing it to be fusiformly enlarged, inhomogeneous and containing a longitudinal split (*arrow*). (*C*) denotes the calcaneus. (**B** and **C**) Transverse sonograms obtained over the Achilles tendon (*T*) at the level of the retrocalcaneal bursa. The needle (*N*) is seen within the bursa and deep into the tendon before (**B**) and during (**C**) injection of anesthetic and corticosteroid. The immiscible nature of these two components can produce small scatterers, which are readily seen during real-time injection. The intrabursal injection of material is confirmed during real-time observation.

well seen. A small amount of anesthetic will help confirm position by active distension of the bursa in real time (Fig. 4).

We similarly approach posterior tibial or peroneal tendons in short axis. Patients with pain in this distribution have been shown to benefit from local tendon sheath injections (2,17). The presence of preexisting tendon sheath fluid can facilitate needle visualization. However, careful scanning prior to the procedure to assess the needle trajectory relative to adjacent neurovascular structures should be undertaken. Use of color or power Doppler imaging can facilitate visualization of the neurovascular bundle. The posterior tibial nerve is closely related to adjacent vascular structures and is usually well seen prior to bifurcating into medial and lateral planter branches. In our experience, fluid is frequently seen in relation to the posterior tibial tendon, in the submalleolar region. The peroneal tendons are less predictable (19). Use of power Doppler imaging in conjunction with real-time guidance can be beneficial in localizing areas of inflammation for directed injection (20,21). In the case of stenosing tenosynovitis, the tendons may only be surrounded by proliferative synovium or scar tissue (Fig. 5) (17). In this latter case, the use of a test injection of local anesthesia to confirm the distribution of the therapeutic agent within the tendon sheath in real time can be invaluable (Fig. 6).

Hand and Wrist

In the hand and wrist, DeQuervain's tendonitis is a frequently encountered tendinopathy involving the abductor pollicus longus tendon, which responds to local administration of

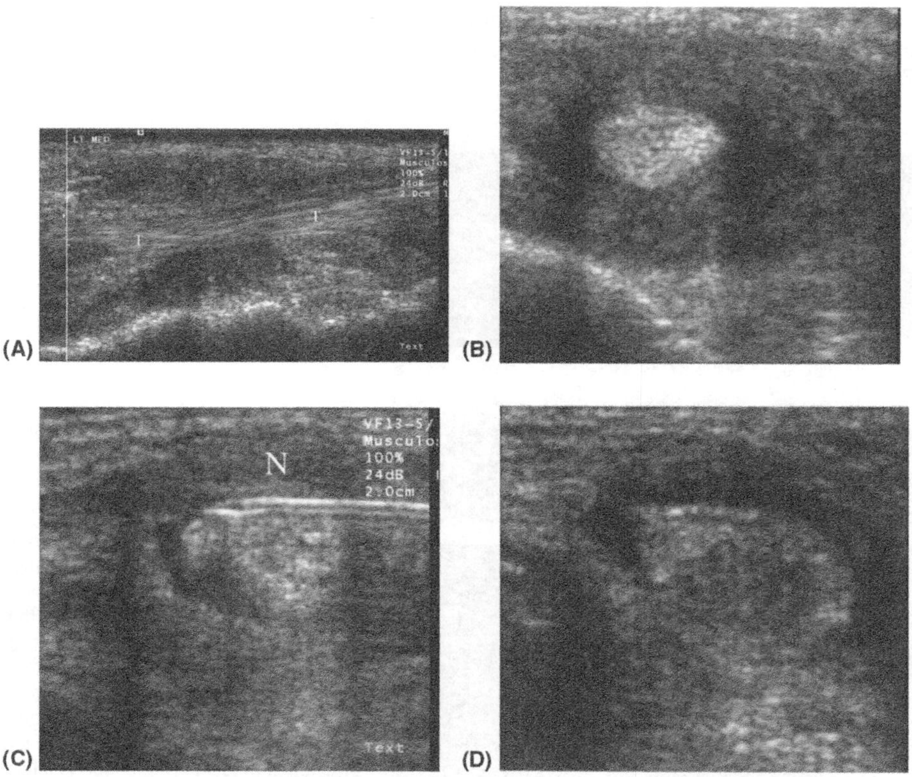

FIGURE 5 Patient with persistent medial ankle pain following deltoid ligament injury and subsequent development of clinical stenosing tenosynovitis. Longitudinal (**A**) and transverse (**B**) sonograms obtained of the posterior tibial tendon shows extensive hypoechoic tissue surrounding the posterior tibial tendon. This was presumed to be related to peritendinous scarring following an injury. The patient complained of pain along the posterior tibial tendon. (**C**) The needle (*N*) is advanced along the superficial margin of the posterior tibial tendon. (**D**) A test injection followed by therapeutic injection of anesthetic and corticosteroid confirms distension of the tendon sheath.

(A)

(B)

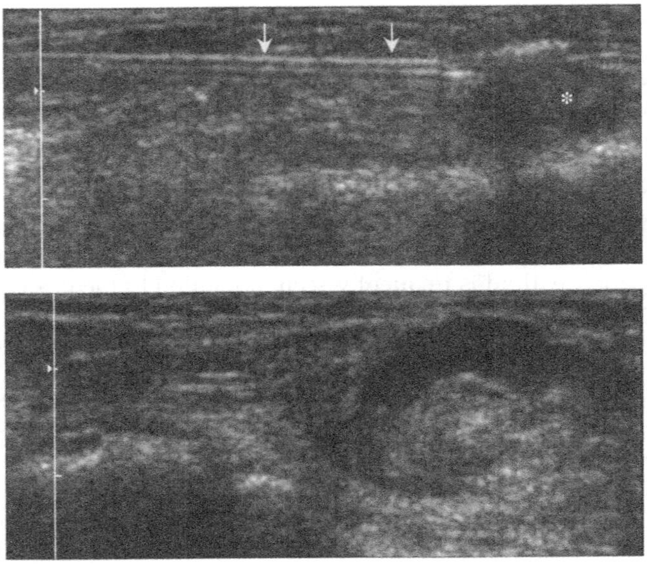

FIGURE 6 Peroneal tendon sheath injection in a patient with peroneus brevis tendonosis. Transverse sonogram of the peroneus brevis tendon in the submalleolar region shows it to be hypoechoic. (**A**) 22-gauge spinal needle (*arrows*) is advanced to the margin of the tendon (∗). (**B**) Following injection of therapeutic mixture, the tendon sheath is distended. The inhomogeneity of the peroneus brevis tendon is better appreciated, due to the surrounding fluid standoff from the distended tendon sheath.

anti-inflammatory agent (5,6). We are also frequently asked to perform injections in patients with rheumatoid arthritis. These patients commonly experience severe tenosynovitis, which can lead to secondary tendon rupture and deformity (18). In either case, the approach is similar to superficial structures in the foot and ankle. A short-axis approach is adopted, in which the surrounding neurovascular structures are avoided, and the corresponding tendon sheaths are injected (Fig. 7).

(A)

(B)

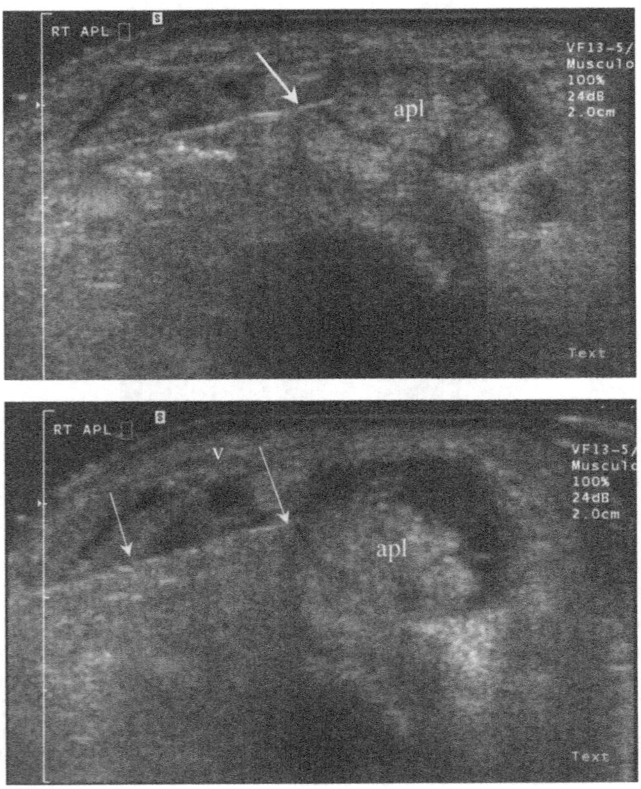

FIGURE 7 Therapeutic injection in a patient with DeQuervain's tendonitis. (**A**) Transverse sonogram obtained over the first extensor compartment of the wrist. Using a transverse approach, a 25-gauge 1.5-inch needle (*arrow*) is advanced to the margin of the tendon sheath of the abductor pollicus longus tendon (*apl*). A small tendon sheath effusion is present. The tendon is inhomogeneous compatible with patient's tendonosis. (**B**) Following injection with anesthetic and corticosteroid under real-time guidance, the tendon sheath is distended. A small vessel (*v*) is noted superficial to the needle (*arrows*).

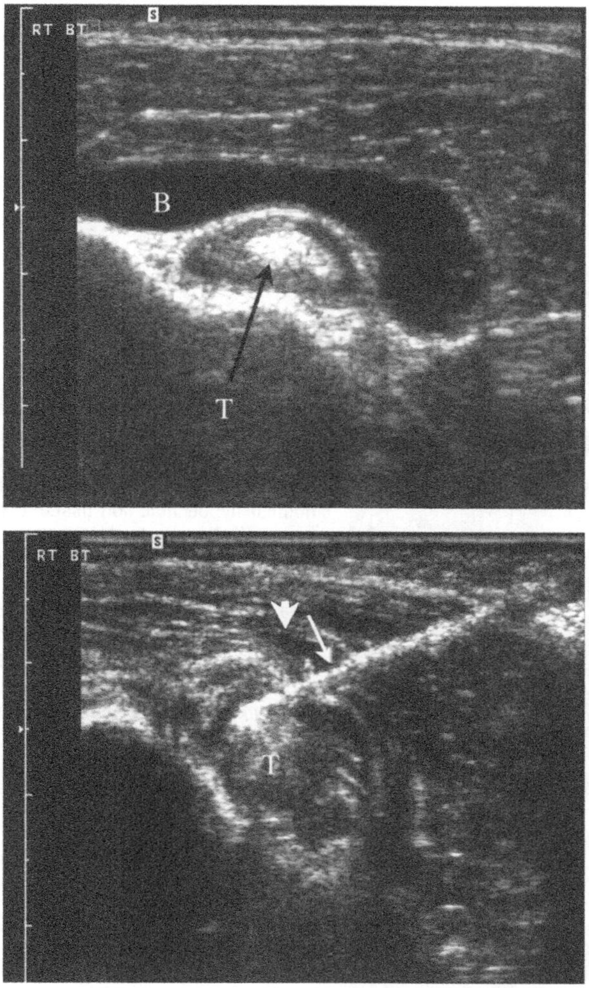

(A)

(B)

FIGURE 8 Bicipital tenosynovitis and distended subdeltoid bursa. (**A**) A transverse photopic image of the long head of the biceps tendon (*T*) shows a distended subacromial subdeltoid bursa overlying the biceps tendon. The tendon (*B*) is surrounded by abnormal soft tissue presumed to relate to synovial proliferation. Of note, the biceps tendon is subluxed medially and is overriding the lesser tuberosity. (**B**) Transverse photopic image of the long head of the biceps tendon during ultrasound-guided injection of the biceps tendon sheath. A 22-gauge spinal needle (thin *white arrow*) is advanced past the overlying bursa with the tip along the superficial margin of the tendon (*T*). The bursa is compressed but evident in the image (*thick white arrow*).

INJECTION OF DEEP TENDONS

The most commonly requested deep tendon injections in our experience include the bicipital tendon sheath, iliopsoas tendon, gluteal tendon insertion onto the greater trochanter, and hamstring tendon origin.

Biceps Tendon

Anterior shoulder pain with radiation into the arm may be secondary to bicipital tendonitis and/or tenosynovitis (22). The biceps tendon can be palpated, but the sheath, if nondistended may offer less than 2 mm clearance to place a needle (22). This is complicated by the caudal extension of the subacromial subdeltoid bursa, which may overlie the bicipital tendon sheath (23). A nonimage-guided injection could therefore result in delivery into an extratendinous synovial space, or possibly result in an intratendinous injection. We have found that ultrasound guidance enables localization of therapeutic agent to the biceps tendon sheath (Figs. 2 and 8).

The patient is placed recumbent with the forearm supinated and the shoulder mildly elevated. The bicipital groove is oriented anteriorly. Using a linear transducer, typically 7.5 MHz, we use a lateral approach with a 22-gauge spinal needle. The long head of the biceps tendon is scanned in short axis. When fluid distends the bicipital tendon sheath, the tip is directed into the fluid. Otherwise the needle is directed along the superficial margin of the

(A)
(B)

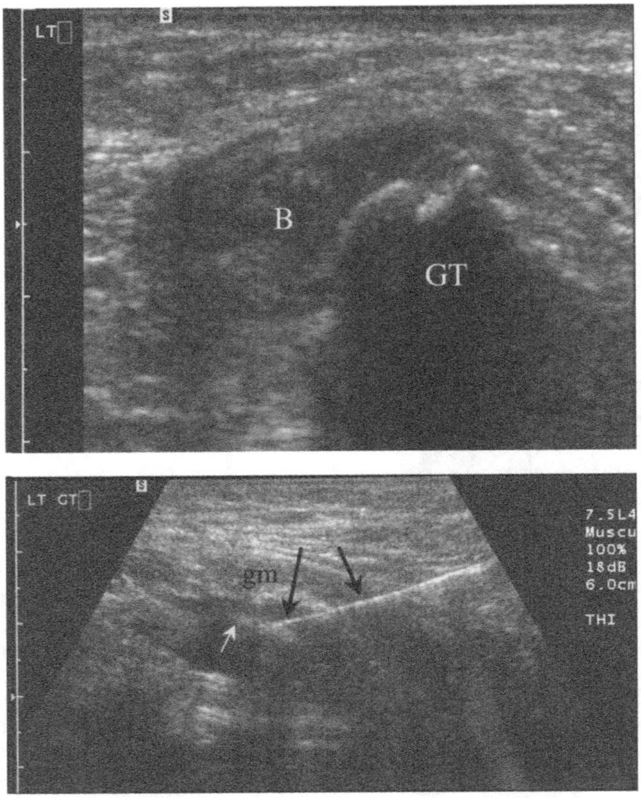

FIGURE 9 Greater trochanteric bursitis. (**A**) Longitudinal sonogram obtained over the greater trochanter in a patient with pain radiating into the buttocks. A complex bursa (*B*) is seen containing nodular hypoechoic soft tissue, contiguous with the greater trochanter (*GT*). The patient complained of localized pain with the transducer positioned over this area. (**B**) Transverse sonogram over the greater trochanteric bursa shows a 22-gauge spinal needle (*black arrows*) positioned deep into the gluteal muscles (*gm*) with its tip within the bursa. The appearance of echogenic microbubbles (*white arrow*) is evident during the injection on real-time observation.

(A)
(B)

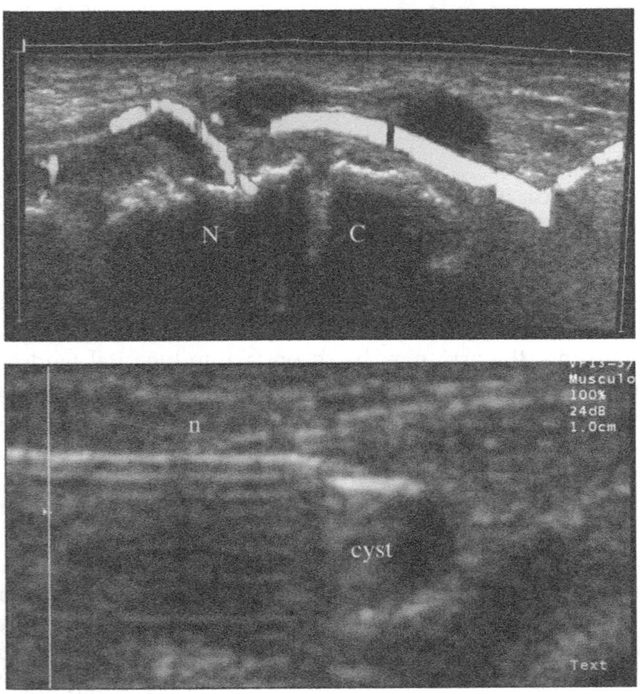

FIGURE 10 Complex ganglion cyst along dorsum of the foot. (**A**) Longitudinal extended-field-of-view sonogram demonstrates a bilobed cyst in relation to the dorsal capsule of the naviculocuneiform joint. A branch vessel of the dorsalis pedis artery is noted on power Doppler imaging, coursing along the deep surface of the cyst. (**B**) Using a short-axis approach, a 22-gauge 1.5-inch needle (*n*) is advanced into the superficial margin of the cyst in order to avoid the deeper arterial branch. The cyst is partially obscured in this image by the strong characteristic reverberation artifact from the needle.

tendon and a test injection of local anesthetic is used to confirm local distension of the sheath, which is then followed by administration of the long-acting corticosteroid. The presence of fluid distension of the sheath with superficially located microbubbles helps to confirm a successful injection.

Iliopsoas Tendon

The iliopsoas tendon lies superficial to and along the medial margin of the anterior capsule of the hip. The tendon inserts into the lesser trochanter (24). A bursa that frequently communicates with the hip is frequently seen in this location and may be distended as the result of underlying joint pathology, or as a primary iliopsoas bursitis (25). Alternatively, one can have an iliopsoas tendinosis in the absence of a preexisting bursitis, in which a peritendinous injection is requested (Fig. 2). A lateral approach to the tendon often requires use of a lower frequency transducer and curved linear or sector geometry. The neurovascular bundle lies medial to the tendon, so that it is advantageous to approach from the lateral margin of the tendon and perform a small test injection to confirm needle position. A successful injection will show the appearance of fluid and/or microbubbles distributed along the long axis of the tendon (Fig. 2).

BURSAL AND GANGLION CYST INJECTIONS

Distended bursae about tendon insertions provide anatomic localization for therapeutic agent. Injections of these areas are often requested in the setting of localized bursitis with abnormality of the adjacent tendon (5–7,15,17,26). Examples would include the retrocalcaneal, greater trochanteric, or ischial burse (Figs. 4 and 9). Alternatively, the presence of a bursitis, distended synovial cyst, or ganglion cyst may cause mechanical impingement of adjacent tendons (Fig. 1). The decompression of these cysts with subsequent administration of therapeutic agent may alleviate these symptoms (15,26). Ultrasound guidance allows one to avoid intratendinous injections as well as adjacent neurovascular structures (Fig. 10). Furthermore, the needle may be redirected as necessary in the presence of a multiloculated cyst.

(A)

(B)

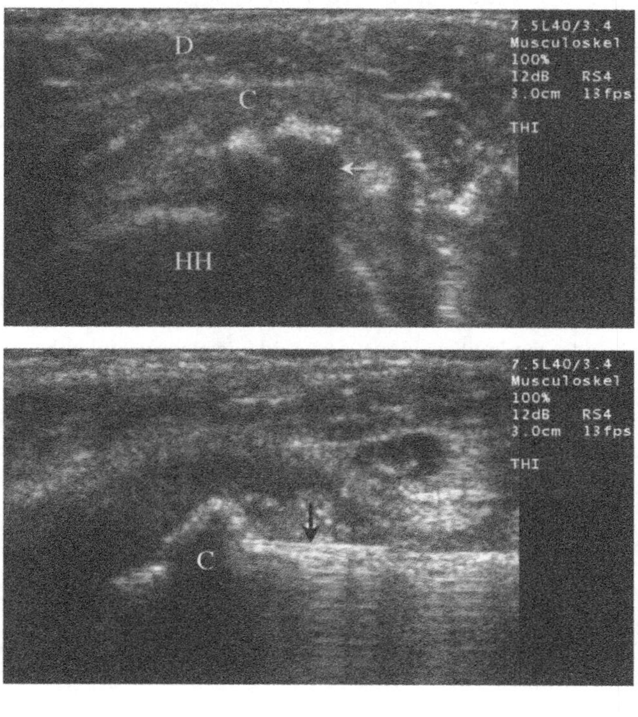

FIGURE 11 Calcific tendonitis. (**A**) Transverse sonogram of the supra spinatus tendon containing two distinct calcifications (*C*) with posterior acoustic shadowing (*arrow*). "HH" and "D" refer to humeral head and deltoid muscle, respectively. (**B**) An 18-gauge spinal needle (*arrow*) is seen within one of the calcifications. The needle displays a characteristic strong reverberation artifact. A second tandem needle was placed into the same calcification and a series of lavages were performed combined with mechanical fragmentation to reduce the level of intratendinous calcification, followed by steroid injection. The patient reported approximately 70% improvement within the next week and has resumed all usual activities.

CALCIFIC TENDONITIS

The presence of symptomatic intratendinous calcification involves the deposition of calcium hydroxyapatite. This often appears as a nodular echogenic mass within the tendon, which may or may not display posterior acoustic shadowing (Fig. 1). This most often occurs in the shoulder, but may occur elsewhere (27,28). Ultrasound-guided fragmentation and lavage have been described as an excellent method to reduce the level of calcification and deposit therapeutic agent (29,30). As described by Farin et al., we employ a dual-needle technique when possible, with one needle acting as inflow for anesthetic and/or sterile saline, and the other acting as an outflow for the calcium solution (Fig. 11) (29). Following multiple lavages, one needle is removed, and the other is used to inject anesthetic and anti-inflammatory mixture. If the placement of a second tandem needle is technically difficult, fragmentation of the calcium with a single needle and peritendinous therapeutic injection has been shown to be effective (4).

SUMMARY

Ultrasound provides several distinct advantages as a method to provide guidance for delivery of therapeutic injections. The most important of these is the ability to visualize the needle and make adjustments in real time to ensure that medication is delivered to the appropriate location. As we have shown through a variety of clinical examples, the current generation of ultrasound scanners provides excellent depiction of the relevant anatomy. The needle has a unique sonographic appearance and can be monitored in real time, as can the steroid-anesthetic mixture. Given these advantages, we believe that ultrasound guidance should become the method of choice for administration of peritendinous injections.

REFERENCES

1. Christensen RA, Van Sonnenberg E, Casola G, Wittich GR. Interventional ultrasound in the musculoskeletal system. Radiol Clin North Am 1988; 26:145–156.
2. Cunnane G, Brophy DP, Gibney RG, FitzGerald O. Diagnosis and treatment of heel pain in chronic inflammatory arthritis using ultrasound. Semin Arthritis Rheum 1996; 25:383–389.
3. Brophy DP, Cunnane G, FitzGerald O, Gibney RG. Technical report: ultrasound guidance for injection of soft tissue lesions around the heel in chronic inflammatory arthritis. Clin Radiol 1995; 50:120–122.
4. Cardinal E, Chhem RK, Beauregard CG. Ultrasound-guided interventional procedures in the musculoskeletal system. Radiol Clin North Am 1998; 36:597–604.
5. Koski JM. Ultrasound guided injections in rheumatology. J Rheumatol 2000; 27: 2131–2138.
6. Grassi W, Farina A, Filippucci E, Cervini C. Sonographically guided procedures in rheumatology. Semin Arthritis Rheum 2001; 30:347–353.
7. Sofka CM, Collins AJ, Adler RS. Use of ultrasonographic guidance in interventional musculoskeletal procedures: A review from a single institution. J Ultrasound Med 2001; 20:21–26.
8. Unverferth LJ, Olix ML. The effect of local steroid injections on tendon. J Sports Med 1973; 1:31–37.
9. Ford LT, DeBender J. Tendon rupture after local steroid injection. South Med J 1979; 72:827–830.
10. Gottlieb NL, Riskin WG. Complications of local corticosteroid injections. JAMA 1980; 243:1547–1548.
11. Oxlund H, Manthorpe R. The biochemical properties of tendon and skin as influenced by long term glucocorticoid treatment and food restriction. Biorheology 1982; 19:631–646.
12. Stapczynski JS. Localized depigmentation after steroid injection of a ganglion cyst on the hand. Ann Emerg Med 1991; 20:807–809.
13. Shrier I, Matheson GO, Kohl HW. Achilles tendonitis: are corticosteroid injections useful or harmful? Clin J Sport Med 1996; 6:245–250.
14. Bouffard JA, Eyler WR, Introcaso JH, van Holsbeeck M. Sonography of tendons. Ultrasound Q 1993; 11:259–286.
15. Breidahl WH, Adler RS. Ultrasound-guided injection of ganglia with coricosteriods. Skeletal Radiol 1996; 25:635–638.
16. Gibbon WW, Cassar-Pullicino VN. Heel pain. Ann Rheum Dis 1994; 53:344–348.
17. Baker KS, Gilula LA. The current role of tenography and bursography. AJR 1990; 154:129–133.
18. Kane D, Greaney T, Bresnihan B, Gibney R, Fitzgerald O. Ultrasonography in the diagnosis and management of psoriatic dactylitis. J Rheumatol 1999; 26:1746–1751.
19. Nazarian LN, Rawool NM, Martin CE, Schweitzer ME. Synovial fluid in the hindfoot and ankle: detection of amount and distribution with US. Radiology 1995; 197:275–278.
20. Newman JS, Adler RS, Bude RO, Rubin JM. Detection of soft-tissue hyperemia: value of power Doppler sonography. AJR 1994; 163:385–389.

21. Breidahl WH, Stafford Johnson DB, Newman JS, Adler RS. Power Doppler sonography in tenosynovitis: significance of the peritendinous hypoechoic rim. J Ultrasound Med 1998; 17:103–107.
22. Middleton WD, Reinus WR, Totty WG, Melson GL, Murphy WA. US of the biceps tendon apparatus. Radiology 1985; 157:211–215.
23. van Holsbeeck M, Strouse PJ. Sonography of the shoulder: evaluation of the subacromial-subdeltoid bursa. AJR 1993; 160:561–564.
24. Stoller DW, Maloney WJ. The hip. In: Stoller DW, ed. Magnetic Resonance Imaging in Orthopaedics and Sport Medicine. Philadelphia, PA: J.B. Lippincott and Co., 1993:45–47.
25. Resnick D, Niyama G. Anatomy of individual joints. In: Resnick D, Niyama G, eds. 2d ed. Diagnosis of Bone and Joint Disorders. Philadelphia, PA: BW Saunders, 1988:713.
26. Chiou HJ, Chou YH, Wu JJ, et al. Alternative and effective treatment of shoulder ganglion cyst: ultrasonographically guided aspiration. J Ultrasound Med 1999; 18:531–535.
27. Farin PU, Jaroma H. Sonographic findings of rotator cuff calcifications. J Ultrasound Med 1995; 14:7–14.
28. Howard CB, Vinzberg A, Nyska M, Zirkin H. Aspiration of acute calcerous trochanteric bursitis using ultrasound guidance. J Clin Ultrasound 1993; 21:45–47.
29. Farin PU, Rasanen H, Jaroma H, Harju A. Rotator cuff calcifications: treatment with ultrasound-guided percutaneous needle aspiration and lavage. Skeletal Radiol 1996; 25:551–554.
30. Farin PU, Jaroma H, Soimakillio S. Rotator cuff calcifications: treatment with US-guided technique. Radiology 1995; 195:841–843.

8 Epidural Steroids

Kirkland W. Davis and James J. Choi
Department of Radiology, University of Wisconsin, Madison, Wisconsin, U.S.A.

INTRODUCTION

For the last century, physicians have employed epidural injections to treat low back and radicular pain. Corticosteroids have been included in this technique for the last 50 years. As we continue to refine this technique and study its effects, it has become clear that epidural steroid injections (ESIs) are an important part of the armamentarium for treating low back pain, sciatica, and, to a lesser degree, neck pain.

Back pain can be quite difficult to treat. The vast majority of patients with acute symptomatic disc herniations will improve without surgery (1); however, they often suffer greatly and experience varying degrees of disability, leading to lost productivity and wages until recovery. Epidural steroids do not cure the underlying cause of a patient's pain; however, they often reduce or eliminate the pain temporarily (2), allowing other rehabilitative efforts to proceed and a more rapid return to a normal lifestyle (3). Although occasionally patients are "cured" by ESIs, it is unrealistic to expect or advertise a complete lasting relief from this procedure alone. The goal should be to decrease pain as much as possible, help the patient cope with the residual pain, and allow maximum function and activity (4).

This chapter reviews the history of epidural injections and the controversy surrounding this technique. A review of the technique's efficacy and a description of the relevant anatomy will be included. This chapter also discusses in detail the practical aspects of the technique including indications, adverse effects, and medications used.

ANATOMY

The epidural space extends from the foramen magnum down through the sacral hiatus. It consists of the space or potential space between the external surface of the spinal dura mater and the inner margin of the spinal canal (5,6). Thus, the outer limit of the epidural space is confined by the peridural membrane, which is demarcated by the periosteum of the osseous canal, the posterior margins of the intervertebral discs, the inner border of the ligamenta flava, and, sometimes, the anterior border of the interspinous ligament (7,8). Noteworthily, some patients have a midline posterior dural reflection, the plica mediana dorsalis, which may divide the epidural space and limit the spread of injectate to one side (5,6,9,10). This variant appears to be more common in the sacral levels.

In the lumbar spine, the epidural space is typically largest in the posterior mid-line, where the triangular posterior recess is usually filled with fat and readily visible on magnetic resonance imaging (MRI) and computed tomography (CT) scans. This recess is the target for epidural injection if the "interlaminar" approach is used (Fig. 1). Postsurgical scarring often obliterates it. Also, the posterior epidural space is commonly only a potential space at L5/S1 or in patients with spinal stenosis. In the cervical spine, the posterior epidural space is often only a potential space, with no focal fat to target. Therefore, reaching the epidural space here relies on feel as much as it does on image guidance.

The sacrum is formed by the fusion of five embryonic vertebrae and is convex dorsally. The coccyx consists of three to five rudimentary vertebrae attached to the base of the sacrum. The sacral hiatus, the portal used for accessing the epidural space, is a natural defect resulting from an incomplete midline fusion of the posterior elements of the lower portion of S4 and the entire S5 vertebra. The hiatus, which is covered by the sacrococcygeal ligament, is bordered laterally by the sacral cornua and its floor is made up of the posterior aspect of S5 (Fig. 2). The thecal sac has a variable termination depending on age: it ends at the lower border of the S1

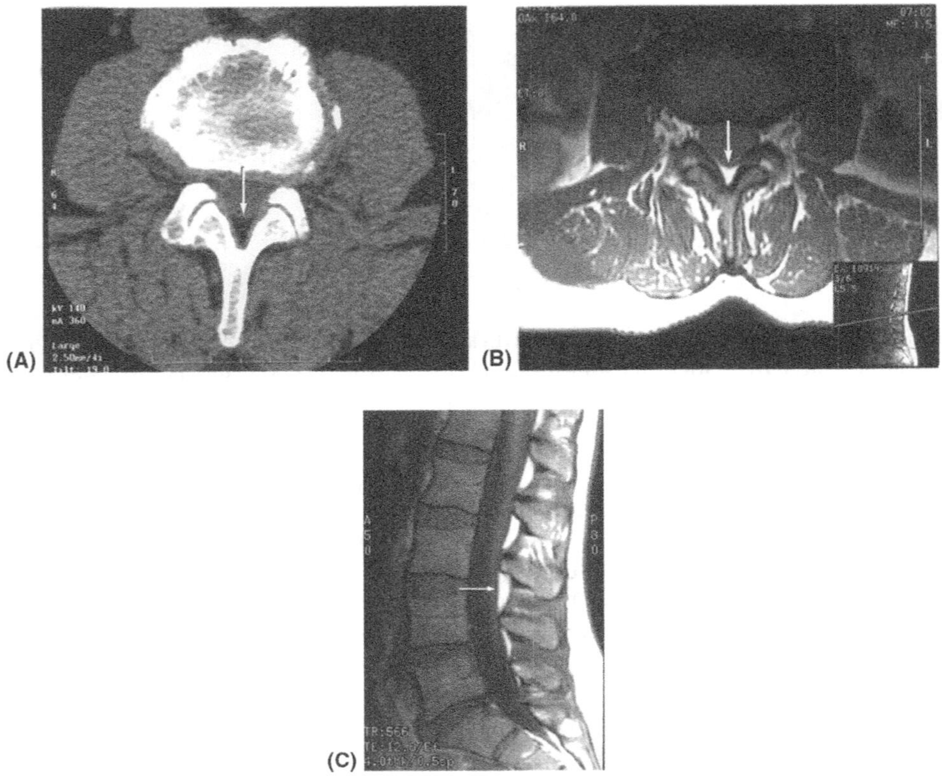

FIGURE 1 Posterior epidural space. (**A**) Axial computed tomography demonstrating fat in the posterior epidural space (*arrow*). (**B, C**) Axial proton density and sagittal T1-weighted magnetic resonance images of a different patient demonstrating fat in the posterior epidural space at L3/4, providing a suitable target for epidural steroid injection. (Note that the posterior recess at L5/S1 is characteristically negligible.)

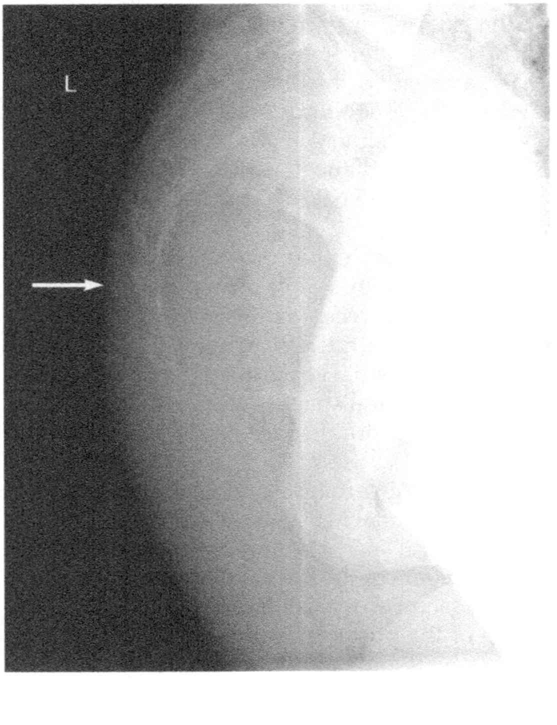

FIGURE 2 The paired sacral cornua (*arrow*), which border the sacral hiatus.

foramen in adults and at the S3 foramen in children, but studies of cadavers suggest that the position of the thecal sac tip is at the middle-third of the S2 body in the average patient (11,12).

HISTORY

A brief review of the history of epidural injections is instructive as to the current status of the technique and some of the surrounding controversy.

Epidural injection was first reported in 1901. At that time, cocaine was injected for lumbago or sciatica. In 1925, Viner described epidurals using 20 cc of 1% procaine in 50 to 100cc of Ringer's solution, normal saline, or petrolatum (13,14). In 1930, Evans described caudal epidural injections in 40 patients with sciatica. Most of them received 1% novocaine, often over 100 cc, although some were treated with saline only. He described some side effects as "abnormal sensations or paresthesiae such as formication. A few [patients] said that they had found it difficult to control a desire to shout or scream." One patient, treated with 2% novocaine, lost consciousness for an hour and was incontinent of urine and feces for 12 hours, implying the need for caution regarding the concentration of anesthetic. Nevertheless, 22 of the patients were "cured," and another five "improved" (15).

Similarly, in 1944, Kelman reported a high rate of improvement or cure, using 50 to 100 cc of saline, 1% novocaine, or 0.75% metycaine in saline administered every other day for four-to-six treatments (16).

These early investigators were often under the impression that benefits of epidural injections not only came from anesthetic effects but that they were also breaking up adhesions around affected nerve roots and displacing nerves away from herniated discs. Thus, the side effects of the procedure, including frequent pain, headache, and dizziness, were thought to be warranted to achieve a reasonably high hydrodynamic effect with large volumes of injectate (1).

In 1952, corticosteroids were added to the local anesthetic mixture for treating acute and chronic back pain (17). The next year, as recounted by Benzon, Lievre et al. administered hydrocortisone into the epidural space via an S1 foraminal injection in 46 patients with sciatica. Of these, 8 had a very good response, 15 good, and 8 mediocre (1). Goebert et al. were the first to report ESI in the United States, using three daily injections of 30 cc procaine and 125 mg hydrocortisone acetate (18).

Since that time, the technique has continued to evolve. Volumes have diminished [from a maximum of 200 cc (19)], and there has been experimentation with numerous steroid preparations and combinations with anesthetics or saline.

Two primary controversies surrounding ESI remain. The first is efficacy. Because most patients with acute radicular pain from disc herniations tend to improve without any or no therapy, initial enthusiasm for the technique has been tempered by conflicting results in subsequent controlled trials. This subject will be examined in the next section.

The second controversial subject is the safety of placing available steroid preparations into the epidural space. The risk is incurred if the injection inadvertently enters the thecal sac or if there is sufficient transit of the materials across the dura into the thecal sac to cause harm (20). Repeated doses of intrathecal steroids have been associated with myelographic signs of arachnoiditis (not necessarily symptomatic), although direct causality is not established (21). The risk of causing arachnoiditis from a single inadvertent intrathecal injection is thought to be negligible (6).

Nelson has campaigned against ESIs because of the purported neurotoxicity of polyethylene glycol, which is contained in two of the steroids popular in ESI, triamcinolone diacetate (Aristocort Intralesional, Fujisawa, Deerfield, Illinois, U.S.A.) and methylprednisolone acetate (Depo Medrol, Pharmacia & Upjohn, Kalamazoo, Michigan, U.S.A.) (20). Noteworthily, the substance that has shown to be potently neurotoxic is propylene glycol. Polyethylene glycol is not neurotoxic at the commercially available concentrations (4,22). As Spaccarelli notes, several studies now refute the risk of arachnoiditis or neurotoxicity of depot steroids in clinically used concentrations (23).

MECHANISMS OF ACTION

Evans's speculation in 1930 that epidural injections achieve their effect by lysis of adhesions and displacement of nerve roots (15) has given way to other theories. In 1983, White, an orthopedic surgeon, made the logical point that adhesions are difficult to dissect with a scalpel, thus throwing considerable doubt on the bulk effect of large volume injections (24). The concept of an injection displacing and stretching nerve roots (25) has never been validated.

It is now generally accepted that epidural steroids give relief by their anti-inflammatory effects. Green et al. noted that inflammation of the involved spinal nerves has often been documented at surgery (26), and there is strong evidence of inflammation in lumbar radiculopathy and disc degeneration. In fact, McCarron et al. observed an intense inflammatory reaction after injection of autologous disc material into the epidural spaces of dogs and a rapid onset of fibrosis (27). Franson et al. demonstrated high levels of phospholipase A2 (PLA2) around symptomatic disc herniations and painful discs. Subsequently, they engendered an inflammatory response by injecting PLA2 into mouse paws (28). PLA2 exerts its effect by liberating arachidonic acid from cell membranes, causing an intense inflammatory reaction in adjacent tissues (4). The inhibition of PLA2 activity is one of the anti-inflammatory actions of corticosteroids (29,30).

Johansson et al. postulated additional effects of corticosteroids in a study of rats by demonstrating that the corticosteroids reduce signal transmission in the unmyelinated C-fibers, which carry nocioceptive information, but not in myelinated fibers (31).

If local anesthetics are combined with the steroids, they give immediate pain relief. This is theorized to have a beneficial psychological effect in that it allows a patient to consider the pain as treatable, and that it breaks the postulated pain-muscle spasm-ischemia-pain cycle (25).

EFFICACY

Lumbar and Caudal Injections

In patients with symptoms emanating from the lumbar spine, interlaminar ("lumbar") and caudal ("sacral") injections have not been shown to have a significant difference in efficacy (32,33). Over the last four decades, numerous uncontrolled studies have touted the results of ESIs, although most studies report a much higher rate of success over the short term (a few months) than the long term (a year or more) (26,34–40). While these studies have often been criticized for their lack of control patients, many demonstrated relief in patients who had not benefited from other nonsurgical treatments, and thus were purported to be valid. In 1996, Spaccarelli reviewed 26 uncontrolled studies and compiled an overall benefit rate of 65% (23).

As greater scientific rigor has been applied to the study of ESI, controlled studies have not uniformly endorsed the benefits of this procedure. Some controlled studies support the use of epidurals (3,19,33,41), whereas others have not shown a statistically significant benefit above placebo (42–45). Noteworthily, the two studies most widely employed to criticize ESI (42,45) were designed in a way that actually minimizes the chance of a successful result given today's understanding of the effects of this procedure (36,46). Both studies evaluated patients first at 48 hours or earlier after the injection, when most patients have not yet fully benefited from the depot steroids, and then not again for at least 8 (45) or 13 (42) months, when the effect has worn off for most patients and many have naturally improved anyway.

Several authors have reviewed and evaluated all of the controlled studies systematically (23,36,46–48), a task that has proven to be quite difficult, given the variability in patient diagnoses, injection techniques, outcome measures, and length of follow-up periods (2,46). Even these reviewers have not agreed, with some citing at least a short-term benefit to ESI (23,48) and others unable to support or refute efficacy of the technique (46,47). All these authors decried the various methodological flaws of each controlled study. One major flaw in all of the studies was that none employed fluoroscopy to document successful positioning of the needle within the epidural space (46). Numerous investigators have shown that blind placement of needles into the epidural space is often unsuccessful (34,49,50).

Finally, one meta-analysis of 11 "suitable" randomized controlled trials of ESI included 907 patients. In this analysis, ESI increased the odds ratio of short-term (up to 60 days) pain relief (greater than 75%) to 2.61 versus placebo. In the long term, this number fell to 1.87. Thus, their results supported the short-term success of this procedure and lent weight to the concept that some, but not as many, patients may experience relief over longer periods (32).

Cervical Injections

ESIs are performed much more sparingly in the cervical spine than via the lumbar or caudal route. Consequently, controlled studies are lacking. This is probably due to an unwillingness to subject patients to this procedure only to have a placebo injected. However, uncontrolled studies have supported this technique (51–55). For instance, Rowlingson and Kirschenbaum performed a retrospective analysis of 25 patients who received a total of 45-cervical ESIs for cervical radiculopathy. Sixty-four percent reported a good (75% pain relief) or excellent response to the injections. Conservative measures had failed these patients (53). Similarly, Cicala et al. found 41% excellent and 21% good results at six months in a study of 58 cervical ESI patients (55). While epidural injections in the cervical spine carry a greater potential risk due to the proximity of the spinal cord, many authors consider it an effective method of therapy in patients who often have few options (52). Alternatively, some of the surgeons, physiatrists, and other physicians who refer patients for lumbar ESI are unwilling to refer cervical spine patients.

INDICATIONS

Most radiologists are not involved in patient selection. Given the relative safety of ESIs and numerous anecdotal successes even in patients whose selection is probably not supported by the literature, we are willing to perform the procedure on all our referrals, provided they do not have contraindications. It would be difficult to deny a patient a chance at relief, even if it is small. Nevertheless, it is prudent that all who perform this procedure be conversant with the accepted indications and thus be able to discuss with patients the likelihood of success.

Lumbar/Caudal

It is widely accepted that the principal indication for lumbar ESI is radicular pain, with or without low back pain (32,38,48,56). There is little support for consistent efficacy of ESI in mechanical or nonradiating pain (23,38). While Rivest et al. reported lower success in patients whose symptoms emanate from spinal stenosis than from herniated discs (57), others have not supported this discrepancy.

Additionally, the majority of investigators have reported greater rates of success in patients with short-term symptoms (less than three or six months) (1,26,37,56,58) than in patients with greater than one year of pain, possibly because inflammation eventually gives way to fibrosis (23). In an analysis of 209 patients who underwent one to three lumbar ESIs each, Hopwood and Abram showed a statistically significant increased risk of failure in patients who were unemployed, smoked, and had nonradicular and long-term pain (59).

Conditions that appear to be recalcitrant to ESI treatment include the post-surgical state (26,56,60) and neurogenic claudication (61). Fredman et al. posit that surgical adhesions and scar tissue often prevent the injectate from reaching the pain generator, an idea supported by their contrast injections on postsurgical patients (60). Fukusaki et al. hypothesized that the lack of efficacy in neurogenic claudication is related to its causation by ischemic neuropathy and not inflammation (61). Back pain related to neoplasm, infection, or spondyloarthropathy is not an indication for epidural injection.

Cervical

ESIs are much less commonly performed in the cervical spine. With the consequent relative paucity of studies compared to the extensive lumbar ESI literature, there is no consensus on indications for cervical ESI. Indications supported by various studies include radicular pain, cervical spondylosis, subacute cervical strain, cervicogenic headache, chronic neck pain (52,54,55,62,63), and a "variety of indications" (52). While some report reduced efficacy in axial neck pain, most do not (62). Hopefully, future work will clarify a set of standard indications for this procedure.

Expectations

Most patients want to know what they can expect after the procedure. A thorough preprocedure discussion addresses these issues and includes the following points: many patients get initial relief from the anesthetic that wears off by the end of the day. The steroids, though, may not exert their effect for two to six days (26). Thus, relief may be bimodal, with a transient return of symptoms over the first few days. Very occasionally, symptoms can be paradoxically worse during this interval. Of course, some patients do not respond, whereas others respond to only the anesthetic or the steroid, without a bimodal effect. Likewise, the degree of relief in responders is highly variable.

Patients should know that initial nonresponders sometimes do benefit from a second or even third injection (19,64), and some referring doctors will preschedule three monthly injections for their patients. This schedule should be revised if the patient maintains complete relief after the first or second injection.

CONTRAINDICATIONS

Several contraindications to ESI are obvious and are common to all invasive procedures: hemorrhagic diathesis, local infection, and a history of reaction to the medications. Systemic or distant infection is also a contraindication, because the steroid will reduce the patient's ability to fight the infection (5,23,38). Cauda equina syndrome, pilonidal cysts, and neurologic disorders that could be masked by this procedure are also included in the list of contraindications (5).

Noteworthily, two studies have described the safety of aspirin and other anti-platelet medications in these patients. These two studies documented no signs or symptoms of epidural hematoma in a total of 637 epidural injections and infusions (64,65). Because we use a 25-gauge needle in our department, patients may continue taking aspirin.

In fact, epidural injection with a 25-gauge needle may be safe in anticoagulated patients. Waldman et al. reported a series of 336 epidural blocks (morphine sulfate and bupivicaine) for intractable pain in cancer patients who were either medically anticoagulated or thrombocytopenic. They documented only two ecchymoses in the superficial tissues and no signs or symptoms of epidural hematoma. While this is not to advocate routine use of ESI in anticoagulated patients, they felt it likely to be safe when absolutely necessary (66). To guard against a cervical epidural hematoma, which can be catastrophic, aspirin use in the last five to seven days should be an absolute contraindication in cervical ESI patients.

While diabetes and cardiac disease are not contraindications to this procedure, they do warrant extra caution. Steroids can increase the lability of blood glucose levels in diabetics, and can cause water retention and thus initiate congestive heart failure in patients with poor cardiac reserve (38,67). Thus, the steroid dose in these patients is reduced by 25% to 50% in our department.

MEDICATIONS

Steroids

There are several depot preparations of corticosteroids available and none of these is more effective than the others. For many years, the most common preparations in use were methylprednisolone acetate (Depo Medrol) and triamcinolone diacetate (Aristo-cort Intralesional). These are said to confer high local tissue levels for at least two weeks (38). Due to lingering, although probably unfounded, concerns about the possible neurotoxicity of polyethylene glycol (see "History" section), Silbergleit et al. employed either triamcinolone acetonide (Kenalog-40, Bristol-Meyers Squibb, Princeton, New Jersey) or a betamethasone preparation (Celestone Soluspan, Schering, Kenilworth, New Jersey) (10). Celestone Soluspan contains both betamethasone sodium phosphate and betamethasone acetate. Betamethasone sodium phosphate is water soluble and expected to have a more rapid onset. Betamethasone acetate is not water soluble and is longer lasting (68).

As with volumes of injectate, steroid doses have diminished as experience with this technique has increased. In 1980, Knight and Burnell recommended limiting the dose of methylprednisolone acetate to 3 mg/kg patient weight, which was considerably lower than some received prior to that time (69). Today, most authors administer the dose equivalent of 80 to 120mg methylprednisolone acetate in an average-size patient (23,38). In our facility, the choice is 80 mg Kenalog-40 for lumbar and caudal ESI, and 12 mg Celeston Soluspan for cervical ESI.

Patients are limited to three injections over a six-month period at our institution, although some referring physicians will not allow more than three per year. Standards for number and timing of injections have not been established yet (23).

Anesthetic

Adding anesthetic to the injectate may provide several benefits: it increases muscle relaxation, it supplies a psychological benefit from immediate pain relief, and it purportedly breaks the pain-muscle spasm-ischemia-pain cycle (23). If a less concentrated dose of the anesthetic is chosen, occasionally there can be partial anesthesia of the dura, but rarely a significant motor or sensory block (5,24).

It is important to note that methyl paraben- or phenol-containing anesthetics can cause flocculation of depot steroid preparations (68). For these reasons, we inject preservative-free 0.5% lidocaine. El-Khoury et al. (5) reported no known interaction or flocculation in mixing contrast with the medications. We eschew bupivicaine and other longer-acting anesthetics to limit the duration of the occasional case of a significant block from a lumbar ESI. If bupivicaine is chosen, it should be 0.25% (24).

TECHNIQUE

The descriptions of these procedures in the following paragraphs reflect the methods and techniques employed in our department. Multiple aspects of these procedures can be varied, and these techniques should be adjusted according to training, experience, facilities, and preferences.

Neither the lumbar/interlaminar nor the caudal approaches has been shown to be more effective. Thus, selection of route depends on operator preference. The lumbar approach has more constant anatomy, often allows placement of the needle closer to the lesion, and carries a lower risk of intravenous placement. The caudal approach is often easier in postsurgical patients, allows better access to the L5/S1 space given typical cephalad flow of the injectate, and carries a lower risk of dural puncture (23,29,70).

The patient's images are reviewed and a route selected. While most prefer to be within two levels of the pathology (ideally at it or below it), Harley showed that 6 cc of contrast injected at the L4/L5 level consistently spread up to L1 and down over the sacrum (37). For lumbar injections, we target a level with a sufficient posterior epidural recess. If the patient does not have a CT scan or MRI immediately available, the caudal approach is preferred to a blind interlaminar attempt. Cross-sectional studies may also reveal anatomy that may preclude or alter the procedure, such as Tarlov cysts or developmental anomalies of the spinal canal.

Informed consent is always obtained, as well as a directed history, including symptoms, previous surgery and injections, allergies to contrast and medications, and relevant medical conditions such as diabetes and cardiac disease.

The patient's cardiovascular status is not routinely monitored during the procedure.

We use a stationary C-arm fluoroscope and contrast (Omnipaque-300), because image guidance affords the best chance of being both within the epidural space and extravascular. Numerous authors report frequent misplacement of needles when inserted blindly (20,49). White et al. documented malposition in 30% of interlaminar approaches and 25% of caudal ones (34). The frequency of thecal sac punctures ranges from 0.6% to 10%, but is usually quoted as 2.5% in blind placements (71,72). Intravascular placement is not unusual, ranging from 0.2% to 11% of cases, and up to three-quarters of these will not be detected by aspiration (9,29,34,50,71,72). The risks of intravascular injections include cardiac toxicity, respiratory arrest, and seizures. In fact, Dawkins reported a case where intravascular injection of lignocaine caused convulsions during an attempted ESI (72). Thus, consistent placement of the medication in its intended location requires fluoroscopy and contrast.

Lumbar/Interlaminar

The patient is placed prone on the fluoroscopy table, with a pillow or bolster under the abdomen. While rotating the C-arm slightly may be necessary to center the relevant spinous processes,

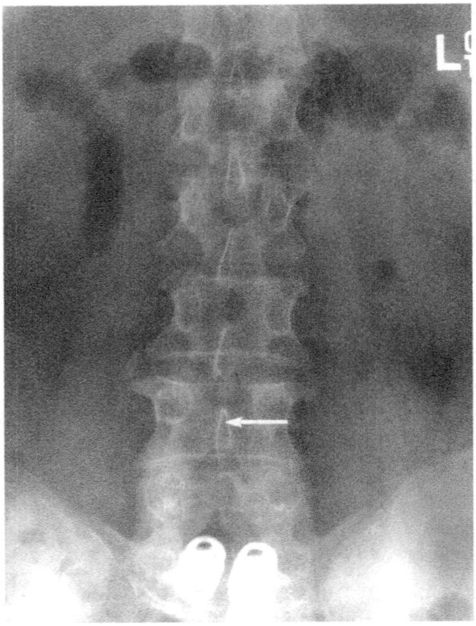

FIGURE 3 *Arrow* denotes a typical skin entry site, if the target is L3–L4.

craniocaudal tilt is not routinely applied. A direct midline approach allows the needle to pass between the spinous processes to reach its target. A skin entry site is marked, usually over the upper portion of the more inferior spinous process (Fig. 3).

Standard sterile preparation and drape are followed by anesthesia of the skin and subcutaneous tissues, with 1% lidocaine buffered with sodium bicarbonate. Under intermittent fluoroscopic guidance, a 25-gauge 3.5-inch spinal needle is advanced, usually with a slight cephalad angulation. Knowing that the needle tends to track in a direction opposite of its bevel, it can usually be steered to keep it in the midline and following the appropriate angle to the target. Once the needle engages the interspinous ligament, it is less likely to deflect laterally, and the needle is advanced with lateral fluoroscopic visualization.

The needle is stopped when it crosses the posterior margin of the facets (Fig. 4). A midline position is ensured by checking the AP view. From this point the needle is advanced into the posterior epidural space utilizing the "air release technique" with anteroposterior (AP) fluoroscopic guidance. The air-release technique is performed by attaching a glass or

(A) (B)

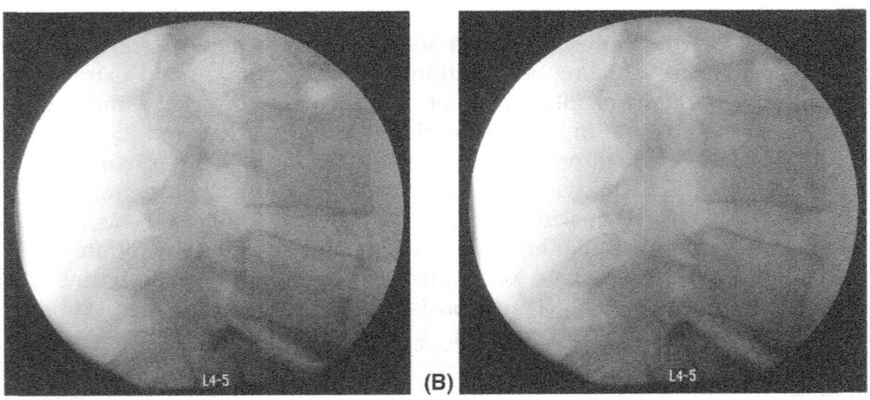

FIGURE 4 (**A**) Needle tip just crossing the back edge of the facets, the depth at which the air-release technique is begun. (**B**) Needle tip advanced a bit deeper, at which point the epidural space is entered. (Note a small amount of contrast.)

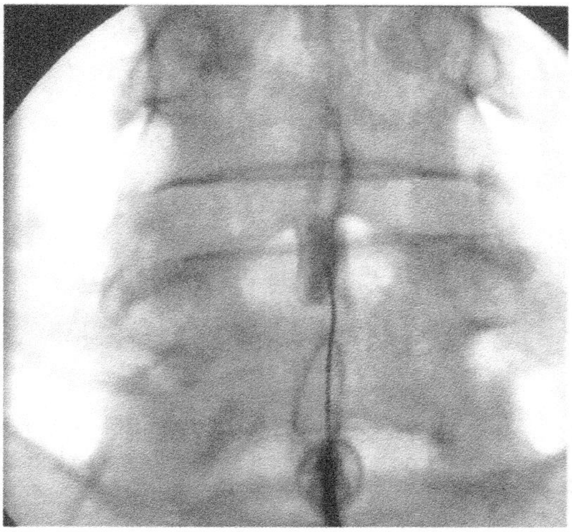

FIGURE 5 Contrast pooling around the needle tip confirming epidural location.

low-resistance plastic syringe (such as a Terumo 3-cc syringe) to the needle (10) and attempting to infiltrate approximately 0.5 cc of air. If the needle tip is still posterior to the epidural space, the plunger will bounce back. Continue advancing a millimeter at a time, checking with an air puff, until there is no bounce back, signaling needle tip entry into the epidural space. Contrast is then instilled via extension tubing, under continuous AP fluoroscopy. Contrast appropriately within the epidural space pools around the tip of the needle (Fig. 5). Alternatively, contrast in the thecal sac quickly diffuses and does not pool at the needle tip. If the tip is in a vein, contrast whisks away into branching or tortuous venous structures (Fig. 6).

Check the lateral projection, which should show contrast collecting along the dorsal margin of the thecal sac (Fig. 7) or extending slightly anterior to this if there is enough contrast to flow around the sides of the thecal sac. Intrathecal contrast will instead layer on the dependent ventral margin of the thecal sac (Fig. 8). Luckily, fluoroscopy helps keep this occurrence down to less than 1% of the time in our department.

If the thecal sac is inadvertently entered, the injection cannot proceed at that site. Proceeding would cause a spinal block, and the steroids would be wasted in the cerebrospinal fluid. While Silbergleit et al. terminate the procedure and reschedule in this situation (10), most authors will make a second attempt at another level and have not reported mishaps from this practice (8,37,60). Our approach is to make a second attempt if the needle is 25-gauge, but reschedule if it is 22-gauge.

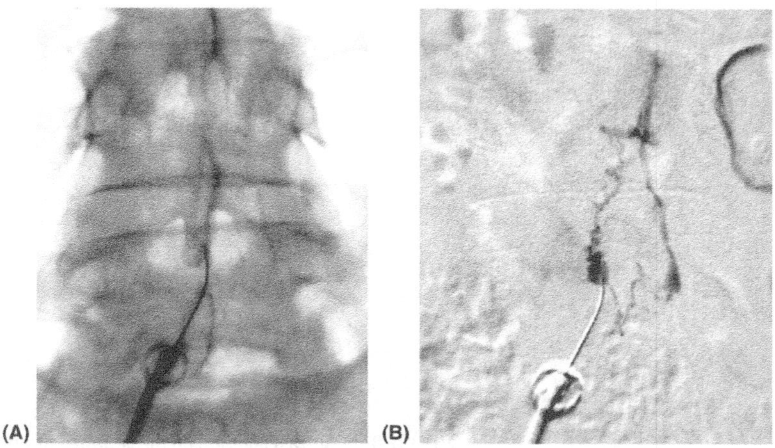

FIGURE 6 (**A**) Contrast in tortuous veins difficult to see on a static image. (**B**) Digital subtraction image more clearly demonstrates intravascular position of the needle tip.

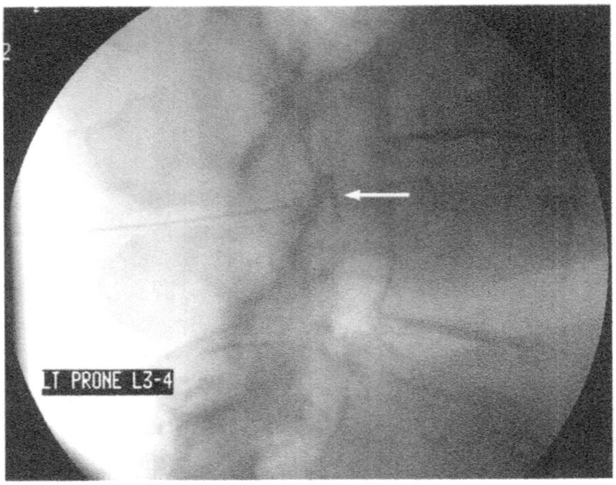

FIGURE 7 Contrast collecting on the dorsal surface (*arrow*) of the thecal sac confirming epidural location.

When proper positioning is assured, 2 cc Kenalog-40 mixed with 5 cc preservative-free 0.5% lidocaine is instilled over one or two minutes, with intermittent visualization in the AP plane. Provided the needle has not dislodged, the contrast pool should increase briefly, from the residual amount left in the needle, and then be dispersed by the radiolucent medication. It is not unusual to elicit discomfort or even pain with the injection. This usually improves with a slower injection. Save the AP and lateral images documenting epidural contrast as well as an AP image after contrast is dispersed by the medication. These images often aid the next operator if the patient returns for another injection and can document extrathecal position if the patient subsequently develops a headache. After the medication is delivered, the needle is removed, the overlying puncture site is cleansed with alcohol, and an adhesive bandage is placed. The patient is assisted while rising slowly.

Given that these patients are almost never sedated and rarely have significant motor nerve anesthesia, they are not detained longer than the time it takes for cleanup and a short postprocedure discussion. While some patients are light headed after the procedure, this is usually due to the time spent prone with an abdominal bolster in place, and the sensation clears shortly. Patients should take it easy that day and then resume normal activities the next day. We insist all patients bring another person to drive them home.

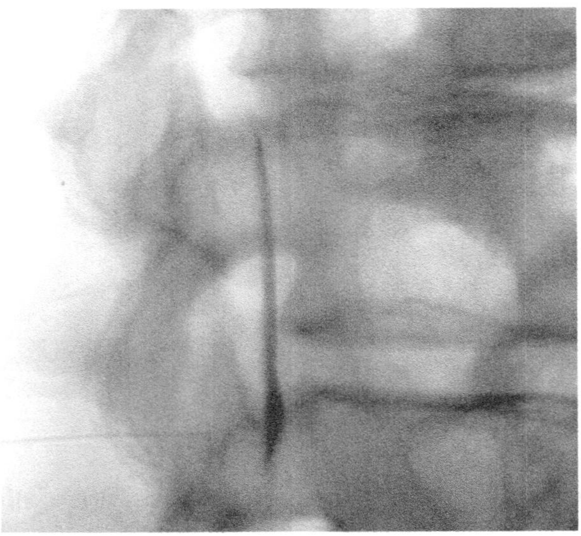

FIGURE 8 Contrast layering on the ventral margin of the thecal sac demonstrating inadvertent entry of the needle into the thecal sac.

Hints

The procedure often must be altered to suit the situation. Be careful to evaluate for adequate space between the spinous processes. If not sufficient initially, the space will often increase significantly with the addition of a second bolster. In this situation, it helps to check the approach, before prep, in the lateral fluoro position, to be certain that the selected skin site will allow a direct path to the target.

Occasionally patients require no needle angulation at all to reach the target, which can affect skin entry site. Again, previewing in the lateral plane will help.

If one is unexpectedly contacting bone and the AP and lateral views do not clearly show which bone is obstructing the path, craniocaudal tube tilt down the barrel of the needle is usually illuminating.

Larger patients may necessitate the use of a 6-inch needle. At this length, 25-gauge needles are usually quite difficult to steer; thus, a 6-inch 22-gauge needle is a better choice for large patients.

Occasionally, as the needle is advanced with the air-release technique, a bounce-back is followed by a partial air release after the next advancement. While this is unsatisfying, this invariably signals entry into the epidural space, and further advancement ("just to be sure") may only send the needle into the thecal sac. Alternatively, if a partial or full air release occurs before expected, the position may be off-center and posterior to the ligamentum flavum, a space that can accept air insufflation (Fig. 9).

(A) 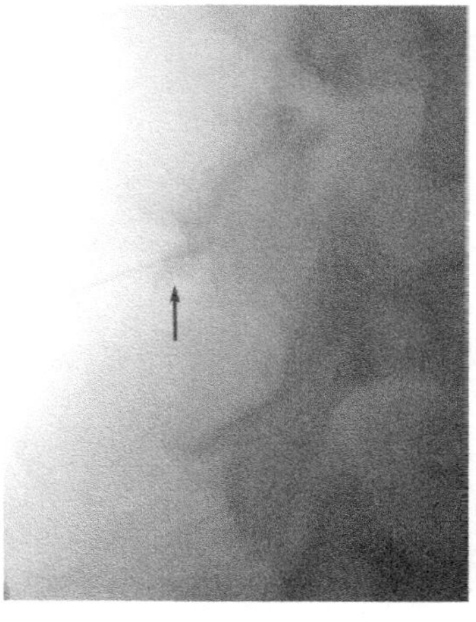 (B) 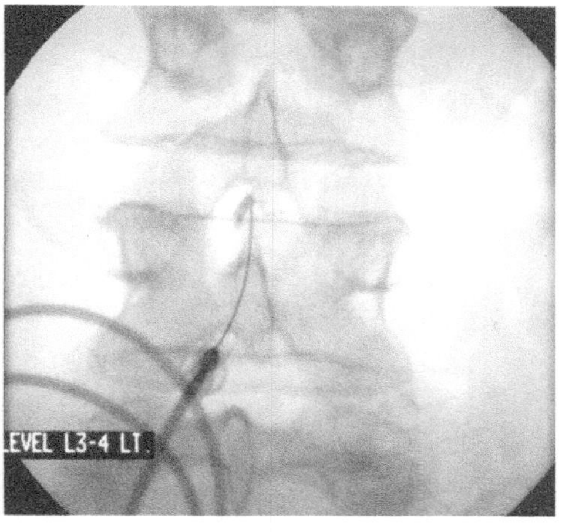

(C) 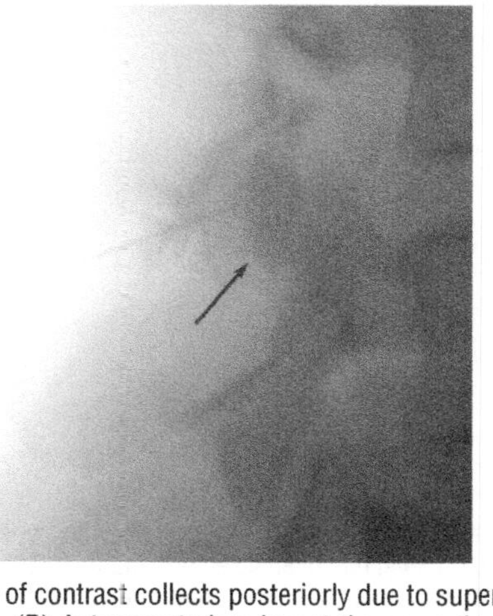

FIGURE 9 (**A**) Initial injection of contrast collects posteriorly due to superficial position of the needle tip, which has not yet entered the epidural space. (**B**) Anteroposterior view at the same time demonstrates unilateral contrast flow to be expected if the tip is posterior to the ligamentum flavum. (**C**) After further advancement epidural location is confirmed.

After each set of four needle advances and tests, reinsertion of the stylet is advisable to clear the needle tip.

Many ESI patients with degenerative disc disease will have close apposition of their spinous processes. Sometimes, this phenomenon will preclude passage of a needle between the spinous processes at the desired level. An alternative approach to the posterior epidural space, the paramedian approach, almost always proves successful. For this approach, tilt the tube 10° oblique from the true AP plane. This will profile a space between the spinous process and the ipsilateral lamina, shaped like an upside-down horseshoe (Fig. 10). One then adjusts craniocaudal tilt of the fluoroscopy tube to place the apex of the horseshoe at the level of the disc. The "horseshoe" is a tunnel for passage of the spinal needle into the posterior epidural space. One should attempt to pass the needle near the margin of the spinous process, rather than more laterally. Once the needle is at the posterior margin of the facet joint on the lateral view, proceed using lateral fluoroscopy as with a standard interlaminar approach. If one looks in the straight anteroposterior plane, the needle will be nearly midline when it reaches the epidural space.

Finally, if the patient is allergic to iodinated contrast, gadolinium will suffice, especially if a digital subtraction run is employed.

Caudal

The prone position is preferred for caudal ESIs, but these can also be done in the lateral decubitus position with the hips flexed. Often, padding is placed under the pelvis to accentuate the

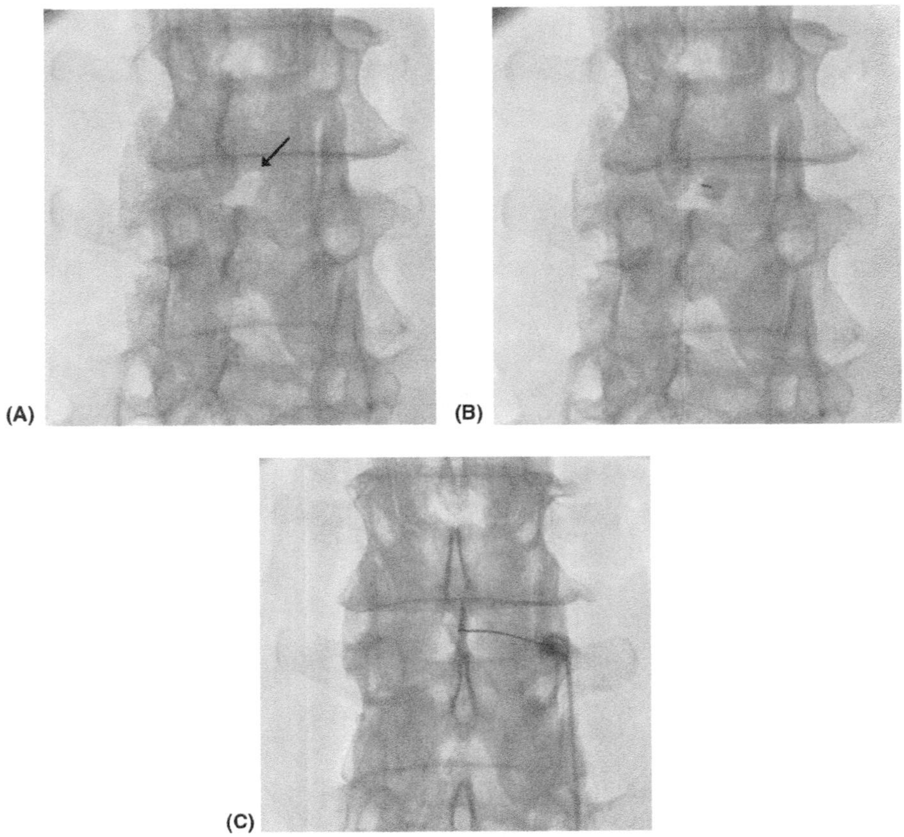

FIGURE 10 (**A**) Oblique spot image (10°) demonstrating the tunnel through which one can access the posterior epidural space from a paramedian approach. The L3/4 level has been selected, and the window resembles an upside-down horseshoe (*arrow*). (**B**) Similar spot image with spinal needle approaching the spinous process, which is to the left on this image. (**C**) Straight anteroposterior spot image demonstrates contrast within the posterior epidural space. The needle tip is directly in the midline when it enters the epidural space.

sacral hiatus. The sacral hiatus is palpated by running a finger down the middle of the sacrum, until the coccyx is identified. The paired cornua are just above this level, off midline, and often palpable. Once the hiatus is identified by palpation, it is verified with lateral fluoroscopy. After marking the hiatus, gauze padding is placed within the gluteal crease to block the spillage of sterilizing agent onto the perineum. Standard sterile preparation and drape are followed by anesthesia of the skin and sacral hiatus region, including periosteum, with 1% lidocaine buffered with sodium bicarbonate.

The choice of epidural needle size is based on individual preference, but patients complain of pain less with a 25-gauge spinal needle (3.5-inch) than a 22-gauge needle. A 20-gauge or larger may cause prolonged superficial sacral pain because of irritation caused by the needle scraping along the periosteum during its advancement. Also, larger needles may not be flexible enough to follow the hiatus contour and may penetrate through the sacrum into the perirectal region or dorsally out the sacral roof to the skin (Fig. 10). Creating a gentle curve at the end of the needle may help the needle to follow the slope of the sacrum.

With a 45° angle of entry and the bevel down, the needle is advanced through the sacrococcygeal ligament and stopped when the underlying bone is encountered. The needle is withdrawn slightly from the periosteum, and the entry through the hiatus is verified fluoroscopically with AP and lateral views. After achieving satisfactory positioning, the bevel is turned up and the needle is laid almost horizontal, until the needle shaft is in the same plane as the sacral canal. The needle is slowly advanced through the canal, often requiring a gentle rotating motion allowing it to follow the curved contour of the canal. Intermittent AP and lateral views should confirm the midline position of the needle. Needle advancement is stopped at the S2/S3 junction or slightly lower (Fig. 11).

Following needle placement, the extradural location of the needle tip is verified, initially by having the patient Valsalva as well as aspirating with a 5 cc syringe, examining for cerebral spinal fluid and blood. Additionally, under fluoroscopy, 1 to 2 cc of nonionic contrast is injected. If the needle is extradural in location, the contrast distribution from the epidurogram resembles a Christmas tree and does not disperse rapidly (Fig. 12). If the contrast has a serpentine course and flows away from the midline, the needle tip is likely in an epidural vein. Often, only a minute adjustment to the tip is necessary to extricate it from the vessel. It may also help to direct the needle slightly dorsally because the epidural veins are more concentrated at the ventral aspect of the epidural space. If the contrast fills only one side of the epidural space, it may be due to a plica mediana dorsalis. Redirecting the needle or repuncture may be needed to ensure satisfactory medication delivery.

If the needle is in the subarachnoid space, the procedure is terminated. Readjusting of the needle is not recommended in this instance. The procedure should be rescheduled about seven days later.

After confirming the needle-tip position and documenting it with AP and lateral spot radiographs, the medication is injected. A typical injection consists of 2 cc of Kanalog-40 followed by 5 to 8 cc of preservative-free 0.5% lidocaine. Recommendations for volume of

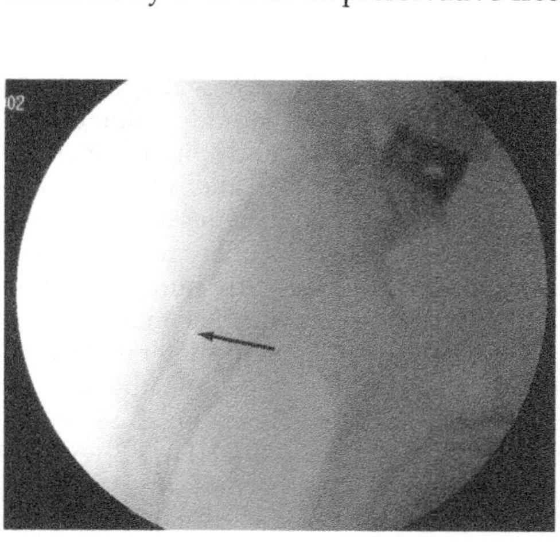

FIGURE 11 Needle tip at the target, which is the midportion of S3.

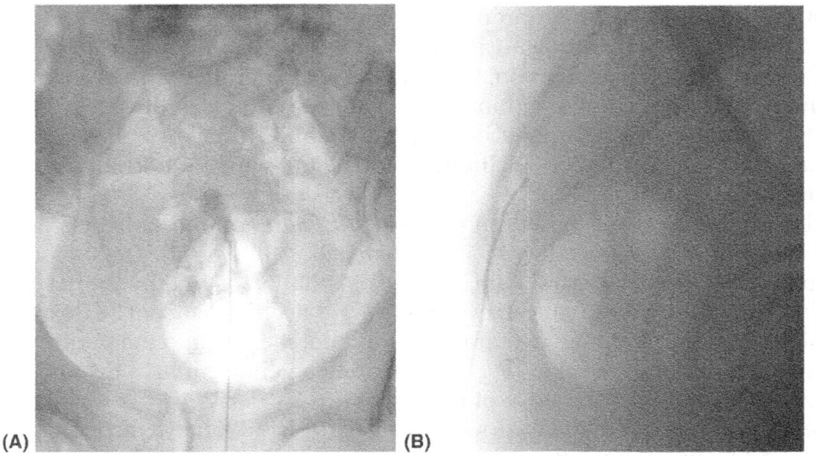

FIGURE 12 (A) Typical "Christmas tree" pattern of epidural contrast in the sacral canal. (B) Lateral view confirms contrast within the canal.

injectate have varied widely. More recent reports (73,74) have shown that a volume of 10 mL consistently reaches the low-to-mid lumbar spine. A volume in this range will not overdilute the steroids.

There should not be any resistance during the injection and if any force is required, needle malposition should be considered. During the injection, patients may complain of pressure at the site of injection or dull or shooting pain down the legs. These symptoms subside with a decreased injection rate. After the medication is delivered, the needle is removed, the overlying puncture site is cleansed with alcohol, and an adhesive bandage is placed. The patient is assisted while rising slowly. The postprocedure routine is the same as for interlaminar injections.

Cervical

The patient is placed in a comfortable, prone position. The patient's head should be neutral and facing straight down on the table. Proper support is necessary to keep the face from being smashed against the table. Aside from general discomfort, some patients may experience claustrophobia in this position. Thus, extra time should be spent positioning the head to alleviate discomfort and minimize patient anxiety. Several firm support pads should be placed under the patient's chest and forehead, elevating the face off the table. This leaves an opening for the patient's face, minimizing the closed-in feeling. Blowing oxygen across the face may also help to decrease claustrophobia. Conscious sedation is not routinely required, but it should be considered for patients who are anxious and less cooperative.

After identifying the target level on the patient's MRI, a target mark is made on the patient's neck utilizing AP fluoroscopy. The initial needle entrance is at the midportion of the lower of the two target vertebral bodies and midway between the pedicle and spinous process. Standard sterile preparation and drape are followed by anesthesia of the skin and subcutaneous tissues with 1% lidocaine buffered with sodium bicarbonate. A variety of needles are available to approach the cervical epidural space, but a 22-guage Tuohy needle is ideal because its blunt tip tends to push firm objects away, decreasing the likelihood of inadvertently puncturing the dura. After obtaining purchase with the Tuohy, the needle is advanced toward the interlaminar space with slight medial angulation, while visualizing the field from a contra-lateral oblique projection. This projection is obtained by rotating the tube about 45° opposite to the side of needle entrance, until the laminae take on a shingled appearance. This allows one to view the needle tip in tangent, and the needle is stopped just posterior to the spinal canal (Fig. 13). As the needle is advanced, its position should be checked in the AP projection periodically, to verify that it is directed toward the midline. After reaching the margin of the spinal canal and

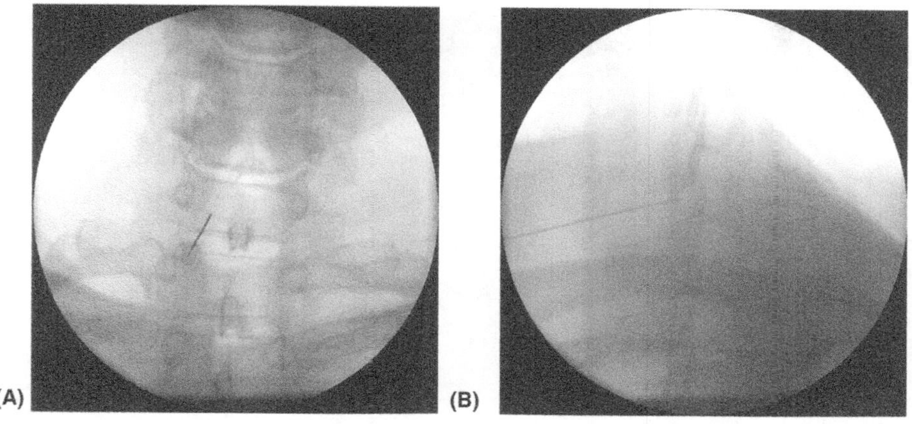

FIGURE 13 Cervical epidural needle positioning. Anteroposterior and oblique fluoroscopic views show the needle heading toward the interlaminar space.

returning to the oblique projection, the remainder of the needle advancement is done with a "hanging drop" technique. A few drops of iodinated contrast are placed on the needle hub ("hanging" there), giving a convex configuration (Fig. 14). As the needle is slowly advanced and finally positioned into the epidural space, the contrast dome will drop down the hub, heralding epidural location.

Once the contrast is imbibed, the needle-tip position is verified by instilling an additional 2 mL of contrast, visualized in real time with both oblique and AP projections. Epidural contrast should circumscribe the outline of the dura (Fig. 15). If the contrast instead has a diluted appearance and wraps around the spinal cord, simulating a myelogram, intrathecal injection is likely. The procedure should be terminated and reattempted in about seven days. Rarely, the contrast conforms to a tubular or serpentine pattern and washes away, due to needle position in an epidural vein. Often a minute adjustment to the needle tip will extricate

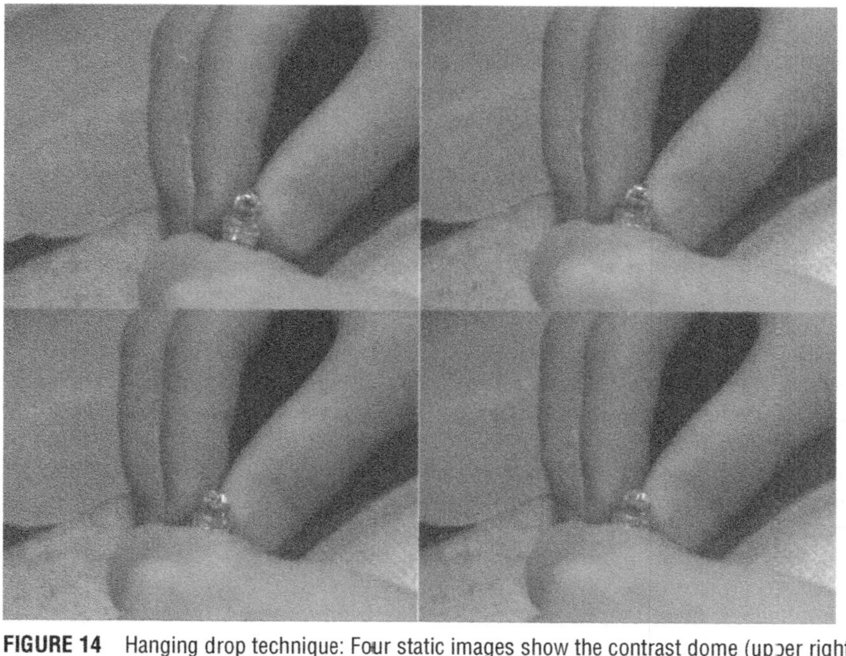

FIGURE 14 Hanging drop technique: Four static images show the contrast dome (upper right image) dropping down the hub, which indicates the epidural positioning of the needle tip.

(A) (B)

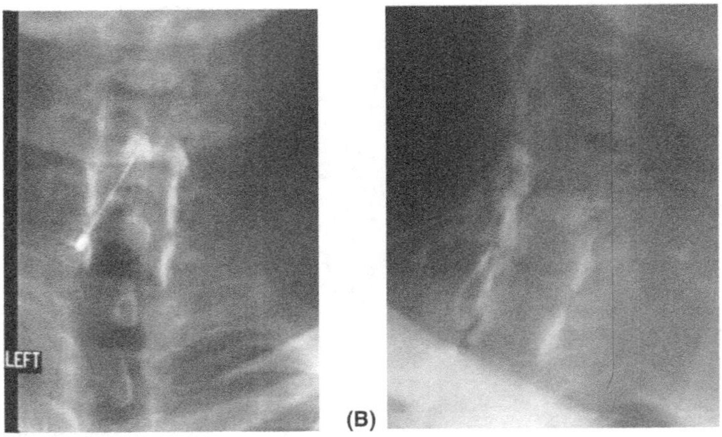

FIGURE 15 Cervical epidural contrast injection. Anteroposterior (**A**) and oblique (**B**) views during contrast injection show the proper location of the contrast circumscribing the dura, thus confirming the epidural location of the needle tip.

it from the vessel. If the repeated needle adjustment is unsuccessful, another level should be accessed.

With satisfactory positioning, 12 mg of Celestone Soluspan or 80 mg Kenalog-40 is instilled for an average-sized adult. An additional 2mL of 0.25% bupivicaine provides immediate pain relief. Higher concentrations of bupivicaine are not used because of the risk of motor block. During the injection, there should not be any resistance; if any force is required, a malposition of the needle should be suspected. During the injection, patients may complain of pressure at the site of the injection or dull or shooting pain. These symptoms should subside with decreasing injection rate. After the injection, the needle is removed, the overlying puncture site is cleansed with alcohol, and an adhesive bandage is placed. The patient is assisted while rising slowly. The postprocedure routine is the same as for lumbar injections.

SIDE EFFECTS AND COMPLICATIONS

In their meta-analysis of controlled trials, Watts and Silagy compiled a list of adverse effects, including subarachnoid tap (2.5%), transient headache (2%), and transient increase in low back or radicular pain (2%). One woman reported irregular menses (32). There also have been reports of transient hypotension or vasovagal response, cardiac angina, and respiratory difficulty with an inadvertent spinal block (56). In their report on 5489 consecutive fluoroscopy-guided ESIs, Johnson et al. found only four complications that required either an emergency department visit or hospitalization. These included one patient with a vasovagal reaction and another with "significant hypotension," both of which resolved uneventfully; a cervical epidural hematoma that required no intervention, with symptoms resolving after 18 hours; and a patient who was admitted for three days of observation for tachycardia and hypertension, thought to be due to an unusual response to the steroids (75).

Insomnia, facial flushing, and nausea are not uncommon side effects of the steroids (67). Other systemic effects of the steroids include Cushingoid effects, which were much more common in the past when higher doses were routine (69,76); and suppression of the hypothalamic-pituitary axis for several weeks, which has been documented in dogs (77) and humans (78). This effect comes into play if the patient has subsequent surgery (77,78), and may preclude the epidural injection or require steroid supplementation at the time of surgery.

We have not found a salient discussion of the risk of osteonecrosis from an ESI.

Local effects, especially long-term ones, are more of a concern. In his review of published complications from ESI, Abram compiled a list that included five epidural abscesses that can be devastating or even fatal; one case of aseptic meningitis in an inadvertent intrathecal injection; two cases of bacterial meningitis, one of which followed an intrathecal injection and the other not documented; and no reports of arachnoiditis. Arachnoiditis has only been established as a complication in patients with multiple intrathecal steroid injections, a former therapy for multiple sclerosis (38). Slucky et al. in a canine study found no significant alterations of the dura after ESI. Also, they found a relative reduction of posterior epidural fat in 17 of 24 specimens (79), arguing against the theory that local steroids may cause epidural lipomatosis.

Overall, Botwin et al. reported a 15.6% rate of minor complications (67). Major complications are the subject of case reports only, and their incidence is uncertain but undoubtedly rare.

Cervical ESI is, of course, subject to the same list of adverse effects. Because the spinal cord is present at this level, neurologic injury with potentially catastrophic sequelae is more of a concern, but the actual risk is not well documented. Presumably this risk is the reason this procedure is less widely practiced. Some side effects, including dyspnea, hypotension, nausea, and neck stiffness, are reportedly more common in cervical injections (80). There have been single reports of reflex sympathetic dystrophy (81) and an epidural hematoma requiring surgical decompression (82,83) after cervical ESI.

Overall, the safety of lumbar and caudal ESI has been established but that of cervical ESI remains controversial.

SUMMARY

ESIs are an important part of the armamentarium of treatment options for patients with low back, neck, and radicular pain. Although these procedures may not alter the causative lesions, they often shorten the clinical course of the disease process, keep patients out of the hospital (34), and provide symptomatic relief and lifestyle improvements onto which it is difficult to place a price tag. These procedures are most accurately performed with fluoroscopic guidance and major complications are rare. While support for cervical ESI is less uniform, a growing body of experience is beginning to support its efficacy and safety.

REFERENCES

1. Benzon HT. Epidural steroid injections for low back pain and lumbosacral radiculopathy. Pain 1986; 24(3):277–295.
2. Rydevik BL, Cohen DB, Kostuik JP. Spine epidural steroids for patients with lumbar spinal stenosis. Spine 1997; 22(19):2313–2317.
3. Bush K, Hillier S. A controlled study of caudal epidural injections of triamcinolone plus procaine for the management of intractable sciatica. Spine 1991; 16(5):572–575.
4. Rowlingson JC. Epidural steroids: do they have a place in pain management? APS J 1994; 3(1):20–27.
5. el-Khoury GY, Ehara S, Weinstein JN, Montgomery WJ, Kathol MH. Epidural steroid injection: a procedure ideally performed with fluoroscopic control. Radiology 1988; 168(2):554–557.
6. Link SC, el-Khoury GY, Guilford WB. Percutaneous epidural and nerve root block and percutaneous lumbar sympatholysis. Radiol Clin North Am 1998; 36(3):509–521.
7. Harrison GR. Topographical anatomy of the lumbar epidural region: an in vivo study using computerized axial tomography. Br J Anaesth 1999; 83(2):229–234.
8. Johnson BA. Image-guided epidural injections. Neuroimaging Clin N Am 2000; 10(3): 479–491.
9. Bryan BM, Lutz C, Lutz GE. Fluoroscopic assessment of epidural contrast spread after caudal injection. J Orthop Med 2000; 22(2):38–41.
10. Silbergleit R, Mehta BA, Sanders WP, Talati SJ. Imaging-guided injection techniques with fluoroscopy and CT for spinal pain management. Radiographics 2001; 21(4): 927–939; discussion 940–942.
11. Ogoke BA. Caudal epidural steroid injections. Pain Physician 2000; 3(3):305—312.
12. Lanier VA, McKnight HE, Trotter M. Caudal analgesia: an experimental and anatomical study. Am J Obstet Gynecol 1944; 47:633–641.
13. Viner N. Intractable sciatica—the sacral epidural injection—an effective method of giving pain relief. Can Med Assoc J 1925; 15:630–634.
14. Hayashi N, Weinstein JN, Meller ST, Lee HM, Spratt KF, Gebhart GF. The effect of epidural injection of betamethasone or bupivacaine in a rat model of lumbar radiculopathy. Spine 1998; 23(8): 887–885.
15. Evans W. Intrasacral epidural injection therapy in treatment of sciatica. Lancet 1930; 2: 1225–1229.
16. Kelman H. Epidural injection therapy for sciatic pain. Am J Surg 1944; 64(2):183–190.
17. Robechhi A, Capra R. L'idrocortisone (composto F). Prime esperienze cliniche in campon reumatologico. Minerva Med 1952; 98:1259–1263.

18. Goebert HW, Jallo SJ, Gardner WJ, Wasmuth CE. Painful radiculopathy treated with epidural injections of procaine and hydrocortisone acetate results in 113 patients. Anesth Analg 1961; 140(1): 130–134.
19. Dilke TF, Burry HC, Grahame R. Extradural corticosteroid injection in management of lumbar nerve root compression. Br Med J 1973; 2(5867):635–637.
20. Nelson DA. Dangers from methylprednisolone acetate therapy by intraspinal injection. Arch Neurol 1988; 45(7):804–806.
21. Roche J. Steroid-induced arachnoiditis. Med J Aust 1984; 140(5):281–284.
22. Delaney TJ, Rowlingson JC, Carron H, Butler A. Epidural steroid effects on nerves and meninges. Anesth Analg 1980; 59(8):610–614.
23. Spaccarelli KC. Lumbar and caudal epidural corticosteroid injections. May Clin Proc 1996; 71(2):169–178.
24. White AH. Injection techniques for the diagnosis and treatment of low back pain. Orthop Clin North Am 1983; 14(3):553–567.
25. Warr AC, Wilkinson JA, Burn JM, Langdon L. Chronic lumbosciatic syndrome treated by epidural injection and manipulation. Practitioner 1972; 209(249):53–59.
26. Green PW, Burke AJ, Weiss CA, Langan P. The role of epidural cortisone injection in the treatment of discogenic low back pain. Clin Orthop 1980; 153:121–125.
27. McCarron RF, Wimpee MW, Hudkins PG, Laros GS. The inflammatory effect of nucleus pulposus. A possible element in the pathogenesis of low-back pain. Spine 1987; 12(8):760–764.
28. Franson RC, Saal JS, Saal JA. Human disc phospholipase A2 is inflammatory. Spine 1992; 17(suppl 6): S129-S132.
29. Cannon DT, Aprill CN. Lumbosacral epidural steroid injections. Arch Phys Med Rehabil 2000; 81(3 suppl 1):S87-S98; quiz S99-S100.
30. Derby R, Kine G, Saal JA. Response to steroid and duration of radicular pain as predictors of surgical outcome. Spine 1992; 17(suppl 6):S176-S183.
31. Johansson A, Hao J, Sjolund B. Local corticosteroid application blocks transmission in normal nociceptive C-fibres. Acta Anaesthesiol Scand 1990; 34(5):335—338.
32. Watts RW, Silagy CA. A meta-analysis on the efficacy of epidural corticosteroids in the treatment of sciatica. Anaesth Intensive Care 1995; 23(5):564–569.
33. Swerdlow M, Sayle-Creer W. A study of extradural medication in the relief of the lumbosciatic syndrome. Anaesthesia 1969; 25(3):341–345.
34. White AH, Derby R, Wynne G. Epidural injections for the diagnosis and treatment of low-back pain. Spine 1980; 5(1):78–86.
35. Hickey RF. Outpatient epidural steroid injections for low-back pain and lumbosacral radiculopathy. N Z Med J 1987; 100(832):594–596.
36. Bowman SJ, Wedderburn L, Whaley A, Grahame R, Newman S. Outcome assessment after epidural corticosteroid injection for low-back pain and sciatica. Spine 1993; 18(10): 1345–1350.
37. Harley C. Extradural corticosteroid infiltration: a follow-up study of 50 cases. Ann Phys Med 1967; 9(1):21–28.
38. Abram SE. Epidural steroid injections for the treatment of lumbosacral radiculopathy. J Back Musculoskel Rehab 1997; 8:135–149.
39. Parisien VM. Conservative treatment of low-back pain with epidural steroids. J Maine Med Assoc 1980; 71(3):83–84, 92.
40. Rosen CD, Kahanovitz N, Bernstein R, Viola K. A retrospective analysis of the efficacy of epidural steroid injections. Clin Orthop 1988; 228:270–272.
41. Ridley MG, Kingsley GH, Gibson T, Grahame R. Outpatient lumbar epidural corticosteroid injection in the management of sciatica. Br J Rheumatol 1988; 27(4):295–299.
42. Cuckler JM, Bernini PA, Wiesel SW, Booth RE Jr, Rothman RH, Pickens GT. The use of epidural steroids in the treatment of lumbar radicular pain. A prospective, randomized, double-blind study. J Bone Joint Surg Am 1985; 67(1):63–66.
43. Beliveau P. A comparison between epidural anaesthesia with and without corticosteroid in the treatment of sciatica. Rheumatol Phys Med 1971; 11(1):40–43.
44. Carette S, Leclaire R, Marcoux S. Epidural corticosteroid injections for sciatica due to herniated nucleus pulposus. N Engl J Med 1997; 336(23): 1634–1640.
45. Snoek W, Weber H, Jorgensen B. Double-blind evaluation of extradural methylpredni-solone for herniated lumbar discs. Acta Orthop Scand 1977; 48(6):635–641.
46. Koes BW, Scholten RJ, Mens JM, Bouter LM. Efficacy of epidural steroid injections for low-back pain and sciatica: a systematic review of randomized clinical trials. Pain 1995; 63(3):279–288.
47. Rozenberg S, Dubourg G, Khalifa P, Paolozzi L, Maheu E, Ravaud P. Efficacy of epidural steroids in low-back pain and sciatica: a critical appraisal by a French Task Force of randomized trials. Critical Analysis Group of the French Society for Rheumatology. Rev Rhum Engl Ed 1999; 66(2): 79–85.

48. Vroomen PC, de Krom MC, Slofstra PD, Knottnerus JA. Conservative treatment of sciatica: a systematic review. J Spinal Disord 2000; 13(6):463–469.
49. Renfrew DL, Moore TE, Kathol MH, el-Khoury GY, Lemke JH, Walker CW. Correct placement of epidural steroid injections: fluoroscopic guidance and contrast administration. Am J Neuroradiol 1991; 12(5):1003–1007.
50. Price CM, Rogers PD, Prosser AS, Arden NK. Comparison of the caudal and lumbar approaches to the epidural space. Ann Rheum Dis 2000; 59(11):879–882.
51. Shulman M, Nimmagadda U, Valenta A. Cervical epidural steroid injection for pain of cervical spine origin. Anesthesiology 1984; 61(3A):A223.
52. Shulman M. Treatment of neck pain with cervical epidural steroid injection. Reg Anesth 1986; 11:92–94.
53. Rowlingson JC, Kirschenbaum LP. Epidural analgesic techniques in the management of cervical pain. Anesth Analg 1986; 65(9):938–942.
54. Martelletti P, Di Sabato F, Granata M. Epidural steroid-based technique for cervico-genic headache diagnosis. Funct Neurol 1998; 13(1):84–87.
55. Cicala RS, Westbrook L, Angel JJ. Side effects and complications of cervical epidural steroid injections. J Pain Symptom Manage 1989; 4(2):64–66.
56. Berman AT, Garbarino JL Jr, Fisher SM, Bosacco SJ. The effects of epidural injection of local anesthetics and corticosteroids on patients with lumbosciatic pain. Clin Orthop 1984; 188:144–151.
57. Rivest C, Katz JN, Ferrante FM, Jamison RN. Effects of epidural steroid injection on pain due to lumbar spinal stenosis or herniated disks: a prospective study. Arthritis Care Res 1998; 11(4): 291–297.
58. Heyse-Moore GH. A rational approach to the use of epidural medication in the treatment of sciatic pain. Acta Orthop Scand 1978; 49(4):366–370.
59. Hopwood MB, Abram SE. Factors associated with failure of lumbar epidural steroids. Reg Anesth 1993; 18(4):238–243.
60. Fredman B, Nun MB, Zohar E. Epidural steroids for treating "failed back surgery syndrome": is fluoroscopy really necessary? Anesth Analg 1999; 88(2):367–372.
61. Fukusaki M, Kobayashi I, Hara T, Sumikawa K. Symptoms of spinal stenosis do not improve after epidural steroid injection. Clin J Pain 1998; 14(2): 148–151.
62. Ferrante FM, Wilson SP, Iacobo C, Orav EJ, Rocco AG, Lipson S. Clinical classification as a predictor of therapeutic outcome after cervical epidural steroid injection. Spine 1993; 18(6):730–736.
63. Cicala RS, Thoni K, Angel JJ. Long-term results of cervical epidural steroid injections. Clin J Pain 1989; 5(2): 143–145.
64. Benzon HT, Brunner EA, Vaisrub N. Bleeding time and nerve blocks after aspirin. Reg Anesth 1984; 9(2):86–89.
65. Horlocker TT, Wedel DJ, Offord KP. Does preoperative antiplatelet therapy increase the risk of hemorrhagic complications associated with regional anesthesia? Anesth Analg 1990; 70(6): 631–634.
66. Waldman SD, Feldstein GS, Waldman HJ, Waldman KA, Allen ML. Caudal administration of morphine sulfate in anticoagulated and thrombocytopenic patients. Anesth Analg 1987; 66(3): 267–268.
67. Botwin KP, Gruber RD, Bouchlas CG. Complications of fluoroscopically-guided caudal epidural injections. Am J Phys Med Rehabil 2001; 80(6):416–424.
68. Maldjian C, Mesgarzadeh M, Tehranzadeh J. Diagnostic and therapeutic features of facet and sacroiliac joint injection. Anatomy, pathophysiology, and technique. Radiol Clin North Am 1998; 36(3): 497–508.
69. Knight CL, Burnell JC. Systemic side-effects of extradural steroids. Anaesthesia 1980; 35(6):593–594.
70. Burn JM. Treatment of chronic lumbosciatic pain. Proc R Soc Med 1973; 66(6):544.
71. Mulroy MF, Norris MC, Liu SS. Safety steps for epidural injection of local anesthetics: review of the literature and recommendations. Anesth Analg 1997; 85(6): 1346–1356.
72. Dawkins CJM. An analysis of the complications of extradural and caudal block. Anaesthesia 1969; 24(4):554–563.
73. Kim KM, Kim HS, Choi KH, Ahn WS. Cephalic spreading levels after volumetric caudal epidural injections in chronic low-back pain. J Korean Med Sci 2001; 16(2):193–197.
74. Manchikanti L, Bakhit CE, Pakanati RR, Fellows B. Fluoroscopy is medically necessary for the performance of epidural steroids. Anesth Analg 1999; 89(5): 1330–1331.
75. Johnson BA, Schellhas KP, Pollei SR. Epidurography and therapeutic epidural injections: technical considerations an experience with 5,334 cases. Am J Neuroradiol 1999; 20(4):697–705.
76. Stambough JL, Booth RE Jr, Rothman RH. Transient hypercorticism after epidural steroid injection. A case report. J Bone Joint Surg Am 1984; 66(7): 1115–1116.
77. Gorski DW, Rao TL, Glisson SN, Chinthagada M, El-Etr AA. Epidural triamcinolone and adrenal response to hypoglycemic stress in dogs. Anesthesiology 1982; 57(5):364–366.

78. Kay J, Findling JW, Raff H. Epidural triamcinolone suppresses the pituitary-adrenal axis in human subjects. Anesth Analg 1994; 79(3):501–505.
79. Slucky AV, Sacks MS, Pallares VS, Malinin TI, Eismont FJ. Effects of epidural steroids on lumbar dura material properties. J Spinal Disord 1999; 12(4):331–340.
80. Bogduk N. Epidural steroids. Spine 1995; 20(7):845–848.
81. Siegfried RN. Development of complex regional pain syndrome after a cervical epidural steroid injection. Anesthesiology 1997; 86(6): 1394–1396.
82. Williams KN, Jackowski A, Evans PJ. Epidural haematoma requiring surgical decompression following repeated cervical epidural steroid injections for chronic pain. Pain 1990; 42(2):197–199.
83. Williams JW, Powell T. Epidural abscess of the cervical spine: case report and literature review. Br J Radiol 1990; 63(751):576–578.

9 Thoracic and Lumbosacral Selective Epidural Injection

William B. Morrison
Department of Radiology, Thomas Jefferson University Hospital, Philadelphia, Pennsylvania, U.S.A.

INTRODUCTION

Demand for epidural injections is rapidly increasing. Patient satisfaction elicited from these procedures is a direct result of imaging guidance, which can shorten and simplify procedures, minimize potential for complications, and verify accurate localization of the needle to selectively provide the lowest necessary dose of analgesic to the optimal area. These procedures can greatly contribute to patient management and surgical planning by determining the sources of pain and treating pain generators. Participation of an experienced radiologist in these procedures optimizes patient care, because radiologists are trained in anatomy and image-guided needle localization procedures. However, this should be considered a "team effort" in conjunction with an experienced clinical diagnostician who can accurately diagnose the patient's problem and recommend the appropriate procedure. The performance of these procedures requires an intimate knowledge of the relevant anatomy, appropriate equipment and facilities, and apprenticeship with an experienced practitioner.

THORACIC AND LUMBAR EPIDURAL INJECTIONS: BACKGROUND AND INDICATIONS

Epidural injections are commonly performed in patients with radicular pain or spinal stenosis. One review of medical records of 25,479 patients presenting with spinal pain or radicular pain showed that epidural steroid injection was recommended in 7.9% (1). In 1999, 482,184 lumbar epidural steroid injections were performed on U.S. Medicare enrollees alone (2). Despite the popularity of these procedures, their efficacy remains somewhat controversial (3). Overall reports of response to injection range from 24% to 91% (4–1 0). This wide range of response is the result of a number of factors. Some of the studies with lower efficacy, such as Carette et al. (6), were performed nonselectively without image guidance. Other studies have documented that there is improved success with image guidance, and that without it there is a significant proportion of injections that miss the target area, either outside the epidural space, or even at a different level than intended (11–17). Therefore, the efficacy studies of injections without image guidance should be viewed in this context. Patient selection is another issue. Patients with true radicular pain respond more frequently to epidural injection than do patients with numbness, weakness, or pseudoradicular symptoms. Also, the definition of efficacy varies between the studies. Some view success as a temporary pain relief or the ability to return to work, and gauge the degree of patient satisfaction with the procedure. These studies generally show a high level of patient satisfaction (4,7,18). Others define efficacy as long-term pain relief or the ability of injections to obviate the need for surgery; however, these studies show that injections do not often take the place of surgery (5,6). However, this latter definition does not describe the intended use of epidural injections. Epidural injections are best used as diagnostic procedures to guide medical or surgical management, or to temporarily relieve pain such that patients decrease analgesic dosage and continue to perform physical therapy in order to avoid muscle deconditioning. Occasionally they may even relieve pain for much longer than the duration of the steroid preparation. In our experience, properly selected patients have a high degree of satisfaction with the procedure, and appreciate the pain relief, even if it is temporary and does not prevent subsequent surgery.

It is important to review the patient's computed tomography (CT) or magnetic resonance imaging studies prior to performing the spinal injection. This can assist in the proper selection of the target area for injection, as well as assure that fracture, infection, or tumor is not the source of symptoms.

EQUIPMENT/MEDICATION

Proper patient positioning and visualization of the area to be injected is of paramount importance for successful completion of these procedures. For thoracic and lumbar injection procedures, fluoroscopy is generally used; a C-arm provides excellent visualization of the anatomy and allows rapid feedback regarding needle positioning in various obliquities while offering an open workspace. A "floating" image intensifier in an angiography suite provides excellent visualization of spinal anatomy in a variety of projections, but offers somewhat restricted workspace compared to a C-arm. CT can also be used for needle localization, but repeated scanning in between the changes in needle positioning results in significant delays and increased radiation dosage. For this reason we prefer fluoroscopic guidance using a C-arm. Another potentially problematic issue regarding CT is that when contrast is injected to verify satisfactory needle positioning, it may be difficult to track (especially if the needle is intravascular and the contrast dissipates during the delay prior to imaging) unless CT fluoroscopy or a combined CT and C-arm is used.

For most diagnostic and therapeutic injections, a combination of anesthetic and steroid is used. A short-acting anesthetic, 1% lidocaine is used, and must be epinephrine-free and preservative-free (the methylparaben preservative can cause flocculation of steroid). This is generally combined with a steroid agent; the three steroid preparations most commonly used in the United States for spinal injections are Depo-Medrol (Methylprednisolone), Kenalog (Triamcinolone) and Celestone Soluspan (Betamethasone). Depo-Medrol has the largest particle size of the three, and the preparation may precipitate (flocculate) more when mixed with lidocaine; therefore, we prefer Celestone, if available. 12 mg Celestone (2 cc of the 6 mg/cc suspension), 40 mg of Kenalog (2 cc of the 20 mg/cc suspension), or 80 mg of Depo-Medrol (2 cc of the 40 mg/cc suspension) mixed with 1 cc of the 1% preservative-free lidocaine (to a total volume of 3 cc) is injected. To verify needle positioning prior to injection, a small amount of iodinated contrast is injected with direct fluoroscopic visualization. In general, for spinal injections only nonionic contrast approved for intrathecal use should be used.

Epidural injection can provide a diagnostic as well as a therapeutic effect. Back pain and radicular symptoms can have numerous etiologies, and imaging findings of disc bulge or herniation may not correspond to the findings on clinical examination. To help localize the source of pain, selective injections have been used. Using a trans-foraminal approach corresponding to an imaging finding or clinically suspected level of pathology, injection of medication can define the pain generator. Anesthetic is the primary medication used for this purpose; if the patient's symptoms improve after the procedure for the duration of the effect of the anesthetic, it can be inferred that the site of pathology is at the level injected. Pain is often evoked during the placement of the needle within the neural foramen; if the pain corresponds to the site of the patient's typical pain, this is also useful information, but is thought to be less diagnostically important than symptom reduction after the injection. Another useful piece of diagnostic information is the delayed response to steroid medication administered at the same time as the anesthetic. One proposed mechanism for generation of radicular pain is irritation of the nerve root due to release of phospholipase A2 from the herniated disc, which induces production of inflammatory mediators (19). Steroids interrupt this pathway, thereby reducing the inflammation. This effect is delayed, with symptom reduction typically after 24 to 48 hours. To observe this effect and control other influences, the patient is instructed to continue the same activity and lifestyle as followed prior to the injection. It can be useful to give the patient a "pain diary" in which they record daily accounts of their level of pain from 0 (no pain) to 10 (the worst pain they can imagine) to document degree of response. Celestone and Depo-Medrol steroid preparations are suspensions of intermediate-acting duration that achieve a slow-release effect, and last approximately 30 days. The effect can last longer or shorter. In general, patients with true radicular pain respond more frequently than those with paresthesia or weakness as their main symptom. A series of three injections spaced a month apart has been employed with the goal of prolonging the therapeutic effect; however, there is no evidence that three ideally placed injections provide any additional beneficial effect. More likely, the "series of three" injection protocol developed in order to assure that at least one injection reached the target area, prior to the use of image guidance. If there is partial relief from the first injection, a second injection at the same site may be beneficial. However, if there is minimal or no relief of pain from the initial injection, a different site (e.g., the level above or below the initial injection) should be targeted if a second injection is requested.

Fluoroscopy or CT may be used for injections. However, there are numerous advantages to using fluoroscopy. First, fluoroscopy provides a better evaluation of intravascular positioning during contrast injection. Even rapid CT-scanning techniques or fluoro-CT could potentially miss intravascular injection of contrast, particularly in the setting of rapid arterial flow. Also, fluoroscopy shortens procedure time; verification of the proper level is rapidly performed fluoroscopically, and changes in needle position can be made quickly. Fluoroscopic equipment is also cheaper than CT equipment, and generally offers more flexible scheduling of procedures than those performed using CT.

TECHNIQUE: SELECTIVE NERVE ROOT INJECTION (TRANSFORAMINAL EPIDURAL INJECTION)

In patients with severe acute radicular pain or chronic radicular pain which does not respond to conservative therapy, selective nerve root injection can be an effective means of therapy. Using image guidance, a small quantity of anesthetic and steroid is injected around the sheath of the nerve thought to be responsible for the symptoms. Patients should be made aware that they may experience recurrence of their pain shortly after the procedure for one to three days in the interval between the short-acting anesthetic and the delayed effect of the steroid. In part, this may be due to a chemical irritation of the nerve root. Patients should also be warned prior to the procedure that they will experience temporary numbness and possibly weakness in the distribution of the nerve root being injected. Potential complications (see below) should be discussed with the patient as part of an informed consent.

Thoracic Selective Epidural Injection

Thoracic selective injections are performed with the patient in prone position. The X-ray beam is tilted laterally such that a "Window" is created allowing access between the facet joint, pedicle, transverse process, and rib (Fig. 1). The beam is tilted cranio-caudally such that it is directed "down the barrel" of the pedicle, the lower margin of which is about halfway between the end plates. The area is prepped and draped, the skin entry site is localized fluoroscopically, and subcutaneous lidocaine is administered. Deeper, lidocaine is administered along the angle of the X-ray beam. Next, a 22- or 25-gauge spinal needle is advanced inferior to the pedicle through the aforementioned "Window" until the posterolateral vertebral body is reached or

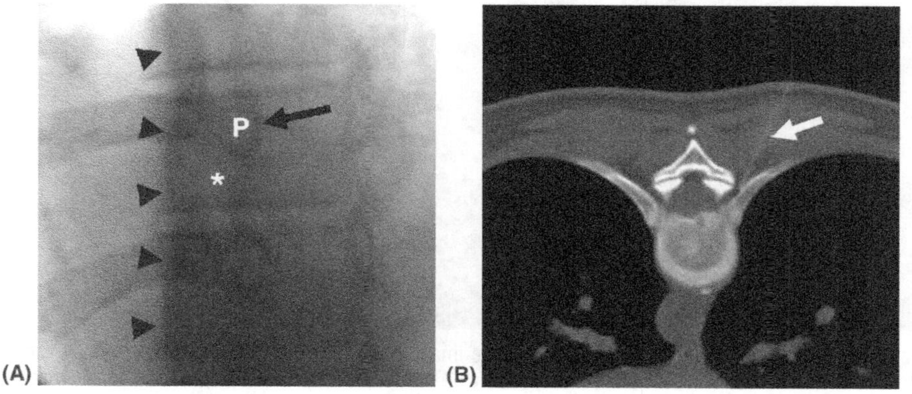

FIGURE 1 Localization for thoracic selective epidural injection: fluoroscopic and computed tomography (CT) methods. (**A**) Fluoroscopic image showing obliquity required for needle placement. Note pleural margin (*arrowheads*) and vertebral pedicle (letter "P"). Optimal needle course between the rib, facet joint, and pedicle is marked with an asterisk. Care must be taken not to pass the needle medial to the medial edge of the pedicle (*arrow*), which demarcates the spinal canal. (**B**) Axial CT image of needle tip (*arrow*) positioned within a thoracic neural foramen. Localization is performed using the same basic technique as for a CT-guided biopsy.

radicular pain is perceived. If there is severe radicular pain, the needle should be repositioned. Care must be taken to avoid the pleural margin more laterally, which is easily visualized fluoroscopically. The needle should not be advanced more medial than the medial margin of the pedicle, which forms the border of the spinal canal.

If there is any blood return from the needle, the tip should immediately be repositioned; in the thoracic spine, especially from T7 to T9, small arterial feeders to the spinal cord can extend through the superior aspect of the neural foramen. Injection of air, lidocaine, or particulate steroid into these branches has the potential to cause cord infarction. For this reason we also position the needle more inferiorly within the neural foramen, because the feeding vessels run just below the pedicle. Next, contrast injection is performed to verify satisfactory positioning. A "Wet" connection is important to assure that no air is injected (in case of intravascular positioning). The injected contrast should extend centrally into the lateral epidural space of the spinal canal, and peripherally around the nerve root. If there is vascular opacification, the needle should be repositioned. If the patient complains of an acute onset of severe radicular pain immediately upon injection, an intraneural injection may have occurred; injection should be stopped immediately and the needle should be repositioned. After proper positioning is verified, the anesthetic and steroid mixture is injected with periodic fluoroscopic observation.

Lumbar Selective Epidural Injection

Lumbar selective injection is also performed with the patient in prone position. The neural foraminal level to be injected is localized fluoroscopically, and the beam is tilted laterally approximately 25° so that the superior articular process is directly below the adjacent pedicle (Fig. 2). Craniocaudal angulation of the image intensifier is performed to "look down the barrel" of the pedicle, the lower margin of which is about halfway between the end plates. A slightly cranial course of the needle tract will help avoid the transverse process. The skin is widely prepped and draped; the skin entry site is localized fluoroscopically and subcutaneous lidocaine is administered. Deeper, lidocaine is administered along the angle of the X-ray beam. The nerve root passes just beneath the medial margin of the pedicle and extends inferiorly, so the needle (22-gauge 3 in. needle for thin-to-average build patients, 6–7 in. needle for obese patients) is advanced to the area just below the inferior margin of the pedicle. The needle is advanced to bone (the posterolateral aspect of the vertebral body) or until radicular pain is elicited. With significant radicular pain, the needle should be repositioned.

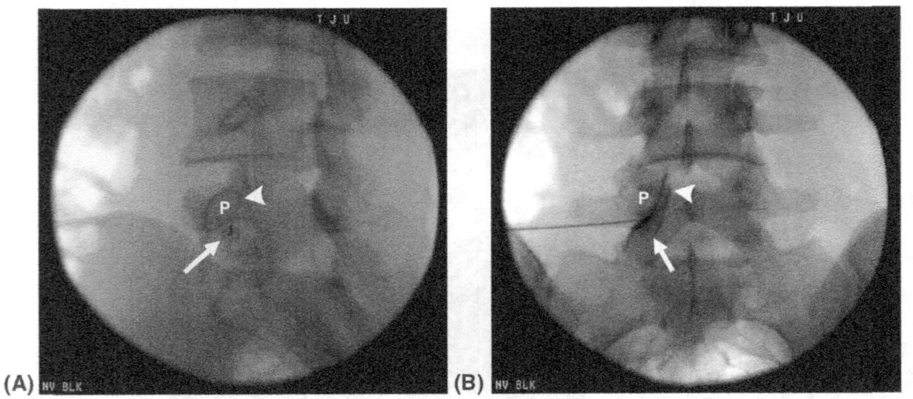

FIGURE 2 Lumbar selective epidural injection. (**A**) Fluoroscopic image showing optimal obliquity for needle placement. Note vertebral pedicle (letter "P") and needle (*arrow*) in place. Care must be taken not to pass the needle medial to the medial edge of the pedicle (*arrowhead*), which forms the margin of the spinal canal. (**B**) Needle tip position within the neural foramen is verified using a frontal projection. Optimally, the needle should be positioned just inferior to the six o'clock point on the pedicle (letter "P"). Nonionic iodinated contrast is injected to verify position; note opacification of the sheath of the exiting nerve root (*arrow*) and extension to the medial epidural space (*arrowhead*). Medication can then be injected with intermittent fluoroscopic observation.

Next, contrast injection is performed to verify satisfactory positioning. The contrast should extend centrally into the epidural space of the spinal canal and peripherally around the nerve root. If there is vascular opacification, the needle should be repositioned. If the patient complains of an acute onset of severe radicular pain immediately upon injection, an intraneural injection may have occurred; injection should be stopped immediately and the needle should be repositioned. After proper positioning is verified, approximately 3 cc of a combination of anesthetic and steroid is injected with intermittent fluoroscopic observation.

A coaxial technique can also be used, with placement of a larger gauge needle with a more straight parasagittal orientation, through which a curved 25-gauge needle is passed to attain access to the neural foramen. This technique is especially helpful in patients with a fusion mass obstructing direct access to the foramen. CT may be required for localization in this setting.

Sacral Selective Epidural Injection

Sacral selective injection is indicated for patients with sciatic radicular pain who have impingement of the S1 nerve root (typically a paracentral disc herniation at L5–S1). This is readily performed fluoroscopically by identifying and distinguishing the dorsal and ventral S1 foramina and tilting the X-ray beam craniocaudally in the anteroposterior projection to attain optimal visualization (Fig. 3). The ventral foramen (broad, ovoid, and sharply defined) is more easily visualized than the dorsal foramen, which is round and often poorly seen. To target the dorsal foramen, it is best to use L5 as a guide: tilt the beam so that the L5 pedicle is seen *en face*,

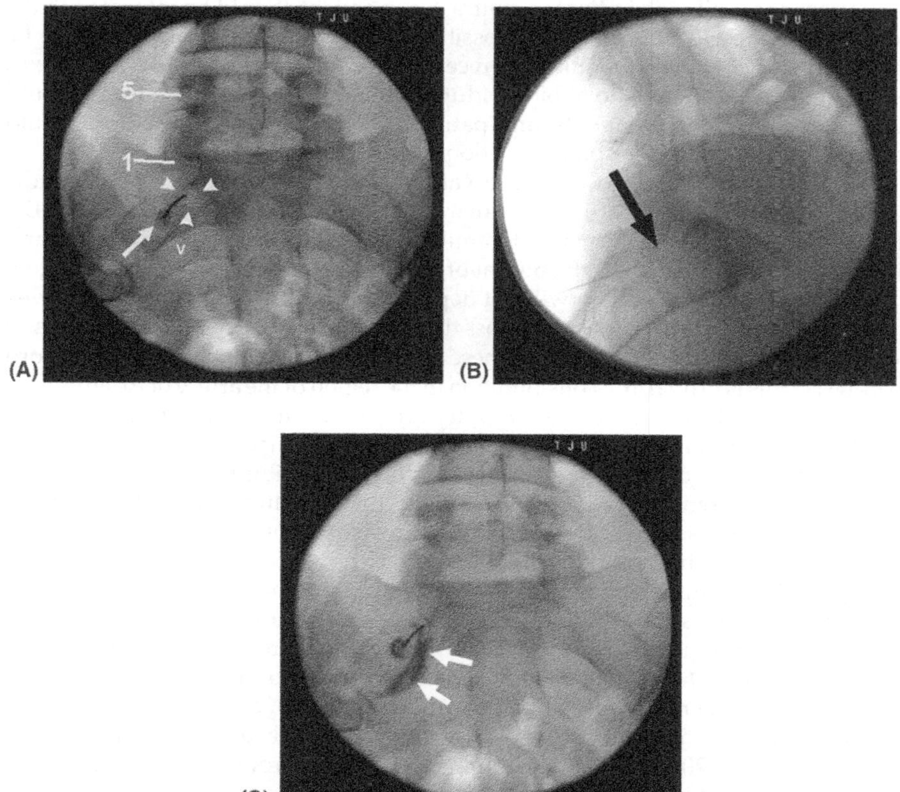

FIGURE 3 Sacral selective epidural injection. (**A**) Fluoroscopic image showing optimal obliquity for needle placement. Note vertebral pedicle at L5 (number "5") and S1 (number "1"). The S1/S2 dorsal foramen is outlined by arrowheads. Note ventral foramen (letter "V"). Needle (*arrow*) is in place. (**B**) Lateral view confirms needle tip (*arrow*) in sacral canal. (**C**) Nonionic contrast injection shows opacification of the exiting S1 nerve root sheath (*arrows*). After confirmation of optimal positioning, medication is injected.

approximately half way between the superior and inferior end plate. More inferiorly, the pedicle of the S1 vertebra will then be in view. Rotate the beam slightly lateral so that the needle will track medially. Advance a 22-gauge needle just inferior to the S1 pedicle until bone is reached or radicular pain is generated. A lateral projection can confirm proper positioning within the foramen. After injection of contrast documents optimal positioning, injection of medication is performed.

Nonselective Epidural Steroid Injection

Anesthetic and steroid can be injected in a nonselective fashion via an interlaminar (transflaval) route, or through the sacral hiatus (caudal injection), with superior infiltration of medication to the lower lumbar levels. This approach results in more diffuse distribution of medication than selective injections described above; however, nonselective injection can be of benefit if there is disease at multiple lumbar levels bilaterally. The main complication of interlaminar injection is intrathecal injection, with spinal anesthesia, headache, or dural leak. This risk is minimized with contrast injection and careful fluoroscopic monitoring.

POTENTIAL COMPLICATIONS

Patients should be made aware that a very small proportion of patients experience increased radicular pain for three to five days after the procedure. This may be related to a chemical sensitivity to the inactive components of the steroid preparation. Patients should be reassured that the pain will resolve; if they require future injections, a different steroid preparation can be used.

As with any procedure utilizing iodinated contrast, the patient should be informed that a contrast reaction could occur. To minimize this possibility, only nonionic contrast should be used. Patients with a history of allergy should first receive a course of oral steroid prophylaxis. Bleeding risk is minimal; nevertheless, to avoid epidural hematoma, these procedures should not be performed on anticoagulated patients, and patients on chronic aspirin therapy should stop the medication a week prior to the spinal injection.

Contrast injection with active fluoroscopic observation verifies proper location of the needle tip and minimizes the risk of subsequent intravascular injection of medication (11,12,14,15,17). Venous collaterals are commonplace in the neural foramina and epidural space of the thoracic and lumbar spine; inadvertent injection of a small amount of medication into these vessels is not generally problematic, although the medication will not deposit in the intended location. However, arterial collaterals of the spinal artery occasionally pass through the thoracic neural foramina, and injection of anesthetic or steroid suspension into these vessels could potentially result in cord infarction and paralysis. This is primarily a consideration in the region of the artery of Adamcowitz at approximally the T7 to T8 level. Injection more inferiorly within the neural foramen could potentially avoid these branches, which course directly beneath the pedicle.

Injection of steroid has its own risks, even though the biodistribution is quite different than a preparation administered orally or intravenously. Risks include avascular necrosis, propagation of infectious processes, elevation of blood glucose in diabetics, suppression of the adrenal-pituitary axis, tendon tears, and deposition of fat in various locations around the body (theoretically including the epidural space, resulting in epidural lipomatosis).

Inadvertent intrathecal or subarachnoid injection is a known complication of epidural injection, but is mainly a consequence of nonselective interlaminar injection (13); intrathecal injection is exceedingly rare when using a proper selective transforaminal approach or a caudal approach through the sacral hiatus. Some potential sequelae of an intrathecal injection include spinal anesthesia (if anesthetic is used), dural leakage, headache, hypotension, meningitis, and arachnoiditis (20,21). Hypotension may also occur from caudal injection if a large amount of fluid is injected in order to "push" the medication superiorly (21). In order to minimize these complications, many prefer the selective transforaminal approach for needle placement. Again, whatever approach is used, verification of needle position with contrast injection is essential to minimize complications. Rarely, intraneural injection can occur, which can injure the nerve. If the needle is placed within the nerve itself, contrast injection will reveal a thin, linear pattern of opacification corresponding to the course of the nerve root, and the patient will experience severe radicular pain. At this point, the injection should be stopped immediately and the needle repositioned.

Vasovagal reaction is occasionally seen, and best treated with supportive measures and termination of the procedure. Facial flushing can also be seen. Finally, as with any procedure, infection is a potential risk that can be minimized by using careful sterile technique.

As outlined previously, lidocaine is typically injected along with steroid for epidural injection procedures; long-acting anesthetics such as Bupivacaine or Marcaine have been used instead of lidocaine by some authors (21,22). With these medications, the extreme numbness (and occasional weakness) after the procedure will last six to eight hours (as opposed to one to two hours with lidocaine). For some patients, especially the elderly, this effect can be temporarily debilitating. This prolonged anesthesia provides no additional diagnostic information compared to injection of a short-acting anesthetic, and anesthetic is not a therapeutic medication; therefore, we do not encourage the use of a long-acting anesthetic for spinal injections. In fact, for elderly patients with spinal stenosis, an argument can be made for avoiding use of epidural anesthetic entirely; in this population, the diagnostic benefit of anesthetic injection is marginal.

Despite this rather long discourse on the potential complications of epidural injection, it should be recognized that these injections are, in actuality, very safe if performed with image guidance and contrast injection. Two large series report few complications, the vast majority of which are mild; this has been our experience as well.

SUMMARY

Selective epidural injections can offer significant diagnostic and therapeutic benefit for patients with radicular pain. Procedures should be performed only after apprenticeship-type training by individuals with extensive knowledge of imaging techniques and relevant anatomy. Attention to proper technique will minimize risk of complications from these procedures while maximizing their benefit.

REFERENCES

1. Fanciullo GJ, Hanscom B, Seville J, Ball PA, Rose RJ. An observational study of the frequency and pattern of use of epidural steroid injection in 25,479 patients with spinal and radicular pain. Reg Anesth Pain Med 2001; 26:5–11.
2. Carrino JA, Morrison WB, Parker L, Schweitzer ME, Levin DC, Sunshine JH. Spinal injection procedures: volume provider distribution and reimbursement in the United States Medicare population from 1993 to 1999. Radiology 2002; 225(3):723–729.
3. Koes BW, Scholten RJ, Mens JM, Bouter LM. Efficacy of epidural steroid injections for low-back pain and sciatica: a systematic review of randomized clinical trials. Pain 1995; 63:279–288.
4. Warfield CA, Crews DA. Work status and response to epidural steroid injection. J Occup Med 1987; 29:315–316.
5. Rosen CD, Kahanovitz N, Bernstein R, Viola K. A retrospective analysis of the efficacy of epidural steroid injections. Clin Orthop 1988; 228:270–272.
6. Carette S, Leclaire R, Marcoux S, et al. Epidural corticosteroid injections for sciatica due to herniated nucleus pulposus. N Engl J Med 1997; 336:1634–1640.
7. Lutz GE, Vad VB, Wisneski RJ. Flouroscopic transforaminal epidural steroids: an outcome study. Arch Phys Med Rehabil 1998; 79:1362–1366.
8. Saal JA, Saal JS. Nonoperative treatment of herniated lumbar intervertebral disc with radiculopathy. An outcome study. Spine 1989; 14:431–437.
9. Watts RW, Silagy CA. A meta-analysis on the efficacy of epidural corticosteroids in the treatment of sciatica. Anaesth Intensive Care 1995; 23:564–569.
10. Schmid G, Vetter S, Gottmann D, Strecker EP. CT-guided epidural/perineural injections in painful disorders of the lumbar spine: short- and extended-term results. Cardiovasc Intervent Radiol 1999; 22:493–498.
11. El-Khoury GY, Ehara S, Weinstein JN, Montgomery WJ, Kathol MH. Epidural steroid injection: a procedure ideally performed with fluoroscopic control. Radiology 1988; 168:554–557.
12. Renfrew DL, Moore TE, Kathol MH, el-Khoury GY, Lemke JH, Walker CW. Correct placement of epidural steroid injections: fluoroscopic guidance and contrast administration. AJNR 1991; 12:1003–1007.
13. Mclain RF, Fry M, Hecht ST. Transient paralysis associated with epidural steroid injection. J Spinal Disord 1997; 10:441–444.

14. Fredman B, Nun MB, Zohar E, et al. Epidural steroids for treating "failed back surgery syndrome": is fluoroscopy really necessary? Anesth Analg 1999; 88:367–372.
15. Furman MB, O'Brien EM, Zgleszewski TM. Incidence of intravascular penetration in transforaminal lumbosacral epidural steroid injections. Spine 2000; 25:2628–2632.
16. Stitz MY, Sommer HM. Accuracy of blind versus fluoroscopically guided caudal epidural injection. Spine 2000; 24:1371–1376.
17. Lui SS, Melmed AP, Klos JW, Innis CA. Prospective experience with a 20-gauge Tuohy needle for lumbar epidural steroid injections: Is confirmation with fluoroscopy necessary? Reg Anesth Pain Med 2001; 26:143–146.
18. Spaccarelli KC. Lumbar and caudal epidural corticosteroid injections. Mayo Clin Proc 1996; 71:169–178.
19. Saal JS, Franson RC, Dobrow R, Saal JA, White AH, Goldthwaite N. High levels of inflammatory phospholipase A2 activity in lumbar disc herniations. Spine 1990; 15:674–678.
20. Abram SE, O'Connor TC. Complications associated with epidural steroid injections. Reg Anesth 1996; 21:149–162.
21. Link SC, El-Khoury GY, Guilford WB. Percutaneous epidural and nerve root block and percutaneous lumbar sympatholysis. Radiol Clin North Am 1998; 36:509–521.
22. Kaplan PA, Dussault RG. Image-guided selective nerve blocks in the spine. Semin Musculoskelet Radiol 1997; 1:231–240.

10 Selective Cervical Nerve Blocks

Philip Lander
Department of Radiology, University of Alabama at Birmingham, Birmingham, Alabama, U.S.A.

CERVICAL FACET BLOCK

Introduction

Chronic neck and shoulder pain as well as pain in the occiput may originate from the cervical zygapophyseal or facet joints (1,2,3). The sequential distention of normal facet joints with saline or contrast produces a segmentally distinctive pattern of pain distribution in the posterior occiput, neck, and shoulder of the injected side (4). Pain receptors in the synovium and capsule of the facets joints from C3 to C7–T1 are innervated by the medial branch of the dorsal ramus of the segmental spinal nerve. These nerves have a constant relationship to the mid-waist of the lateral cortex of the corresponding articular pillars (5), and are easily accessible to local anesthesia by needle placement under fluoroscopic or computed tomography (CT) guidance. Local anesthetic blocks into the facet joint or of the corresponding medial branch may afford complete relief of pain in the segment supplied by these joints, thus attributing this joint as the source of pain (1). The effect of local anesthetic applied to the medial branches has a similar validity to an intra-articular injection of local anesthesia (6).

The prevalence of chronic neck pain at levels below C3 following whiplash was estimated to be at least 49% by Lord et al. (7) and 36% by Speldewinde et al. (8), utilizing medial branch blocks to localize a specific diagnosis of the side and level of the pain generator. The prevalence of third occipital nerve headache following whiplash injury is 27% diagnosed by double blind, controlled diagnostic blocks of the third occipital nerve, but may be as high as 52% in patients presenting with predominant headache (3).

Cervicogenic headache as well as referred pain to the upper neck and posterior occiput may originate from the C2–C3 facet joint innervated by the third occipital nerve (TON) (5). Less commonly, the C1–C2 facet joint and rarely the atlanto-occipital joint are responsible for this pain (9). The C1–C2 facet joint is innervated by the C2 or greater occipital nerve (5), and a selective nerve block or intra-articular block will confirm the diagnosis of cervicogenic headache from this segment.

The use of controlled, comparative selective local anesthetic blocks of the medial branches or the TON-utilizing lignocaine and subsequently bupivacaine is advocated to exclude the high rate of false-positive response in these patients and has the same validity as a placebo-controlled block (10). The duration of action of local anesthetic alone or with corticosteriods injected into the cervical facet joints has been shown to not be greater than three days (11). However, in a recent systematic review of facet joint interventions in chronic spinal pain observational studies of cervical facet joint injections, pain relief was achieved for up to three months. Short term and long term relief of medial branch blocks achieved pain relief up to 12 months (12).

Patient Indications and Selection

Patients with chronic occipital, neck, shoulder, or interscapular pain are candidates for intra-articular or medial branch block, if their pain follows the typical distribution as has been shown from previous clinical studies on normal volunteers and patients with head, neck, and shoulder pain (4,9,13). In patients with whiplash, the prevalence of chronic neck pain and headache was found to be as 54%, most commonly arising at the C2–C3 and C5–C6 levels (7).

Posterior occipital pain is most likely referred from the C2–C3 joints (9). The pain pattern of capsular distention of the lateral atlantoaxial joint and the atlanto-occipital joints is referred to the ipsilateral posterolateral occiput and upper neck (13). Upper posterior or posterolateral cervical regional cervical pain is likely to be referred from the C2–C3 or C3–C4 joints. Middle posterior cervical pain is likely to be referred from the C3–C4 or C4–C5 joints, and lower posterior cervical pain is likely to be referred from the C4–C5 or C5–C6 joints. Suprascapular shoulder pain may be referred from the C4–C5 or C5–C6 joints and superior scapular pain from the C6–C7 joints. Pain in the mid-scapular region may be referred from the C7–T1 joint (4,9).

Routine imaging should exclude other possible etiologies such as tumor or infection before the patient undergoes a block, because it is primarily a test to provide a diagnosis in patients who do not show any evident cause for their chronic neck pain and whose pain map indicates a segmental innervation of their referred pain. However, it must be noted that the localization of cervical referred pain is confounded by the fact that the pattern of pain distribution of cervical discogenic origin is similar to that of the zygapophyseal joints (14).

The easy access to the medial branches of the dorsal rami to anesthetic block makes this the preferred route to establish the diagnosis of cervical facetogenic pain. The choice of the side and level of the intra-articular or medial branch block is determined by the pain distribution. The nerve supply to the facet joints from C3–C7 arises from the medial branches of the dorsal rami of the level above and below the joint (5). Thus two medial branch blocks are required for a complete block of that segment (1). The nerve supply to the C2–C3 joint arises from branches of the TON, and local anesthetic block to this nerve and the ganglion is administered lateral to the C2–C3 joints. In patients whose headache is dominant over neck pain, origin from the TON appears to be the most prevalent (3). Intra-articular blocks performed at C3–C4 and C1–C2 have been shown to block occipital headaches less frequently. Pain of cervicogenic origin may be unilateral or bilateral or involve more than one segment. Therefore, bilateral blocks will be required if the pain is bilateral, or if there is a multiple segmental distribution serial blocks may be required at consecutive sessions.

Radicular upper extremity pain will not be diagnosed or relieved with these blocks. If there is no clear segmental referred pain pattern attributable to one or more facet joint levels, a medial branch or facet block would unlikely be positive and these patients should be investigated for other sources of their pain.

Patients should fill out a pain map (15) and a visual analogue score (VAS) (16) prior to the blocks in order to determine the levels to be blocked and to assess the effectiveness of the block by comparison to the pain map, following the block. If the pain intensity is less than 60 on a 100-point scale, the evaluation may be compromised and consideration should be given to postponing the procedure. The execution of certain activities of daily living such as bending or turning may be prevented by the pain and should be recorded. The evaluation by a separate assessor such as a nurse or technologist rather than the procedure physician is desirable to remove possibility of outcome bias. As well, patients should utilize a pain diary or VAS to record their response to the block over the ensuing days.

Intra-articular Facet Joint Block Technique

A fluoroscope, preferably with a "C" arm, is required to image the needle position. A 22-gauge, or more optimally a 25-gauge, 90 mm spinal needle is utilized without minimal volume extension tubing and a 2 or 3 cc syringe. The injection of 0.3 mL of nonionic contrast before the local anesthetic block will confirm that the needle tip is extravascular or in an intra-articular position. Injection of preservative-free lidocaine 1% to 2% or bupivacaine 0.25% to 0.5% in volumes of 0.3 mL for the medial branch block and 0.5 mL for intra-articular injection is sufficient for anesthetic block.

INTRA-ARTICULAR BLOCK

Both posterior and lateral approaches have been described for joints C3 to C7 (17,18). The plane of the joint is oblique anteroposteriorly, so that the posterior joint is accessible via a puncture two or more segments distal to the target in the prone patient. An angled fluoroscope into the plane of the joint (a pillar view) facilitates this projection. The needle should be directed to hit the end plate bone adjacent to the joint and determine the depth before intra-articular entrance of the needle. Lateral fluoroscopy will also confirm the needle depth. Following the injection of

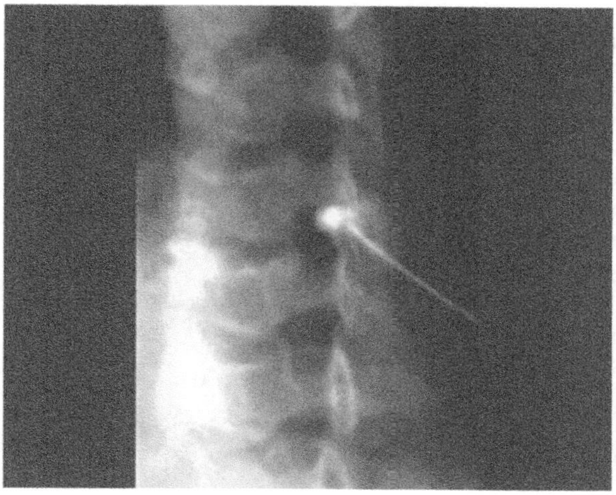

FIGURE 1 Intra-articular C5–C6 block. Oblique projection of the cervical spine. A 22-gauge needle is within the C5–C6 facet joint introduced by a posterolateral approach. Contrast outlines the capsule in the posterior intervertebral foramen.

0.3 mL of contrast, a volume of no more than 1.0 mL of local anesthetic, with or without a corticosteroid, is injected intra-articularly so that capsular rupture does not occur (Fig. 1).

A lateral intra-articular block is performed, with the patient lying on the side with the head supported parallel to the table and the target joint up. To identify the target pillar and joint, the C-arm is rotated slightly anteroposteriorly before and after the needle is inserted into the bone margin of the joint. The needle can then be readjusted so that its tip is felt to enter the joint capsule but not advanced into the joint to prevent entering the epidural or subarachnoid space or the spinal cord. The injection of 0.3 mL of contrast medium will provide confirmation of the intra-articular position (Fig. 2). Intra-articular blocks at C2–C3 can be performed via a lateral approach with a straight tube or with angulation of the tube caudally until the joint is parallel to the beam (Fig. 3) (18).

The lateral atlantoaxial joint can be entered via posterior approach with the needle directed into the lateral aspect of the joint to avoid penetration of the C2 ganglion and ventral ramus. The needle is introduced with 5 mm increments and intermittent fluoroscopy to touch the articular bone before entering the joint, so that the depth of the needle is established. The needle tip should only just enter the joint capsule and the intra-articular position is confirmed with the injection of 0.3 mL of contrast (18). CT scan can also be used to guide a 22- or 25-gauge needle into the posterior C1–C2 joint (Fig. 4) (19).

A lateral fluoroscopic approach to the atlantoaxial joint has also been described by Dreyfuss et al. (20).

The patient is placed on the fluoroscopy table in the lateral position and the "C" arm fluoroscope is rotated until the C1–C2 articular pillars are aligned. Then the C arm is angled cephalocaudad to image the joint parallel to the beam. A 25-gauge needle is positioned in the lateral suboccipital neck to reach the junction of the anterior one-third and posterior two-thirds of the joint bone margin before intra-articular placement. This position is confirmed by anteroposterior (AP) open mouth fluoroscopy, and the needle is then introduced into the joint followed by the intra-articular injection of 0.3 mL of contrast (Fig. 5).

EXTRA-ARTICULAR MEDIAL BRANCH BLOCK

Medial branch blocks at the levels of C3 to C6 are performed most readily in the lateral projection with the affected side up. The target is the center of the articular pillar for branches C3 to C6, with the needle tip abutting the bone at the intersection of the two diagonals of the diamond-shaped articular pillar (Fig. 6). A fluoroscopic true lateral projection of the articular pillars is essential to avoid inserting the needle deep to the pillars and possibly into the spinal canal. Fluoroscopy in the AP plane will confirm the position of the needle against the lateral cortex at the waist of the articular pillar. The injection of 0.3 mL of contrast will

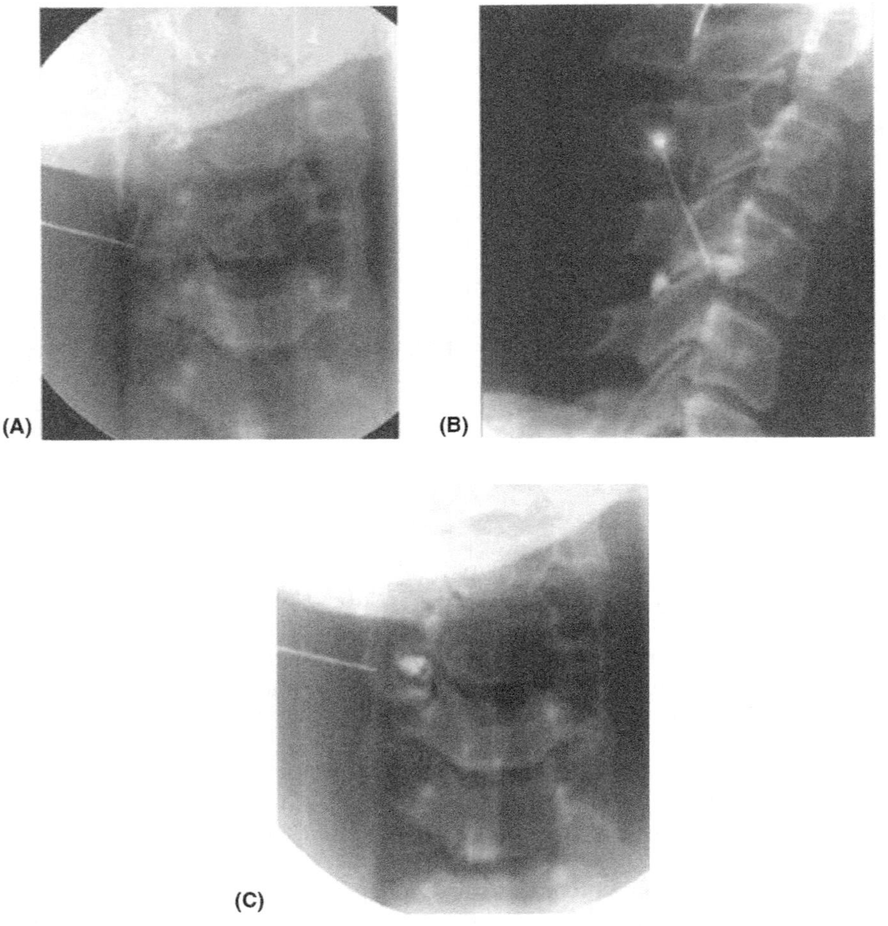

FIGURE 2 Intra-articular block at C6–C7; lateral approach. A 25-gauge needle is advanced to the superolateral cortex of the C6–C7 facet joint, with the patient in the lateral decubitus position. The C arm is rotated 90°, and the needle-tip position is confirmed in the anteroposterior (AP) projection (**A**). The needle tip is adjusted to just enter the capsule, and 0.3 to 0.5 mL of nonionic contrast is injected under fluoroscopy in the lateral projection (**B**) and imaged in the AP projection (**C**).

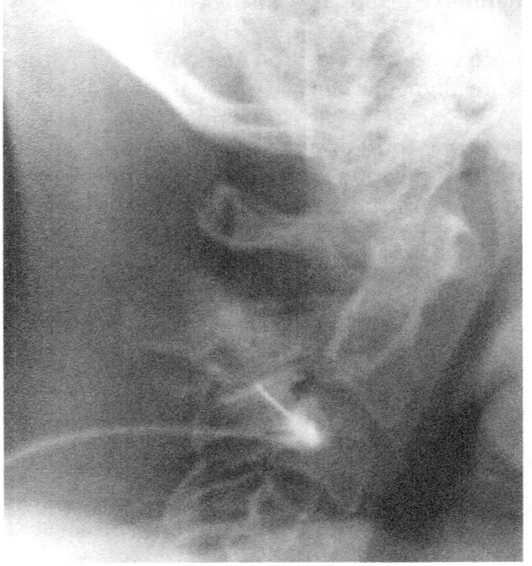

FIGURE 3 Lateral projection. Intra-articular C2–C3 block. A 25-gauge needle placed on the C2 articular cortex to just enter the joint capsule. A small bore, short-extension tube is attached to the needle to allow the exchange of syringes without disturbing the needle position. This is followed by the injection of 0.3 cc of contrast, followed by 1.0 cc of local anesthesia with or without a corticosteroid.

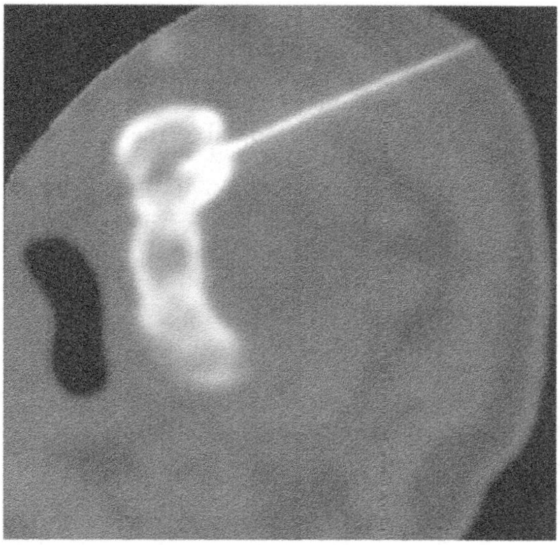

FIGURE 4 Axial computed tomography scan. Intra-articular block, C1–C2; posterior approach. A 25-gauge needle has been localized to the posterior lateral-third of the joint to avoid the C2 spinal nerve dural sac or C2 root sleeve. The periphery of the joint is outlined by contrast.

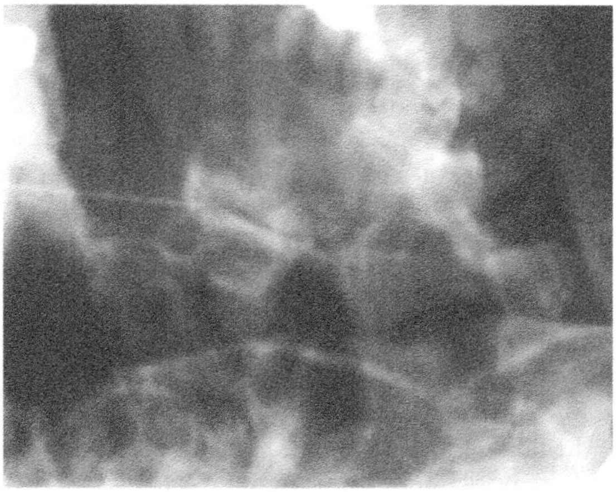

FIGURE 5 Anteroposterior projection. Intra-articular block, C1–C2; lateral approach. Contrast arthrogram outlines the joint capsule. Notice the needle tip remains within the lateral aspect of the joint to avoid inadvertent passage into the spinal canal.

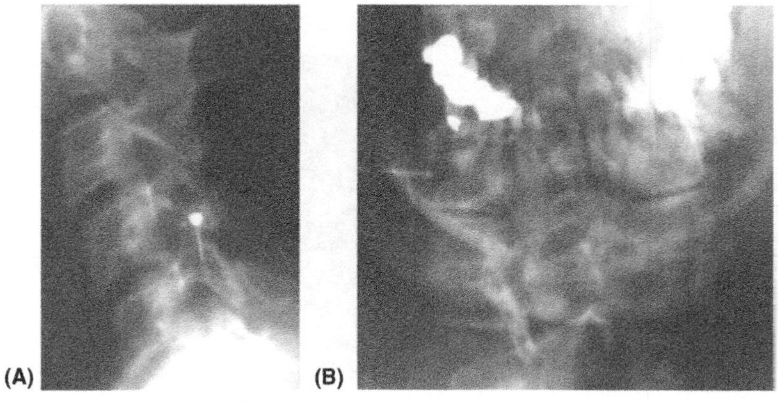

FIGURE 6 (**A**) Lateral and (**B**) anteroposterior projection. Extra-articular C5 medial branch block, lateral approach. The needle is aimed to reach the midpoint or waist of the facet pillar. The injection of 0.5 mL of contrast excludes intravenous washout and is followed with injection of 0.3 to 0.5 mL of local anesthetic.

exclude an intravenous position and washout. This is followed by the administration of 0.3 mL of local anesthesia to the target nerve.

The medial branch block of C7 requires placement of the needle superior to the base of the transverse process within the middle of the upper portion of the pillar. Once the needle contacts the bone, an AP view is obtained to be certain that the needle tip is not lodged on the transverse process but superior to base of the transverse process on the lateral cortex of the superior articular pillar (Fig. 7). An injection of 0.3 mL of local anesthesia is given, and the needle is withdrawn 0.5 cm and a second. 0.3 mL injected to anesthetize a possible variation in the position of the C7 medial branch lateral to the process and within a small muscle tunnel in the seimispinalis capitis muscle, which covers the pillars (5). CT scan can also be utilized to guide the needle tip in large patients whose shoulders prevent a safe fluoroscopic approach (19).

The TON is less constant in location and thicker than the medial branches. It is lateral to the C2–C3 facet joint within the pericapsular fascia (5). This nerve can be anesthetized by the lateral approach, with the needle directed to lie on the bone at three sites along a plane perpendicular to the middle of the C2–C3 joint, cranial to the joint at the level of the tip of the superior articular process, adjacent to the superior articular cortex, and caudal to the superior articular cortex. The needle is withdrawn slightly before injecting 0.3 mL of local anesthetic to each site to avoid an intracapsular injection (Fig. 8). The posteroanterior approach requires the needle to be placed lateral to the C2–3 joint and its final position to be observed in the lateral projection (Fig. 9). Patients frequently experience temporary ataxia with this block and should be instructed to focus on objects along the horizontal until they adapt to this sensation.

Contraindications

The presence of bacterial infection in the skin overlying the needle target or a systemic bacterial infection and a bleeding diathesis are absolute contraindications. Allergy to contrast media or local anesthesia can be compensated by pretreatment with corticosteriods and use of a different class of anesthetic agents. The patients should stop nonsteroidal anti-inflammatory drugs for one to two days prior to the block and anticoagulant treatments for five days prior to the blocks.

Evaluation

Pain relief that is complete within the expected region of the block for the span of the local anesthetic constitutes a positive response. If 100% of the pain in the target segment is relieved, but there is persistence of other segmental pain, this block is still classified as a positive response for that segment. If all the pain is not relieved then additional levels or other pain generators should be sought. Patients with pain apparently mediated by upper and lower segments such as C2, C3 and C5, C6 should be blocked at only two consecutive levels at a time to separate the response to the blocks. Patients with headache as the predominant symptom should have a C2–C3 TON

(A) 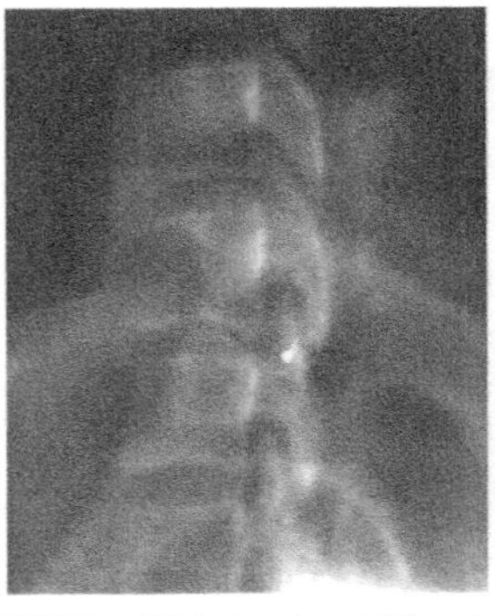 (B) 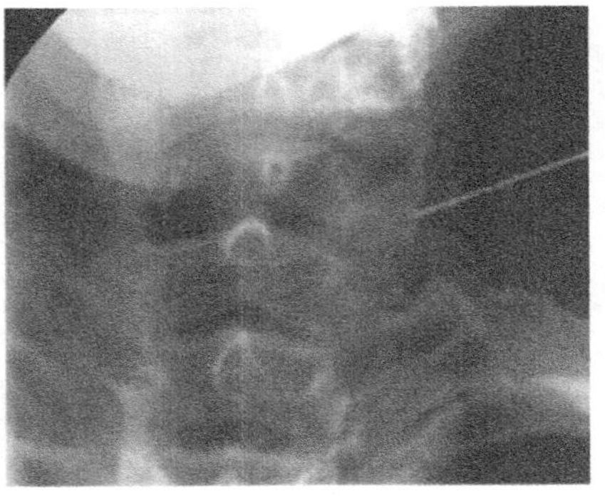

FIGURE 7 (**A**) Lateral and (**B**) anteroposterior projection. Extra-articular C7 medial branch block; lateral approach. The needle abuts the superior lateral C7 articular pillar cortex above the base of the C7 transverse process.

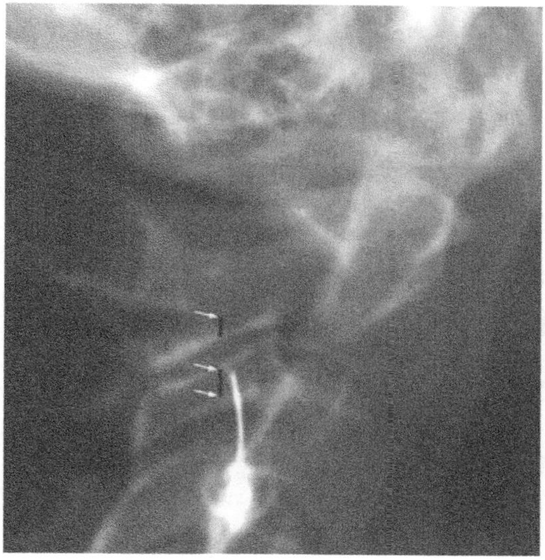

FIGURE 8 Lateral projection. Extra-articular third occipital nerve block. Lateral approach. The needle target is the lateral cortex superior to the joint, adjacent to the superior articular cortex, and then below the joint near the midline on lateral view. The course of this nerve is slightly variable and requires these additional sites to fully anesthetize it. The tip of the needle is withdrawn 1 to 2 mm before the injection of contrast and local anesthesia to avoid an intra-articular puncture.

block because this is the most likely source in 54% of such patients (3). This block should produce cutaneous anesthesia in the posterior occiput to be judged effective. In the case of an unsuccessful block at C2–C3, a C3 and C4 medial branch block would be the next most likely to yield a diagnosis, followed by a C1–C2 intra-articular block, if necessary. Lower neck pain is most likely to be referred from the segments C5 and C6 and less frequently from C6, C7 or C4, C5 (7).

The pain intensity response and the change in the daily activities are recorded immediately following the block, and the patient is then followed by a clinical or a telephone follow-up until pain has returned to the original level. A repeat confirmational block with another type of local anesthetic or block of additional levels is scheduled for a separate session.

Evidence of positive short term (relief lasting less than six weeks) and long term pain relief (six weeks or longer) of chronic cervical pain with intra-articular facet joint injections is limited, while in cervical medial branch blocks the evidence for short- and long-term pain relief blocks is moderate in multiple observational studies (21).

CERVICAL EPIDURAL BLOCK

Introduction

The use of epidural steroids in the cervical spine to treat chronic neck and upper extremity pain has been reported since 1984 (22). The translaminar route without fluoroscopic control has been utilized in most series (22–26). A criticism of this technique is the difficulty in assuming that the needle placement is in the epidural space as well as the risk of intradural puncture. More recently, the use of fluoroscopy and contrast media epidurography have been advocated by Johnson et al. to improve the safety and efficacy of the cervical epidural steroid injection (27). They performed 669 cervical epidural injections, mostly at C7–T1 over a 5.5-year-period in patients suffering neck pain with or without radiculopathy in a cohort study of 5334 patients receiving lumbar, cervical, and thoracic epidural steroid injections and reported only four complications in all the epidural injections. These consisted of transient hypotension; a small dorsal epidural hematoma, which resolved without intervention at 18-hours postinjection; and tachycardia and hypertension occurring 12 hours following a lumbar epidural injection. Only five patients were given intravenous (IV) conscious sedation due to severe anxiety before the procedure.

Direct cervical epidural needle puncture under fluoroscopy control with nonionic contrast epidurography for confirmation will allow a nonselective injection of steroid into the epidural space. The use of corticosteroid injection into the epidural space has been advocated to treat disc protrusion as well as chronic neck pain. Injectates typically ascend to the upper cervical spine and transverse the cervicothoracic region to T2 or lower (27).

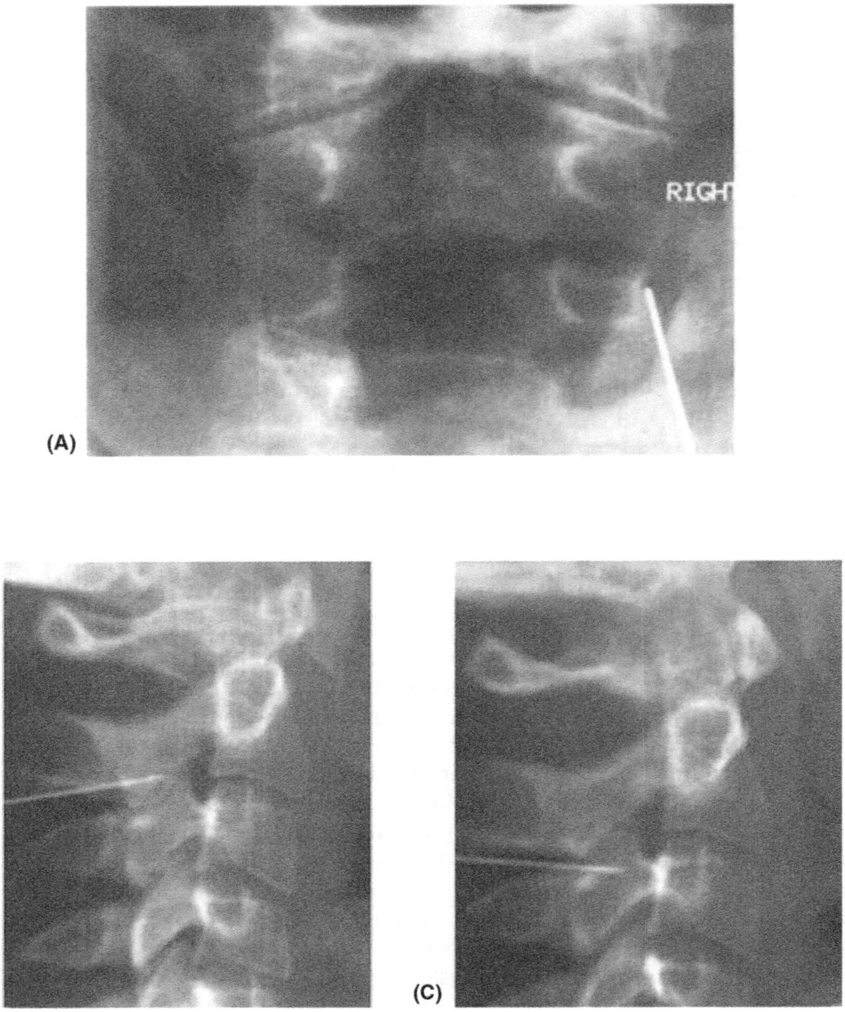

FIGURE 9 (**A**) Posteroanterior and (**B** and **C**) lateral projection. Third occipital nerve block; posterior approach. The needle target is the waist of the joint of C2, and once bone contact is made, the needle is advanced under lateral fluoroscopic control to positions above and below the joint.

TRANSLAMINAR EPIDURAL INJECTION

Technique

The patient is placed in a prone position, with the neck slightly flexed and the forehead supported on a small sponge. The target site is the lamina at C7–T1 because there is less likely to be central stenosis at this level and a larger epidural space. The lamina of C7 is easily seen with C-arm fluoroscopy. No injection is attempted at the site of previous posterior surgical laminectomy due to epidural scarring. Following aseptic preparation and sterile draping, a 22-gauge spinal needle is directed under fluoroscopy to impact the posterior lamina using a paramidline approach (Fig. 10). The depth of the needle is marked with a small forceps attached to the needle at the junction of the skin, while injecting 2 to 3 cc of local anesthetic into the adjacent lamina and the posterior cervical muscles as the needle is withdrawn. A 20- or 22-gauge epidural needle is then inserted into the anesthetized skin and down to the bone margin of the lamina adjacent to the interlaminar space. The needle depth is corrected to the predetermined length so as to bypass the lamina and then directed toward the midline at the interlaminar gap under fluoroscopic control. Because the ligamentum flavum is thin and affords little resistance to the injection, a small volume of contrast in a 3 or 5 cc syringe with a minimum volume

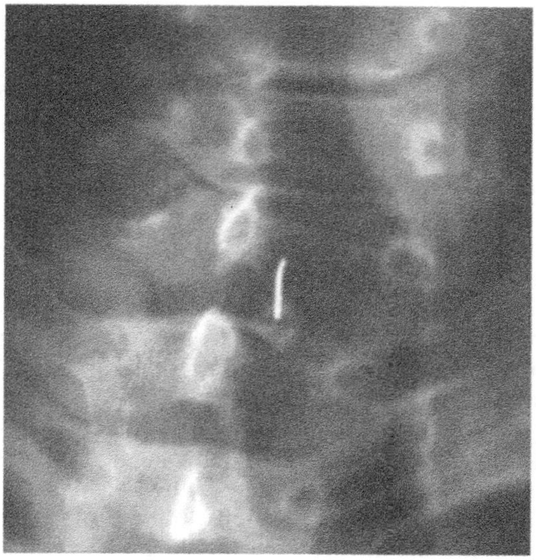

FIGURE 10 Posteroanterior projection. Translaminar C7–T1 epidural block; posterior approach with the patient prone. The needle target is the posterior lamina of C7. A 22- or 25-gauge needle is inserted to touch the posterior cortex of C7 via a paramedian approach and a small forceps is used to grasp the needle above the skin before needle removal to mark the depth of the lamina.

short-extension tubing is used to observe the contrast flow into the epidural space by keeping a small constant pressure on the syringe until the contrast flows freely away from the needle tip into the epidural space. The position is checked with fluoroscopy in the true PA and both oblique projections to show a linear configuration of the contrast against the dural sac and exclude a subarachnoid injection. Usually no more than 3 to 4 cc of contrast is necessary to confirm this position (Fig. 11). A subsequent injection of 6 to 12 mg of a preservative-free corticosteriod such as betamethasone is administered into the epidural space. The use of a short-acting local anesthetic such as lignocaine is optional because there is a risk of intradural injection.

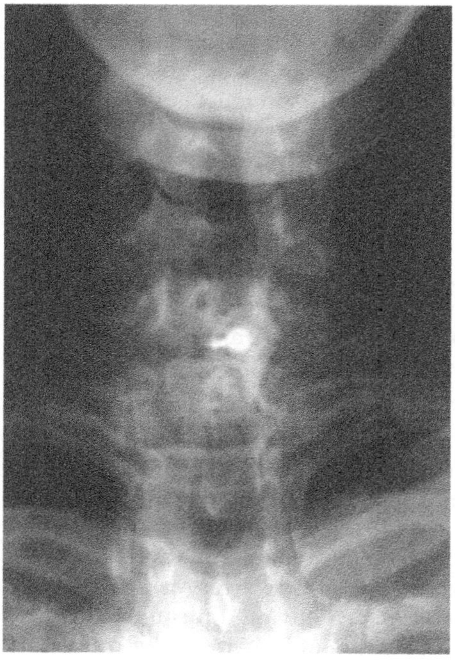

FIGURE 11 Posteroanterior projection. Translaminar C7–T1 epidural block; posterior approach with the patient prone. Contrast outlines the epidural space bilaterally from C6–T2.

Monitoring devices for blood pressure, EKG, and pulse oximetry are recommended. There should be oxygen, suction, and a ventilator available to manage cardiac and respiratory emergencies.

Complications

Inadvertent placement of the needle into the dural sac and the spinal cord is the most important complication. The use of C arm fluoroscopy and careful attention to the needle-tip position are paramount to avoid the injection of contrast or corticosteroid into the spinal cord. Also, there is a plethora of epidural veins, and intravenous injection of contrast will show these as tubular structures in the epidural space. Intrathecal or intravascular positioning of the needle following contrast injection is an immediate indication to withdraw the needle and abandon the procedures. Serious complications have recently been reported, including brainstem stroke and guadriparesis following routine translaminal cervical epidural steroid injections (28,29).

The injection of methylprednisolone acetate (MPA) and its vehicle-containing polyethylene glycol has been reported to cause archnoiditis, and this steroid has been shown to be neurotoxic in laboratory animal-tissue studies (30). However, a recent study of the histopathological effect following intrathecal injection of betamethasone showed no effect with small volumes of injectate into the thecal sac of sheep (31). These authors indicate that volumes of up to 2 mL injected intrathecally in humans are unlikely to cause arachnoiditis. There is no evidence that injection of steroids into the epidural space is associated with arachnoiditis (30,31).

No adhesive arachnoiditis has been reported in patients receiving only epidural corticosteroids (30,31). However, inadvertent dural puncture and injection of MPA will result in arachnoiditis (30,31).

An epidural abscess has been reported in a diabetic patient receiving epidural triamcinolone acetonide, and an epidural hematoma in a patient who concurrently received indomethacin. Reported side effects include allergic reaction to triamcinolone diacetate, transient cushingoid side effects, and flushing. Suppression of plasma cortisol levels has been documented for up to three weeks following the epidural injection of 80 mg of MPA. However the literature supports the safety of one to three epidural injections of low doses of steroids such as 40 to 80 mg of MPA or 50 mg of triamcinolone diacetate (32).

Outcome

The results of epidural cervical steroid injection have been reported to give pain relief in 38% of 96 patients with chronic neck pain over a period of one month or longer (22), and up to 41% excellent and 29% good pain relief at six months in 58 patients with chronic neck pain (23). Klein et al. reported that of 62 patients with cervical radiculopathy, pain relief persisted in 71% of patients for up to 18 months following one or two epidural steroid injections (33).

CERVICAL TRANSFORAMINAL INJECTION

Introduction

The placement of an image-guided needle into the cervical foramen for the diagnostic delivery of local anesthetic and the therapeutic injection of corticosteroid has been advocated in the diagnosis and treatment of severe cervical radicular pain and disc protrusion or cervical spondylosis and stenosis that are resistant to conservative medical and physical therapy (34).

Patient Selection

The selection of patients for transforaminal block is usually reserved for those patients with radicular pain who do not respond to conservative management and who are likely candidates for surgical treatment.

These injections are technically demanding and potentially dangerous, and the inadvertent intra-arterial injection of particulate steroids into a radicular artery is a postulated mechanism for spinal cord infarction (29,30,35). Therefore patient monitoring with pulse oximetry and EKG as well as the availability of resuscitation equipment in the procedure room is necessary. An IV is recommended before the procedure to ensure venous access, and the injectionist must ensure availability of personnel trained in these resuscitative procedures. Conscious sedation with short-acting IV Midazolam (Versed) may be used in small incremental doses of 0.5 to 5 mg to reduce anxiety and improve the patient's tolerance if required. Anticoagulants including aspirin, coumadin, and nonsteroidal anti-inflammatory drugs should be stopped five days before the procedure.

Absolute contraindications for transforaminal injections include untreated localized infection, bleeding diathesis, and inability of the patient to remain immobile during the procedure. A relative contraindication is a known systemic infection or an immunosupressed patient. An anatomical bone derangement such as a previous fusion with bone graft or marked osteophyte formation may prevent needle entry into the foramen.

The use of contrast is mandatory to exclude intravascular or subarachnoid injection. Digital subtraction angiography in the frontal plane is advocated to exclude inadvertent radicular artery injection (36). Only nonionic contrast media such as omnipaque (iohexol) or isovue (iopamidol) and local anesthetic agents that are preservative- and methylparabene-free such as lidocaine HCl should be used.

Procedure and Technique for Transforaminal Injection

A high-resolution C arm fluoroscopy is the preferred imaging device so that AP, oblique, and lateral imaging planes can be assessed while the needle is advanced.

Under fluoroscopy the patient is positioned in a supine position or a steep oblique lateral decubitus position, so that the cervical intervertebral canal and foramina are seen in multiple projections. The ideal position of the target canal is at the widest diameter with the fluoroscope angled slightly caudally along the axis of the canal to achieve this. After accurate count to the target canal and sterile skin preparation, a 25-gauge or 22-gauge short-beveled spinal needle is positioned, so that the target is the anterior cortex of the superior articular pillar adjacent to the middle of the posterior wall of the canal. In the latter instance, the needle position must be changed by withdrawal and reentry at a slightly different plane.

The needle is then introduced along the central ray in small increments and directed to the bone cortex of the mid-posterior canal. Because the exit zone of the canal is quite superficial and the length of the canal is short, frequent small-needle increments with interval fluoroscopic screening are necessary before the needle reaches the bone. Fluoroscopy in the true AP position will determine the accurate depth of the needle, which can be then repositioned to see the needle position in the oblique and lateral view abutting the anterior cortex of the articular pillar posterior to the intervertebral canal (Fig. 12). Progressive entry of the needle into the canal is performed under AP fluoroscopic control, so that the tip does not pass beyond the

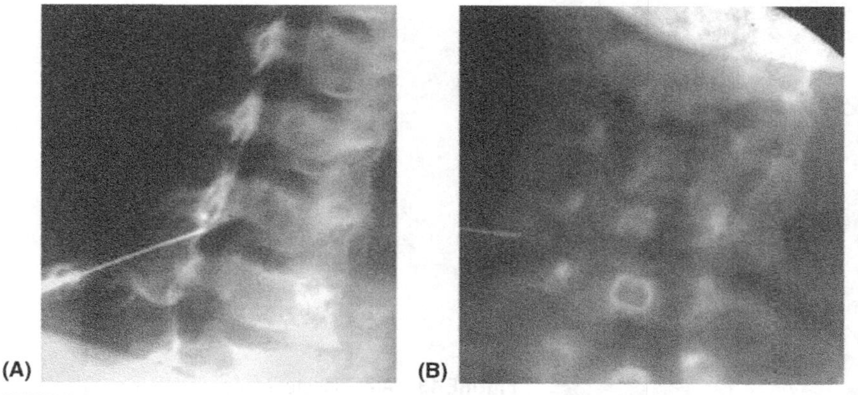

FIGURE 12 (**A**) Oblique projection and (**B**) anteroposterior (AP) projection. Selective C8 epidural nerve block. The target is the posterior cortical wall of the C7–T1 intervertebral foramen in the oblique projection. The needle tip should not project anterior to the articular pillar to avoid piercing the vertebral artery. The needle is then advanced under AP fluoroscopic guidance lateral to the pedicles and uncinate processes to avoid piercing the dura.

lateral aspect of the intervertebral canal. Induction of severe radicular pain or intense paresthesia indicates the needle must be repositioned before any injection is performed. Spot film or image capture in true AP, oblique and lateral projections should document the position of the needle tip before the injection of contrast. To prevent needle movement, a short-length of minimal volume extension tubing is attached to the needle hub. Syringe aspiration will infrequently demonstrate blood or cerebraspinal fluid because of the small needle size. Therefore injection of 1.0 to 1.5 cc of contrast in a 3.0 cc syringe should be performed under real-time fluoroscopy with digital subtraction angiographic imaging in the antero-posterior plane to exclude inadvertent intra-arterial radicular arterial filling into the spinal canal (36). The contrast will typically outline the spinal nerve and flow into the adjacent epidural space (Fig. 13). Contrast may also follow the anterior spinal ramus distally. Rapid upward clearance indicates vertebral artery filling, and the needle must be repositioned without further injection. Venous filling of the radicular, epidural, and paraspinous veins leads to a slow washout of contrast usually directed caudally. Rotation of the needle and withdrawl to reposition the needle tip should be followed by a repeat digital subtraction angiogram to exclude intravascular filling. The procedure should be aborted with contrast opacification of a radicular artery. Injection of a small volume of lidocaine 2% (0.5–1.0 ml) requires observation of the non sedated patients for adverse neurological effect for several minutes and serves as a diagnostic block.

The dural sleeve extends along the nerve roots to the entrance zone of the foramen. A rapid dilution of contrast usually indicates that the needle tip has punctured the dura and is within the subarachnoid space. Withdrawal of the needle is then performed without further injection and consideration should also be given to abandoning the procedure and rescheduling it.

Bupivacaine and longer-acting anesthetics should be avoided because of the possibility of an unintended subarachnoid injection and possible respiratory arrest. Corticosteroids are given for therapeutic purposes in doses of 3.0 to 6.0 mg of betamethasone or 20 to 40 mg of triamcinolone for a single level. Dreyfuss et al. have demonstrated comparatively similar effectiveness of nonparticulate and particulate steriod preparations. They employed 12.5 mg dexamethasone sodium phosphate and found similar therapeutic responses to an injectate of 60 mg of triamcinolone (37).

CT scan or fluoroscopic CT-guided cervical transforaminal blocks have been advocated by some authors (38), and are useful in large or obese patients, especially at the lower levels (C6–C7; C7–T1), which may be partially obscured by the shoulders or if there is marked bone

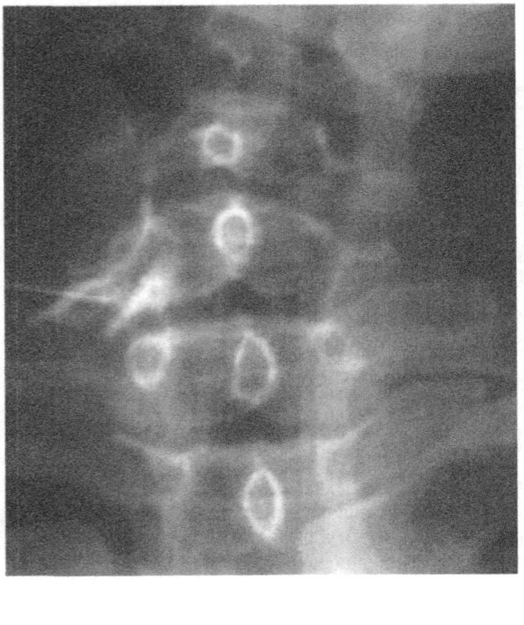

FIGURE 13 Anteroposterior view. Selective C8 epidural nerve block. A 22-gauge needle tip within the midportion of the C7–T1 intervertebral foramen. Contrast outlines the C8 spinal nerve within the canal and the adjacent medial epidural space.

hypertrophy or bone spurs from the adjacent zyagapophyseal joint. The principle of targeting the posterior wall cortex and the use of nonionic contrast to exclude intravascular injection must still be adhered to (Fig. 14A and B). The alignment of the carotid vessels and the vertebral artery to the foramen will be variable and may be in the path of the needle (Fig. 14C and D), particularly at C4–C5 and C5–C6. The use of a 25-gauge spinal needle decreases the risk of intravascular damage.

A diagnostic block with local anesthesia is achieved by injecting a small volume (0.5–1.5 cc) of preservative-free (single dose) lidocaine 1% into the perineural epidural space. Marcaine and longer-acting anesthetics should be avoided because of the possibility of an unintended subarachnoid injection and possible respiratory arrest. Corticosteroids are given for therapeutic purposes in doses of 3.0 to 6.0 mg of betamethasone or 20 to 40 mg of triamcinolone for a single level.

Complications

It is important for the injectionist to be aware that the vertebral artery lies anterior to the spinal nerve and anterior ramus within the foramen (Fig. 14). A radicular artery is present anterior to the nerve and may have a communication with the anterior spinal artery supplying the spinal cord.

The most serious complication to avoid is the injection of particulate corticosteriod into the anterior spinal artery, which can lead to irreversible and sudden central cord infarction (39,40). The inadvertent injection of local anesthetic in the subarachnoid space may induce high spinal anesthesia with potential pulmonary arrest. The perforation of an epidural vein can also lead to epidural hematoma (21).

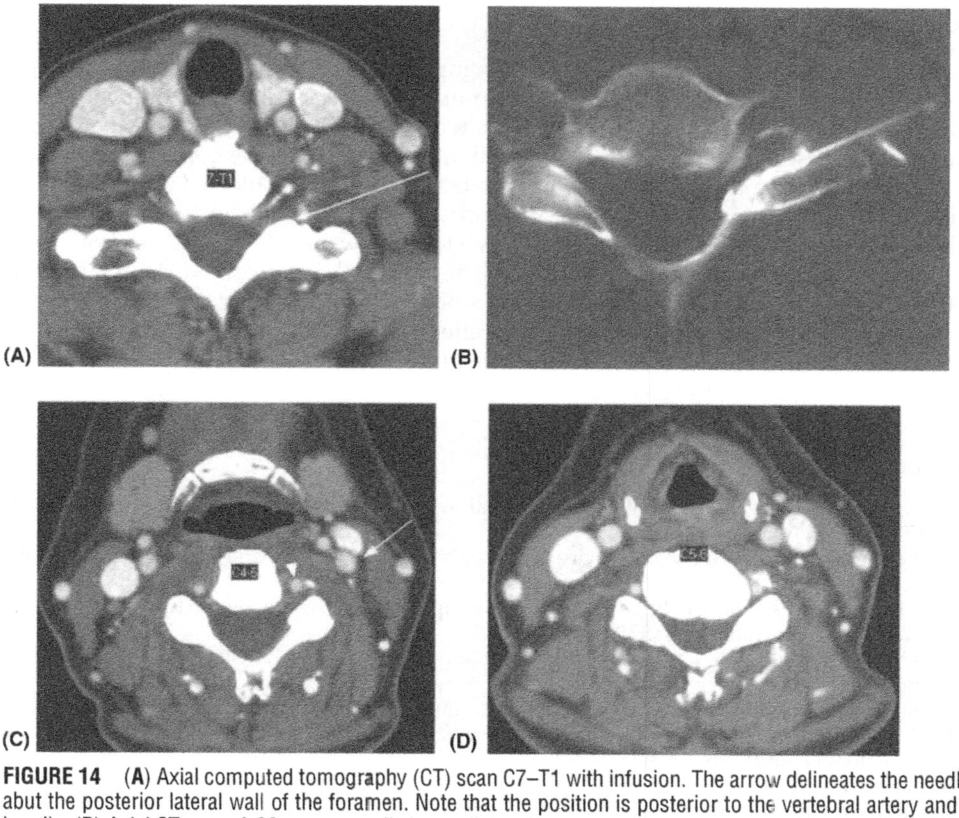

FIGURE 14 (**A**) Axial computed tomography (CT) scan C7–T1 with infusion. The arrow delineates the needle course to abut the posterior lateral wall of the foramen. Note that the position is posterior to the vertebral artery and the carotid bundle. (**B**) Axial CT scan. A 22-gauge needle is parallel to the posterior cortex of the foramen. Contrast outlines the C8 spinal nerve before administration of the corticosteroid. (**C**) Axial CT scan C4–C5. The carotid vascular bundle (*arrow*) is lateral to the foramen, and the vertebral artery (*arrowhead*) is anterior to the foramen. (**D**) Axial CT scan at C5–C6. Severe foraminal stenosis. Note the close proximity of the vertebral artery to the posterior wall cortex.

Direct puncture of the spinal cord has been reported by Hodges et al. (41). In their report, the needle tip was placed into the spinal canal without the injectionist being aware of its initial position prior to the injection of a corticosteroid due to incomplete fluoroscopic assessment, AP fluoroscopic assessment. In a series of 1036 fluoroscopically guided extraforaminal cervical nerve blocks in 844 patients Ma et al. reported only 1.66% minor complications with this procedure without catastrophic complications (42).

Outcome

A prospective study by Bush and Hillier (34) in 68 patients with cervical monoradiculopathy receiving intraforaminal periradicular epidural steroids noted that of 42 patients treated by selective epidural nerve blocks, 76% were completely asymptomatic at average of seven months postinjection of 40 mg of triamcinolone acetonide. All these patients had radicular pain of the upper extremity with associated paresthesia of the fingers or thumb and positive neurological signs, and 56 patients had an MRI or a CT confirmation of nerve compression by disc hernia or bone stenosis.

Vallee et al. reported that in 32 patients unresponsive to medical therapy for at least two months (mean 9.5), fluoroscopic image-guided periradicular injection of 50 mg of prednisolone to treat chronic upper extremity radicular pain in the distribution of C6, C7, and C8 produced good or excellent pain relief in 18 patients (56%) and was unsuccessful in 14 patients (43%) (43).

Good or excellent results in 60% of 20 patients with cervical spondylotic radicular pain receiving one to two selective fluoroscopic-guided nerve root blocks was reported by Slipman et al. (44–47).

Post-procedure Care

The patient should be observed immediately following the procedure and the pain intensity response should be reassessed with a VAS and pain map. A successful block will completely relieve upper extremity pain. There may be associated sensory changes to the upper extremity, which usually resolve in a few hours. An evaluation of the pain response is best made by a practitioner other than the injectionist immediately following the block and at a delayed reevaluation in one or two days. A significant improvement which is short term or incomplete can lead to a repeat steroid injection within two to three weeks. If there is no short-term response to the corticosteroid, a repeat injection should not be performed. However, the absence of pain for the expected duration of the anesthetic is diagnostic for the site of the pain generator. A failure to provide complete symptom relief in the distribution of the spinal nerve implies that other pain generators should be considered with diagnostic cervical medial branch blocks or cervical discography.

REFERENCES

1. Bogduk N, Marsland A. The cervical zygapophysial joints as a source of neck pain. Spine 1988; 13:610–617.
2. Aprill C, Bogduk N. The prevalence of cervical zygapophyseal joint pain: a first approximation. Spine 1992; 17:744–747.
3. Lord SM, Barnsley L, Wallis BT, Bogduk N. Third occipital nerve headaches: a prevalence study. J Neurol Neurosurg Psychiatr 1994; 57:1187–1190.
4. Duyer A, Aprill C, Bogduk N. Cervical zygophyseal joint pain patterns 1: a study in normal volunteers. Spine 1990; 15:453–457.
5. Bogduk N. The clinical anatomy of the cervical dorsal rami. Spine 1982; 7:319–330.
6. Barnsley L, Bogduk N. Medical branch blocks are specific for the diagram of cervical zygopophyseal joint pain. Regional Anesthesis 1993; 18:343–350.
7. Lord S, Barnsley L, Wallis BJ, Bogduk N. Chronic cervical zygopophyseal joint pain whiplash: a placebo-controlled prevalence study. Spine 1996; 21:1737–1745.
8. Speldewinde GC, Bashford GM, Davidson IR. Diagnostic cervical zygaphysial joint blocks for chronic cervical pain. Med J Aust 2001; 174:174–176.
9. Fukui S, Ohsetok, Shiotani M, et al. Referred pain distribution of the cervical zyagapo physeal joints and cervical dorsal rami. Pain 1996; 68:79–83.

10. Lord SM, Barnsley L, Bogduk N. The utility of comparative local anesthetic blocks versus placebo-controlled blocks for the diagnosis of cervical zygapophysial pain. Clin J Pain 1995; 11:208–213.
11. Barnsley L, Lord SM, Wallis BJ, Bogduk N. Lack of effect of intraarticular corticosteroids for chronic pain in the cervical zygapophyseal joints. N Engl J Med 1994; 330:1047–1050.
12. Boswell MV, Colson JD, Sehgal N, Dunbar EE, Epter R. A systematic review of therapeutic facet joint interventions in chronic spinal pain. Pain Physician. 2007; 10:229–253.
13. Dreyfuss P, Michaelson M, Fletcher D. Atlanto-occipital and lateral atlanto-axial joint pain patterns. Spine 1994; 10:1125–1131.
14. Grubb SA, Kelley CK. Cervical discography: clinical implication from 12 years of experience. Spine 2000; 25:1382–1389.
15. Margolis RB, Tait RC, Krause SJ. A rating system for use with patient pain drawings. Pain 1986; 24:57–65.
16. Jensen MP, Karoly P, Braver S. The measurement of clinical pain intensity: a comparison of six methods. Pain 1986; 27:117–126.
17. Dory MA. Arthrography of the cervical facet joints. Radiology 1983; 148:379–382.
18. Bogduk N, Aprill C, Derly R. Diagnostic blocks of spinal synovial joints. In: Spine Care Diagnosis and Conservative Treatment. Vol. 1. Ed. White, Mosby, St. Louis 1995:298–321.
19. Silbergleit R, Mehta BA, Sanders WP, Talati SJ. Imaging guided injection techniques with fluoroscopy and CT for spinal pain management. Radiographics 2001; 21:927–942.
20. Dreyfuss P, Rogen J, Dreyer S, Fletcher D. Atlanto-occiptal joint pain. A report of three cases and description of an intraarticular joint block technique. Reg Anesth 1994; 19:344–351.
21. Slipman CW, Lipetz JS, Jackson HB, Rogers DP, Vresilovic EJ. Therapeutic selective nerve root block in the nonsurgical treatment of atraumatic cervical spondylotic radicular pain: a retrospective analysis with independent clinical review. Arch Phys Med Rehabil 2000; 81:741–746.
22. Cathelove RF, Bratia R. The use of cervical epidural nerve block in the management of chronic head and neck pain. Can Anesth Soc J 1984; 31:188–191.
23. Shulman M. Treatment of neck pain with cervical epidural steroid injection. Reg Anesth 1986; 11:92–94.
24. Cicala RS, Thoni K, Angel J. Long term results of cervical epidural steroid injection. Reg Anesth 1986; 5:143–145.
25. Rowlingson JC, Kirschenbaum LF. Epidural analgesic techniques in the management of cervical pain. Anesth Analg 1986; 65:938–942.
26. Ferrante FM, Wilson SP, Iacoba C, et al. Clinical classification as a prediction of therapeutic outcome after cervical epidural steroid injection. Spine 1993; 18:730–736.
27. Johnson BA, Schellhar KP, Pollie SR. Epidurography and therapeutic injection: technical considerations and experience with 5334 cases. AJNR 1999; 20:697–705.
28. Bose B, Quadriparesis following cervical epidural steroid injections: case report and review of the literature. The Spine Journal 2005; 5:558–563.
29. Ziai W, Ardelt A, Llinas R, Brainstem Stroke Following Uncomplicated Cervical Epidural Steroid Injection. Archives of Neurology 2006; 63:1643–1646.
30. Abram SE, O'Connor TC. Complications associated with epidural steroid injection. Reg Anesth 1996; 21:149–162.
31. Klein GR, Vaccaro AR, Curk J, et al. Efficacy of cervical epidural steroids in the treatment of cervical spine disorders. Am J Anesthesiol 2000; 27:547–550.
32. Bush K, Hillier S. Outcome of cervical radiculopathy treated with periradicular epidural corticosteroid injections: a prospective study with independent clinical review. Eur Spine J 1996; 5:319–325.
33. Vallee JN, Feydy A, Cartier RY. et al. Chronic cervical radiculopathy: lateral approach periradicular corticosteroid injection. Radiology 2001; 218:886–892.
34. Cyteval C, Thomas E, Decoux E, et al. Cervical Radiculopathy: Open study on percutaneous periradicular foraminal steroid infiltration performed under CT control in 30 patients. AJNR 2004. 25: 441–445.
35. Ludwig M, Burns S. Spinal cord infarction following cervical transforaminal epidural injection. Spine 2005; 30:E266–E268.
36. Hodges SD, Castleberg RL, Miller T, et al. Cervical epidural steroid injection with intrinsic spinal cord damage. Two case reports. Spine 1998; 23(19); 2137–2142.
37. Rathmell J, Aprill C, Bogduk N. Cervical transforaminal injection of steroids. Anesthesiology 2004; 100:1595–1600.
38. Williams KN, Jackowski A, Evans PJ. Epidural hematoma requiring surgical decompression following repeated cervical epidural steroid injections for chronic pain. Pain 1990; 42:198–219.
39. Hodges SD, Carteberg RL, Miller T, et al. Cervical epidural steroid injection with intrinsic spinal cord damage. Spine 1998; 23:2137–2142.
40. Suresh S, Berman J, Connell DA. Cerebellar and brainstem infarction as a complication of CT-guided transforaminal cervical nerve root block. Skeletal Radiol 2007; 36:449–452.
41. Baker R, Dreyfuss P, Mercer S, et al. Cervical transforaminal injection of corticosteroids into a radicular artery: a possible mechanism for spinal cord injury. Pain 2003; 103:211–215.

42. Ma D, Gilula L, Riew K. Complications of Fluoroscopically Guided Extraforaminal Cervical Nerve Blocks. J Bone Joint Surg Am 2005; 87:1025–1030.
43. Brouwers PS, Kohink ET, Simon MA, et al. Anterior spinal artery syndrome after diagnostic blockade of the right C6 nerve root. Pain 2001; 3:397–399.
44. Aprill C, Axinn MJ, Bogduk N. Occipital headaches stemming from the lateral axial (C1–C2) joint. Cephalalgia 2002; 22:15–22.
45. Boswell MV, Colson JD, Sehgal N, et al. A systematic review of therapeutic facet joint interventions in chronic spinal pain. Pain Physician 2007; 10:229–253.
46. Nelson D. Intraspinal therapy using methyprednislone acetate. Spine 1993; 18:278–286.
47. Latham SM, Fraser RD, Moore JR, et al. The pathologic effects of intrathecal betamethasone. Spine 1997; 22:1558–1562.
48. Boswell MV, Colson JD, Sehgal N, Dunbar EE, Epter R. A systematic review of therapeutic facet joint interventions in chronic spinal pain. Pain Physician 2007; 10:229–253.
49. Dreyfuss P, Baker R, Bogduk N. Comparative effectiveness of cervical transforaminal injections with particulate and nonparticulate corticosteroid preparations for cervical radicular pain. Pain Med 2006; 7:237–242.

11 Selective Sacroiliac Joint and Facet Joint Injections

John A. Carrino
Department of Radiology and Radiological Science, Johns Hopkins University School of Medicine, Baltimore, Maryland, U.S.A.

SACROILIAC JOINT INTERACTIONS
Introduction

The sacroiliac (SI) joint can be a primary source of low back pain. More often, it is a secondary site or part of a multifactorial syndrome from dysfunction elsewhere in the spine. SI joint injections provide diagnostic information and potential therapy in certain circumstances.

The exact prevalence of SI joint dysfunction is unknown because of difficulty in establishing a reference (criterion) standard. There is a high degree of mechanical interdependence in this region of the body and this concept is reflected in the term "lumbo-pelvic-hip complex." Etiologies other than biomechanical dysfunction affect the SI joint and include trauma (diastasis), inflammation (infectious or noninfectious arthropathies), or neoplasia. Tumor transgression may occur across the fibrous portion of the SI joint, but similar to other synovial joints the articular cartilage is a relative boundary to neoplastic invasion. Abnormalities of adjacent structures such as the sacrum (stress fractures) may alter SI joint biomechanics. Iatrogenic causes such as a large bone defect from a graft harvest site may also cause SI joint–mediated pain.

Anatomy

The sacrum is a block of bone at the base of the vertebral column. The sacrum is the largest vertebral element, consisting of five fused vertebrae and is the place where the spine meets the pelvis vis-a-vis the SI joint. It supports the spine and transmits load to the lower extremities. It has a triangular shape, which is broader cephalad and tapers caudally. There are foramina anteriorly and posteriorly. It provides two roles with respect to dispersing forces longitudinally and transversely. The lateral margins of the sacrum have two regions. One area is an auricle-shaped (ear-like) surface, which is smooth and provides the articulation for the SI joint. The other area is ligamentous, which is rough and provides the interosseous portion of the SI joint. The SI joint itself has a small range of motion. There is no muscle that executes active movements across the SI joint. All of the movements are passive and thus it is considered a "stress relieving" type of articulation. Nevertheless, it is a complex set of movements. The axis of movement passes obliquely across the pelvis. In flexion, the axis passes backward from the pubic symphysis to the sciatic notch. In extension, the axis passes from the pubic symphysis through the pelvis between the ischium and coccyx. There are normal age-related (senescent) changes of the SI joint. In the embryo, the SI joint starts as a strip of mesenchyme that then goes on to cavitate. During the first decade of life the joint enlarges and the surface is flat. During the second decade there begins corrugation of the joint surfaces. During the fifth and sixth decades, osteophytes develop. By the eighth decade, large interdigitizing osteophytes may be identified. These imaging findings may be part of the normal aging process and there is no correlation between degenerative type findings of the SI joints and symptomatology (particularly over the age of 50).

Indications/Rationale

Primary SI joint pain is noted to be maximal below the level of L5, but on palpation sacral sulcus tenderness is present. However clinical findings are not reliable (1). There are not any specific history or physical examination findings that can accurately identify the SI joint as a source of pain.

Using diagnostic intra-articular blocks producing temporary symptomatic relief as the reference (criterion) standard, the prevalence of primary SI joint pain and chronic low back pain is in the range of 18% to 30% based on two studies (2,3). The SI joint has a diffuse innervation pattern without a fixed course for the efferent nerves. Therefore, there is no effective nerve block for the SI joint and only intra-articular injections can selectively anesthetize the SI joint. One confounder may be the degree of pain the patient is in at the time of the injection. If the patient is not in a high level of pain, then the opportunity for demonstrating dramatic improvement is lessened. Control injections (control blocks) are useful in mitigating the placebo effect. Because of the risks of a false positive response, the placebo injection of normal saline at another time would be useful to show no improvement. Alternatively, anesthetic agents of different time durations (lidocaine vs. bupivacaine) on separate occasions provide comparative data. The duration of pain relief should be comparable to the anticipated duration of the agent. For example, a lidocaine block should last a couple of hours, whereas a bupivacaine block should last several hours and a saline injection should have minimal effect. It has also been postulated that some of the pain mediated from the SI joint may be via communication with adjacent nerve structures. Fortin et al. (4) found five principal patterns of extracapsular contrast extravasation, with three of these patterns having a potential pathway of communication between the SI joint and nearby neural structures. These included posterior extension into the dorsal sacral foramina, superior recess extravasation at the sacral alar level to the fifth lumbar epiradicular sheath, and ventral leakage to the lumbosacral plexus. They purport the concept that inflammatory mediators could leak from the SI joint along these pathways and cause irritation of adjacent neurological structures. This then could in part explain how SI joint dysfunction may manifest as lower extremity symptoms. Nevertheless, this does not negate the potential for pain mediation via direct subchondral bone irritation through cartilage loss, altered biomechanics, and marrow edema as it occurs in other synovial-type articulations. Pain referral maps have been successfully produced using intra-articular injections in asymptomatic people (5). In this study, a physical examination immediately after SI joint injection revealed an area of buttock hyperesthesia extending about 10 cm caudally and 3 cm laterally from the posterior superior iliac spine. This map would explain the referral distribution based on primary SI joint pathology.

Therefore, the indications for SI joint injection are to identify, diagnose, and treat pain putatively if it is of SI joint origin. Absolute contraindications to the procedure include infection (unless this is done as a diagnostic arthrocentesis), uncontrollable bleeding diathesis, and pregnancy. Relative contraindications include patients on anticoagulant therapy and history of severe contrast reaction.

Technique

The SI joint may be injected under fluoroscopic or cross-sectional imaging guidance. Computed tomography (CT) is the most common cross-sectional modality used, however magnetic resonance (MR)-guided SI joints have been described in a technical note (6). The SI joint can be one of the most difficult joints to access, and therefore there have been numerous technical descriptions, and the preferred modality is often determined on the basis of operator experience and equipment availability.

Several fluoroscopic techniques have been described. In general, the patient is placed prone on a radiolucent table. The target zone is the inferior, posterior aspect of the SI joint, about 1 to 2 cm proximal to the most caudal margin of the joint. Because the sacrum is broader anteriorly than posteriorly, two joints may be visualized as defined by a thin cortical opaque line on each side of the articulation. Under direct posteroanterior (PA) projection the more medial joint is posterior and the lateral joint is anterior. Some operators use this projection and target the medial joint often hitting the sacrum and then creeping along the margin until the needle encounters the joint space (Fig. 1). In the oblique approach (7), the C-arm is rotated medially to profile the joint with supraimposition of all the joint margins. One would then simply target these superimposed joints at their inferior aspect (Fig. 2). However, this technique often fails to allow entry into the SI joint.

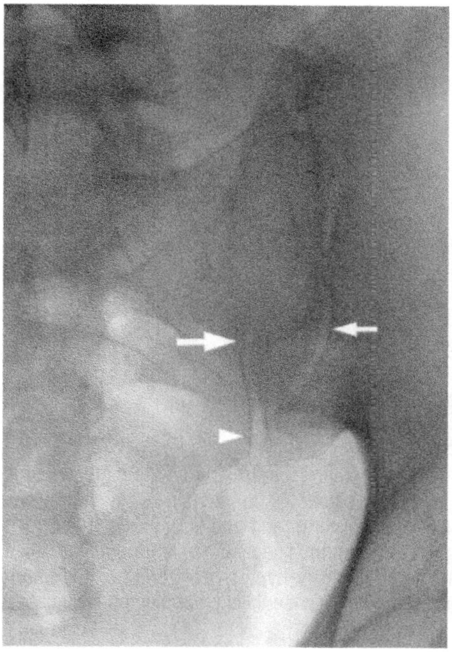

FIGURE 1 Sacroiliac (SI) joint anatomy. The target zone is the inferior, posterior aspect of the SI joint about 1 to 2 cm proximal to the most caudal margin of the joint (*arrowhead*). Because the sacrum is broader anteriorly then posteriorly, two joints may be visualized. Under a direct posteroanterior projection, the more medial joint is posterior (*large arrow*) and the more lateral joint is anterior (*small arrow*).

Others have developed modifications using the advantage of multidirectional C-arm fluoroscopy. The technique described by Dreyfuss et al. (8) is a modification of the oblique approach. The C-arm is rotated medially (contralateral to the affected side or side of interest) to profile the joint. This orients the anterior and posterior planes to be parallel to the central ray. A single lucent zone is produced inferiorly, which reflects the confluence of anterior and posterior joints. The C-arm is then rotated back in the opposite direction. Rotation is continued until "optimal separation" is produced. This is defined by a well-visualized posterior (medial) joint with thin sclerotic margins and an inferior hyperlucent zone. Typically it requires about 5° to 20° of rotation to produce this configuration (Fig. 3). Another modification has been described by Dussault et al. (9), again using a rotating C-arm. Initially the X-ray tube is perpendicular to the table and the skin is marked over the distal 1 cm of the SI joint. The tube is then angled approximately 20° to 25° in a cephalic direction to displace the posteroinferior portion of the

(A) 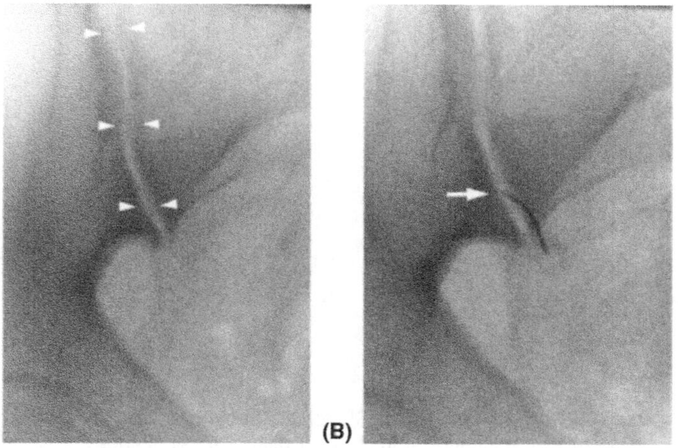 (B)

FIGURE 2 Sacroiliac joint oblique approach: (**A**) the C-arm is rotated medially to profile the joint with superimposition of all the joint margins (*arrowheads*); (**B**) the needle is placed in the target zone at the inferior aspect or mid-portion of the joint (*arrow*).

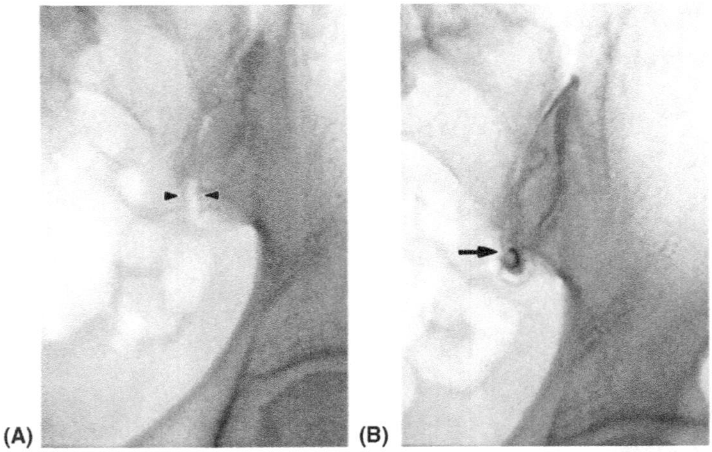

FIGURE 3 Sacroiliac joint modified oblique approach. The C-arm is rotated medially (contralateral to the affected side or side of interest) to profile the joint. This orients the anterior and posterior planes to be parallel to the central ray. A single lucent zone is produced inferiorly, which reflects the confluence of the anterior and posterior joints. The C-arm is then rotated back in the opposite direction, and (**A**) rotation is continued until "optimal separation" is produced defined by a well-visualized posterior (medial) joint with thin sclerotic margins and an inferior hyperlucent zone (*arrowheads*). Typically it requires about 5° to 20° of rotation to produce this configuration. (**B**) The needle is placed in the target zone at the inferior aspect of the joint (*arrow*).

joint in a caudal direction. The needle is advanced through the original site marked on the skin, but the tube is maintained in the cephalic position. The needle is advanced toward the posterior SI joint without angling of the needle (i.e., maintained perpendicular to the table). The needle is advanced until it encounters the SI joint.

In all of these techniques, SI joint arthrography is performed to confirm an intra-articular location (Fig. 4). Often a small inferior caudal recess is opacified with a variable portion of the synovial joint. It is uncommon to opacify the entire SI joint. Also common in most techniques is that the needle encounters the sacral side, avoiding being too low or too lateral because of the proximity of the sciatic notch (sciatic nerve). From here the needle is migrated into the joint until one feels a depression or a "pop," then advancing it 1 to 2 mm. When the inferior joint is

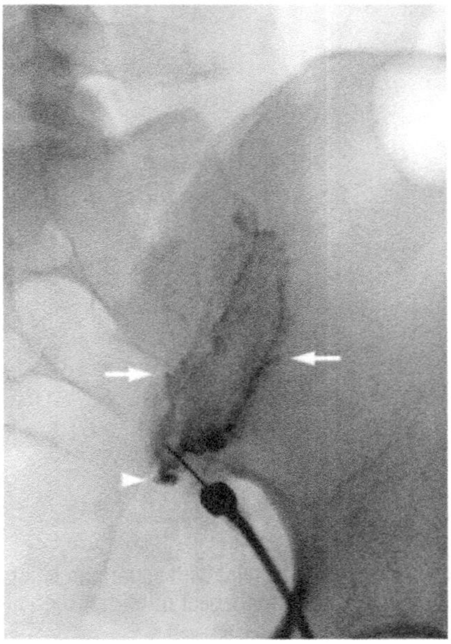

FIGURE 4 Sacroiliac (SI) joint arthrography. It is uncommon to opacify the entire SI joint. Most often a small inferior caudal recess is opacified (*arrowhead*). This arthrogram also shows synovitis with interdigitating of contrast between proliferative fronds of synovium (*arrows*).

not accessible under fluoroscopy, then most would advocate injecting the mid-portion of the joint (more cephalad). Alternatively, a cross-sectional imaging technique can be used if attempts using fluoroscopy are not adequate.

The cross-sectional imaging technique that is used most often is CT (Fig. 5). The patient is placed prone on the gantry and axial images are obtained through the SI joint. Again the inferior or mid-portion of the joint is targeted. The needle is advanced until the tip is in the synovial portion of the joint (10). The injection of contrast material is not mandatory if one feels that there is an adequate position based on the imaging. A similar technique can be used with MR imaging. Despite the absence of ionizing radiation, the lesser availability and higher cost of MR precludes its widespread use. In addition, special consideration is needed with respect to MR-compatible equipment and needles. While many needles may be safe (i.e., no significant migration) for use in the magnet environment, imaging can be problematic unless special needles are used. In terms of modality preference, it is the author's experience to first attempt SI joint injections using fluoroscopy. If this is unsuccessful then CT is employed. Overall, the majority of SI joints should be fluoroscopy accessible if one uses one of the modified C-arm techniques described above.

The SI joint may be injected with a 22-, 23-, or 25-gauge needle. The needle length depends on the patient's body habitus. Larger needles (20 gauge) may be used for arthrocentesis if there is a suspected infection. No intravenous line, premedication, or physiologic monitoring is required. The contrast material should be suitable for intrathecal administration. The volume of contrast injected should be between 0.3 and 1.0 mL maximum. This is to allow enough residual volume within the joint for anesthetic injection. In a small number of patients, it may be impossible to obtain an arthrogram. Other flow patterns to be cognizant of are venous and extracapsular opacification. With venous opacification, there are small tributaries identified flowing away from the SI joint. Extracapsular injections are characterized by irregular collections or a "blob" pattern. These may occur concurrently with an intra-articular injection. The needle tip may be advanced or rotated slightly in this circumstance. To best visualize the arthrogram, an anteroposterior or slight oblique projection can be used to profile the structures of interest. If the needle tip is too anterior it may hit an artery but this is extremely uncommon. Degenerative joint disease can limit the amount of contrast injected. Several extravasation patterns have been described by Fortin et al. (4). They are characterized as fascia, piriformis, sacral foramina (anterior and posterior), and lumbosacral plexus extravasation patterns. Similar to other spine injections, the injectate consists of a short-acting anesthetic and/or a corticosteroid. Injection is performed until there is a firm end point or extracapsular extravasation. The total volume should not exceed 2.5 to 3 cc.

The interpretation of this procedure is based primarily upon pain relief and not the provocative aspect. Concordant pain production during SI joint injection is not validated as a diagnostic test confirming pain of SI joint etiology. Practically, the assessment should be before the injection, immediately subsequent, and approximately 15 to 30 minutes after the injection. The anesthetic response is most often assessed using a visual analog scale (VAS) and or percent pain relieved. The result is considered positive (good) if greater than 75% of pain is relieved and negative if less than 50% of the pain is relieved. Pain reduction in the range of 51% to 74% is

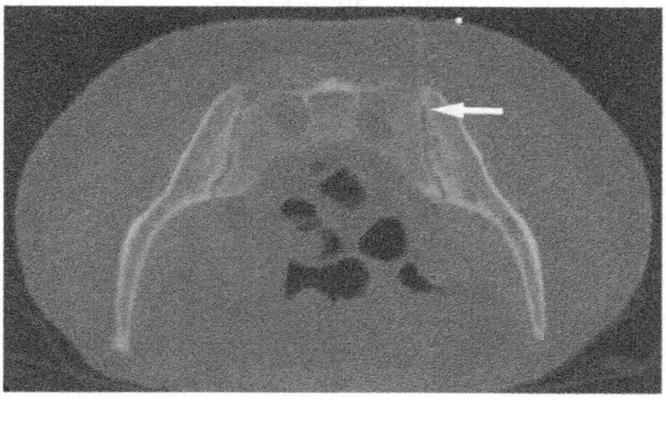

FIGURE 5 Sacroiliac joint computed tomography (CT) guidance. CT image shows a needle positioned in the synovial portion of the sacroiliac joint (*arrow*).

considered equivocal and may be a placebo response. Pain relief may not be expected to be total because the SI joint is usually only one component of a multifactorial reason for back pain. However, there may be complete relief regionally in an area that corresponds to the SI joint distribution (posterior low back below L5 and over the buttocks), establishing the SI joint as a nociceptor.

Patients should be observed for about 30 to 60 minutes postinjection. It is recommended that a companion be present to drive the person home in case of a complication. The patient may resume normal activity immediately in the absence of a complication. The most common complication is the subjective sensation and weakness of the lower extremities (about 10%). This may occur without extravasation and is felt to be related to loss of proprioception in the pelvic girdle. In addition, the patient may temporarily experience a wobbly or unsteady gait. Lower extremity numbness and true weakness may be secondary to anesthetic leakage around the sciatic nerve.

CLINICAL EXPERIENCE

Anesthetic injections into the SI joint are considered a diagnostic procedure, and when technically successful are valid indicators of the SI joint as a nociceptor. However, once the SI joint is identified as a substantial pain generator, the question then is the therapeutic benefit of intra-articular cortical steroid injection. There is good evidence from several trials (11–14) for treating inflammatory spondyloarthropathies that have SI joint involvement. This makes sense because these are true causes of sacroiliitis. However, the data on the efficacy of steroid intra-articular injections for mechanical somatic dysfunction are conflicting. In one series of 58 patients, Pulisetti and Ebraheim (15) found that the effects wore off within two weeks in 90% of the patients and felt that intra-articular injection is primarily for diagnostic purposes with little therapeutic benefit. A more recent study by Slipman et al. (16) retrospectively reviewed their experience with 31 patients and found a significant reduction in disability score and VAS, combined with a significant improvement of work status. They concluded that SI joint injections are clinically effective. However, prospective controlled studies would provide better evidence.

FACET (ZYGAPOPHYSEAL) JOINT INJECTIONS

Introduction

The facet joint is more appropriately termed zygapophyseal joint (z-joint) because the facet actually only represents one portion, that is, the articular cartilage surface, which also lines numerous other joints. The z-joints have been implicated as nociceptors in the cervical, thoracic, and lumbar spine to varying degrees. The z-joints are formed by the articulation of the inferior articular process of the posterior elements of one vertebra with the superior articular process of the posterior elements of the next vertebra. It is a synovial-type joint with a capsule, synovium, and reciprocating surfaces lined by a hyaline (articular cartilage). Similar to other joints in the spine, the imaging manifestations of degenerative joint disease (i.e., osteoarthritis for synovial joints) do not correlate well with symptomatology, and thus intra-articular injections or nerve blocks have served as a diagnostic modality for "facet syndrome." In this context, Mooney and Robertson in 1976 (17) produced low back pain by injecting hypertonic saline into lumbar z-joints and reported symptomatic relief with corticosteroid injection.

Anatomy

In the cervical spine, the z-joints are coronally oriented at angles of approximately 30° to 45° relative to the horizontal. In the thoracic spine they are coronally oriented at about a 60° angle relative to the horizontal (steeper angle than the cervical spine). In the lumbar spine, the z-joints are oblique relative to the sagittal plane. Each z-joint receives dual innervation, with the superior portion innervated by branches from the nerve one level cephalad and the inferior portion innervated by branches from the exiting nerve at that level (immediately above the joint of interest). This has implications for the number of intra-articular injections that should be performed at a single session.

Indications/Rationale

The indications for z-joint injections are axial pain, referred pain, or putative implication that this joint is contributing to a spine pain syndrome. Pain maps have been generated in the cervical, thoracic, and lumbar regions. Imaging has no clear role other than documenting degenerative joint disease and associated findings, or excluding other pathologies such as neoplasm or infection. However, particularly in the cervical spine in the postwhiplash patient, normal appearing z-joints can be symptomatic. Osteoarthritis of the z-joints is often part of the degenerative cascade where the anterior column articulation (intervertebral disc) is first affected with subsequent subluxation and degeneration of the posterior element articulations. There are only a few contraindications to intra-articular z-joint injections. The absolute contraindications include infection (unless a needle placement is part of a diagnostic arthrocentesis), pregnancy, or an uncontrollable bleeding diathesis. The relative contraindications include patients on anticoagulants and patients with contrast allergies or local anesthetic allergy. These relative contraindications can be worked around by removing a patient from anticoagulation medication, providing steroid preparation for contrast reaction, or obviating contrast use and determining which class of local anesthetics the patient has an allergy to and avoiding these agents. Informed consent is a process that includes discussion of risks, benefits, and alternatives to the procedure. Potential risks include spinal block, pneumothorax (in the thoracic spine), infection, and other organ injury. The most common "complication" is that the patient receives no relief or is slightly worsened for a period of time. While this is not a true complication, it is prudent to discuss this during the consent process so as not to provide false expectations. The patient's level of pain should be assessed immediately before the injection, immediately after the injection, and within 15 to 30 minutes after the injection. Typically patients are observed for a period of 30 minutes for any potential reactions. There is no special postprocedure care other than observing for immediate complication and for subsequent infection. For injections in the lumbar region there is no requirement for premedication, sedation, intravenous access, or physiologic monitoring. However, in the cervical and thoracic spines, intravenous access should be considered to treat hypotension related to vasovagal reaction or dehydration. In addition, the procedure may be facilitated by light sedation using an agent such as midazolam. In most institutions, this would require physiologic monitoring, a nursing assessment, and a longer postprocedure observation (i.e., an intravenous conscious sedation protocol).

The complications of z-joint injections are typically minimal. Because of its posterior location, there is less likelihood of collateral damage to the exiting nerve roots and spinal cord. Spinal blocks from intrathecal needle positioning are uncommon. Infection is possible and there have been case reports of a paraspinal abscess complicating z-joint injections (18,19). As with other spinal injections, there may be no symptomatic relief or increased pain for a short duration, which usually dissipates back to baseline in one to two weeks.

There is a concept that pain from a nociceptor may be "blocked" either by intra-articular injection or administration of an anesthetic to the afferent pain fibers. Single time blocks are problematic because of the high false positive rate and/or placebo effect. Therefore, the practice of controlled diagnostic blocks has developed, whereby after a putatively symptomatic joint has been successfully blocked from the first injection, subsequent injections are with agents of different effect duration. The duration of effect should be comparable to the type of agent injected. Algorithms may include repeat injections with bupivacaine and/or isotonic saline. Symptoms should be relieved for several hours with the intermediate-acting anesthetic (bupivacaine) and not at all with the placebo (saline) injection. In terms of evaluation, a provocative response is not a valid measure for determining whether the z-joint is a nociceptor; it is the pain relief level and duration that are important. If a single joint or a set of joints is causing the pain, then greater than 90% relief should be obtained but typically greater than 75% relief is considered a positive (good) response, because most spine pain is multifactorial. Once a good response is obtained, serial blocks can be performed to see if the patient is a candidate for a more definitive procedure (e.g., neurotomy).

Cervical Technique

For cervical spine z-joint injections, the approach may be posterior, lateral, or by using a pillar-type projection (10,20). The patient is positioned decubitus or supine with the head

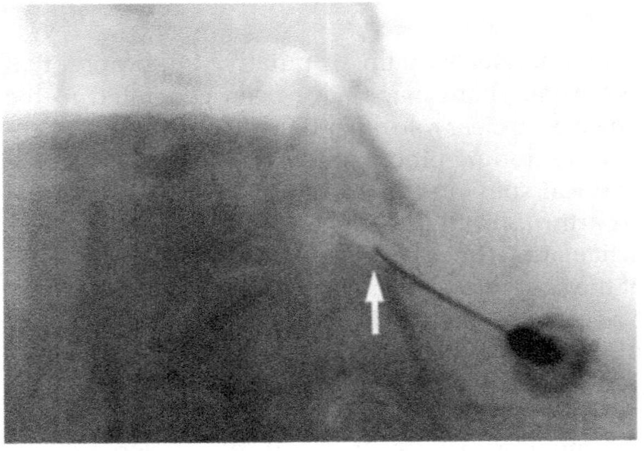

FIGURE 6 Cervical zygapophyseal joint injection lateral approach: For the C3–C6 level a lateral projection is obtained profiling the joint margins. Under fluoroscopy the mid-point of the joint is targeted (*arrow*).

turned away from the affected side. These injections are best performed under C-arm fluoroscopy. At the C2–C3 level the z-joint slopes inferiorly at its posterior and medial aspect; therefore particular or specialized projections are used. One option is to rotate the X-ray tube cephalad until the joint is profiled. Another option is to rotate the X-ray tube to the patient's rear (posteriorly), to bring the posterolateral aspect of the joint into profile. For the C3–C6 levels, a lateral approach is advocated (Fig. 6). The target zone is the mid-portion of the joint along its lateral margin. If the X-ray tube is rotated close to a true lateral, this will profile the joint margins. The left and right joints might be nearly superimposed and slight rotation of the C-arm may be performed to determine which is the appropriate side (e.g., closest). However, if truly superimposed, then the targeting should be similar despite the side and not be problematic. The volume of contrast injected is typically 0.1 to 0.3 mL, so as not to rupture the joint capsule prior to anesthetic/corticosteriod administration (Fig. 7).

Thoracic Technique

Thoracic spine z-joint injections are a relatively recent phenomena and uncommon. Although fluoroscopic techniques have been described, the thoracic spine is one area where CT may be advantageous for demonstrating the joint. The difficulty with using CT guidance is identifying the precise level, and typically more than one joint is injected during a single session. Therefore, CT fluoroscopy would be advantageous in this setting, but there is increased radiation exposure to the patient and the operator.

The fluoroscopic technique is as follows (21). The patient is placed prone on a radiolucent table with a C-arm fluoroscope. Posteroanterior imaging is employed intially. The skin entry zone overlying the inferior vertebral end plate of the same numbered vertebral body for T1–T5

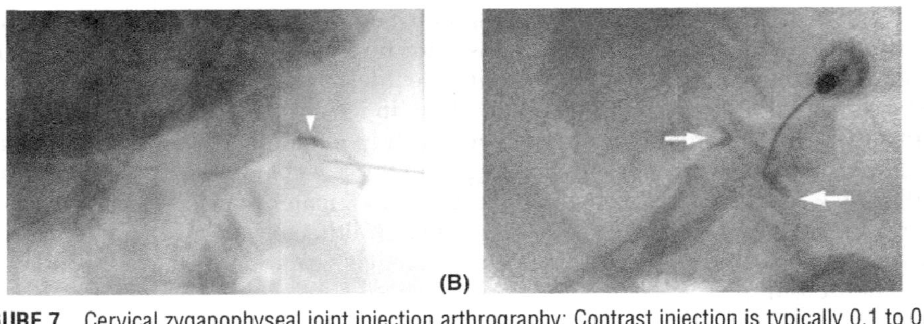

(A) (B)

FIGURE 7 Cervical zygapophyseal joint injection arthrography: Contrast injection is typically 0.1 to 0.3 mL so as not to rupture the joint capsule prior to anesthetic and corticosteriod administration. (**A**) A ring-like configuration is seen on the anteroposterior projection (*arrowhead*); (**B**) the superior recess (*small arrow*) and inferior recess (*large arrow*) are identified on the lateral projection.

z-joints or just superior to the pedicle of the vertebral body below the level of interest for T6–T10 z-joints is marked (Fig. 8A). The inferior aspect of a z-joint will be located at or slightly above the superior aspect of the pedicle of the lower vertebral body forming that articulation on PA imaging. The skin just above the superior portion of the pedicle below the level of interest is marked. The spinal needle is inserted through the skin mark angling about 60° toward the target joint. Proximal segments (T1–T5) may require a more obtuse angle. Using intermittent PA imaging, the needle is advanced cephalad toward the superior articular process of the articulation of interest. The needle should not stray medial to the medial aspect of the pedicle on PA imaging in order to prevent epidural or subarachnoid puncture. After a few centimeters of

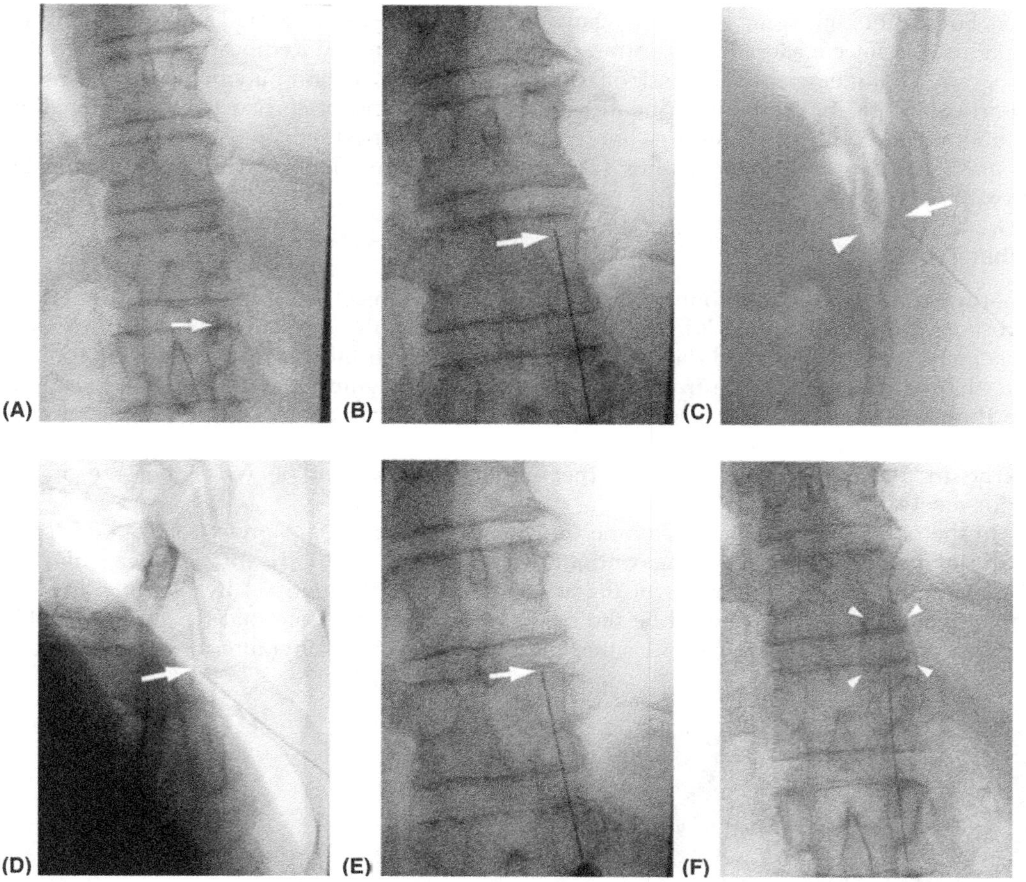

FIGURE 8 Thoracic zygapophyseal joint (z-joint) injection (**A–F**). Posteroanterior imaging is utilized first, in this example, the T11–T12 z-joint is to be blocked, so the inferior aspect of that z-joint will be located at or slightly above the superior aspect of the T12 pedicle on posteroanterior imaging. The skin just above the superior portion of the pedicle below (L1) is marked (*arrow* in **A**). The spinal needle is inserted through the skin mark angling about 60° toward the target joint. Proximal segments (T1–T5) may require a more obtuse angle. Using intermittent posteroanterior imaging, the needle is advanced cephalad toward the superior articular process of T12. The needle should not stray medial to the medial aspect of the pedicle on PA imaging to prevent epidural or subarachnoid puncture. After a few centimeters of needle insertion, the tip should be projected over the mid-aspect of the T12 pedicle (*arrow* in **B**). The C-arm intensifier is then rotated away from the side being injected until the outline of the joint is first clearly visible. This requires almost full lateral imaging (typically 20–30° from true lateral and also with minor rotational changes until the X-ray beam is aligned with the plane of the joint). The tip of the needle should be seen at or near the inferior aspect of the target joint. If the angle is not suitable (*arrow* in **C** shows that the tip is too high; *arrowhead* in **C** shows region of the inferior recess) then the trajectory may be readjusted, being careful not to deviate too much medially or laterally. The needle should be advanced such that the tip projects in the inferior recess on the lateral projection (*arrow* in **D**). This will correspond to the cephalad aspect of the T12 pedicle (*arrow* in **E**). Injection of contrast should show an arthrogram with ovoid or ring-like collection on the posteroanterior projection (*arrowheads* in **F**) and/or opacification of the superior and inferior recesses on the lateral projection.

needle insertion, the tip should be projected over the mid-aspect of the pedicle of the lower vertebral body (Fig. 8B). The C-arm intensifier is then rotated away from the side being injected until the outline of the joint is first clearly visible. This requires slightly off-lateral imaging; typically 20° to 30° from true lateral and also with minor rotational changes until the X-ray beam is aligned with the plane of the joint. The tip of the needle should be seen at or near the inferior aspect of the target joint. If the angle is not suitable (Fig. 8C), then the trajectory may be readjusted, being careful not to deviate too much medially or laterally. The needle should be advanced such that the tip projects in the inferior recess on the lateral projection (Fig. 8D). This will correspond to the cephalad aspect of the pedicle of the lower vertebral body (Fig. 8E). Injection of contrast should show an arthrogram with ovoid or ring-like collection on the PA projection (Fig. 8F) and/or opacification of the superior and inferior recesses on the lateral projection. For the thoracic spine, a 23- or 25-gauge needle may be used. A small amount of contrast (0.3–0.5 mL) is injected to confirm intra-articular location by the demonstration of a z-joint arthrogram. Obliqueing the tube and/or using the lateral projection may be needed to verify intra-articular opacification. Subsequently, local anesthetic and or corticosteroid combination may be injected. Typically, a higher concentration of local anesthetic is used given the small intra-articular volumes of the z-joints (2% or 4% lidocaine). The total injectate should be approximately 1.5 to 2 cc.

Lumbar Technique

Numerous technical descriptions exist for lumbar z-joint injections performed under CT or fluoroscopy guidance (10,22–26). Some prefer the CT technique because the posterior joint space can be readily identified (Fig. 9). The anatomy of the lumbar z-joints is such that there is a C-shaped auricular curve from anterior to posterior, which is variable. There are two main fluoroscopic techniques used for accessing the lumbar z-joints—oblique and posterior. In both approaches the patient is prone on a radiolucent table with pillows or cushions underneath the abdomen to decrease the lumbar lordosis and increase the size of the inferior recess.

In the oblique approach (22,25), the C-arm is started in the PA projection and centered over the joint of interest. While using continuous fluoroscopy, the tube is rotated toward the lateral direction until the first profile of the articular margins (as defined by sclerotic lines) is encountered (Fig. 10). This should be the posterior joint space. Note that the traditional 45° oblique position ("Scotty dog") profiles the anterior aspect of the joint (anterior joint space) but

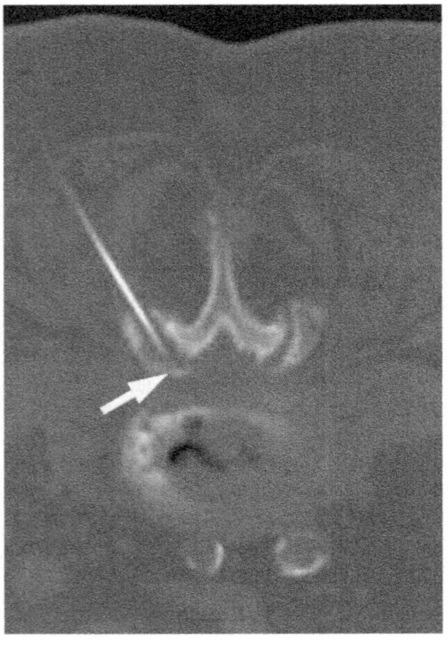

FIGURE 9 Lumbar zygapophyseal (facet) joint injection lumbar [computed tomography (CT) guidance]. CT image shows a needle positioned in a lumbar zygaphophyseal joint.

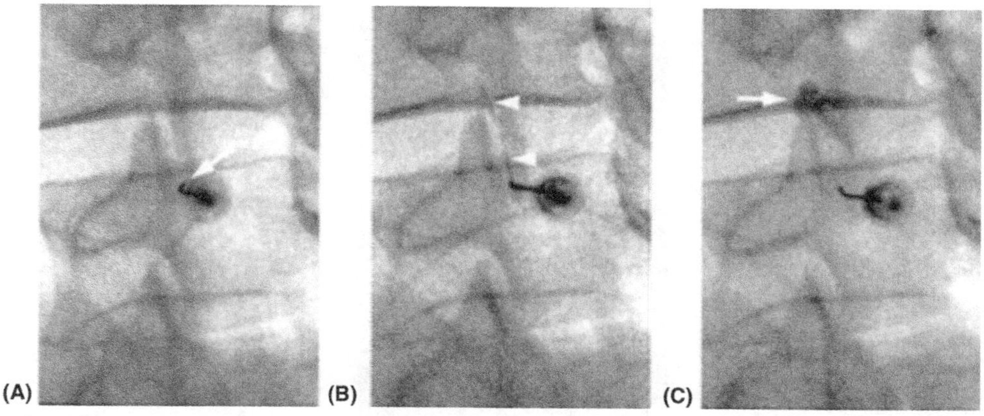

FIGURE 10 Lumbar zygapophyseal joint injection oblique approach: In the oblique approach the C-arm is started in the PA projection and centered over the joint of interest. While using continuous fluoroscopy, the tube is rotated toward the lateral direction until the first profile of the articular margins is encountered. This should be the posterior joint space and may or may not be similar to a "Scotty dog" projection. Note that in this example that the projection is similar to the traditional 45° oblique position. (**A**) shows the needle tip projected over the inferior aspect of the joint (*arrow*). The tip is slightly bent, which often indicates that it is intra-articular. (**B**) shows an early arthrogram as a linear region of intra-articular opacification (*arrowheads*). (**C**) shows a later arthrogram as ovoid contrast accumulating in the superior recess (*arrow*).

is not necessarily the correct target in many patients. If one simply just positions at this designated angle without observing for the most posterior portion of the joint, then the needle will not be able to access this area because of bone. This approach is also problematic for lumbar z-joints with prominent osteophytosis.

The posterior approach is also known as the European approach (26). In this approach, the X-ray tube is left in a vertical position centered over the level of interest. The goal is to access the inferior recess. The target zone is just below the superior articular process along its inferomedial aspect. Often the landmarks of the superior articular process are not well visualized and an alternative target zone is just above the pedicle at its superomedial aspect (Fig. 11). The skin is marked and the needle is placed straight down until the bone is encountered. For the L5–Sl (lumbosacral junction) the target zone is just below (inferior to) the superior aspect of the

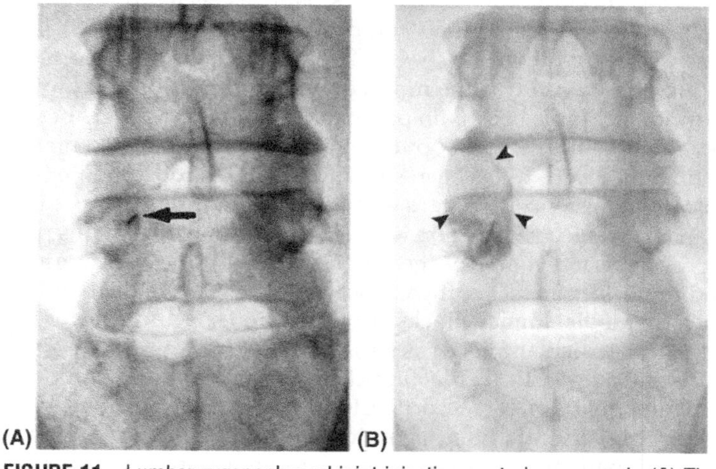

FIGURE 11 Lumbar zygapophyseal joint injection posterior approach. (**A**) The target zone is just below the superior articular process along its inferomedial aspect. However, often the landmarks of the superior articular process are not well visualized and an alternative target zone is just above the pedicle at its superomedial aspect (*arrow*). (**B**) shows an arthrogram as ovoid area of contrast accumulating in the expected region of the zygapophyseal joint (*arrowheads*).

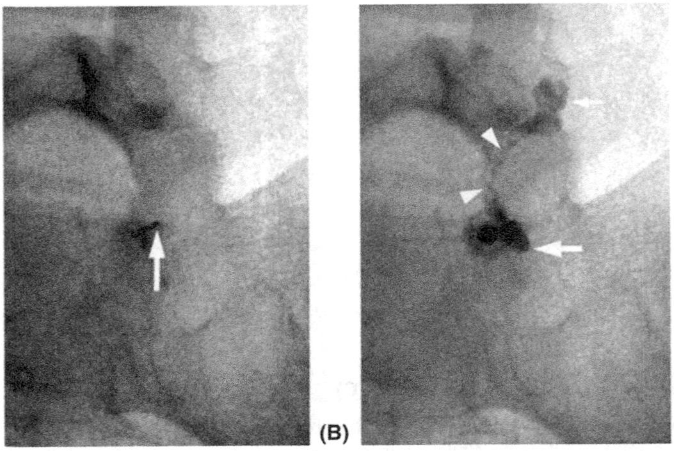

(A) (B)

FIGURE 12 Lumbar zygapophyseal joint injection posterior approach for the lumbosacral junction. (**A**) For the L5–S1 level (lumbosacral junction), the target zone is just below (inferior to) the superior aspect of the sacrum (*arrow*). (**B**) shows an arthrogram as a curvilinear ringlike configuration (*arrowheads*) with contrast accumulating in the superior (*small arrow*) and inferior joint recesses (*large arrow*).

sacrum (Fig. 12). Once the needle encounters the bone, passage into the joint is often perceived as a "pop" or loss of resistance as the tip passes through the capsule.

With either the oblique or posterior technique, a z-joint arthrogram should be obtained with the installation of a small amount (0.5 cc) of contrast material. The arthrogram may be circular (ring-like) and there is often flow into the superior recess (Figs. 10B and C, 11B, and 12B). Often the PA projection is sufficient and no other deviation of the fluoroscope is needed to visualize the arthrogram. However, if flow is not perceived but rather a "blob" sign is identified, then rotating the C-arm may be useful. In the lumbar spine, the z-joint arthrogram may be quite variable with contrast passing to the contralateral side and either around or communicating with the contralateral z-joint. Alternatively, contrast may proceed cephalad to the z-joint above. If a pars defect (spondylolysis) is present, it is often opacified by injecting the z-joint.

Cervical Clinical Experience

On the basis of using intra-articular injections or nerve blocks as the reference standard, the prevalence of the z-joint as a nociceptor in chronic axial-type neck pain has been reported from about 35% in a specialist referral setting (27) to about 50% in postwhiplash patients (28). The level most commonly involved in either group was C5–C6. The C3–C4 level was the most common in the nonwhiplash group and C2–C3 was the most common in the whiplash group. Cervical z-joint pain is common among chronic neck pain patients after whiplash and is of significant clinical importance. In evaluating this patient population, placebo controlled comparative blocks should be the algorithm employed. It has been shown that uncontrolled diagnostic blocks are compromised by substantial false positive rates and diminish the specificity of this modality (29). An alternative to an intra-articular injection is a medial branch block. In this procedure, a needle is placed along the expected course of the medial branch of the dorsal ramus that innervates the facet joint and multifidus muscles, then a local anesthetic is instilled and an assessment is made of pain reduction. Medial branch blocks are considered specific tests for the diagnosis of z-joint pain similar to intra-articular injections (30).

Both intra-articular injections and anesthetic medial branch blocks are considered primarily as diagnostic tests. Intra-articular injection of a corticosteroid is not an effective pain management therapy for postwhiplash cervical z-joint symptoms. In a double-blinded, randomized, placebo-controlled trial, 41 patients were randomized to either an intra-articular injection of anesthetic or corticosteroids. In both groups less than half the patients reported relief of pain for more than one week and less than 20% reported relief for more than one month (31). Therefore, in terms of therapy, procedures encompassing nerve ablations are useful to provide more lasting reduction of symptomatology. A randomized, placebo-controlled, doubleblinded trial studying 24 patients found that percutaneous radio frequency neurotomy creating multiple lesions at each level can provide lasting relief (32). In this investigation, the control group underwent a sham procedure with needle placement but no radio frequency current applied. Both groups were followed until they reported that their pain had returned to about 50% of the preprocedure level. There was a significant difference between the active treatment group (263 days) versus the control group (8 days).

Thoracic Clinical Experience

There is a paucity of literature regarding thoracic z-joint injections. Pain patterns were studied in normal volunteers in a preliminary investigation confirming that thoracic z-joints can cause both local and referred pain (33), and a referral pain diagram has been constructed based on these data from intra-articular injections. However, innervation to the z-joints in the thoracic spine is different than that anticipated for the cervical and lumbar spine. In an anatomic investigation, Chua and Bogduk (34) found that the medial branch upon leaving the inter-transverse space typically crossed the supralateral corners of the transverse processes and then passed medially and inferiorly across the posterior surfaces of the transverse processes, before dividing to innervate the multifidus muscles. Therefore the supralateral corners of the transverse processes may be more accurate target points if considering medial branch blocks for thoracic z-joints. A nonrandomized trial evaluating 40 patients who underwent thoracic medial branch neurotomy for a variety of reasons found this percutaneous denervation procedure to be safe and beneficial in patients with chronic pain thought to be of z-joint origin (35). The short-term results (within two months) had approximately one-half the patients pain free and a third with greater than 50% pain relief. However, the long-term results show that less than half (44%) were pain free with just above a third (39%) maintaining more than 50% pain relief.

Lumbar Clinical Experience

While there is no doubt that the z-joint can contribute to low back pain, there are no unique identifying features in the history, physical examination, and imaging of these patients (36) and therefore the existence of a primary "facet syndrome" or "zygapophyseal joint syndrome" is dubious (37). Nevertheless a prospective randomized trial was performed to try and elucidate clinical characteristics that support the z-joint as a primary or significant nociceptor. Although there is no specific syndrome that discriminates between lower back pain caused by z-joint and other etiologies, there were several predictors identified. The factors of age above 65 years and pain that was not exacerbated by coughing, not worsened by hyperextension, not worsened by forward flexion, not worsened when rising from flexion, not exacerbated by extension-rotation, and relieved by recumbent positioning were found to be five clinical characteristics indicating pain related to the z-joint (38). This investigation was performed using placebo-controlled intra-articular injections. Similar to the SI and other z-joints, the lumbar z-joint evaluations after injections are predominantly for pain relief. The criterion of pain provocation from an intra-articular injection is not considered valid (39).

There are sufficient data to support that the definitive diagnosis of lumbar z-joint–mediated pain is based on selective analgesic injections of these joints or their nerve supply (the medial branch of the dorsal ramus). Notwithstanding, one must be cautious of the false positive rate of uncontrolled diagnostic blocks (40). The false positive rate of uncontrolled blocks approaches 38% with a positive predictive value of only 31%, when compared to the reference (criterion) standard of using comparative and confirmatory blocks. Given that the prevalence of predominantly lumbar z-joint pain is likely to be less than 50% in all patients with low back pain, this is felt to be an unacceptably high false positive rate. There are numerous uncontrolled trials evaluating the therapeutic effect of lumbar z-joint injections. The preponderance of evidence is that intra-articular lumbar z-joint injections and anesthetic nerve blocks are diagnostic tests, but neither is satisfactory for chronic low back pain (41,42) similar to the experience in the cervical spine. Randomized, placebocontrolled trials support also that intra-articular z-joint injections of corticosteroid do not have sustaining therapeutic effects (43,44).

Due to the ineffectiveness of anesthetic and corticosteroid injections, healthcare providers have turned to medial branch neurotomy for providing therapy. Initial uncontrolled trials have demonstrated efficacy (45,46). The lumbar medial branch neurotomy has been shown to be an effective means of reducing pain in patients who are selected on the basis of controlled diagnostic blocks (either intra-articular or medial branch blocks). Technically, adequate coagulation of the target nerve is achieved by carefully placing the electrode in the correct position as judged by osseous landmarks using imaging (most commonly fluoroscopy). Randomized controlled trials have had somewhat conflicting results. In a study by Leclaire et al. (47), the investigators found that neither functional disability nor pain level showed any significant difference in the treatment or the control group at 12 weeks. They felt that denervation provided some short-term improvement, but the efficacy of the treatment is not established at the three-month follow-up point. In another randomized controlled trial by van Kleef et al. (48), the investigators found significant differences in pain relief and functional disability between the treatment and the control group both on a short-term and long-term basis. This trial had a longer follow-up and the significant differences were primarily noted at 6 and 12 months. Therefore, this procedure, in addition to its immediate neuroablative effects, may also facilitate rehabilitation such that improvement may not be identified until a longer interval follow-up is obtained.

Synovial Cysts

As previously mentioned, the z-joints are diarthrodial synovial joints characterized by an articular surface covered by hyaline cartilage and surrounded by a joint capsule, which is lined by synovium. Similar to other synovial joints, diverticula may form, known as synovial cysts. These cysts may emanate from the inferior recess and protrude posteriorly (extracanalicular). In this location they typically do not cause symptoms in and of themselves, but serve as a marker of underlying osteoarthritis. When synovial cysts emanate from the superior recess, they tend to protrude anteriorly into the vertebral canal or medially underneath the ligamentum flavum (these are often referred to as intraspinal or intracanalicular). When large, these cysts compromise the lateral recess or neural foramen and can produce radicular symptoms mimicking a disc herniation clinically. Cross-sectional imaging (CT or MR) will show a rounded lesion in the lateral recess or neural foramen in close proximity to the z-joint (49,50). On MR imaging synovial cysts typically follow fluid signal intensity. However, they may be complicated by hemorrhage (51) and the wall may calcify (Fig. 13). The lumbar spine is the most common location, but these have been described in the thoracic spine (52,53).

The synovial cyst is a lesion that can be treated by z-joint intra-articular injections as an alternative to surgery. This technique was described in three patients who were treated with complete relief in two and partial relief in the third (54). No complications were identified. A retrospective series (55) treating 30 patients with radiculopathy secondary to lumbar z-joint synovial cysts found that about one-third of the patients had long-term relief greater than six months. A more recent series of 12 patients (56) with clinical and imaging follow-up found a correlation between excellent pain relief and resolution of the synovial cyst by imaging. In this

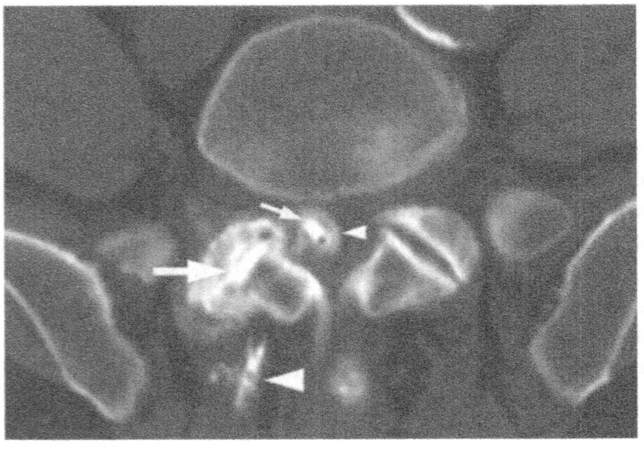

FIGURE 13 Lumbar synovial cyst. Postinjection computed tomography image shows intra-articular contrast in an osteoarthritic zygapophyseal joint (*large arrow*) and opacification of a synovial cyst that is causing right lateral recess stenosis (*small arrow*). Note that the cyst wall is also hyperattenuating, but not as dense as the contrast reflecting that it is calcified (*small arrowhead*). Note also the contrast extravasation into the posterior musculature (*large arrowhead*).

study, 75% (9 of 12 patients) had excellent pain relief, which makes this procedure an attractive treatment option prior to surgical intervention. The three patients with no or only partial relief showed a persistent synovial cyst at followup imaging. As mentioned previously, the wall of the synovial cyst may calcify and while not addressed in the literature, it is this author's experience that cysts with calcified walls are more resilient to injection treatment, but this is not an absolute contraindication. Given the simplicity and low morbidity of this technique, it is considered a suitable first-line therapy despite the possibility of symptomatic recurrence. Also, in the author's experience, if temporary or partial relief is achieved with a single injection, then a repeat in a couple or a few weeks may enhance or augment the therapeutic effect similar to a partial response to an epidural steroid injection. Postprocedure CT imaging is not routinely performed but may be useful if one is uncertain on fluoroscopic imaging that an intra-articular injection was performed (Fig. 13).

CONCLUSION

In this chapter, the indications, contraindications, rationale, technique, and clinical experience for performing SI joint and zygapophyseal (facet) joint intra-articular injections was reviewed. Intra-articular injections have an established role for identifying these articulations as nociceptors. However, intra-articular injections with anesthetic and/or corticosteroid are often not sufficient for a long-lasting therapeutic effect. Once a z-joint has been implicated as a substantial or significant nociceptor, then targeted therapy options exist that may include a neuroablative procedure (medial branch neurotomy) in conjunction with functional restoration via physical therapy. SI joint treatment can be more problematic given the diffuse innervation of the articulation; however SI joint fusion is a technique practiced by some orthopedic surgeons. Two areas where intra-articular injections can successfully treat patients are in the following settings. For true inflammatory sacroiliitis related to a spondyloarthropathy, intra-articular corticosteroid is proven to be an effective component of treatment. For a z-joint–related synovial cyst that is causing lateral recess narrowing and radicular symptoms intra-articular injection with corticosteroids assists in decreasing the perineural inflammation, reducing the size of the cyst, and alleviating the symptoms. In summary, intra-articular SI and z-joint injections are an important component of a comprehensive management approach to chronic spine pain syndromes, for establishing a diagnosis, directing therapy, and facilitating rehabilitation.

REFERENCES

1. Dreyfuss P, Michaelsen M, Pauza K, McLarty J, Bogduk N. The value of medical history and physical examination in diagnosing sacroiliac joint pain. Spine 1996; 21(22): 2594–2602.
2. Schwarzer AC, Aprill CN, Bogduk N. The sacroiliac joint in chronic low back pain. Spine 1995; 20(1):31–37.
3. Maigne JY, Aivaliklis A, Pfefer F. Results of sacroiliac joint double block and value of sacroiliac pain provocation tests in 54 patients with low back pain. Spine 1996; 21(16): 1889–1892.
4. Fortin JD, Washington WJ, Falco FJ. Three pathways between the sacroiliac joint and neural structures. AJNR 1999; 20(8):1429–1434.
5. Fortin JD, Dwyer AP, West S, Pier J. Sacroiliac joint: pain referral maps upon applying a new injection/arthrography technique. Part I: Asymptomatic volunteers. Spine 1994; 19(13):1475–1482.
6. Pereira PL, Gunaydin I, Trubenbach J, et al. Interventional MR imaging for injection of sacroiliac joints in patients with sacroiliitis. Am J Roentgenol 2000; 175(1):265–266.
7. Hendrix RW, Lin PJ, Kane WJ. Simplified aspiration or injection technique for the sacroiliac joint. J Bone Joint Surg Am 1982; 64(8):1249–1252.
8. Dreyfuss P, Cole AJ, Pauza K. Sacroiliac joint injection techniques. Phys Med Rehabil Clin North Am 1995; 6(4):815–823.
9. Dussault RG, Kaplan PA, Anderson MW. Fluoroscopy-guided sacroiliac joint injections. Radiology 2000; 214(1):273–277.
10. Silbergleit R, Mehta BA, Sanders WP, Talati SJ. Imaging-guided injection techniques with fluoroscopy and CT for spinal pain management. Radiographics 2001; 21(4): 927–939.
11. Maugars Y, Mathis C, Berthelot JM, Charlier C, Prost A. Assessment of the efficacy of sacroiliac corticosteroid injections in spondylarthropathies: a double-blind study. Br J Rheumatol 1996; 35(8): 767–770.

12. Braun J, Bollow M, Seyrekbasan F, et al. Computed tomography guided corticosteroid injection of the sacroiliac joint in patients with spondyloarthropathy with sacroiliitis: clinical outcome and followup by dynamic magnetic resonance imaging. J Rheumatol 1996; 23(4):659–664.
13. Bollow M, Braun J, Taupitz M, et al. CT-guided intraarticular corticosteroid injection into the sacroiliac joints in patients with spondyloarthropathy: indication and followup with contrast-enhanced MRI. J Comput Assist Tomogr 1996; 20(4):512–521.
14. Hanly JG, Mitchell M, MacMillan L, Mosher D, Sutton E. Efficacy of sacroiliac corticosteroid injections in patients with inflammatory spondyloarthropathy: results of a 6 month controlled study. J Rheumatol 2000; 27(3):719–722.
15. Pulisetti D, Ebraheim NA. CT-guided sacroiliac joint injections. J Spinal Disord 1999; 12(4):310–312.
16. Slipman CW, Lipetz JS, Plastaras CT, et al. Fluoroscopically guided therapeutic sacroiliac joint injections for sacroiliac joint syndrome. Am J Phys Med Rehabil 2001; 80(6): 425–432.
17. Mooney V, Robertson J. The facet syndrome. Clin Orthop 1976; (115):149–156.
18. Magee M, Kannangara S, Dennien B, Lonergan R, Eramett L, Van der Wall H. Paraspinal abscess complicating facet joint injection. Clin Nucl Med 2000; 25(1): 71–73.
19. Cook NJ, Hanrahan P, Song S. Paraspinal abscess following facet joint injection. Clin Rheumatol 1999; 18(1):52–53.
20. Chevrot A, Chevrot AL, Drape J, et al. Image-guided spinal steroid injections. Semin Musculoskelet Radiol 1997; 1(2):221–230.
21. Dreyfuss P, Tibiletti C, Dreyer S, Sobel J. Thoracic zygapophyseal joint pain: a review and description of an intra-articular block technique. Pain Digest 1994; 4:46–54.
22. el-Khoury GY, Renfrew DL. Percutaneous procedures for the diagnosis and treatment of lower back pain: diskography, facet-joint injection, and epidural injection. AJR 1991; 157(4):685–691.
23. Carrera GF. Lumbar facet joint injection in low back pain and sciatica: description of technique. Radiology 1980; 137(3):661–664.
24. Destouet JM, Gilula LA, Murphy WA, Monsees B. Lumbar facet joint injection: indication, technique, clinical correlation, and preliminary results. Radiology 1982; 145(2): 321–325.
25. Agorastides ID, Kumar N. The oblique needle technique in lumbar facet joint injection. Eur J Radiol 2001; 40(3):240–243.
26. Sarazin L, Chevrot A, Pessis E, et al. Lumbar facet joint arthrography with the posterior approach. Radiographics 1999; 19(1):93–104.
27. Speldewinde GC, Bashford GM, Davidson IR. Diagnostic cervical zygapophyseal joint blocks for chronic cervical pain. Med J Aust 2001; 174(4):174–176.
28. Lord SM, Barnsley L, Wallis BJ, Bogduk N. Chronic cervical zygapophysial joint pain after whiplash. A placebo-controlled prevalence study. Spine 1996; 21(15):1737–1744; discussion 1744–1745.
29. Barnsley L, Lord S, Wallis B, Bogduk N. False-positive rates of cervical zygapophyseal joint blocks. Clin J Pain 1993; 9(2):124–130.
30. Barnsley L, Bogduk N. Medial branch blocks are specific for the diagnosis of cervical zygapophyseal joint pain. Reg Anesth 1993; 18(6):343–350.
31. Barnsley L, Lord SM, Wallis BJ, Bogduk N. Lack of effect of intraarticular corticosteroids for chronic pain in the cervical zygapophyseal joints. N Engl J Med 1994; 330(15):1047–1050.
32. Lord SM, Barnsley L, Wallis BJ, McDonald GJ, Bogduk N. Percutaneous radiofrequency neurotomy for chronic cervical zygapophyseal-joint pain. N Engl J Med 1996; 335(23):1721–1726.
33. Dreyfuss P, Tibiletti C, Dreyer SJ. Thoracic zygapophyseal joint pain patterns. A study in normal volunteers. Spine 1994; 19(7):807–811.
34. Chua WH, Bogduk N. The surgical anatomy of thoracic facet denervation. Acta Neurochir (Wien) 1995; 136(3–4):140–144.
35. Stolker RJ, Vervest AC, Groen GJ. Percutaneous facet denervation in chronic thoracic spinal pain. Acta Neurochir (Wien) 1993; 122(1–2):82–90.
36. Dreyer SJ, Dreyfuss PH. Low back pain and the zygapophysial (facet) joints. Arch Phys Med Rehabil 1996; 77(3):290–300.
37. Schwarzer AC, Aprill CN, Derby R, Fortin J, Kine G, Bogduk N. Clinical features of patients with pain stemming from the lumbar zygapophyseal joints. Is the lumbar facet syndrome a clinical entity? Spine 1994; 19(10):1132–1137.
38. Revel M, Poiraudeau S, Auleley GR, et al. Capacity of the clinical picture to characterize low back pain relieved by facet joint anesthesia. Proposed criteria to identify patients with painful facet joints. Spine 1998; 23(18):1972–1976; discussion 1977.
39. Schwarzer AC, Derby R, Aprill CN, Fortin J, Kine G, Bogduk N. The value of the provocation response in lumbar zygapophyseal joint injections. Clin J Pain 1994; 10(4): 309–313.
40. Schwarzer AC, Aprill CN, Derby R, Fortin J, Kine G, Bogduk N. The false-positive rate of uncontrolled diagnostic blocks of the lumbar zygapophyseal joints. Pain 1994; 58(2): 195–200.
41. Marks RC, Houston T, Thulbourne T. Facet joint injection and facet nerve block: a randomised comparison in 86 patients with chronic low back pain. Pain 1992; 49(3): 325–328.
42. Lau LS, Littlejohn GO, Miller MH. Clinical evaluation of intra-articular injections for lumbar facet joint pain. Med J Aust 1985; 143(12–13):563–565.

43. Carette S, Marcoux S, Truchon R, et al. A controlled trial of corticosteroid injections into facet joints for chronic low back pain. N Engl J Med 1991; 325(14):1002–1007.
44. Lilius G, Laasonen EM, Myllynen P, Harilainen A, Gronlund G. Lumbar facet joint syndrome. A randomised clinical trial. J Bone Joint Surg Br 1989; 71(4):681–684.
45. Dreyfuss P, Halbrook B, Pauza K, Joshi A, McLarty J, Bogduk N. Efficacy and validity of radiofrequency neurotomy for chronic lumbar zygapophyseal joint pain. Spine 2000; 25(10):1270–1277.
46. Tzaan WC, Tasker RR. Percutaneous radiofrequency facet rhizotomy—experience with 118 procedures and reappraisal of its value. Can J Neurol Sci 2000; 27(2):125–130.
47. Leclaire R, Fortin L, Lambert R, Bergeron YM, Rossignol M. Radiofrequency facet joint denervation in the treatment of low back pain: a placebo-controlled clinical trial to assess efficacy. Spine 2001; 26(13):1411–1416; discussion 1417.
48. van Kleef M, Barendse GA, Kessels A, Voets HM, Weber WE, de Lange S. Randomized trial of radiofrequency lumbar facet denervation for chronic low back pain. Spine 1999; 24(18):1937–1942.
49. Jackson DE Jr, Atlas SW, Mani JR, Norman D. Intraspinal synovial cysts: MR imaging. Radiology 1989; 170(2):527–530.
50. Apostolaki E, Davies AM, Evans N, Cassar-Pullicino VN. MR imaging of lumbar facet joint synovial cysts. Eur Radiol 2000; 10(4):615–623.
51. Kaneko K, Inoue Y. Haemorrhagic lumbar synovial cyst. A cause of acute radiculopathy. J Bone Joint Surg Br 2000; 82(4):583–584.
52. Fritz RC, Kaiser JA, White AH, DeLong WB, Gamburd RS. Magnetic resonance imaging of a thoracic intraspinal synovial cyst. Spine 1994; 19(4):487–490.
53. Lopes NM, Aesse FF, Lopes DK. Compression of thoracic nerve root by a facet joint synovial cyst: case report. Surg Neurol 1992; 38(5):338–340.
54. Bjorkengren AG, Kurz LT, Resnick D, Sartoris DJ, Garfin SR. Symptomatic intraspinal synovial cysts: opacification and treatment by percutaneous injection. Am J Roentgenol 1987; 149(1):105–107.
55. Parlier-Cuau C, Wybier M, Nizard R, Champsaur P, Le Hir P, Laredo JD. Symptomatic lumbar facet joint synovial cysts: clinical assessment of facet joint steroid injection after 1 and 6 months and long-term follow-up in 30 patients. Radiology 1999; 210(2): 509–513.
56. Bureau NJ, Kaplan PA, Dussault RG. Lumbar facet joint synovial cyst: percutaneous treatment with steroid injections and distention—clinical and imaging follow-up in 12 patients. Radiology 2001; 221(1):179–185.

12 Selective Therapeutic Injections of Axial Skeleton Joints (Spine and Sacroiliac Joints Excluded)

Marc Wybier
Department of Bone and Joint Radiology, Lariboisière Hôpital, Assistance Publique–Hopîtaux de Paris, Paris, France

INTRODUCTION

A radiologically guided approach is commonly necessary for therapeutic injections of axial extraspinal joints, due to the small size or the deep location of these joints. In this chapter, selective injections of the acromioclavicular, sternoclavicular, and costovertebral joints, as well as that of the pubic symphysis, will be detailed. Injection of the transversosacral neoarthrosis, which may be encountered in cases of transitional lumbosacral anomaly, will also be mentioned.

GENERAL TECHNICAL CONSIDERATIONS

Therapeutic injections into axial extraspinal joints are performed on an outpatient basis in the fluoroscopic suite. Radiological guidance first consists in opacifying the joint space prior to the therapeutic injection to check for the accurate intraarticular needle placement. The patient is placed in the supine (or, if necessary, prone) position to avoid or to reduce the effects of vagal discomfort. Prescription of a sedative treatment just before the procedure is usually not necessary. However, as for any radiological invasive procedure, it is necessary to inform and reassure the patient. The skin is prepped with an iodine solution. Further procedure varies according to the superficial (acromioclavicular and sternoclavicular joints) or deep (pubic symphysis, costovertebral joint, and transversosacral neoarthrosis) situation of the joint. For superficial joint puncture, skin anesthesia is not always required: actually, the direct puncture of the joint is not more painful than the puncture for skin anesthesia. A 20-gauge, 9-cm needle may be used as for appendicular joints. This type of needle is flexible enough to be curved after having entered the joint, so that further contrast medium injection may be controlled safely under fluoroscopic monitoring. It also allows easy aspiration of fluid from the joints. However, due to the superficial location of these joints, a short intradermic-type (25 gauge, 16 mm) needle may also be used, avoiding any risk of injury to the underlying soft tissues. In addition, this type of needle is too thin to allow release of contrast medium from the syringe if the needle is not correctly inserted within the joint cavity. After removal of the needle, gentle compression and massage of the skin at the puncture site help avoid fluid extravasation.

For deep joint puncture, anesthesia of the skin and underlying soft tissues is appropriate, as well as the use of a 20-gauge, 9-cm needle.

Prior to contrast medium injection, aspiration of joint fluid is recommended. Joint effusion would actually dilute the active principle and potentially reduce its effectiveness. It would also lead to deterioration of the image quality of the arthrogram—even if the joint opacification is not requested for diagnostic purposes, a rudimentary arthrogram (one or two films) should be obtained in every case, at least for legal purposes. If present, joint fluid must be inspected immediately to rule out the possibility of unexpected infection, which would contraindicate a planned steroid injection. The injection of a small amount (1–2 mL) of contrast medium is followed under fluoroscopic control. Additional contrast material may be injected for diagnostic purposes. However, joint distension may be responsible for extravasation and may thus compromise the efficiency of the therapeutic injection. To avoid excessive extravasation of the

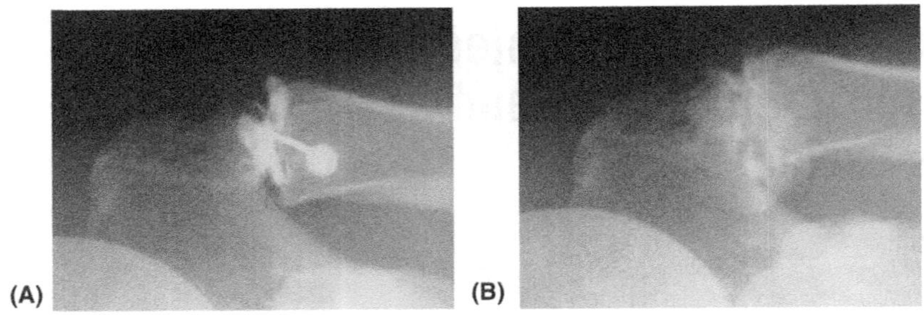

FIGURE 1 Arthrography during acromioclavicular joint therapeutic injection. Arthrography as the first step prior to steroid injection (**A**). Attenuated opacification of the joint after injection of the steroid (**B**).

steroid after removal of the needle, the finger is released periodically from the plunger to verify that no injected material returns into the syringe. At least two radiographs are taken. The first radiograph demonstrates the correct opacification of the joint, and the second one shows the dilution of the contrast material by the injected therapeutic fluid (Fig. 1) and the absence of extravasation through the site of joint puncture.

ACROMIOCLAVICULAR JOINT

Anatomy

The acromioclavicular joint is a diarthrodial joint. Both articular surfaces have an ovoid shape. The clavicular surface lies on the acromial one, with the joint space oriented ventrally and medially. Articular surfaces are covered with fibrocartilage, which suggests a poor range of movement between them. A meniscus may arise from the superior part of the capsule (Fig. 2). In some cases, it separates the joint into two compartments, but communications between them may be found that are related to aging (1). Joint stability mainly relies on the coracoclavicular ligaments and the trapezius, deltoid, and sternocleidomastoid muscles. During abduction–adduction movements of the shoulder, the acromioclavicular joint allows the scapula to rotate around the distal end of the clavicle. Maximal tension on the coracoclavicular ligaments is obtained in forced adduction of the arm, such as in tennis backhand preparation.

Indications

Acromioclavicular degenerative osteoarthritis is a common finding, encountered in about 90% of patients over 60 years of age, especially in manual workers (2). It is however frequently asymptomatic. Painful cases may be efficiently treated with selective steroid injections. Septic osteoarthritis of the acromioclavicular joint is uncommon except in cases of drug addiction or

FIGURE 2 Meniscus-like lucency (*arrows*) in an acromioclavicular joint arthrogram.

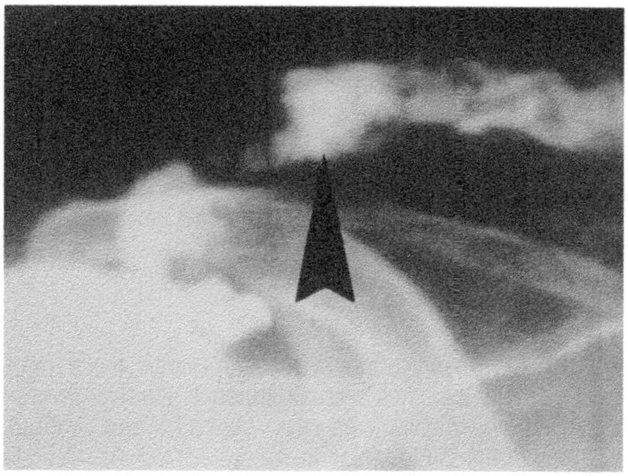

FIGURE 3 Synovial cyst (*arrowhead*) associated with degenerative acromioclavicular osteoarthritis.

immunodeficiency (3). Aspiration of joint space fluid under radiological guidance may be indicated for bacteriological purposes. Acute acromioclavicular arthritides may be encountered in ankylosing spondylarthritis, rheumatoid arthritis, and chondrocalcinosis. As a rule, selective steroid injection is rapidly effective in these unusual conditions.

Infrequently, a bulging synovial cyst may be associated with acromioclavicular osteoarthritis (Fig. 3). Such a cyst is generally associated with an abnormal communication between the acromioclavicular and glenohumeral joints through a wide rotator cuff tear with disruption of the subacromial bursa. Aspiration and steroid injection of the cyst may be of interest, but the cyst may recur if the rotator cuff tear is not repaired.

Technique of Injection

The patient lies supine with the symptomatic shoulder on the operator side. The radiological approach is anterior and superior (Fig. 4). Opacification immediately outlines the lower pole of the joint. On the anteroposterior view, the joint cavity has a vertical orientation (Fig. 5).

STERNOCLAVICULAR JOINT

Anatomy

The sternoclavicular joint is a diarthrodial joint. The articular surface of the proximal end of the clavicle has a coronal convexity and a slight sagittal concavity, whereas that of the sternal manubrium has a coronal concavity and a slight sagittal convexity. The cartilage of the first rib also articulates with the clavicle. Both articular surfaces are covered with fibrocartilage, which is thicker on the clavicle than on the sternum. A circular meniscus extending from above the articular surface of the clavicle to the junction between the sternal manubrium and

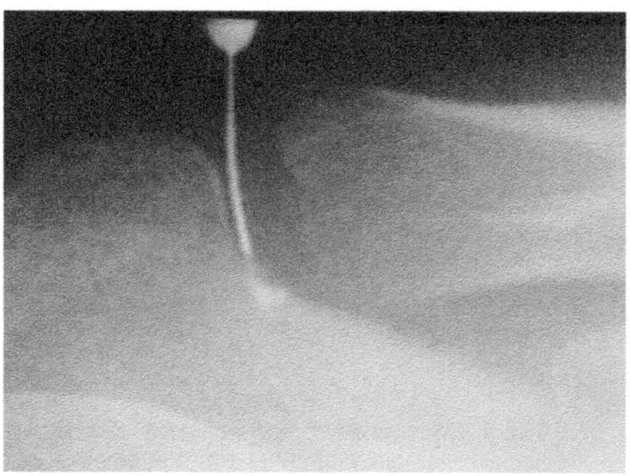

FIGURE 4 Needle placement into an acromioclavicular joint.

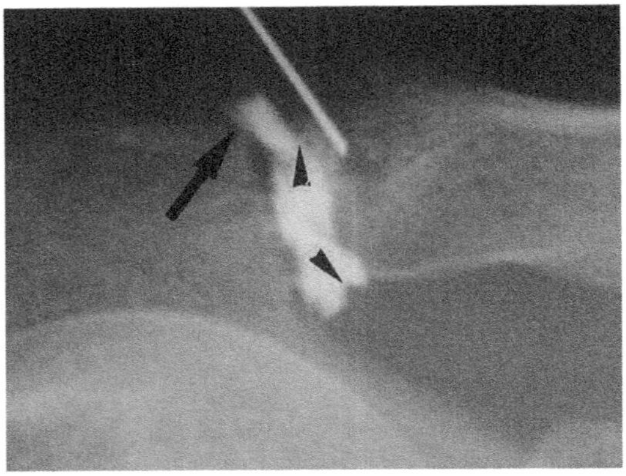

FIGURE 5 Opacification of the acromioclavicular joint cavity. *Arrow* indicates a small superior synovial diverticle. *Arrowheads* indicate the superior and inferior portions of the joint meniscus.

the cartilage of the first rib separates the joint cavity into two compartments, the clavicular one being larger than the sternal one (Fig. 6) (4). Communications between these compartments are related to aging (5). The meniscus also acts as a checkrein that prevents the proximal end of the clavicle from medial subluxation. The joint space is anteriorly and inferiorly oriented. During the movements of the shoulder, both the end of the clavicle and the meniscus rotate around the longitudinal axis of the clavicle and simultaneously shift vertically on the steady surface of the sternum. In addition, the end of the clavicle also slightly rotates against the meniscus.

Indications

The frequency of degenerative changes in the sternoclavicular joint may be underestimated using plain films. Such changes are frequently found at computed tomography (CT) (6) or pathologic (5) examinations in asymptomatic people over 60 years of age. Discomfort and soft-tissue swelling about the proximal end of the clavicle may, however, commonly occur in females of 50 years of age. Long-lasting limitation of the glenohumeral joint due to resistant rotator cuff tendonitis or frozen shoulder might favor overuse, and thus degenerative changes of the sternoclavicular joint in these patients (7). Symptomatic overuse of the sternoclavicular joint is also encountered in manual workers having frequent sustained shoulder abduction of more than 90° or those commonly wearing heavy loads on their shoulder. In some patients, pain from the sternoclavicular joint may project onto the shoulder. Uncommonly, sternoclavicular synovitis or synovial cyst may be found at magnetic resonance (MR) examination (Fig. 7). Discomfort may also occur in both homolateral acromioclavicular and sternoclavicular joints in some

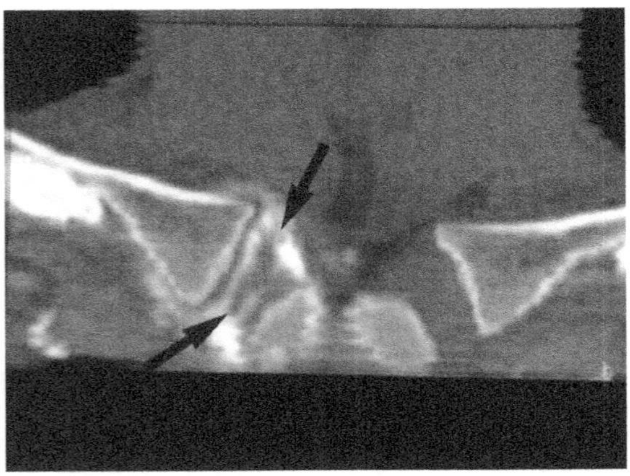

FIGURE 6 Meniscus (*arrows*) in a coronal view of a sternoclavicular joint computed arthrography.

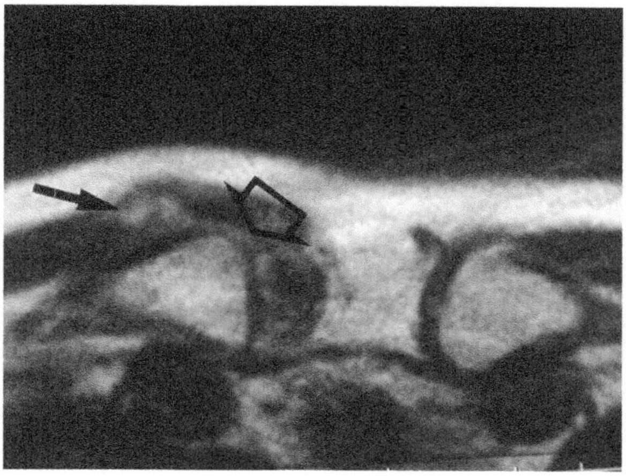

FIGURE 7 Axial T2-weighted magnetic resonance scan of a synovial cyst (*solid arrow*) connected with the articular surface (*open arrow*) of the head of the clavicle in a degenerative sternoclavicular joint.

patients who underwent long-lasting extended cervicotomy due to the mechanical stress of spacers on these joints.

Steroid injection under radiological guidance may also be indicated in some patients with sternoclavicular arthritis associated with seronegative spondylarthropathy or Synovitis, Acne, P. cestulosis, Hyperostosis and Osteitis (SAPHO) syndrome (8), when nonsteroid anti-inflammatory drugs are not effective. Twenty percent of the cases of septic osteoarthritides occurring in drug addicted people involve the sternoclavicular or sternocostal joints (9). In this situation, management of samples should also be appropriate to peculiar microorganisms. Permanent subclavian catheter is another potential cause of sternoclavicular joint infection. Joint space aspiration under radiological guidance may be indicated for bacteriological culture purposes.

Technique of Injection

The patient lies supine with the symptomatic joint at the operator side. The skin is punctured about 5 mm above the middle of the cartilaginous surface of the manubrium. The needle is inserted according to a descending route toward the inferior border of the proximal end of the clavicle (Fig. 8). A lower site of skin puncture would lead the needle to enter the meniscus or the cartilage of the first rib. Opacification first occurs in the subclavicular synovial recess. The joint cavity has a crescent shape with a thin interosseous part and a thick subclavicular part corresponding to the synovial recess (Fig. 9). Small synovial diverticles may be present at the superior part of the joint (Fig. 10).

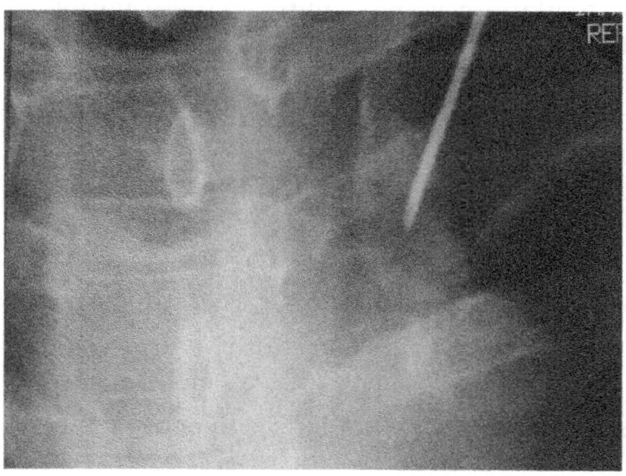

FIGURE 8 Needle placement into a sternoclavicular joint.

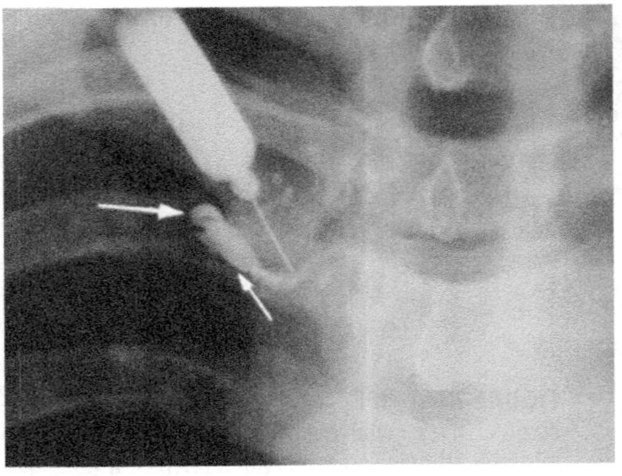

FIGURE 9 Subclavicular synovial recess (*arrows*) in an opacified sternoclavicular joint.

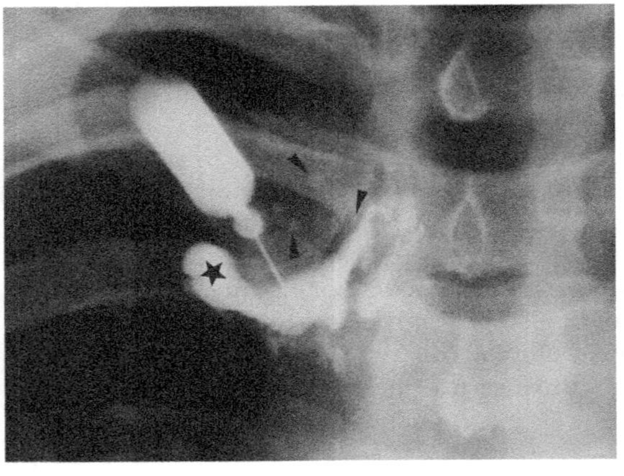

FIGURE 10 Small synovial recess diverticles (*arrowheads*) in a degenerative sternoclavicular joint with enlarged subclavicular synovial recess (*star*).

COSTOVERTEBRAL JOINT

Anatomy

Thoracic spine and ribs articulate at two areas, namely between the heads of the ribs and vertebral bodies (joints of the heads of the ribs) and between the tubercles of the necks of the ribs and the transverse processes of the vertebrae (costotransverse joints), except for T11 and T12 vertebrae, which have no costotransverse joints. The head of each rib articulates with the facets of both the overlying and underlying vertebral bodies through diarthroses, each containing a synovial membrane, separated by a thick ligament extending from the intervertebral disc to the head of the rib (Fig. 11). Communications between the diathroses frequently occur through this ligament. The 1st, 10th, 11th, and 12th ribs articulate, however, with a single vertebral body. Articular surfaces are covered with fibrocartilage. The costotransverse joint is a trochoid joint, the articular surface of the costal tubercle being convex and that of the transverse process being concave. The costotransverse ligaments separate the joints of the heads of the ribs and the costotransverse joints (Fig. 12). Communicating rami of the sympathetic nervous system are located close to the anterior part of the 11th and 12th costovertebral joints. Clinical signs and symptoms of costovertebral arthropathies may thus simulate intrathoracic or intra-abdominal visceral diseases.

Technique of Injection

The patient is in the prone position. The approach is parallel to the head of the rib and slightly ascending in order to avoid the transverse process of the vertebra (Fig. 13). Injected contrast material immediately flows in a linear, semicircular fashion between the head of the rib and the

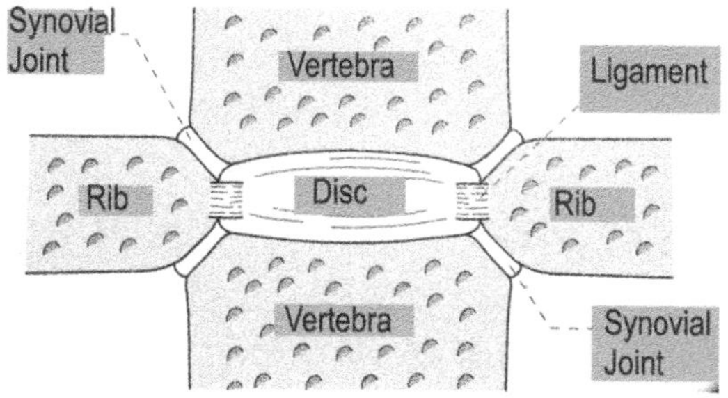

FIGURE 11 Diagrammatic coronal representation of the costovertebral joint.

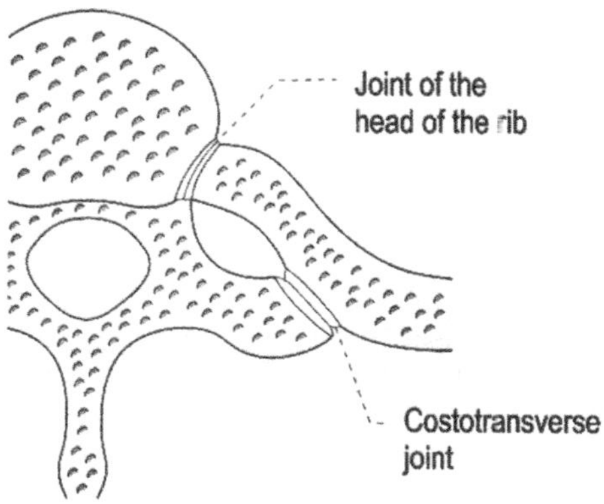

FIGURE 12 Diagrammatic axial representation of the joint of the head of the rib and the costotransverse joint.

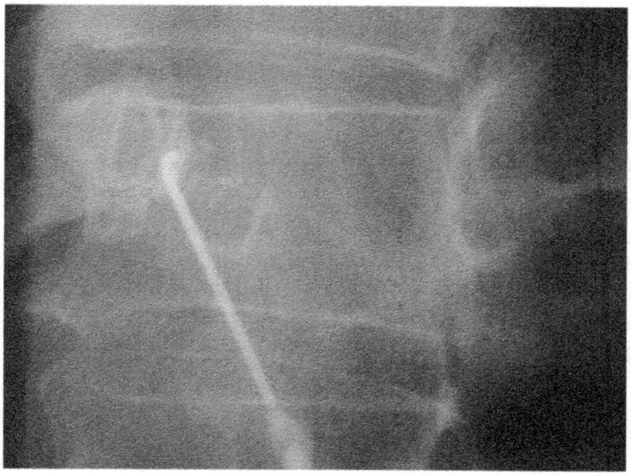

FIGURE 13 Needle placement into a joint of the head of a rib (oblique view).

vertebral body facet. A small synovial recess is present at the inferior aspects of the neck and head of the rib (Fig. 14). Use of nonionic contrast material is recommended to avoid complications of unexpected injection into the thoracic spinal canal.

Indications

Steroid injection into the clinically or arthrographically selected costovertebral joint is indicated in patients who do not respond to usual anti-inflammatory and analgesic therapy.

In some cases, costovertebral pain may be easily recognized. Pain is located at the posterior chest wall, increasing with the rotation of the trunk, deep inspiration, or coughing. It is also

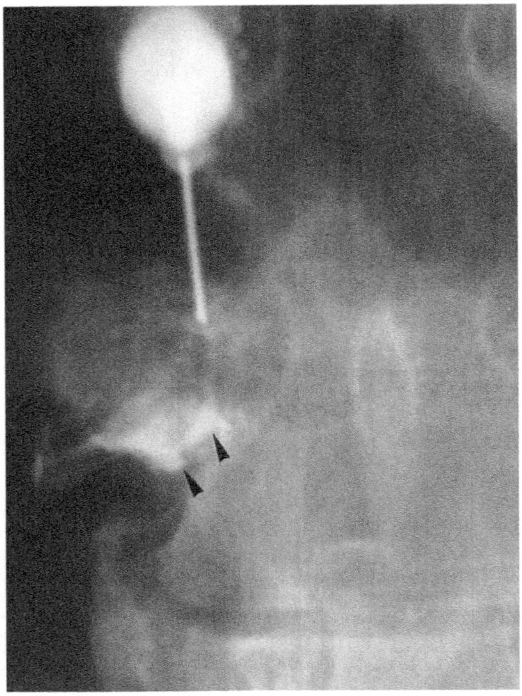

FIGURE 14 Anteroposterior view of the opacified joint of the head of a rib with an inferior synovial recess (*arrowheads*).

provoked by the firm palpation of the abnormal costovertebral joint, as well as the horizontal or vertical passive displacement of the rib. In other patients, especially those with involvement of the 11th or 12th costovertebral joints, pain may simulate a visceral disease, such as coronaropathy, renal colic, or gastrointestinal tract disease (10,11). In these cases, Benhamou et al. recommend early costovertebral arthrography as a diagnostic test, the pain being reproduced by capsular puncture and distension and relieved by a local anesthetic injection (10,12). According to these authors, pain relief should be obtained with injection of one costovertebral level, while injection of an adjacent level is inefficient.

The radiological appearance of the costovertebral joints is of poor diagnostic value, because degenerative changes are age related and may be found in asymptomatic individuals.

Costovertebral joint inflammation is common in seronegative spondyloarthropathy. Rheumatoid costovertebral arthritis has also been reported (13). Steroid injection may be indicated in rheumatic patients with intractable pain.

PUBIC SYMPHYSIS

Anatomy

The pubic symphysis is a diarthroamphiarthrosis joining the medial aspects of each pubic bone. The main axis of the joint space is caudally and ventrally oriented. The bone surfaces are covered with a layer of cartilage and surrounded by a strong interosseous ligament (Fig. 15), the central part of which is soft with frequently a rudimentary joint cavity originating from sagittal splitting of ligament fibers. The ventral aspect of the interosseous ligament is covered with fibers of the adductor muscles interlacing with those of the anterior abdominal wall muscles. The normal pubic joint is almost completely motionless, except during pregnancy.

Indications

Painful pubic joint instability is the main indication of selective steroid injection of the pubic symphysis. It occurs mainly in multiparae (Fig. 16). It is however wise to first rule out pubic

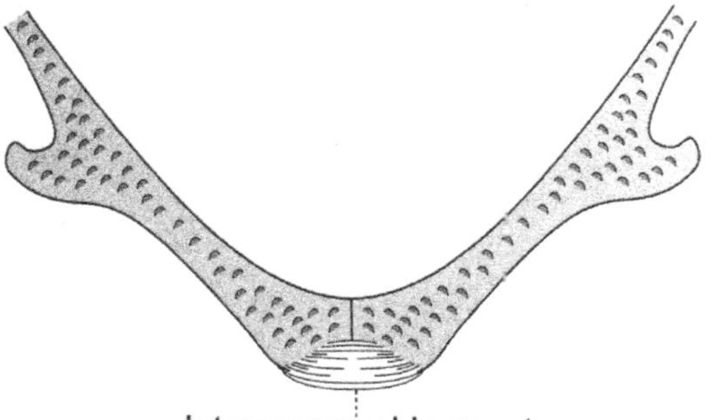

FIGURE 15 Diagrammatic axial representation of the pubic symphysis.

joint infection. Pubic joint infection may complicate intrapelvic surgery or interventional procedure, as well as delivery or abortion. A spontaneously abortive pubic joint infection might be the first step of chronic pubic joint instability, because both diseases share the same radiological changes with subchondral bone erosion and joint space widening. In some cases, however, changes due to infection may be rapidly progressive on plain films or CT examinations. A soft tissue mass at CT or MR examinations would also help recognize infection. Finally, it is sometimes necessary to obtain bacteriological and pathological studies of pubic joint lavage or pubic subchondral bone samples prior to injecting steroid into pubic symphysis.

Technique of Injection

The patient is supine. The approach is anterior on the midline, sagittal, slightly ascending, (Fig. 17). The needle is inserted carefully enough to avoid intra-abdominal penetration of the needle in cases of loose degenerative pubic joint. In cases of obvious pubic diastasis, it is recommended to use a more coronal approach from the ventral aspect of one pubic lamina to the opposite articular surface (Fig. 18). In these cases, the puncture may be safer under CT guidance. Pubic joint space arthrogram is linear and vertically oriented, with the interosseous ligament impinging into the joint cavity as a caudal meniscus-like defect (Fig. 19).

TRANSVERSOSACRAL NEOARTHROSIS

Lumbosacral transitional anomaly is a frequent finding on plain films. In some individuals, it is associated with a neoarthrosis between the transverse process of the lower lumbar vertebrae and the homolateral sacral or, less commonly, iliac wings. The neoarthrosis is frequently unilateral. In a few cases, pain may be clinically suspected to originate in the neoarthrosis. In these uncommon cases, selective injection of a local anesthetic and steroid into the neoarthrosis may be of diagnostic and therapeutic interest. Prior to performing the injection, it is wise to rule out

(A) 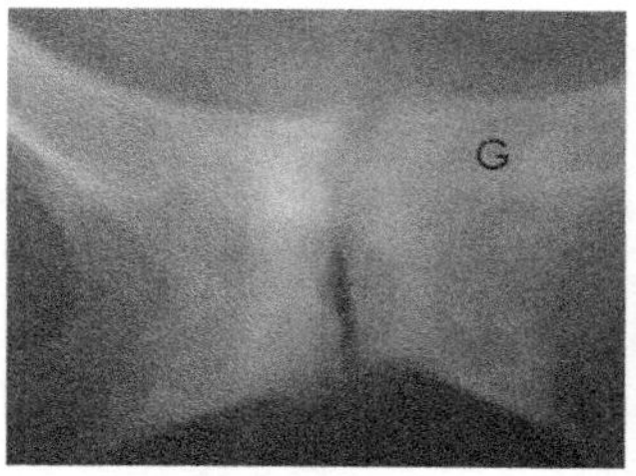

(B) 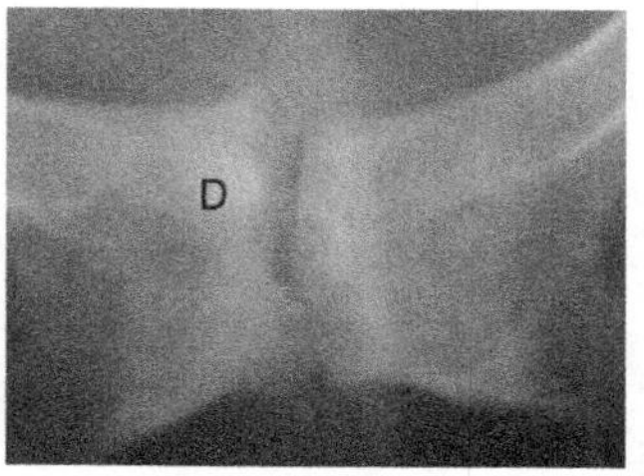

FIGURE 16 Instability in a degenerative pubic symphysis with abnormal (**A**) vertical displacement of each pubic bone according to left (*G*, **A**) or right (*D*, **B**) unipodal standing of the patient. Note the vacuum phenomenon within the pubic joint (**A**).

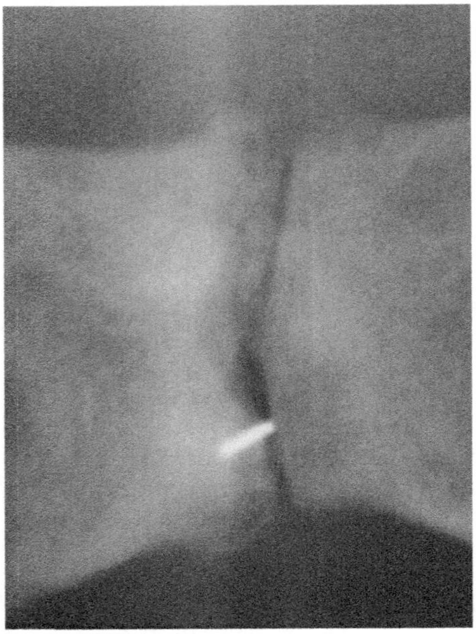

FIGURE 17 Needle placement into the pubic symphysis.

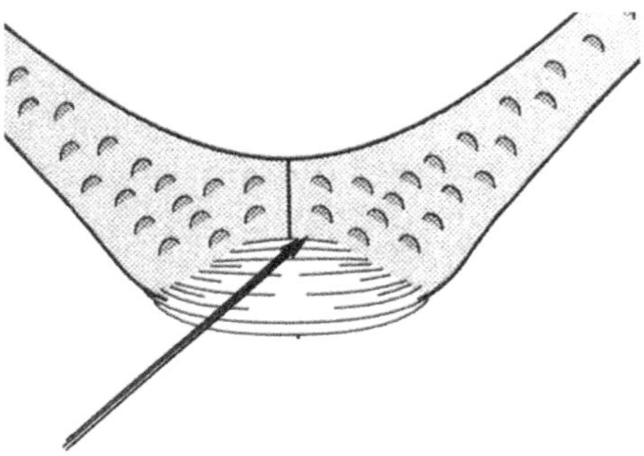

FIGURE 18 Diagrammatic representation of the needle approach in case of pubic diastasis.

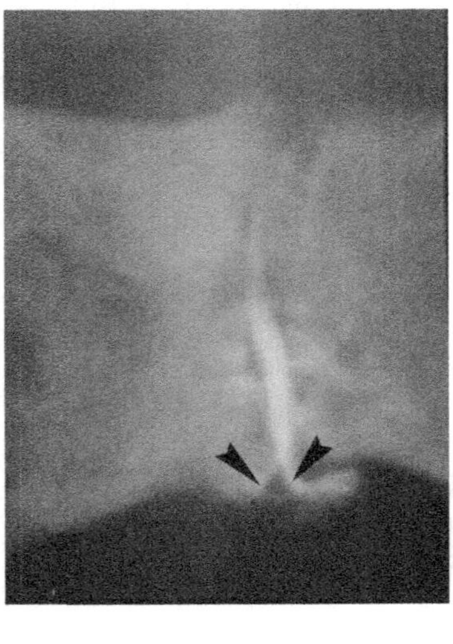

FIGURE 19 Anteroposterior arthrographic view of the pubic symphysis with a meniscus-like lucency (*arrowheads*) at the distal end of the joint cavity.

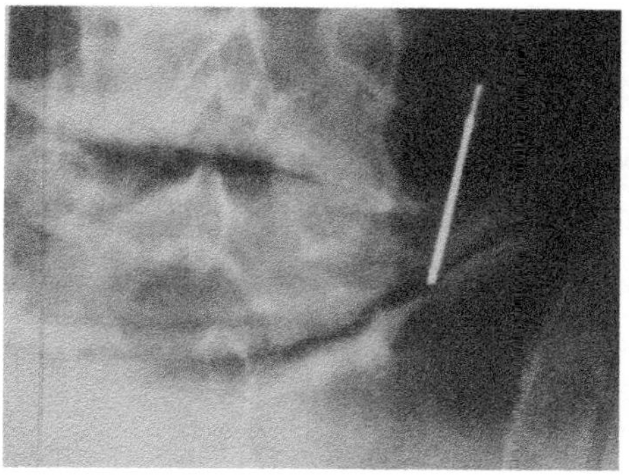

FIGURE 20 Needle placement into a transversosacral neoarthrosis.

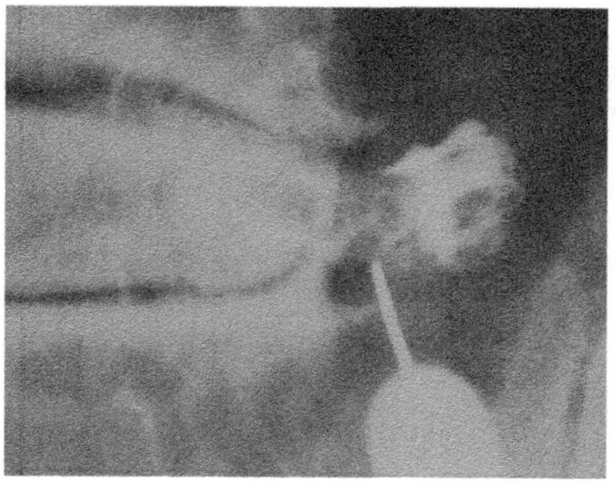

FIGURE 21 Opacified synovial pouch in a transversosacral neoarthrosis.

a simple radiographic superimposition of the transverse process on the sacral wing. In this case, the space between these bones would increase on additional views with appropriate X-ray beam inclination. Selective injection is performed through a posterior approach with the patient prone. The needle is inserted along an oblique direction optimally determined by the X-ray beam orientation that best delineates the neoarthrosis joint space (Fig. 20). Contrast injection may disclose a true synovial pouch surrounding the neoarthrosis (Fig. 21) (14).

REFERENCES

1. Petersson CJ. Degeneration of the acromioclavicular joint. Acta Orthop Scand 1983; 54:434–438.
2. Stendlund B, Goldie I, Marions O. Diminished space in the acromioclavicular joint in forced arm in adduction as a radiographic sign of degeneration and osteoarthrosis. Skeletal Radiol 1992; 21:529–533.
3. Zimmermann B, Erickson AD, Mikolich DJ. Septic acromioclavicular arthritis and osteomyelitis in a patient with acquired immunodeficiency syndrome. Arthritis Rheum, 1989; 32:1175–1178.
4. Sarrafian SK. Gross anatomy and kinesiology. In: Post M. The Shoulder. Surgical and Nonsurgical Management. Philadelphia: Lea G Febiger, 1988:35–53.
5. Kier R, Wain SL, Apple J, et al. Osteoarthritis of the sternoclavicular joint: radiographic features and pathologic correlation. Invest Radiol 1986; 21:227–233.
6. Lucet L, Le Loet X, Menard JF, et al. Computed tomography of the normal sternoclavicular joint. Skeletal Radiol 1996; 25:237–241.
7. Ryckewaert A. Arthroses. In: Rhumatologie–Pathologie Osseuse et Articulaire. Paris: Médecine-Sciences Flammarion, 1987:374–393.
8. Kahn MF, Khan MA. The SAPHO syndrome. Baillère's Clin Rheumatol 1994; 8:333–362.
9. Tsouderos Y, Audisio F. Pathologie ostéoarticulaire des toxicomanes. In: Les Actualités Rhumatologiques. Paris: L'Expansion Scientifique Française, 1982:126–131.
10. Benhamou CL, Roux C, Tourliere D, Gervais T, Viala JF, Amor B. Pseudovisceral pain referred from costovertebral arthropathies. Twenty eight cases. Spine 1993; 18:790–795.
11. Raney FL Jr. Costovertebral–costotransverse joint complex as the source of local or referred pain. J Bone Joint Surg 1966; 48A:1451–1452.

12. Benhamou CL, Roux C, Gervais T, Viala JF. Costovertebral arthrography. diagnostic and therapeutic value of arthrography. Clin Rheum 1988; 7:220–223.
13. Bywaters EGL. Rheumatoid discitis in the thoracic region due to costovertebral joint involvement. Rheumatol Int 1981; 1:83–97.
14. Resnick D. Degenerative disease of the spine. In: Diagnosis of Bone and Joint Disorders. Philadelphia: WB Saunders, 2002:1382–1475.

13 Anesthetic Joint Injections

Anthony C. Toppins
Section of Musculoskeletal Radiology, Baylor University Medical Center, Dallas, Texas, U.S.A.

INTRODUCTION

Percutaneous injections of joints with anesthetic agents have long been employed in the diagnosis and treatment of multiple joint disorders. The mechanism of action of intra-articular anesthetic injections is variable because structures including the joint capsule, synovium, intrinsic ligaments, and exposed subchondral bone underlying cartilage defects are in direct contact with the administered agents (1). Injections utilized primarily in the symptomatic treatment of patients affected by osteoarthrosis or inflammatory synovial processes are referred to as therapeutic injections. Anesthetic joint injections may also be requested during the preoperative evaluation of patients especially during the evaluation for joint arthrodesis or arthroplasty. These are known as diagnostic injections. This chapter will describe the general techniques involved in the injection of a variety of articular compartments with anesthetic agents and the objective evaluation of diagnostic injections.

HIP INJECTIONS

Anesthetic injection of the hip is one of the most commonly requested and performed articular injection procedures in the modern interventional musculoskeletal radiology practice. This is primarily due to the superior accuracy of the fluoroscopically guided needle positioning and the proximity of the joint to the femoral neurovascular structures. Verification of intra-articular needle positioning can also be difficult without fluoroscopy due to the depth of the joint capsule. Fluoroscopic guidance is therefore crucial to determine the proper needle approach and to verify intra-articular needle position. Although many techniques and approaches have been described, an anterolateral approach to the hip joint is preferred. This technique is both effective and safe due to the relative avoidance of the femoral neurovascular bundle.

All injection procedures should begin with a brief patient interview to elicit pertinent clinical history including contraindications for medications such as anticoagulants, medication allergies, and the patient's current pain level. A verbal pain scale from 0 through 10 (0, no pain; 10, worst imaginable pain) is useful for objective assessment (2).

The anterolateral approach to the hip joint is performed with the patient supine, utilizing single plane posterior-anterior (PA) fluoroscopy with the hip internally rotated. The skin overlying the greater trochanter of the femur is marked with an indelible marker prior to skin preparation and draping. Following local anesthesia, a 20 or 22 gauge spinal needle is directed in a posteromedial direction toward the femoral neck at or near its junction with the femoral head. The needle is advanced under fluoroscopic guidance until the cortex of the femoral neck is contacted. If under fluoroscopic monitoring, the needle tip passes medial to the femoral head–neck junction without cortical contact, a steeper entry angle is needed. The needle is then pulled back partially and redirected. If cortical contact is made significantly lateral to the head–neck junction, a shallower angle of the approach is needed and the needle tip can be "walked" medially along the femoral neck until the capsule is entered. After the desired needle tip position is obtained, a small amount of contrast should be injected to document capsular filling (Fig. 1). This should be followed immediately by injection of the desired anesthetic solution.

The injection solution depends on personal preference and availability of medications. The author prefers a mixture of equal parts of lidocaine (2%) and bupivicaine (0.75%). Steroid solutions may also be added for sustained anti-inflammatory effect if desired. It should be noted that some studies have found intra-articular steroid injections to be harmful in some

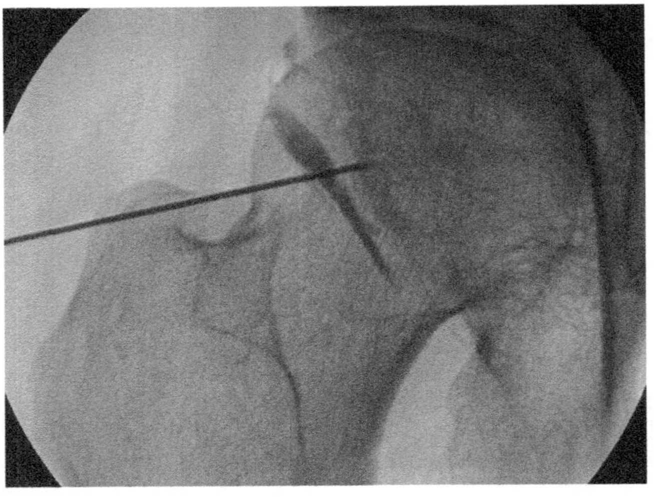

FIGURE 1 Pulmonary artery fluoroscopic image depicting early capsular filling with a typical "ring-like" contrast opacification during anterolateral approach hip injection.

joints (3). Please see Chapter 2 for detailed information regarding the biochemistry of anesthetic agents for selective intra-articular injections.

After the procedure, a repeat pain level assessment is performed to gauge response to the injection. This is typically assessed following a short walk by the patient to assure adequate intra-articular anesthetic disbursement.

Procedural complications with this technique have been minimal in the author's experience. Obvious risks include bleeding from arterial puncture, joint or soft tissue infection, allergic reactions, and inadvertent femoral nerve anesthesia due to infiltration of the local anesthetic. These risks are reduced by a strict aseptic technique, careful needle positioning, and minimization of the overall procedure time. The risk of neurovascular injury is minimized by the anterolateral approach, which is especially helpful in obese patients in whom palpation and isolation of the femoral artery may be difficult. This approach also obviates the need for lateral fluoroscopy, which is variably employed during the lateral approach to the hip joint (4). Ultimately, the operator's familiarity and comfort level should dictate the selected approach to the hip.

Anesthetic agents can also be added to the contrast solution during arthrographic evaluation of the postarthroplasty hip to determine intra- or extra-articular origin of recurrent hip pain (5).

KNEE INJECTIONS

Fluoroscopically guided anesthetic knee injections are uncommonly requested. This is likely due to the relative ease in which the joint can be injected without fluoroscopic guidance. Additionally, excellent noninvasive imaging of the knee exists to complement the clinical assessment. If anesthetic knee injection is requested, the purpose is usually therapeutic rather than diagnostic. The technique usually involves needle placement into the patellofemoral joint from a medial or lateral approach. The margins of the patellofemoral joint can usually be palpated with the patient supine and knee relaxed (firing of the quadriceps muscle narrows the patellofemoral joint space). The skin is marked and anesthetized. A 20- to 22-gauge hypodermic needle is then advanced at an angle appropriate for the medial or lateral patellar facet (steeper angle for medial entry) (Fig. 2). The knee joint can easily accommodate 40 cc or more of fluid. Therefore, dilution of the anesthetic/steroid solution with sterile, nonbacteriostatic saline to a 10 cc volume is useful to assure adequate disbursement. Drainage of the preexisting joint fluid is useful to avoid uncontrolled dilution of the anesthetic agents. It should be noted that osteoarthrosis in any joint and in particular the knee may significantly worsen following intra-articular anesthetic injection likely due to the elimination of the feedback inhibition pathway and cartilage "softening" due to steroid affects (6).

Recent reports have been presented in the literature indicating that intra-articular anesthetic and steroid injection in the knee may reduce pain and the time of convalescence

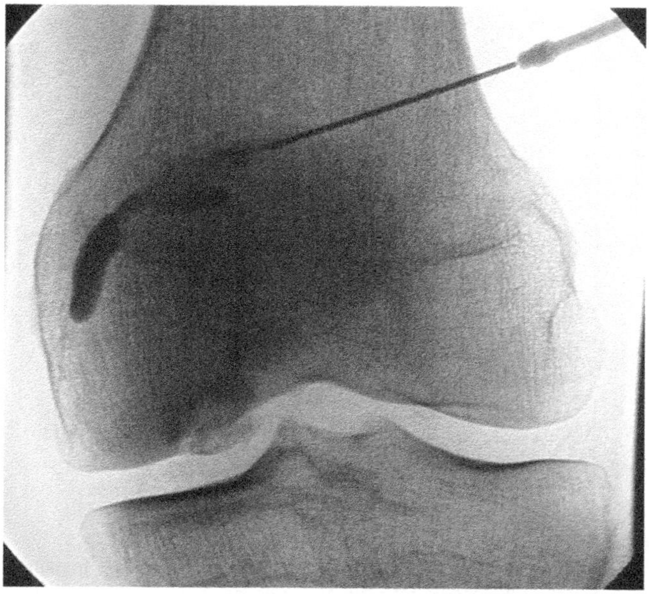

FIGURE 2 Pulmonary artery fluoroscopic image during knee injection from lateral approach into the patellofemoral joint.

following arthroscopic knee surgery (7). This has not been widely accepted by orthopedic surgeons at our institution. Finally, diagnostic injections are occasionally requested to evaluate postoperative pain following knee arthroplasty.

FOOT AND ANKLE INJECTIONS

Selective anesthetic injections of the foot and ankle are also commonly requested of the musculoskeletal radiologist. These are often diagnostic in nature to assess the anatomic location of foot or ankle pain, because the closely associated osseous and nonosseous structures often make clinical and radiographic assessment difficult (8). Often these injections are performed in the preoperative assessment for foot or ankle arthrodesis or to assess symptomatic tarsal coalitions. Anterior subtalar joint injections are also useful in the assessment of the sinus tarsi syndrome. Less commonly, these injections are performed on a purely therapeutic basis with the addition of steroid medication. Objective assessment of the degree of pain relief following an injection is crucial for diagnostic injections. The pain level assessment is identical to that for the hip injection described previously.

Ankle and Hindfoot Injections

Anesthetic injections of the ankle and subtalar joints can be performed under fluoroscopic or computed tomography (CT) guidance. Needle choice is by personal preference. A 22-gauge, 1.5 in needle works well for all of the ankle injections. The author prefers fluoroscopic guidance for the ankle and subtalar joints.

The ankle joint is entered by first marking the desired needle entry location at the medial third of the tibiotalar joint in the anterior projection to avoid the location of the dorsalis pedis artery. This is best accomplished using a vertical mark with an indelible marker. The skin entry site is then determined in the lateral projection where the exact location of the anterior margin of the tibiotalar joint can be visualized rather than the talar dome, which is visualized in the anterior projection (9). A horizontal skin mark is made, which will intersect the vertical line previously drawn, indicating the exact point of skin entry (Fig. 3). The test injection with a water-soluble contrast material is then performed in the lateral projection (Fig. 4). Reasonable injection volume is 10 to 12 cc.

The posterior subtalar joint can be entered with fluoroscopic guidance. Difficult cases such as patients suffering from advanced osteoarthrosis of the hind foot may benefit from CT guidance. For fluoroscopic guidance, the patient is positioned with the contralateral side down

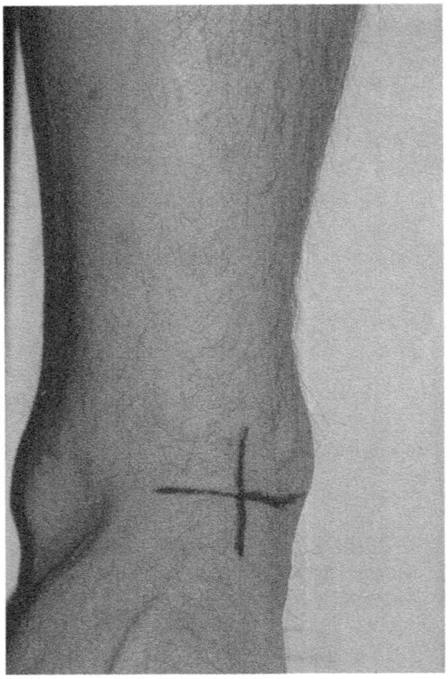

FIGURE 3 Intersecting vertical and horizontal skin markings depicting the needle entry site for ankle injection, based on location of medial third of joint from pulmonary artery fluoroscopy and anterior margin of joint from lateral fluoroscopy, respectively.

and the ankle of interest positioned medial side down (Fig. 5). The desired approach to the posterior subtalar joint is from the lateral aspect of the ankle anterior or posterior to the lateral malleolus (Fig. 6) (10). Abnormal communication between tarsal compartments should be noted because the passage of anesthetic into adjacent compartments can lead to diagnostic uncertainty, that is, communication between the posterior subtalar and ankle joints, which occurs in roughly 10% of patients (Fig. 6).

Due to its communication with the talonavicular joint, the anterior subtalar joint is readily accessible from a medial approach under fluoroscopic guidance with patient supine and foot placed lateral side down (Figs. 7 and 8). Significant navicular curvature requires an oblique

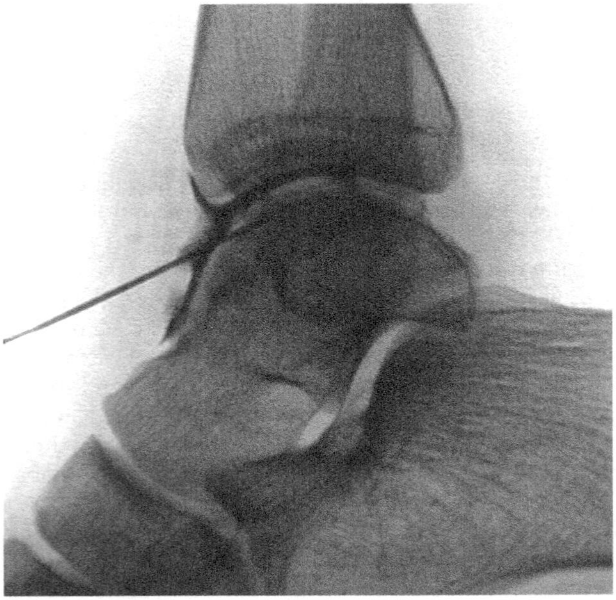

FIGURE 4 Lateral fluoroscopic image during ankle injection. *Source*: Courtesy of Guerdon Greenway M.D., Baylor University Medical Center, Dallas, Texas, U.S.A.

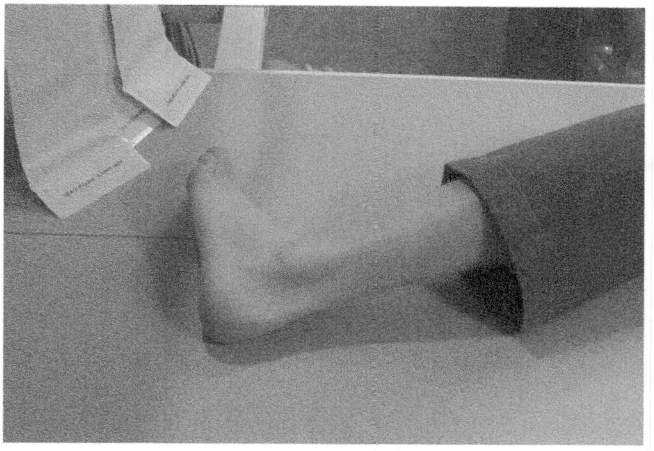

FIGURE 5 Patient positioning for posterior subtalar injection. *Source*: Courtesy of Drew Small M.D., Baylor University Medical Center, Dallas, Texas, U.S.A.

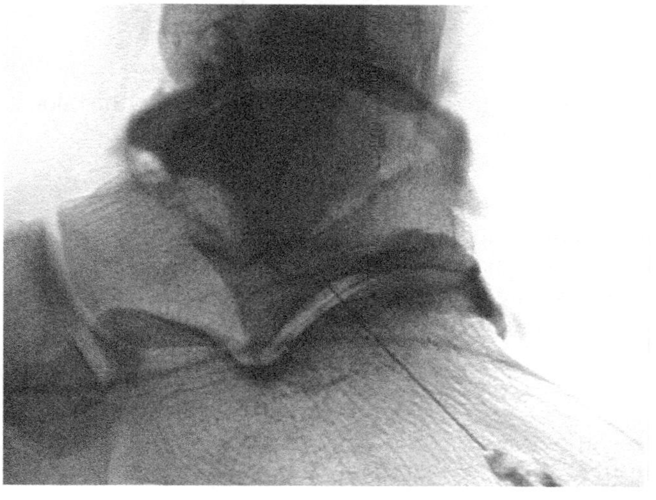

FIGURE 6 Lateral fluoroscopic image demonstrating the posterolateral approach to the posterior subtalar joint. Note the passage of contrast into the ankle joint, which occurs in 10% of patients.

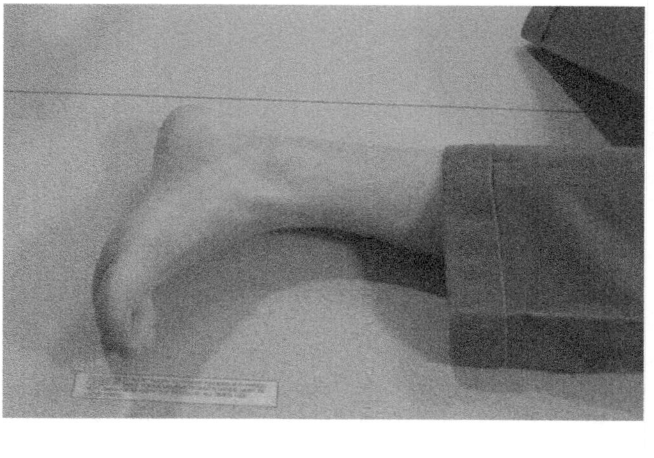

FIGURE 7 Patient positioning for anterior subtalar (talonavicular) injection.

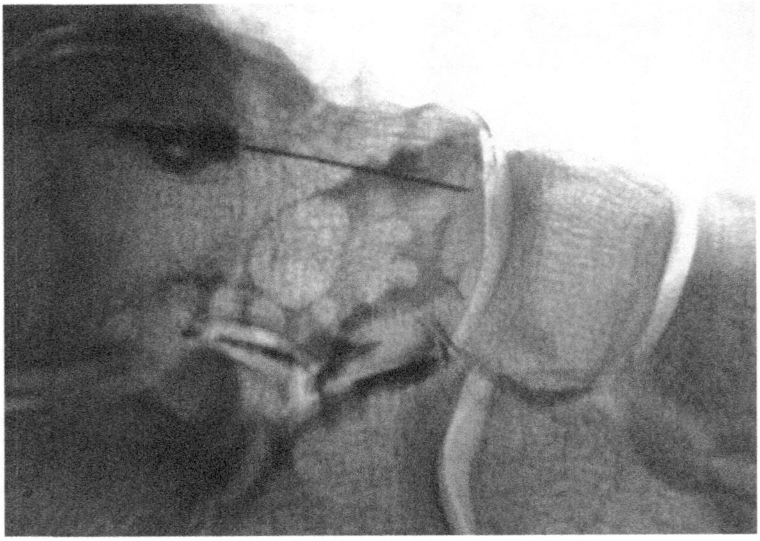

FIGURE 8 Lateral fluoroscopic image during talonavicular joint injection with expected opacification of the anterior subtalar joint.

needle path, as shown on this PA fluoroscopic image (Fig. 9). A dorsal approach to the talonavicular joint can be used but is often difficult due to dorsal osteophyte formation.

Other Injections of the Foot

Additional, requested fluoroscopically guided injections of the foot include injections of tarsometatarsal, metatarsophalangeal, and interphalangeal joints. These injections are facilitated by a dorsal approach utilizing the fine-needle technique. These injections, similar to tarsal injections, are usually employed in the evaluation of possible sites of surgical arthrodesis.

SHOULDER INJECTIONS

Percutaneous injections of the glenohumeral joint are commonly performed without fluoroscopic guidance by a variety of clinicians to treat shoulder pain. However, as with hip injections,

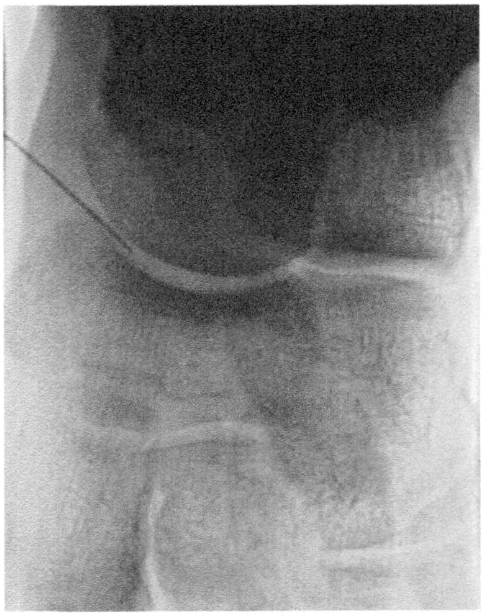

FIGURE 9 Pulmonary artery fluoroscopic image demonstrating oblique needle course required due to arcuate configuration of tarsal navicular joint.

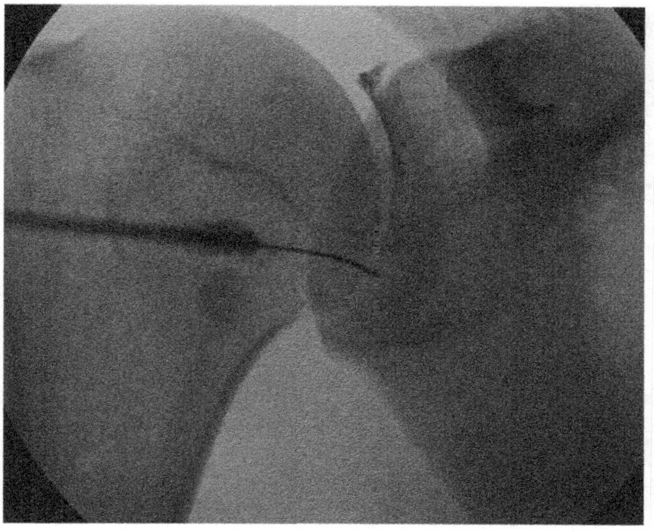

FIGURE 10 Pulmonary artery fluoroscopic image during glenohumeral injection using anterior approach. *Source*: Courtesy of Guerdon Greenway M.D., Baylor University Medical Center, Dallas, Texas, U.S.A.

fluoroscopy provides the ability to definitively assess intra-articular needle position. Glenohumeral joint injections are most commonly performed from an anterior approach, utilizing 20- to 25-gauge spinal needles. A 22-gauge needle is preferred due to its relatively small diameter and maneuverability. The technical aspects are identical to those of glenohumeral arthrography and typically involve entering the joint capsule from an anterior approach at the level of the lower one-third of the glenoid (Fig. 10). Occasionally, a posterior approach will be required due to osteophytic ridging along the anterior margin of the joint. This technique involves positioning the patient in a prone-oblique position on a conventional fluoroscopic table (Fig. 11) or utilizing an oblique orientation with a C-arm device to visualize the glenohumeral joint in profile (Fig. 12). The needle can then be advanced between the humeral head and the glenoid. This technique has been recently described as the preferred method in magnetic resonance arthrography, particularly in the assessment of anterior instability of the glenohumeral joint (11).

Anesthetic injection of the glenohumeral joint can be useful diagnostically when dilemmas exist between joint-related pain and cervical radicular symptoms. Objective assessment of the patient's response to anesthetic injection is identical to that for hip and ankle injections described previously in this chapter.

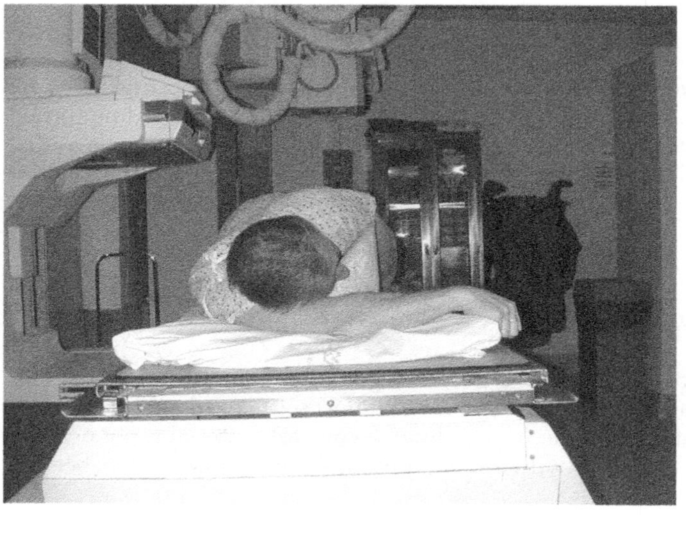

FIGURE 11 Prone-oblique positioning used during posterior approach to the glenohumeral joint. *Source*: Courtesy of Christine Chung M.D., VA San Diego Healthcare System and University of California San Diego with special thanks to Troy Gonzalez M.D.

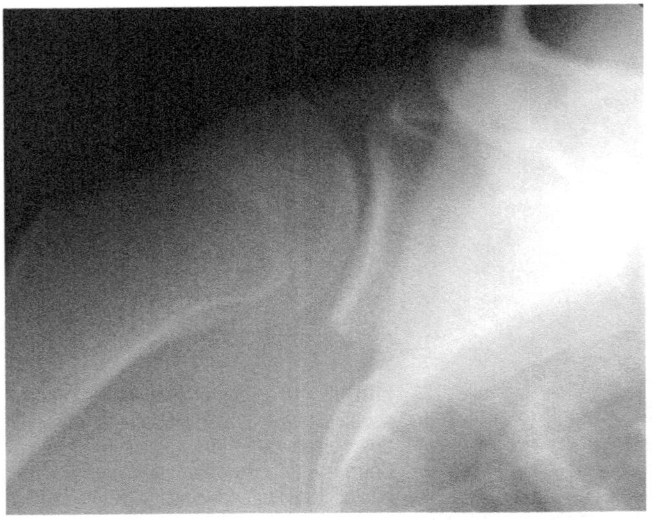

FIGURE 12 Pulmonary artery fluoroscopic image depicting glenohumeral joint in profile during posterior approach injection. *Source*: Courtesy of Christine Chung M.D., VA San Diego Healthcare System and University of California San Diego.

The distention shoulder arthrogram involves progressive distention of the glenohumeral joint capsule with a combination of water-soluble contrast material and anesthetic agents such as bupivicaine. This procedure has been advocated in the diagnosis and treatment of glenohumeral adhesive capsulitis or "the stiff and painful shoulder" (12). It is a minimally invasive alternative to surgical capsular release. Objective assessment of response to this procedure can be performed by comparing passive and active range of motion before and after anesthetic capsular distention. Often the administered anesthetic agents combined with the capsular distention will provide enough symptomatic relief to facilitate increased exercise capacity. This can ultimately end the cycle of the stiff and painful shoulder, allowing the patient to gradually increase the duration and frequency of range of motion exercises.

WRIST INJECTIONS

Anesthetic injections of the wrist are not commonly requested as a separate procedure. However, the addition of anesthetic agents such as lidocaine and/or bupivicaine to the contrast material during tricompartmental wrist arthrography can be useful. This modification to standard wrist arthrography adds the additional element of localizing symptomatic compartments. This is especially useful when no abnormal communication is documented between compartments, suggesting intra-articular causes of wrist pain other than intrinsic ligament injuries.

ELBOW INJECTIONS

Anesthetic injections in and around the elbow are occasionally requested. Elbow injections are performed with the patient prone, arm over head, with elbow flexed at 90°, and thumb turned upward (Fig. 13). A 22- to 25-gauge hypodermic needle is then advanced from a radial approach entering the humeroradial joint. The needle position is verified using a small amount of contrast material followed by injection of desired anesthetic/steroid agent(s) (Fig. 14). Periarticular injections of the olecranon bursa and humeral epicondyles can also be performed to treat bursitis and epicondylitis, respectively.

ULTRASOUND-GUIDED INJECTIONS

Certain articular compartments are amenable to ultrasound-guided anesthetic injections. This technique is especially useful in satellite imaging centers or clinical settings where fluoroscopy units are unavailable. It is also cost effective because iodinated contrast material is not needed. In addition, the lack of ionizing radiation with sonographic guidance is an added benefit.

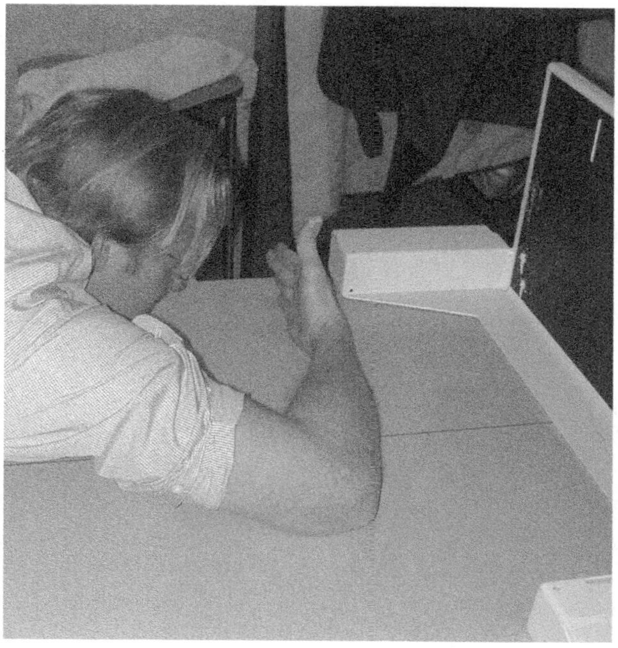

FIGURE 13 Patient positioning for elbow injection.

Sonographically guided injections of the knee and hip have been described (13). An effusion within these joints facilitates verification of intra-articular needle position by providing a sonographic window through which the hyperechoic needle tip is visualized. Small amounts of air can also be injected to demonstrate hyperechoic bubbles within the joint fluid. This guidance technique is more difficult to perform in the absence of a joint effusion.

Sonographically guided injections of the knee are performed by localizing fluid (Fig. 15A) within the suprapatellar synovial bursa typically with a transverse orientation (Fig. 15B) of a linear, high-frequency transducer, that is, 13 MHz. The needle is then placed into the joint fluid entering the skin superolateral to the patella (Fig. 15C). If no fluid is present within the knee, the needle is placed into the lateral patellofemoral compartment by palpation alone. Intra-articular placement can then be assessed by injecting small amounts of air through a syringe attached to the needle. The echogenic bubbles of air should then be visualized by sonographic interrogation of the suprapatellar bursa (Fig. 16).

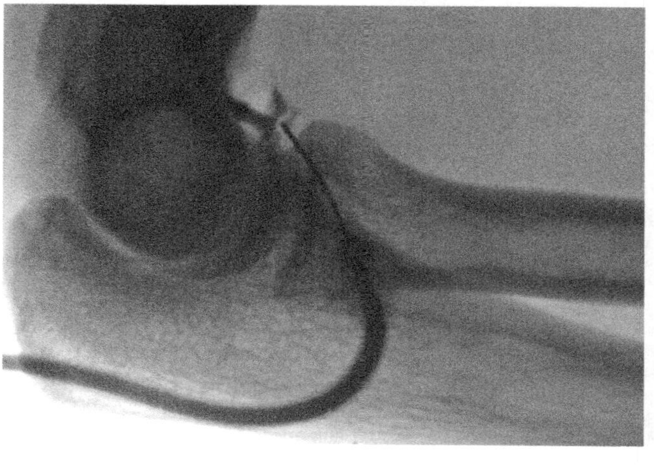

FIGURE 14 Lateral fluoroscopic image during elbow injection.

(A) (B)

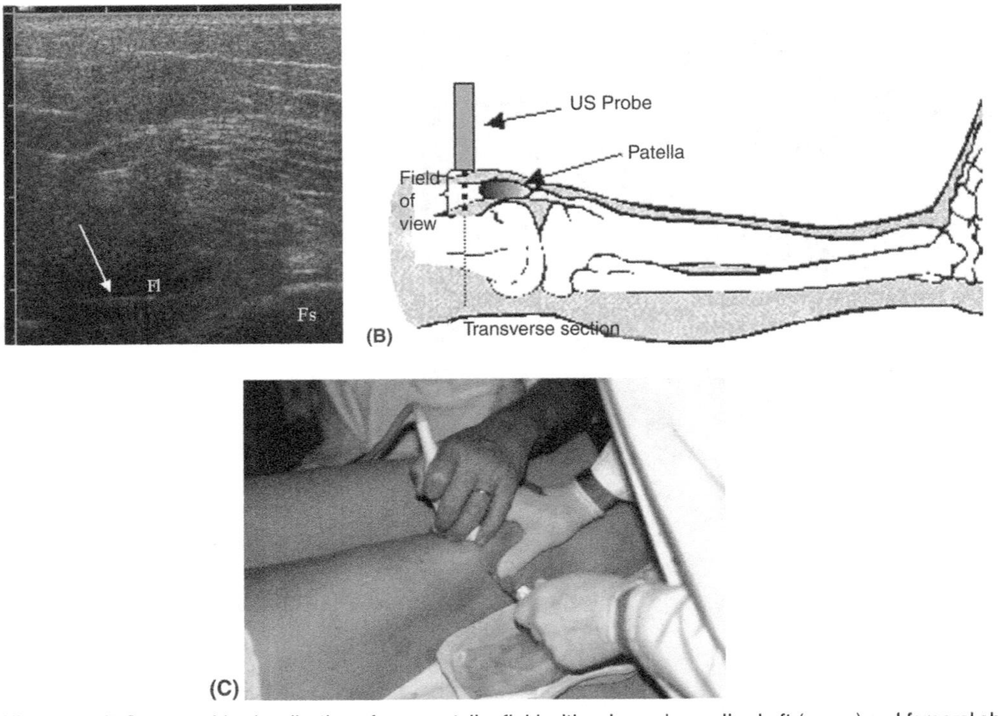

(C)

FIGURE 15 (**A**) Sonographic visualization of suprapatellar fluid with echogenic needle shaft (*arrow*) and femoral shaft. (**B**) Transducer positioning for knee injection. (**C**) Needle/transducer positioning for knee injection. Abbreviations: Fl, suprapatellar fluid; Fs, femoral shaft. *Source*: Courtesy of Henning Bliddal M.D., The Parker Institute, H:S Frederiksberg Hospital, Copenhagen, Denmark.

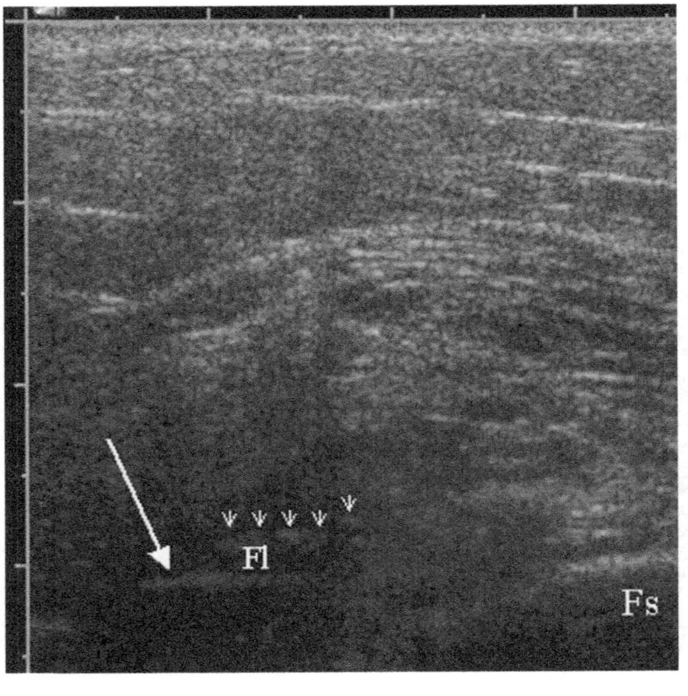

FIGURE 16 Sonographic image demonstrating layering air bubbles (*small arrows*), needle shaft (*large arrow*) and femoral shaft, and joint fluid. *Abbreviations*: Fl, joint fluid; Fs, femoral shaft. *Source*: Courtesy of Henning Bliddal M.D., The Parker Institute, H:S Frederiksberg Hospital, Copenhagen, Denmark.

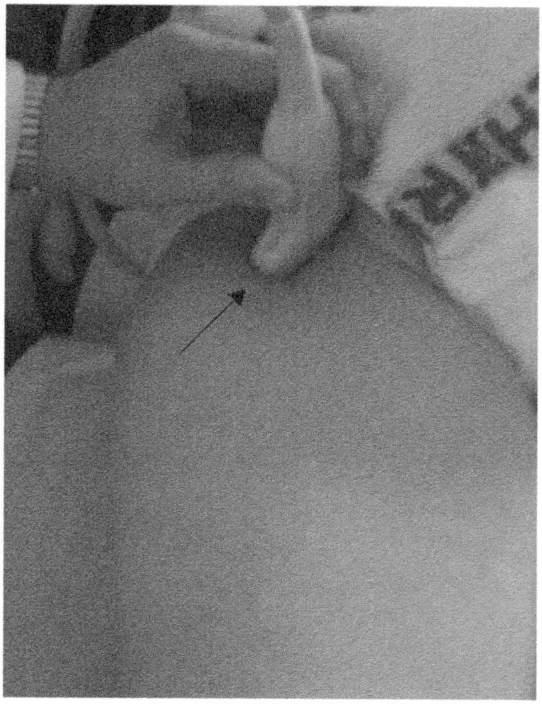

FIGURE 17 Transducer positioning for hip injection. *Source*: Courtesy of Henning Bliddal M.D., The Parker Institute, H:S Frederiksberg Hospital, Copenhagen, Denmark.

Hip injections are performed with the patient supine and with a lower frequency linear transducer such as 8 MHz, which is needed for deeper penetration of the thicker soft tissues overlying the hip. The transducer is placed parallel to the femoral neck (Fig. 17) and the needle is inserted 8 to 10 cm below the inguinal ligament to enter the anteroinferior joint capsule just below the femoral head (Fig. 18A and B). As with knee injections, this technique is facilitated by a joint effusion but can be performed successfully without an effusion. Injection of a small amount of air verifies the intra-articular needle position (Fig. 19A and B). A distinct advantage of sonography during hip injections is the real-time visualization of the femoral artery and vein, utilizing color Doppler evaluation, thus increasing safety and reducing the risk of vascular complications.

(A)

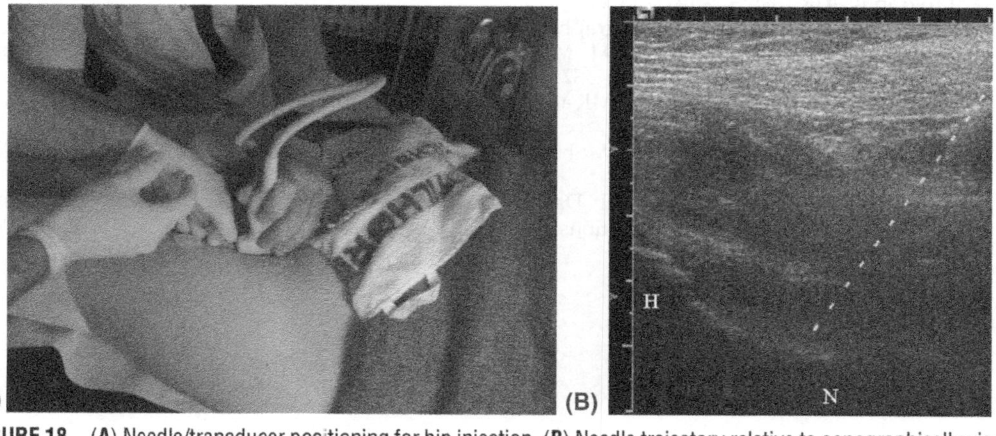

(B)

FIGURE 18 (**A**) Needle/transducer positioning for hip injection. (**B**) Needle trajectory relative to sonographically visualized femoral neck and femoral head. *Source*: Courtesy of Henning Bliddal M.D., The Parker Institute, H:S Frederiksberg Hospital, Copenhagen, Denmark.

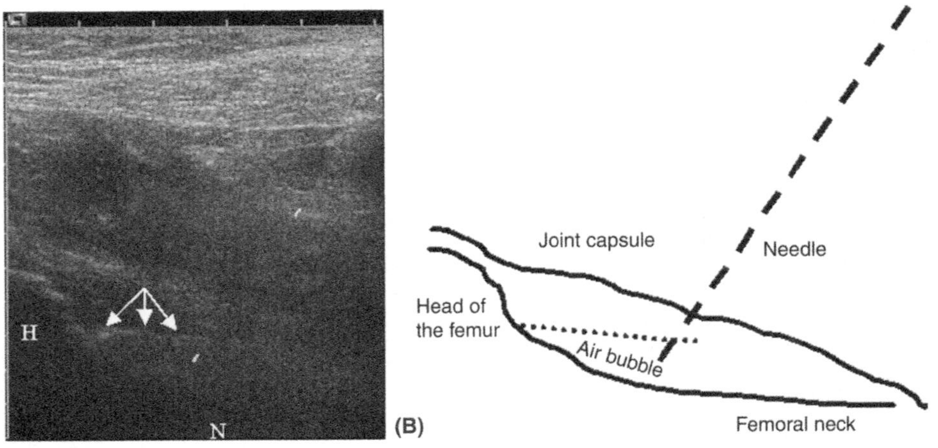

(A) (B)

FIGURE 19 (**A**) Layering echogenic air bubbles within hip joint capsule (*arrows*). (**B**) Diagram. *Source*: Courtesy of Henning Bliddal M.D., The Parker Institute, H:S Frederiksberg Hospital, Copenhagen, Denmark.

The choice of imaging guidance should be based on personal preference and operator experience.

REFERENCES

1. Creamer P, Hunt M, Dieppe P. Pain mechanisms in osteoarthritis of the knee: effect of intraarticular anesthetic. J Rheum 1996; 23(suppl 6):1031–1036.
2. Lucas P, Hurwitz S, Kaplan P, Dussault R, Maurer E. Fluoroscopically guided injections into the foot and ankle: localization of the source of pain as a guide to treatment prospective study. Radiology 1997; 204:411–415.
3. Flanagan J, Casale F, Thomas T, Desai K. Intra-articular injection for pain relief in patients awaiting hip replacement. Ann Royal Coll Surg Engl 1988; 70:156–157.
4. Kilcoyne R, Kaplan P. The lateral approach for hip arthrography. Skeletal Radiol 1992; 21:239.
5. Braunstein E, Cardinal E, Buckwalter K, Capello W. Bupivicaine arthrography of the post-arthroplasty hip. Skeletal Radiol 1995; 24:519–521.
6. Wright V, Hartfall S, Chandler G. Efficacy of intra-articular steroid. Br Med J 1965; 1:1065.
7. Rasmussen S, Larsen A, Thomsen S, Kehlet H. Intra-articular glucocorticoid, bupivicaine and morphine reduces pain, inflammatory response and convalescence after arthroscopic meniscectomy. Pain 1998; 78:131–134.
8. Mitchell M, Bielecki D, Bergman A, Kursunoglu-Brahme S, Sartoris D, Resnick D. Localization of specific joint causing hindfoot pain: value of injecting local anesthetics into individual joints during arthrography. AJR 1995; 164:1473–1476.
9. Pavlov H. Ankle and subtalar arthrography. Clin Sports Med 1982; 1(1):47–69.
10. Ruhoy M, Newberg A, Yodlowski M, Mizel M, Trepman E. Subtalar joint arthrography. Semin Musculoskeletal Radiol 1998; 2(1):433–437.
11. Chung C, Dwek J, Feng S, Resnick D. MR arthrography of the glenohumeral joint: a tailored approach. AJR 2001; 177(1):217–219.
12. Gilula L, Schoenecker P, Murphy W. Shoulder arthrography as a treatment modality. AJR 1978; 131:1047–1048.
13. Qvistgaard E, Kristoffersen H, Terslev L, Danneskiold-Samsoe B, Torp-Pedersen S, Bliddal H. Guidance by ultrasound of intra-articular injections in the knee and hip joints. Osteoarthritis Cartilage 2001; 9:512–517.

14 Arthrography—Techniques and Indications

David Karasick
Department of Radiology, Thomas Jefferson University Hospital, Philadelphia, Pennsylvania, U.S.A.

INTRODUCTION

Arthrography is a well-recognized diagnostic tool in the evaluation of joint disorders. The technique, indications, and contraindications are discussed. The opinions rendered in this chapter reflect my personal experience with arthrography over the last 25 years. My interest in arthrography began in the winter of 1978 when I attended Dr. Freiberger's course at the Hospital for Special Surgery in New York City. The lectures by Dr. Freiberger, Kaye, Ghelman, Schneider, Goldman, and Pavlov peaked my interest, which has never waned ever since. The chapter will primarily deal with arthrography of the wrist, elbow, shoulder, hip, knee, and ankle, in that order. With the advent of magnetic resonance (MR), the number of conventional arthrograms of these joints has steadily decreased over the last decade. MR arthrography steadily replaced the "classic" arthrogram in the last five years, and the performance of these procedures still requires a basic understanding of arthrographic techniques, indications, and contraindications.

The general indications for all arthrographic procedures are firstly to visualize intra- and para-articular structures that cannot be adequately identified on conventional radiographs. This includes the capsule, synovium, cartilaginous structures, and intra-articular and collateral ligaments. Secondly, there is often a need to obtain synovial fluid for diagnostic laboratory evaluation. The arthrographic contraindications, on the other hand, include superficial skin infections and previous severe allergic reaction to contrast medium. Patients at risk for contrast media reaction can be given one of two premedication regimens. The first is a corticosteroid/antihistamine option: prednisone 50 mg orally at 13 hours, 7 hours, and 1 hour before contrast media injection plus diphenhydramine (Benadryl) 50 mg orally one hour before the contrast media injection. The second regimen option is corticosteroids alone: methylprednisolone (Medrol) 32 mg orally 12 hours and 2 hours before contrast media injection. Either ionic (diatrizoate) or nonionic (Iopamidol) contrast agents are acceptable. Nonionic contrast is usually preferable because of its less-irritating effect on the synovial lining (1) The amount of contrast injected per joint depends on the size of the joint capsule. Another option in patients with allergy history is to perform a pneumoarthrogram, only injecting room air as a contrast agent. Over the years, I have salvaged many a knee, wrist, and shoulder arthrogram study this way.

EQUIPMENT

The arthrogram procedures are performed in the fluoroscopic suite, utilizing a 0.3 or 0.6 focal spot. I prefer to use one of the commercial arthrogram trays such as Pharmaseal (Allegiance Healthcare Corp, Texas, U.S.A.). This tray contents include two locktip syringes (10 cc), one locktip syringe (60 cc), four needles ranging between 18, 20, 21, and 25 gauge, an extension tube, three swab sticks, three gauze pads, fenestrated drape, a towel, lidocaine hydrochloride (United States Pharmacopeia) 1% (5 mL), and bandage. An additional lidocaine hydrochloride for deep joints and 3.5 in. spinal needles, either of 20 or 22 gauge should be also on hand. Epinephrine (0.3 mL 1:1000 strength) may be injected to retard contrast absorption in joints when there may be a delay in visualizing structures. However, a well-performed study takes 10 to 15 minutes, thereby obviating the need for epinephrine.

SHOULDER ARTHROGRAM

Scout anteroposterior (AP) radiographs of the patient's shoulder are done in internal and external rotation with 10° to 15° caudal angulation. The injection is performed with the patient lying supine on the fluoroscopic radiographic table, with the arm minimally abducted and the thumb

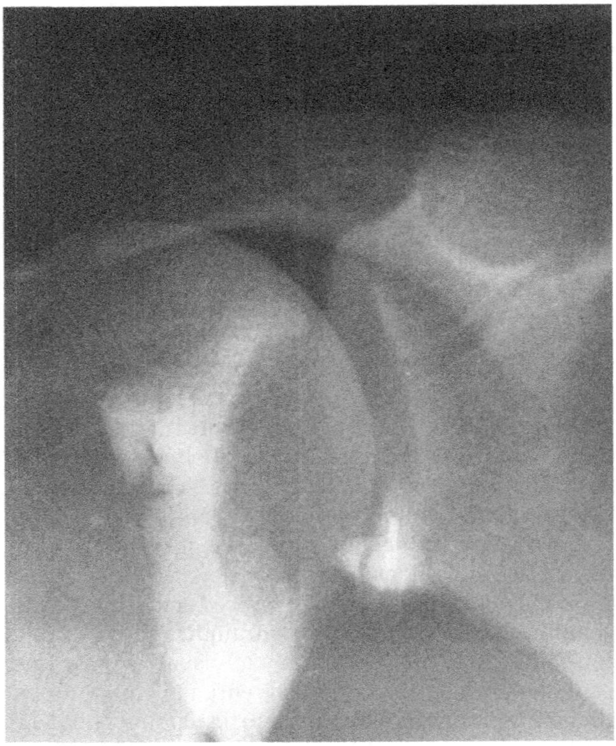

FIGURE 1 Anteroposterior view of shoulder shows early contrast filling within the joint capsule following proper needle placement in the glenohumeral joint.

in the up or 12 o'clock position. There is no need to oblique the patient to view the joint in tangent because this will only cause the fibrocartilaginous labrum of the glenoid to be positioned over the joint space, causing difficult needle placement. Under fluoroscopic guidance, a lead marker or BB is centered on the midpoint of the glenohumeral joint, and this point is then marked on the skin for the site of injection. Following marker placement, sterile preparation of the shoulder and administration of a local anesthetic, a 20- or 21-gauge 3.5 in. spinal needle is directed straight down into the center of the glenohumeral joint. With proper placement, a test injection will result in contrast flowing away from the tip of the needle and covering the humeral neck, humeral head, and subcoracoid (subscapular) recess (Fig. 1). Following successful needle placement, 4 to 5 cc of iodinated or noniodinated contrast is followed by an injection of 10 cc of room air. The needle is removed, and supine radiographs identical to the scout internal and external views are obtained (Fig. 2). Additional radiographs obtained include erect films with the arm

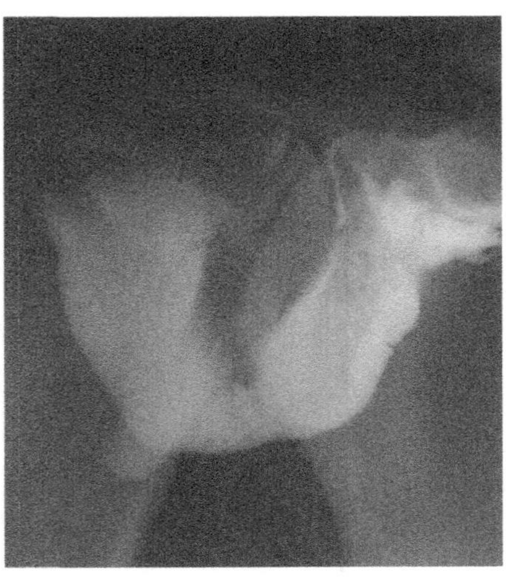

FIGURE 2 Anteroposterior view of shoulder shows contrast-filled joint following needle removal.

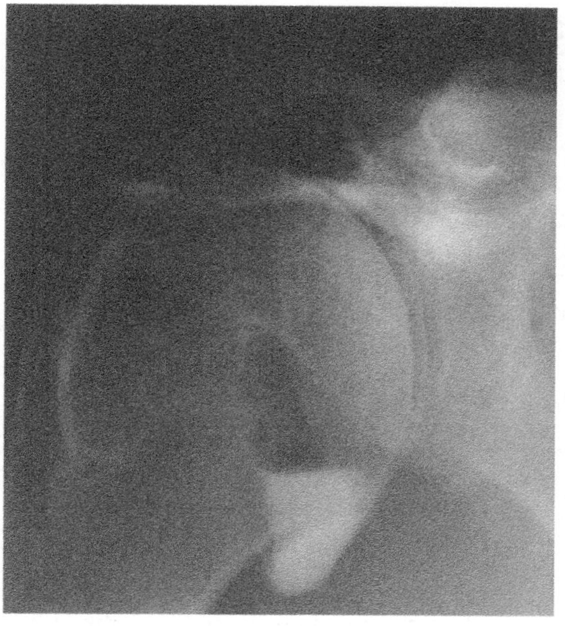

FIGURE 3 Erect anteroposterior of shoulder view shows double-contrast effect within joint capsule.

holding a 5 to 10 lb. weight or sandbag; these are done again with caudal angulation, internal and external rotation, and pre- and postexercise (Fig. 3). Exercise involves the patient rotating the shoulder for several minutes. If necessary, fluoroscopic spot radiographs are obtained to visualize a particular area such as suspected undersurface partial rotator cuff tears. There is no need to obtain additional views like axillary, bicipital groove, or Grashey projections.

Indications for shoulder arthrography include detection of partial or complete rotator cuff tears, adhesive capsulitis, loose bodies, bicipital tendon rupture, and steroid therapeutic injections. For evaluation of prosthetic loosening, 5 to 10 cc of contrast media is instilled at the level of the metallic neck (Fig. 4) (2,3).

Subacromial bursography is rarely performed, except for steroid therapy of symptomatic bursitis.

ELBOW ARTHROGRAPHY

Scout radiographs are obtained in the flexed lateral and straight AP projections. While the patient can sit next to the X-ray table, it is preferable to lie prone on the table with the elbow

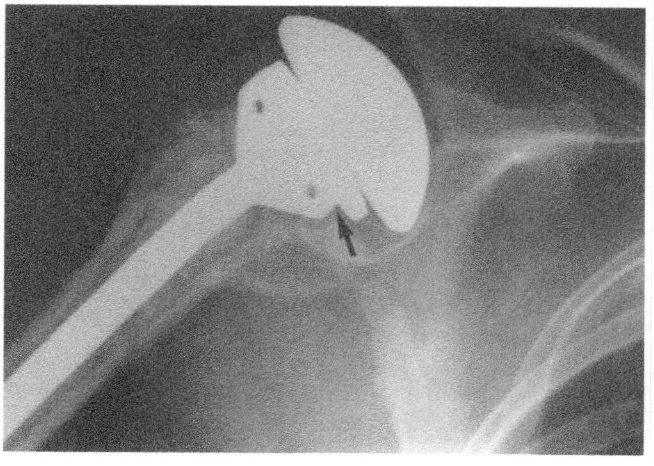

FIGURE 4 Anteroposterior view shows hemiarthroplasty. Proper site of injection is at the level of the prosthetic neck (*arrow*).

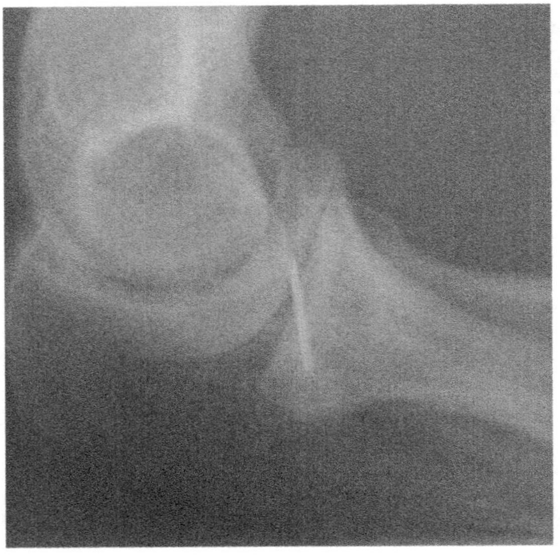

FIGURE 5 Needle placement at level of radiocapitellar articulation.

flexed in front of the head. With the elbow flexed 90°, a marker is positioned over the radiocapitellar joint space. This point is marked over the skin as a site of injection. Following sterile preparation and administration of local anesthetic, a 21- or 22-gauge 1.5 in. needle is directed down into the radiocapitellar joint (Fig. 5). A test injection is performed with a small amount of contrast or air. Successful needle-tip placement occurs when the contrast or air extends anteriorly to the humeral shaft, elevating the anterior fat pad or seen running in the joint space between the radius and capitellum. Once needle position is confirmed, single or double-contrast injection can be performed. With single contrast, the elbow joint holds approximately 8 to 10 cc of contrast (Figs. 6 and 7). With double contrast, 1 to 2 cc of contrast can be injected with 7 to 8 cc of room air.

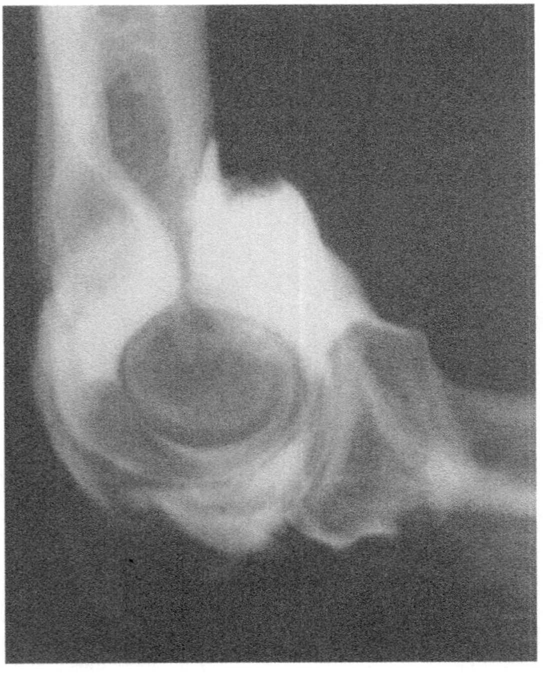

FIGURE 6 Lateral elbow view shows contrast-filled joint space.

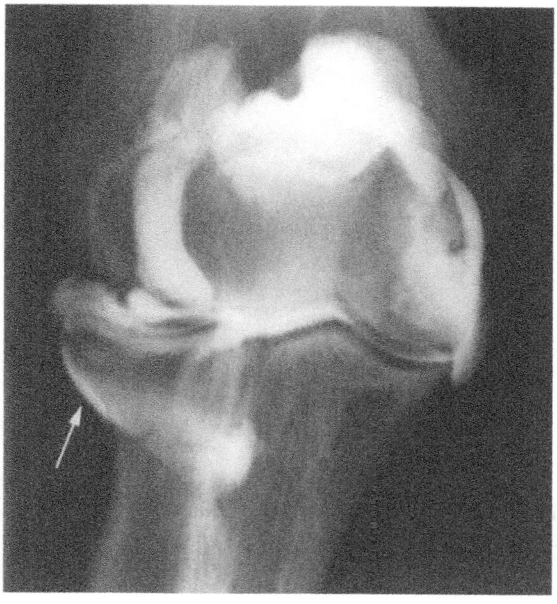

FIGURE 7 Anteroposterior elbow view shows contrast-filled joint space including periradial recess (*arrow*).

Limited elbow flexion and extension suffices, whereas postinjection exercise is not advisable as it may rupture the joint capsule. Postinjection radiographs include a lateral view of the humerus with 90° of flexion, a lateral view in full extension, an AP view, and both oblique views. Computed tomography arthrography may be utilized with double-contrast technique for evaluation of loose bodies, if it is desirable to locate all loose bodies, determine their size and position, as well as to evaluate cartilage thickness and degree of degenerative joint disease (4).

Indications for elbow arthrography include detection of loose bodies, osteochondritis dissecans, ligament and capsule tears, and steroid therapeutic injections (5,6).

WRIST ARTHROGRAPHY

Scout radiographs include AP and lateral views of the wrist. The patient is placed supine on the X-ray table, with the hand placed palm down at the side and the wrist flexed over a small triangular rubber pad or pillow; this position widens the radiocarpal joint, making easier needle placement. Fluoroscopic observation of carpal motion may be useful with a history of a "clicking" disorder (7).

Under fluoroscopic guidance, a lead marker is placed over the radiocarpal joint at the level of the mid-scaphoid, and this point is marked on the skin at the site of injection (Fig. 8). It is important not to place the needle too close to the scapholunate articulation as this may result in erroneous injection into the mid-carpal row. Following administration of local anesthetic and under sterile technique, a 21-gauge 1.5 in. needle is passed directly down into the radioscaphoid joint. A test injection will result in contrast flowing away from the needle outlining the radiocarpal joint. Three cubic centimeters of contrast will successfully fill the radiocarpal joint at which point the needle is removed (Fig. 9). Following injection, AP and lateral views of the wrist are obtained followed by postexercise radiographs in the AP projection supplemented by adduction and abduction AP views. The carpal deviation views help evaluate the spaces between the scapholunate, lunate, and triquetrum for tears of the corresponding ligaments. It is preferable to fluoroscopically monitor the initial injection as this is the best way to visualize and identify intercarpal ligament tears as contrast flows into the mid-carpal row or into the distal radioulnar joint through a tear of the triangular fibrocartilage.

The so-called "triple joint" injection requires initially the radiocarpal injection and films, followed, four hours later, by injecting 1 cc of contrast into the distal radioulnar joint (25-gauge needle) and then immediately followed by a 3 cc midcarpal contrast injection (21- or 22-gauge needle) (Figs. 10 and 11) (8).

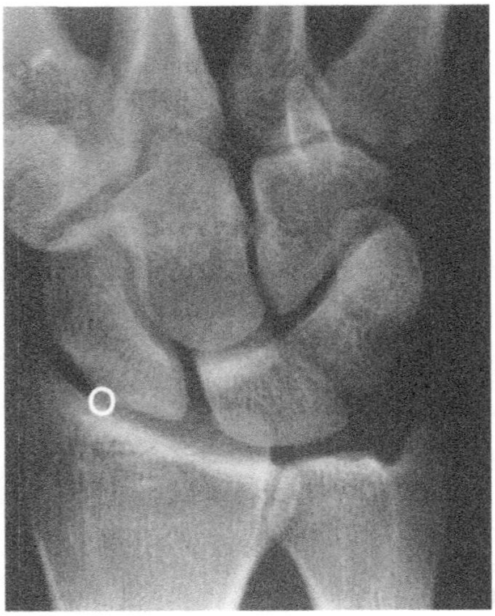

FIGURE 8 Posteroanterior wrist view shows site (O) of contrast injection between the scaphoid and radius.

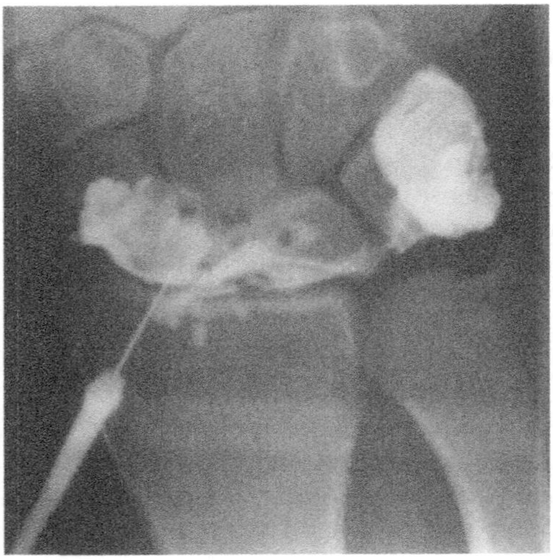

FIGURE 9 Posteroanterior wrist view shows proper needle location and contrast filling the radiocarpal articulation.

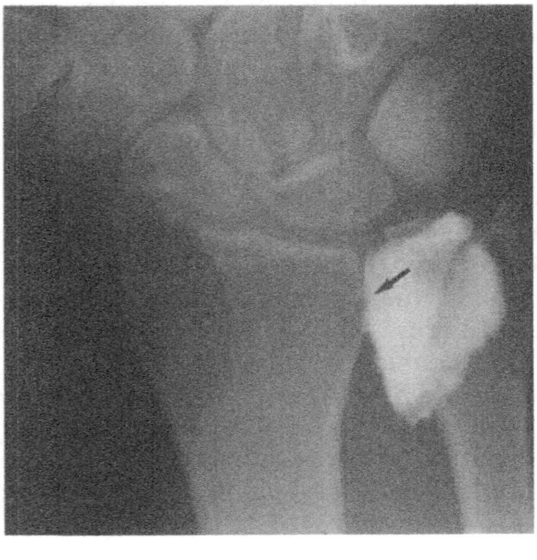

FIGURE 10 Posteroanterior wrist view shows selective contrast injection of the distal radioulnar joint. *Arrow* points to site of injection.

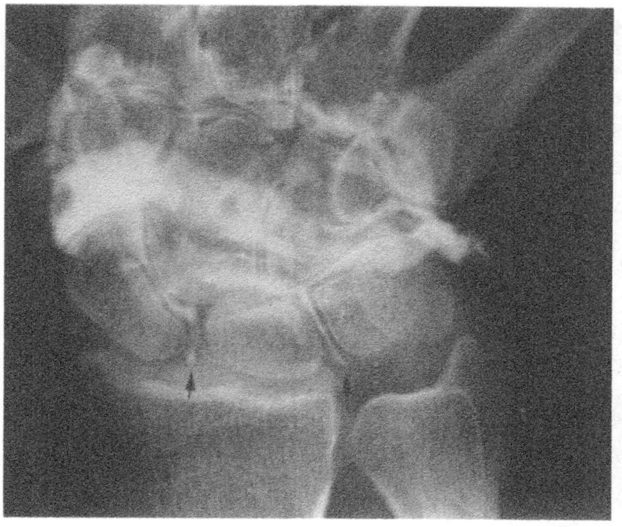

FIGURE 11 Posteroanterior wrist view shows selective mid-carpal joint contrast injection, with opacification of scapholunate and triquetrolunate articulations down to site of normal ligament attachments (*arrows*).

Indications for wrist arthrography include detection of intercarpal ligament tears, triangular fibrocartilage tears, adhesive capsulitis, and communicating ganglia (9).

HIP ARTHROGRAPHY

Scout radiographs are obtained in the AP and lateral projections. The patient is placed supine on the X-ray table, with the hip held in neutral or internal rotation position for optimal visualization of the femoral neck. The femoral artery is palpated and avoided during injection. Under fluoroscopic guidance, a lead or BB is placed over the center or lateral margin of the mid-femoral neck, and this point is marked on the skin to indicate the site of injection (Fig. 12). Under sterile technique and following the administration of local anesthetic, a 20-gauge 3.5 in. spinal needle (5.5 in. spinal needle for obese patients) is directed downward until it touches the femoral neck preferably at the level of the mid-neck or lateral edge. When the tip of the needle touches the bone, it is slightly withdrawn and aspiration of synovial fluid is attempted. If an aspiration study and no fluid is obtained, 10 mL of sterile saline (nonbacteriostatic) may be

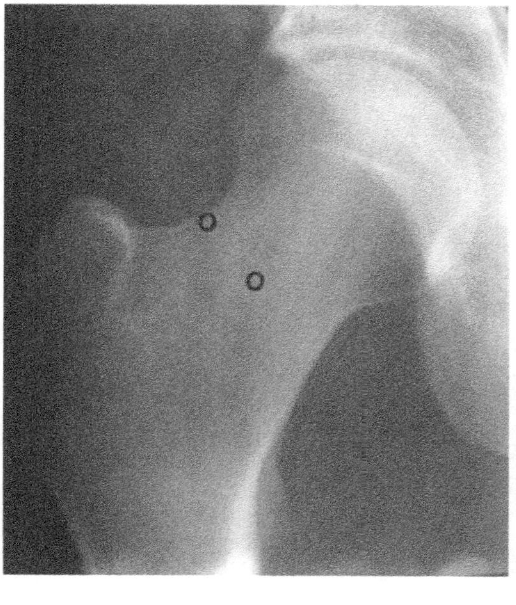

FIGURE 12 Anteroposterior view shows possible injection sites (O) for hip arthrography.

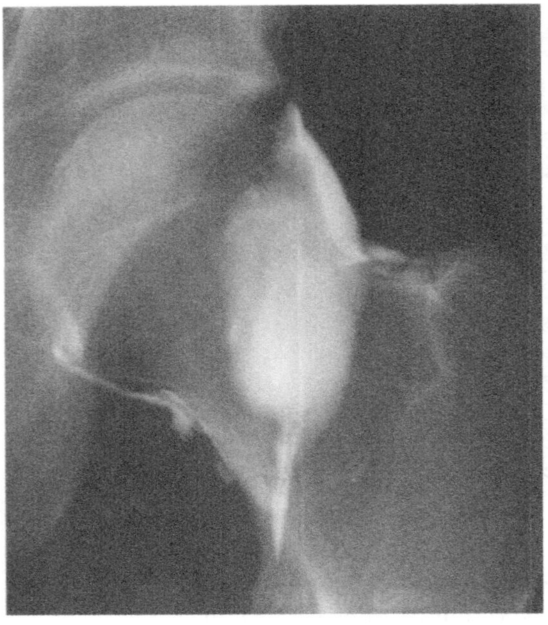

FIGURE 13 Anteroposterior hip view shows contrast-filled joint space.

injected and reaspirated. If the needle is intra-articular, the contrast will flow freely away from the tip of the femoral neck filling the joint capsule (Fig. 13). Six to ten cubic centimeters of contrast medium can be injected and the needle removed. Routine postinjection radiographs include AP and lateral views. Postexercise films are not usually necessary.

For prosthetic replacements, the needle should be directed onto the metal neck and a small amount of contrast injected after aspiration. A pseudocapsule will fill if the needle is placed correctly. Up to 10–15 cc of contrast may be injected when evaluating for prosthetic loosening.

Indications for hip arthrograms include joint aspiration for septic-joint diagnosis, steroid therapy, and detecting arthroplasty complications (Fig. 14) (10–13).

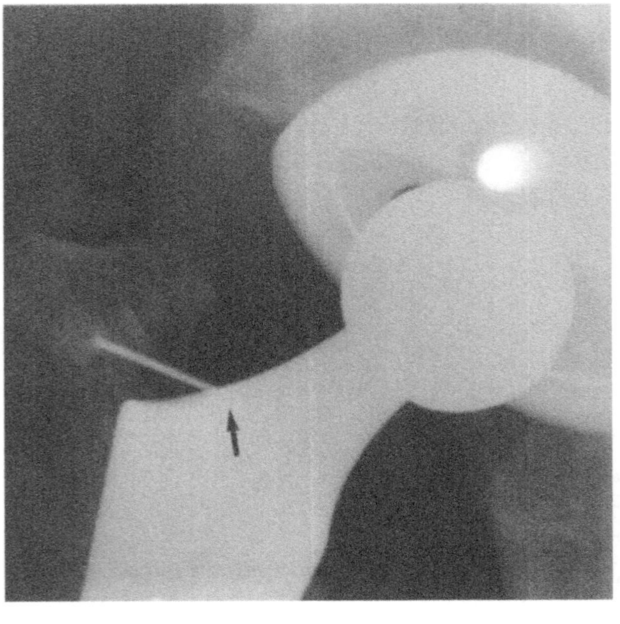

FIGURE 14 Anteroposterior view shows total hip prosthesis with proper needle (*arrow*) placement onto the metallic neck.

KNEE ARTHROGRAPHY

Scout radiographs of the knee are taken in the AP and lateral view with slight flexion. The patient is placed supine on the fluoroscopic radiographic table, with the leg in extension. Following sterile preparation of the knee, local anesthesia may be administered although I personally find it unnecessary. A 20-gauge 1.5 in. needle is directed between the undersurface of the patella and the top of the femoral condyle (Fig. 15). This space can be enlarged by grabbing the patella and pulling it toward oneself. With placement of the needle and following aspiration of all possible joint fluid, a small amount of test-contrast dose should be evaluated for intra-articular location. With intra-articular location, the contrast moves away from the needle tip and runs into the lateral recesses of the suprapatellar pouch and into the tibiofemoral joint (Figs. 16 and 17). If the needle is extra-articular, contrast will pool around the needle

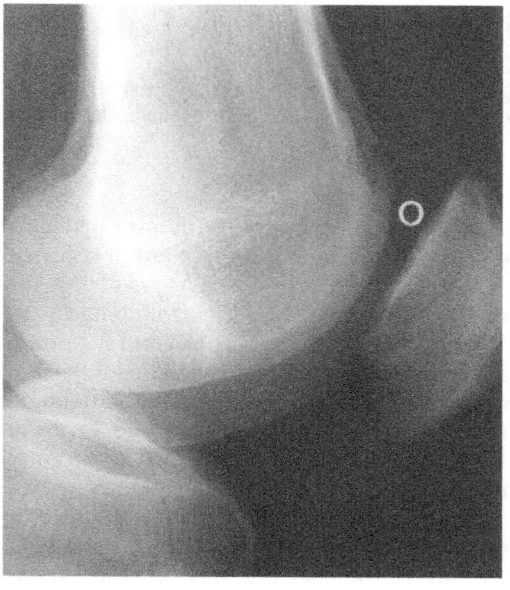

FIGURE 15 Lateral view of knee shows site (O) of proper needle placement.

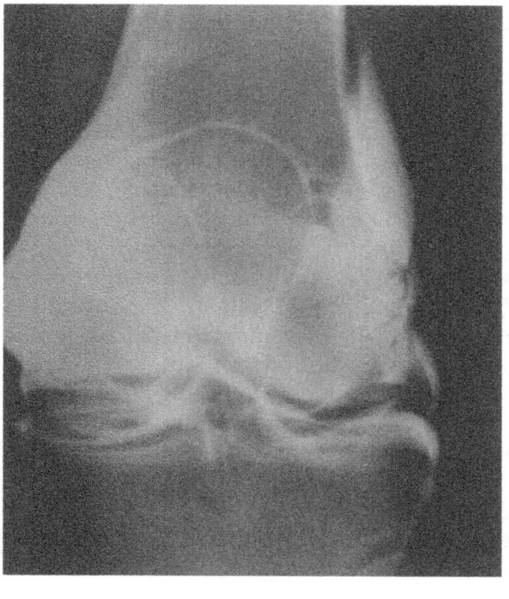

FIGURE 16 Anteroposterior knee view shows normal double-contrast-filled joint space.

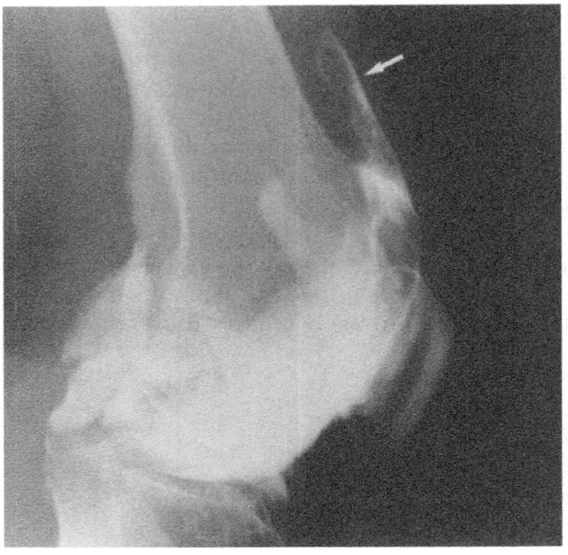

FIGURE 17 Lateral space knee view shows double-contrast-filled joint with filling of suprapatellar bursa (*arrow*).

tip and not flow off to the sides of the smooth-walled suprapatellar bursa. With intra-articular location of the needle, there should be no resistance felt during the injection of either contrast or air.

For double-contrast examination of the knee, 5 cc of contrast is injected followed by an injection of 30 cc of room air. Following removal of the needle, the patient is placed in the prone position, the knee flexed several times, and then a stress device applied to the knee in such a way as to allow the arthrographer to stress the medial and lateral joint sides. With adequate stress, there is proper distraction and visualization of each meniscus. A series of radiographs of each meniscus is done as the patient is rotated under stress from the posterior horn to the anterior horn (14). The subsequent radiographs include a minimum of 12 images per meniscus. Following meniscal examination, a lateral flexed view of the knee is performed with anterior stress (drawer test) on the tibia to visualize the anterior cruciate ligament, which is then spot radiographed. Subsequent AP and lateral extension films of the knee are obtained for evaluation of joint anatomy and possible communicating popliteal cyst.

Indications for knee arthrography include detection of meniscal tears, osteochondritis dissecans, chondromalacia, communicating popliteal cysts, and anterior cruciate tears (15,16).

ANKLE ARTHROGRAPHY

Scout radiographs of the ankle are obtained in AP and lateral projections. Before performing an injection, the dorsalis pedis artery is palpated to avoid its injection. The patient is placed on the side and the ankle is placed in lateral position on the fluoroscopic radiographic table, with the foot held in plantar flexion to open the anterior aspect of the tibiotalar joint. Under fluoroscopic guidance, a lead marker is moved either up or down so it is at the same level as the anterior tibiotalar joint. This point is marked on the skin as the site of injection. Following sterile preparation and administration of a small amount of local anesthetic, a 22-gauge 1.5 in. needle is advanced into the anterior ankle joint capsule. Following needle placement, a test injection of contrast is performed. If the tip of the needle is intra-articular, the contrast will flow into the anterior joint capsule and enter the joint at the level of the talar dome (Figs. 18 and 19). Injection of 6 to 10 cc of contrast media will result in full distention of the joint capsule. Plantar and dorsiflexion will disperse the contrast evenly. Postinjection radiographs include AP view, lateral view, and both oblique views.

Indications for ankle arthrography include detection of medial and lateral collateral ligament tears, loose bodies, osteochondritis dissecans, syndesmosis rupture, and adhesive capsulitis (17,18).

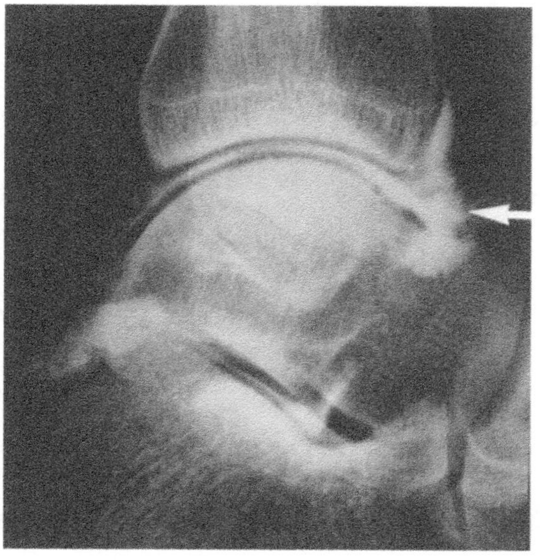

FIGURE 18 Lateral ankle view shows contrast-filled joint space. Proper needle placement requires entry into the anterior joint capsule (*arrow*).

POSTINJECTION COMPLICATIONS

Complications following arthrography are few in number (19,20). The patient may experience pain at the injection site, which may be related to the number of needle passes necessary to obtain intra-articular location. Patients may develop syncopal reaction or vasovagal reactions during the examination, which require standard medical treatment. With the use of nonionic contrast, there has been a decrease in the incidence of synovitis of the involved joint, which may begin two to four hours after the study and last for up to 12–24 hours. The incidence of nerve blocks and intraosseous injections are very rare, utilizing the techniques described above. Very rarely, lidocaine skin reactions may occur.

Bursography and other joint injections occasionally are performed for therapeutic steroid treatment. Bursas that can be injected include subacromial subdeltoid, iliopsoas, and trochanteric bursae (Fig. 20). Joints that can be injected include temporomandibular, acromioclavicular, proximal tibiofibular, sacroiliac, symphysis pubis, subtalar, and facet joints (Figs. 21–24) (2,21).

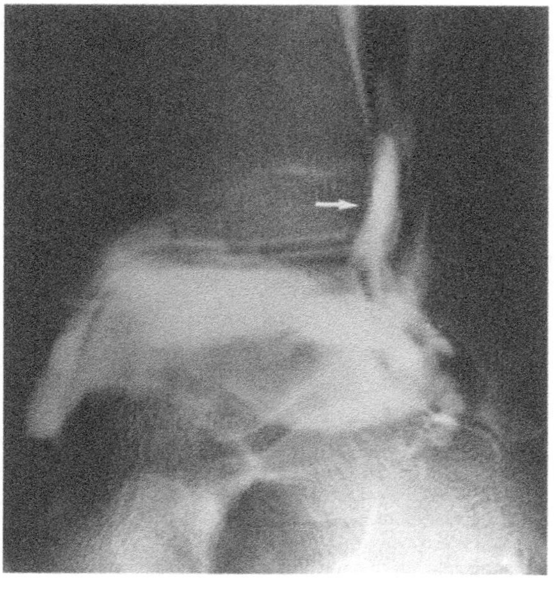

FIGURE 19 Anteroposterior internal oblique view of ankle shows contrast outlining joint margins, including the syndesmotic recess (*arrow*).

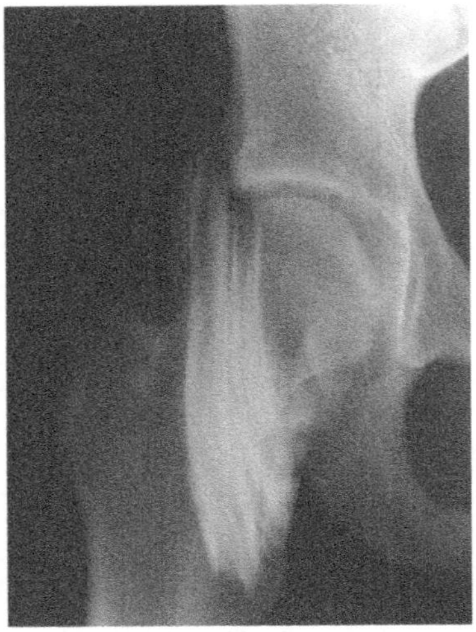

FIGURE 20 Contrast-filled iliopsoas bursa visualized from lesser trochanter cephalad.

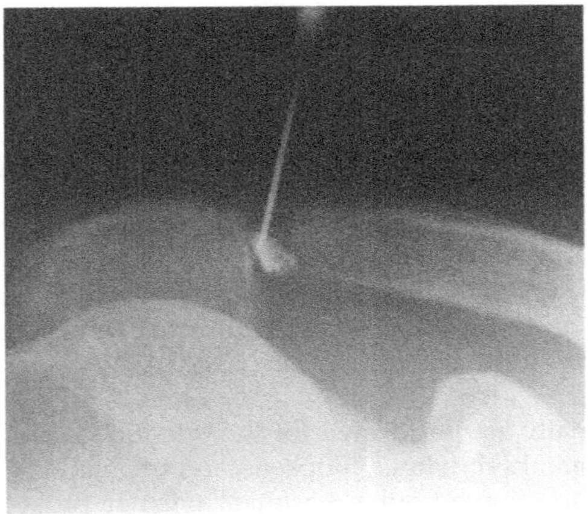

FIGURE 21 Proper needle placement and contrast injection within the acromioclavicular joint.

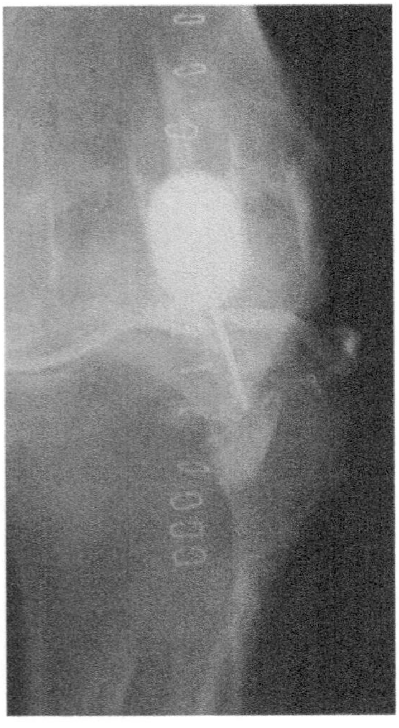

FIGURE 22 Proper needle placement and contrast injection within the proximal tibiofibular joint.

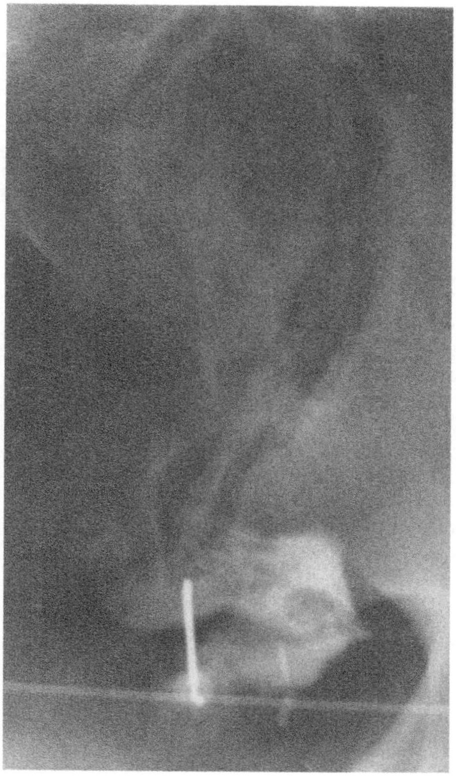

FIGURE 23 Proper needle placement and contrast injection within the inferior aspect of the sacroiliac joint.

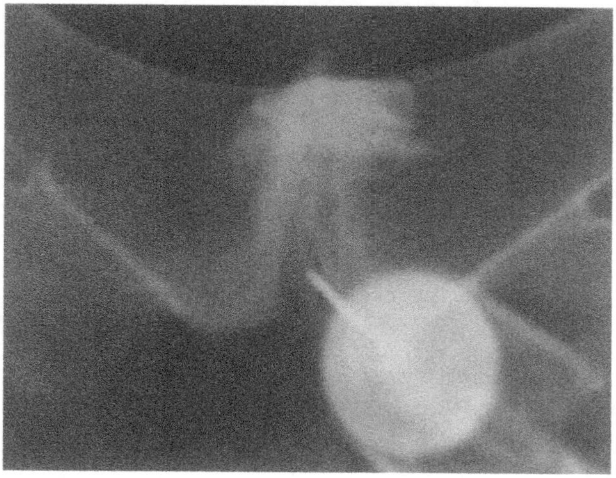

FIGURE 24 Proper needle placement and contrast injection in the symphysis pubis.

REFERENCES

1. Pasternak S, Resnick D, Niwayama G, Danzig L, Haghighi P. The effects of watersoluble contrast media on the synovial membrane. Radiology 1982; 143:331–334.
2. Resnick D. Arthrography, Tenography and Bursography. Philadelphia: W.B. Saunders, 1996:113–135.
3. Goldman A. Shoulder Arthrography. In: Technique, Diagnosis and Clinical Correlation. Boston: Little Brown, 1982:164–167.
4. Pavlov H, Ghelman B, Warren R. Double contrast arthrography of the elbow. Radiology 1979; 130:87–91.
5. Goldman A. Procedures in Skeletal Radiology. Orlando: Grune and Stratton, 1984:295–338.
6. Steinbach L, Schwartz M. Elbow arthrography. Radiol Clin N Am 1998; 36:635–649.

7. Protas J, Jackson W. Evaluating carpal instabilities with fluoroscopy. Am J Radiol 1980; 135:137–140.
8. Levinsohn M, Rosen I, Palmer A. Wrist arthrography: value of the three-compartment injection method. Radiology 1991; 179:213–219.
9. Linkous M, Gilula L. Wrist arthrography today. Radiol Clin N Am 1998; 36:651–672.
10. Tehranzadeh J, Schneider R, Freiberger R. Radiologic evaluation of painful total hip replacement. Radiology 1981; 141:355–362.
11. Berquist T, Bender C, Maus T, Ward E, Rand J. Pseudobursae: a useful finding in patients with painful hip arthroplasty. Am J Radiol 1987; 148:103–106.
12. Aliabadi P, Baker N, Jaramillo D. Hip arthrography, aspiration, block and bursography. Radiology Clin N Am 1998; 36:673–690.
13. Weissman B. Imaging of the total hip replacement. Radiology 1997; 202:611–623.
14. Freiberger R, Pavlov H. Knee arthrography. Radiology 1988; 166:489–492.
15. Tegtmeyere C, McCue F, Higgins S, Ball D. Arthrography of the knee: a comparative study of the accuracy of single and double contrast techniques. Radiology 1979; 132:37–41.
16. Coumas J, Palmaer W. Knee arthrography: evaluation and current status. Radiol Clin N Am 1998; 36:703–728.
17. Olson R. Ankle arthrography. Radiol Clin N Am 1981; 19:255–268.
18. Dory M. Arthrography of the ankle in chronic instability. Skeletal Radiol 1986; 15: 291–294.
19. Hall F. Morbidity from shoulder arthrography: etiology, incidence and prevention. Am J Radiol 1981; 136:59–62.
20. Hugo P, Newberg A, Newman J, Wetzner S. Complications of arthrography. Semin Musculoskel Radiol 1998; 2:345–348.
21. Baker K, Gilula L. The current role of tenography and bursography. Am J Radiol 1990; 154:129–133.

15 The Brisement Procedure for Adhesive Capsulitis

Caroline Parlier-Cuau, Marc Wybier, and Jean-Denis Laredo
Department of Bone and Joint Radiology, Lariboisière Hôpital, Assistance Publique–Hopîtaux de Paris, Paris, France

INTRODUCTION

Frozen shoulder syndrome (FSS) is of uncertain etiology. It is characterized by the spontaneous onset of pain in the shoulder, with insidious progressive restriction of both active and passive motion in every direction, mainly external rotation and anterior elevation. Pain is often very severe and disturbs sleep. After several weeks and months, the painful phase gradually abates and is followed by a period of stiffness. This period of stiffness without improvement lasts between 4 and 12 months (1). Spontaneous gradual recovery of motion then follows over a period of several months. The total duration of FSS may be difficult to evaluate because the exact times of onset and resolution are frequently not clear: the initial pain may be confused with that of a shoulder tendonitis or trauma, which, not infrequently, often proceeds to FSS. Complete recovery from FSS is also difficult to define: some patients who consider their range of motion (ROM) still restricted are found to have no restriction on objective testing at long-term follow-up and conversely many other patients who regard their ROM as normal are found to have significant restriction at clinical examination (1–4). Distention arthrography (DA) involves the injection into the joint of contrast media, steroids, lidocaine, and a large volume of fluid in order to obtain joint distension.

TECHNIQUE

DA is usually performed on an outpatient basis. The patient is placed in the supine position, with the arm at the side and, as far as possible, the palm supported to hold the shoulder in external rotation in order to facilitate needle placement into the anterior aspect of the glenohumeral joint. The X ray beam is vertically oriented over the joint. The skin is prepped with an iodine solution. The skin and superficial planes are anesthetized with 1% lidocaine. A 20-gauge 9-cm needle is vertically inserted under fluoroscopic control into the anterior aspect of the joint through the anesthetized skin. First, 2 to 3 mL of contrast material (meglumine loxaglate) is injected to outline the glenohumeral space, confirming correct intra-articular needle position. The joint capacity must be measured and is typically dramatically reduced (1,4–10). Most patients have not more than 2- to 3-mL joint space volume together with no more than 8 to 10 mL with loss of distensibility of the capsule. In most cases, the arthrogram is very suggestive of adhesive capsulitis (Fig. 1): there is a marked reduction in size of all joint recesses, the inferior (axillary) and internal (subcoracoid and subscapularis) recesses. The bicipital tendon sheath is inconstantly opacified. In addition, the synovial pouch around the humeral head is "tight looking" and has an irregular indented outline. Early lymphatic filling is not uncommon. However, arthrography may be surprisingly normal in some patients with full clinical criteria of FSS (4,5). A rotator cuff tear is present in 10% (1,3,5,7) to 31.3% (9) of cases.

Next, 3 mL of 2% lidocaine and 1.5 mL of cortivazol are injected intra-articularly. This injection often causes a severe exacerbation of shoulder pain. Characteristically, releasing the finger from the plunger of the syringe leads to immediate return of fluid into the syringe and immediate decrease of shoulder pain. Third, distension of the capsule is then performed using 30 to 40 mL of chilled sterile saline solution according to Fareed and Gallwan's technique (8). The maximum volume injected depends on the distensibility of the joint capsule.

Joint distension, the "brisement procedure" (11), requires slow, gradual, intermittent injection of a larger and larger volume of chilled sterile saline solution. To avoid excessive fluid reflux from the needle while settling a new syringe of saline solution, a lockable three-way

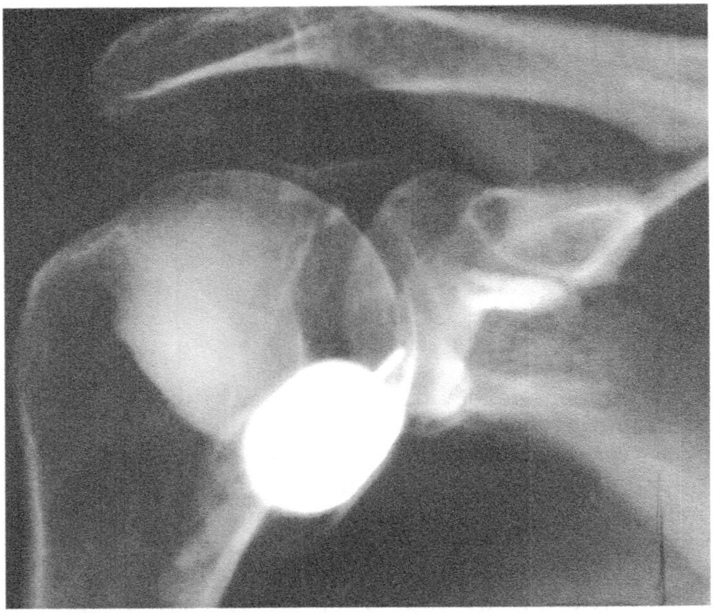

FIGURE 1 Arthrogram of frozen shoulder syndrome; inferior and internal joint recesses are filling up insufficiently.

stopcock could be placed over the needle. The aim of the procedure is to distend the joint with the largest volume of fluid (usually up to 40–50 mL) (Fig. 2) without fluid extravasation. Fluid extravasation may occur at the subscapular recess or the bicipital tendon sheath, causing a sudden fall in the joint resistance to distension. Further injection would be then ineffective and the procedure should be stopped.

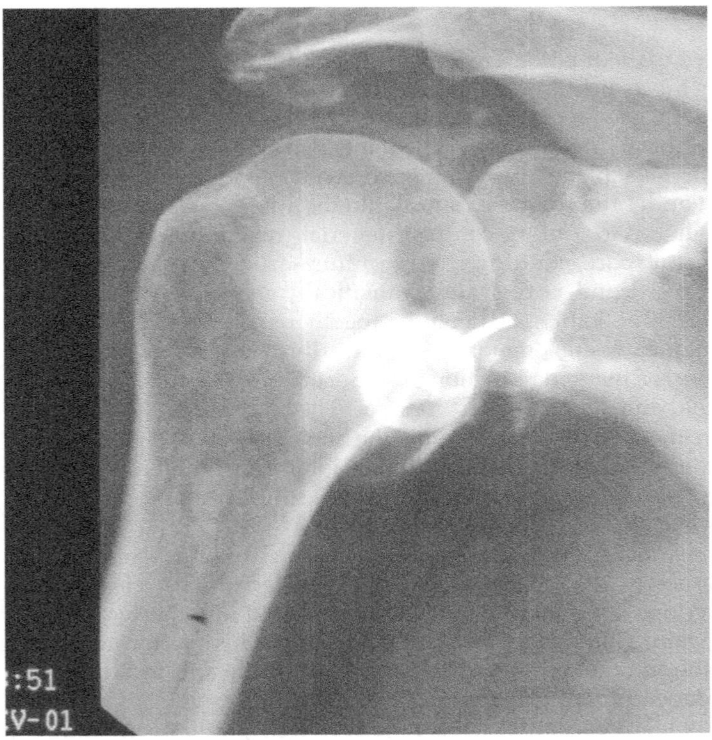

FIGURE 2 Arthrogram of distended shoulder after injection of a large amount of fluid.

Arthrographic distension is immediately followed by active assisted ROM exercises under the supervision of a physical therapist. The following days, the patient continues with regular home physical therapy exercises.

AUTHORS' EXPERIENCE

The authors' experience is a noncontrolled study including 30 glenohumeral joints in 29 patients with FSS (17 female and 12 male patients; age range 41–65 years; mean age 49 years) treated with DA. Patients were assessed for pain (Huskinsson scale) and shoulder ROM in internal and external rotation, anterior and posterior elevations, and abduction; first prior to DA and then at 15-day-and 45-day-follow-up after treatment. At 15-day follow-up, 80% of patients considered that they had very good (53%) or good (27%) results. At 45-day follow-up, 90% of patients considered that they had very good (80%) or good (10%) results; none of the patients considered that they had no benefit at all at 45-day follow-up.

DISCUSSION

A wide variety of treatments have been investigated in FSS. Physiotherapy is widely recommended in most of the published reports, including home exercises consisting of pendulum exercises and resisted exercises performed several times daily. Oral (2) and local (12) steroids are also frequently prescribed. Manipulations of the shoulder performed under general anesthesia (13) or under interscalenic brachial plexus block anesthesia (14,15) have also been proposed. More recently, arthroscopic release (14,16–19) or surgical excision (10,20) of the coracohumeral ligament has been reported to be highly effective in patients who did not improve under conservative therapy: the coracohumeral ligament normally restrains external rotation of the shoulder with the arm at the side, and its contracture due to the chronic fibrotic thickening present in FSS (10,20) acts as a check-rein against external rotation resulting in loss of both active and passive movement (20).

The therapeutic value of DA has been studied by many authors (6–9,13,21–25) since Andren and Lundberg first reported that joint distension occurring during diagnostic arthrography could be effective for shoulder restriction (11). Most studies evaluating DA were noncontrolled series (7–9,13,23,25). Satisfactory (good and excellent) results were obtained in 68% (13 cases) and 96% (22 cases). Data provided by controlled studies (6,22,24) is more limited with less convincing results. Corbeil et al. found no statistical difference in ROM at three-month follow-up between patients having a nondistensive arthrography with intra-articular injection of steroids and those having additional intra-articular injection of 20 mL of lidocaine (6). Also, Jacobs et al. reported no statistical difference in ROM at 16-week followup between patients treated with intra-articular injection of steroids alone and those treated with intra-articular steroids combined with joint distension (22). However, in the later group, distension consisted in intra-articular injection of only 6 mL of lidocaine plus 3 mL of room air (22). Gam et al. reported significant improvement in ROM at 12-week follow-up for the patients treated with intra-articular steroids combined with distension (24). We believe that additional controlled studies are necessary to confirm the value of DA in FSS. DA should also be used in combination with other treatments (intra-articular glucocorticoids, gentle mobilization under local anesthesia, and sedation or general anesthesia) (25).

Considering the technique of DA, Rizk et al. found, in an open trial including 16 patients, that good results were achieved only when capsular rupture was obtained during the procedure (1); our own results do not confirm this statement as, in most of our patients, capsular rupture did not occur during joint filling.

As mentioned above, physiotherapy is widely recommended as an early therapy either alone or in association to other treatments including DA. The goal of physiotherapy is to relieve pain, improve motion, and restore function. Physiotherapy is widely used and generally held to be beneficial. However, no convincing evidence of efficacy has been found in well-designed studies (26). Also, to our knowledge, the advantage of physiotherapy in association with DA over DA alone has not been studied. In contrast, DA was found to be effective in patients unsuccessfully treated by previous physiotherapy in many reports (7,8,13) suggesting DA has advantages over physiotherapy alone in FSS. In our institution, patients treated with DA are instructed to self-perform home physical therapy exercises in the days following the procedure. Finally, mid–long-term assessment of treatment results in FSS is challenging because FSS is a self-limiting condition with spontaneous recovery after several months or years. Bulgen et al. reported that various treatment regimens, including intra-articular injection of steroids, ice therapy, and mobilizations, have little long-term advantage over no treatment in FSS (12). However, the aim of procedures such as DA in FSS is not to modify the overall course of the disease, but rather to shorten the most disabling phase of this condition.

CONCLUSION

In noncontrolled studies, DA, which includes intra-articular injection of steroids followed by hydraulic joint distension, provides good and excellent results in 90% of the cases. DA obtains rapid improvement in pain and joint stiffness and shortens the disabled period. However, the respective role of intra-articular injection of steroids and hydraulic distension in the achievement of good results remains to be studied in controlled studies.

REFERENCES

1. Reeves B. The natural history of the frozen shoulder syndrome. Scand J Rheumatol 1975; 4:193.
2. Blockley NJ, Wricht J, Kellgren JH. Oral cortisone therapy in periarthritis of the shoulder. Br Med J 1954; 4:1455.
3. Sany J, Caillens JP, Rousseau JR. Evolution lointaine de la rétraction capsulaire de l'épaule. Rev Rhum 1982; 49:815.
4. Shaffer B, Bone JE, Kerlan RK. Frozen shoulder. A long-term follow-up. J Bone Joint Surg 1992; 74-A:738.
5. Binder AI, Bulgen DY, Hazleman BL, et al. Frozen shoulder: an arthrographic and radionuclear scan assessment. Ann Rheum Dis 1984; 43:365.
6. Corbeil V, Dussault RG, Leduc BE, et al. Capsulite rétractile de l'épaule: étude comparative de l'arthrographie avec corticothérapie intra-articulaire avec ou sans distension capsulaire. Can Ass Radiol J 1992; 43:127.
7. Ekelund AL, Rydell N. Combination treatment for adhesive capsulitis of the shoulder. Clin Orthop 1992; 282:105.
8. Fareed DO, Gallivan WR Jr. Office management of frozen shoulder syndrome Clin Ortho 1989; 242:177.
9. Mulcahy KA, Baxter AD, Oni Ooa, et al. The value of shoulder distension arthrography with intra-articular injection of steroid and local anaesthetic: a follow-up study. Br J Radiol 1994; 67:263.
10. Ozaki J, Nakagawa Y, Sakurai G, et al. Recalcitrant chronic adhesive capsulitis of the shoulder. Role of contracture of the coracohumeral ligament and rotator interval in pathogenesis and treatment. J Bone Joint Surg 1989; 71-A:1511.
11. Andren L, Lunberg BJ. Treatment of rigid shoulder by joint distension during arthrography. Acta Orthop Scand 1965; 36:45.
12. Bulgen DY, Binder AI, Hazleman BL, et al. Frozen shoulder: prospective clinical study with an evaluation of three treatment regimens. Ann Rheum Dis 1984; 43:353.
13. Sharma RK, Bajekal RA, Bhan S. Frozen shoulder syndrome. A comparison of hydraulic distension and manipulation. Int Orthop 1993; 17:275.
14. Pollock RG, Duralde XA, Flatow EL, et al. The use of arthroscopy in the treatment of resistant frozen shoulder. Clin Orthop 1994; 304:30.
15. Dodenhoff RM, Levy O, Wilson A, Copeland SA. Manipulation under anesthesia for primary frozen shoulder on early recovery and return to activity. J Shoulder Elbow Surg 2000; 9:23–26.
16. Andersen NH, Sojbjerg JO, Johannsen HV, Sneppen O. Frozen shoulder: arthroscopy and manipulation under general anesthesia and early passive motion. J Shoulder Elbow Surg 1998; 7:218–222.
17. Pearsall AW, Osbahr DC, Speer KP. An arthroscopic technique for treating patients with frozen shoulder. Arthroscopy 1999; 15:2–11.
18. Klinger HM, Otte S, Baums M, Haerer T. Early arthroscopic release in refractory shoulder stiffness. Arch Orthop Trauma Surg 2002; 122:200–203.
19. Holloway GB, Schenk T, Williams GR, Ramsey ML, Iannotti JP. J Bone Joint Surg Am 2001; 83(A): 1682–1687.
20. Bunker TD, Anthony PP. The pathology of frozen shoulder. J Bone Joint Surg 1995; 77-B:677.
21. Hsu SYC, Chan KM. Arthroscopic distension in the management of frozen shoulder. Int Orthop 1991; 15:79.

22. Jacobs LGH, Barton MAJ, Wallace WA, et al. Intra-articular distension and steroids in the management of capsulitis of the shoulder Br Med J 1991; 302:1498.
23. Rizk TE, Gavant ML, Pinals RS. Treatment of adhesive capsulitis (frozen shoulder with arthrographic capsular distension and rupture). Arch Phys Med Rehabil 1994; 75:803.
24. Gam AN, Schydlowsky P, Rossel I, Remvig L, Jensen EM. Treatment of "frozen shoulder" with distension and glucorticoid compared with glucorticoid alone. A randomized controlled study. Scand J Rheumatol 1998; 27:425–430.
25. Laroche M, Ighilahriz O, Moulinier L, Constantin A, Cantagrel A, Mazieres B. Adhesive capsulitis of the shoulder: an open study of 40 cases treated by joint distention during arthrography followed by an intraarticular corticosteroid injection and immediate physical therapy. Rev Rhum [Engl Ed] 1998; 65:313–319.
26. Noel E, Thomas T, Schaeverbeke T, Thomas P, Bonjean M, Revel M. Frozen shoulder. Joint Bone Spine 2000; 67:393–400.

16 Selective Radiosynoviorthesis

Leon Rybak
Department of Radiology, New York University/Hospital for Joint Diseases, New York, New York, U.S.A.

Robert G. Dussault and Phoebe A. Kaplan
Musculoskeletal Radiology, Laënnec Radiology, Montreal, and Reso-Concorde MRI, Laval, Quebec, Canada

DEFINITION

Radiosynoviorthesis or radiation synovectomy is a form of treatment for recurrent synovitis that involves ablating the synovium by intra-articular administration of radionuclides. The purpose is to destroy the synovium before it can cause cartilage destruction and result in a chronic, debilitating arthropathy. This treatment was developed as an alternative or adjunct to surgical and chemical forms of synovectomy. Ansell et al. (1) were the first to publish their results in 1963, and Delbarre et al. (2) first coined the term "radiosynoviorthesis" to describe the treatment.

In this chapter, we will review the basics of performing this technique with an emphasis on recent advances in dose calibration and implications of imaging for patient selection and disease monitoring. The major indications for radiosynoviorthesis, including rheumatoid arthritis (RA), hemophilia, pigmented villonodular synovitis (PVNS), and osteoarthritis (OA), will be discussed separately.

GOALS

Chronic recurrent synovitis, whether resulting from an inflammatory arthropathy such as RA, from intra-articular bleeding secondary to a coagulopathy, or from a synovial tumor, will cause pain, limitation of motion, and, eventually, joint destruction. The goal of radiosynoviorthesis, as with the other forms of synovectomy, is to arrest this process at an early stage, decrease joint pain, improve mobility, and prevent irreversible joint damage.

MAJOR INDICATIONS

The major indication for radiosynoviorthesis is chronic synovitis, particularly in patients with RA and hemophilia. Synovial proliferation in the form of PVNS has also been successfully treated. Although not considered a primary feature of OA, concomitant synovitis seen with this condition has been treated with less impressive results (3–7). The procedure has also been used with variable results in patients with psoriasis, ankylosing spondylitis, reactive arthritis, postoperative synovitis, and synovitis in the setting of joint replacements (5,8–11). For the purposes of this review, further discussion will be limited to RA, hemophilia, PVNS, and OA.

Rheumatoid Arthritis

In the past, patients with RA, who have failed medical treatment and gone on to develop persistent debilitating synovitis, have been treated with surgical synovectomy. This procedure entails anesthesia, a postoperative hospitalization period, extensive rehabilitation, and a significant incidence of decreased range of motion (12). Intra-articular steroid injections have also been used for many years in this patient population. More recently, radiosynovectomy has become part of the therapeutic armamentarium.

Numerous investigators have reported good results in 50% to 80% of patients using radiosynoviorthesis with success defined as improved range of motion and decreased pain, effusion, and synovitis (5,6,8,10,13,14). Though there are conflicting reports on the relative efficacy of intra-articular steroids and radionsynoviorthesis, an additive effect has been noted by several investigators and their use in combination may be most beneficial (15–17). Despite reports suggesting little benefit of a repeat injection in the setting of initial therapeutic failure, some investigators believe that patients with particularly thick synovium or refractory synovitis may require more than one treatment to achieve a response. Patients demonstrating a recurrence of symptoms after initial response may also benefit from additional treatments (4,6,18–21).

Most authors agree that younger patients treated earlier during the course of the disease stand a greater chance of benefiting (3,6,8,12,19,22–24). Advanced bone or cartilage destruction, as manifested by loss of joint space, subchondral cysts, erosions, or instability on plain film radiography, was a consistent indicator of poor response (6,25–29).

Hemophilia

The inciting event in the debilitating arthropathy associated with hemophilia is intra-articular hemorrhage. The mechanism by which joint damage then ensues is still a subject of debate. One hypothesis is that hemosiderin and iron deposition in the synovial membrane results in synovitis with subsequent progression as in the other inflammatory arthropathies. The reactive, edematous synovium is more susceptible to injury from minor trauma resulting in additional episodes of hemorrhage. Thus, a vicious cycle of hemorrhage-synovitis-hemorrhage is established (30). The other hypothesis is that hemorrhage alone may lead to cartilage degeneration, independent of the effect on the synovium (28). The first line of treatment in patients with a coagulopathy is factor replacement. Failing response, patients with repeated hemorrhages can be treated with surgical synovectomy or synoviorthesis. The surgical approach, whether open or arthroscopic, though resulting in slightly higher long-term success rates, involves extensive perioperative factor replacement and prolonged recovery time. A certain percentage of patients may also develop postoperative arthrofibrosis leading to decreased range of motion (30).

Ahlberg and Pettersson were the first to treat hemophiliacs with radiosynovectomy in 1971 (31). Using decreased frequency of intra-articular hemorrhage as the measure of success, many publications with excellent results in the range of 60% to 80% have followed (25,32–38). As opposed to surgery, no decreased range of motion seems to result from radiosynovectomy. In addition, radiosynovectomy has proven much more cost effective. Given these facts, some proponents feel that radiosynovectomy should always be attempted prior to surgical synovectomy (39).

As with RA, patients treated at an earlier stage of disease seem to have a better response. Though many initial reports cited young age as a relative contra-indication to the procedure, attempts to halt disease progression in the early stages has led some investigators to advocate the use of radiosynoviorthesis in children. Citing good results with few complications and a substantial cost saving as compared to medical prophylaxis, they propose its use in young hemophiliacs who have repeated hemorrhage involving one joint in particular or in those whose disease is refractory to medical management. In patients with multiple joint involvement, they continue to advocate prophylaxis with factor replacement (30,34,36).

Curiously, several investigators, though demonstrating decreased rates of hemorrhage with radiosynovectomy, noted no apparent alteration in the long-term progression of joint destruction as evidenced on both plain film and unenhanced MRI (27,28).

Pigmented Villonodular Synovitis

Intra-articular PVNS may present in localized or diffuse form. The diffuse form can resemble the chronic synovitis of an inflammatory arthritis. Without therapy, the process will cause joint destruction. Both open and arthroscopic synovectomy have been performed in joints with PVNS with some success. The recurrence rates after this procedure are high, however, ranging from 8% to 46% (40). In light of this, several investigators have chosen to attempt radiosynovectomy both alone and in combination with surgery. The initial reports, though few, have been enouraging (41–44). The most promising results came from a group in Germany that reported a 100% success rate with regard to both clinical assessment and decreased early uptake on Tc99m bone scans utilizing a combined treatment of surgical synovectomy followed within a six-month period with Y90 radiosynovectomy (40). In all of these studies, patients who were treated earlier in the disease process seemed to fare better.

Osteoarthritis

Radiosynoviorthesis has not traditionally been considered a first-line treatment for OA and is not directly listed as an indication in the recently published guidelines of the European Association of Nuclear Medicine (11). However, at many centers in Europe, patients with OA who demonstrate evidence of concomitant active synovitis or persistent effusion have been treated. Results have varied with response rates ranging anywhere from 35% to 78% (5,6,10). One group demonstrated good short-term response with relapse at six months (45). Almost all of the studies, which included both RA and OA have demonstrated significantly greater responses in the former.

More research is necessary to determine which patients with OA may respond to this treatment. At present, it can be considered somewhat controversial at best.

TECHNIQUE

Radionuclide Selection

Several factors become important in selecting the most appropriate radiopharmaceutical for radiosynoviorthesis. The ideal radionuclide is a pure beta(B)-emitter that has a large particle size, a short half-life, and a penetration range of less than 1 cm. Beta particles are electrons with varying energies that typically deposit all of their energy locally, causing biological damage through excitation and ionization of atoms and molecules within the exposed medium. Gamma rays do not have as limited a range and lead to higher total body doses. Unfortunately, extra-articular leakage with a pure beta emitter cannot be imaged and precisely quantified with a gamma camera and requires the measurement of Bremsstrahlung radiation or a Geiger–Muller counter (12).

The particle size and half-life of the agent also become vital with respect to leakage. The particles should be small enough to permit pinocytosis by the macrophages in the synovial membrane, but not so small that they escape the joint space before phagocytosis occurs (14). Half-life should be short enough to prevent sustained activity and damage in the event of leakage, but not so short that the use of the agent becomes impractical from a technical standpoint (Table 1).

Much of the earliest experience was with gold-198 (^{198}Au). Although found to be effective from a treatment perspective, its use has become less popular due to coincident 4.11 meV gamma emission, a maximum tissue penetration range of 4 mm, and a small particle size, resulting in leakage (12,46). Early studies showed a concentration of radioactivity in local lymph nodes of anywhere from 0% to 60% of the dose (8).

Yttrium-90 (^{90}Y) has been advocated as an ideal agent by many, particularly in Europe, where it is available. It is a pure beta emitter with a higher energy of 2.27 MeV and a maximum tissue penetration of 11 mm. The half-life of three days makes it easier to use. It has become the agent of choice in Europe for treating larger joints like the knee and is recommended for this purpose by the European Association of Nuclear Medicine (11). Rhenium-186 (^{186}Re), a beta

TABLE 1 Radiopharmaceuticals Used for Intra-articular Therapy

Radiocolloid	Half-life (days)	Type	Energy B (meV)	Penetration range in soft tissue		Particle size (μ)
				Max.	Mean	
Gold-198	2.7	B, Y	0.96	3.6	1.2	20–70
Phosphorus-32	14.0	B	1.7	7.9	2.6	500–2000
Rhenium-186	3.7	B, Y	0.98	3.6	1.2	5–10
Yttrium-90	2.7	B	2.2	11.0	3.6	100
Dysprosium-165	0.1	B, Y	1.29	5.7	1.8	3000–8000

Source: From Ref. 12.

emitter with a maximum energy of 1.07 meV, an average soft tissue range of 1.1 mm and a half-life of 3.7 days, is recommended for use in medium-sized joints (hip, shoulder, elbow, wrist, and ankle) and Erbium-169 (^{169}Er) with a maximum energy of 0.34 meV, an average soft tissue penetration of 0.3 mm and half-life of 9.4 days is suggested for the small joints of the hands and feet (11).

Another agent that has demonstrated promise is Dysprosium-165 (^{165}Dy), a beta emitter with an energy of 1.29 meV, a short half-life of 2.3 hours, a maximum tissue penetration of 5.7 mm, and a very large particle size. The large size and short half-life minimize the risk of radiation damage with studies reporting average leakage to the regional lymph nodes of only 0.12% and to the liver of 0.64% of the total dose administered (12,26). A gamma emission of 3.6% makes it possible to use a gamma camera for detection of possible leakage. The use of this agent has proven impractical however, due to the short half-life and inherent technical limitations (26).

As several of the agents discussed above are not available for use in the United States, Phosporus-32 (^{32}P) chromic phosphate has become the most widely utilized. This is a pure beta emitter with a large particle size, a half-life of 14 days, and a maximum penetration of 8 mm (12,46).

The selection of the most appropriate agent and dose should be tailored toward achieving the desired effect of penetrating the synovium while minimizing risk of adverse side effects. The normal synovium is composed of a layer of fibroblasts and macrophages, one to two cells thick. In the actively inflamed joint, the synovium may have a thickness of 10 to 12 cell layers or greater than 5 mm (12). Optimally, the radionuclide selected should penetrate the desired thickness of synovium and be delivered in a dose sufficient to treat the entire joint lining (Tables 1 and 2). Absorption beyond this depth may result in undesirable effects on other components of the joint. Most of the effects of the radionuclide are mediated through oxidation and free-radical formation. Fortunately, cartilage being oxygen poor is, thus, relatively radioresistant.

The radioactive emissions undergo many interactions as they penetrate tissue, with a resultant loss of energy. The absorbed dose at the maximum penetration range is, thus, much less than that at a shallower depth. Several groups of investigators have utilized models of the synovial joint and computer analysis to study several radionuclides and determine the effective absorbed dose penetration range (46,47). Using this data, it is possible to select a radionuclide and the amount of activity to be delivered, which will achieve the desired dose at the desired depth of inflamed synovium (Table 2).

Patient Selection

Patients with recurrent synovitis are the prime candidates for radiosynoviorthesis. Rheumatoid patients are considered eligible after a period of medical treatment of at least six months. Hemophiliacs with multiple episodes of intra-articular hemorrhage are also excellent candidates. Patients with the diffuse form of PVNS can benefit from the procedure, usually following surgical synovectomy.

Unfortunately, many of the patients selected by these criteria may have already reached an advanced stage of disease at which point the benefits from the procedure are not as great. There is a poor response in patients with evidence of advanced arthropathy, as manifested in particular

TABLE 2 Comparison of × 90 and the Maximum and Average Ranges of Penetration

Radionuclide	Maximum range (mm)	Average range (mm)	× 90 (mm)
Phosphorus-32	7.9	2.6	2.2
Yttrium-90	10.8	3.8	2.8
Dysprosium-165	5.6	1.4	1.3
Holmium-166	8.7	2.2	2.1
Rhenium-186	4.5	0.9	1.0
Gold-198	3.9	0.8	0.9

Note: This table compares the average and maximum range of penetration achieved by the beta particles of the various radiocolloids with the distance from the source at which 90% of the absorbed dose is deposited (×90).
Source: From Ref. 46.

by cartilage loss. Much of the current research with regard to radiosynoviorthesis centers around the most appropriate method of screening for patients who are most likely to benefit.

Monitoring for Leakage

Although to date, no correlation between this procedure and an increased incidence of malignancy has been established, it is reasonable to incorporate a method of monitoring leakage into the basic technique. This depends largely on the agent utilized and the type of emission involved. Pure beta emitters with no gamma radiations such as ^{90}Y and ^{32}P must be monitored through the use of a Geiger–Muller counter for Bremstrahlung radiation. Radioisotopes emitting gamma radiation may be imaged for leakage by using a gamma camera. Measurements have been made at either one or multiple intervals usually within the first 72 hours following treatment (18,26,29).

Basic Technique

Given the number of radionuclides available and their differing properties, some variations in technique are to be expected. The authors provide a method utilizing ^{32}P as an example below (Table 3). The basic principles, however, are similar. In most cases, with the exception of young children and particularly anxious patients, therapy may be offered as an outpatient procedure with no need for general anesthesia or conscious sedation. A 20- or 22-gauge needle should suffice in most cases. Although some investigators advocate the use of anatomic landmarks for "blind" needle placement, the authors prefer to use fluoroscopic guidance. An attempt at joint aspiration is made to prevent dilution of the radiopharmaceutical and reduce the risk of leakage from overdistention of the joint. A small amount of radiographic contrast is then injected to confirm appropriate intra-articular needle position and avoid soft tissue injection. A radiograph is obtained to document the intra-articular position of the needle. The radionuclide mixture is then injected. (For information on the dosage to be administered, the readers are referred to Table 3 and Ref. 11.)

TABLE 3 Radiosynoviorthesis

Indications	Chronic synovitis (i.e., rheumatoid arthritis) Recurrent hemarthrosis (i.e., hemophilia) Synovial tumor (i.e., pigmented villonodular synovitis)
Radiopharmaceuticals	Phosphorus-32 (^{32}P) chromic phosphate B emitter Penetration 8 mm Half-life 14 days
Dose	Large joint (i.e., knee) = 1 mCi (37.0 MBq) of P32 diluted[a] Small joint (i.e., elbow) = 0.5 mCi (18.5 MBq) of P32 diluted[a]
Method	Sterile technique Local anesthesia Place 22-gauge needle into joint under fluoroscopy Aspirate any fluid from joint Confirm intra-articular position with contrast (1–2 cc) Hook up three-way stopcock and tubing. Inject P32 through tubing and flush with contrast Inject joint and needle tract with a mixture of 1 cc of methylprednisolone and 1 cc of lidocaine 1% Ace wrap for 48 hr after therapy and limit activities. Then resume normal activities Monitor for leakage with a Geiger counter immediately and 2 hr after the injection of P32
Risks	Mild inflammation at injection site Soft-tissue necrosis from extravasation
Bleeding prophylaxis	Von Willebrand's Desmopressin acetate (dD AVP) nasal spray qd × 3 days prior to procedure day Desmopressin acetate (dD AVP) nasal spray day of procedure within 60 min of and at least 30 min before the procedure Hemophilia-8 Factor 8 IV replacement within 30 min of procedure

[a]Must dilute the dose with sterile saline to a total volume of 1 cc.

The joint and needle tract are flushed with a solution of anesthetic agent and corticosteroid as the needle is withdrawn. Though some practitioners argue against the routine use of steroids citing possible long-term adverse effects on joint cartilage, this practice serves several purposes. It results in immediate reduction in inflammation and decreases vascularity of the synovium promoting intra-articular retention of the agent and preventing leakage into the local soft tissues. It also helps to prevent burns along the needle tract and reduces the incidence of postprocedure radiation-induced synovitis. Lastly, as described earlier in the chapter, some investigators believe that there is an additive benefit to the steroid in combination with the radionuclide (15–17).

The joint is then wrapped with an ace bandage or other pressure bandage, splinted, and the patient instructed to limit activities for at least 48 hours. No correlation was found between efficacy of treatment and intra-articular distribution pattern (48,49). Leakage is monitored with a Geiger counter immediately and two hours following the procedure.

Special Considerations in Hemophilic Patients

We utilize a single dose of factor replacement within 30 minutes of the procedure. Other investigators have utilized a more conservative approach with more frequent administration of factors, either before or after the procedure. Patients with inhibitors were given special consideration and monitored overnight in the hospital (32).

Risks

One of the obvious concerns about adverse effects in radiosynovectomy has to do with possible chromosomal damage and mutagenesis resulting in increased risk of cancer. Leakage to regional lymph nodes and the liver is the primary consideration. Most reports have shown minimal risk of leakage with some variability, depending on the agent utilized (12). Several authors have even studied the incidence of chromosomal aberrations in patients after radiosynovectomy. One investigator did find a low incidence of chromosomal aberrations in hemophiliacs treated with radiosynovectomy; all but one subsequently tested HIV positive, which has been associated with chromosome damage (50). Other reports noted no correlation between leakage, as measured by detectable lymph node activity, and chromosomal abnormalities (12). To date, no correlation between this procedure and an increased incidence of malignancy has been established.

Other complications including skin and soft tissue necrosis, though rare, have arisen as a result of leakage of radioactive material along the needle tract (51). This can be prevented with proper technique, by confirming the intra-articular position of the needle with contrast, by flushing the needle as it is removed and by temporarily immobilizing the joint in the immediate postprocedure period (25).

CONTROVERSIES AND FUTURE RESEARCH

A review of the current literature on radiosynovectomy reveals some encouraging, although somewhat inconsistent, results. Clearly, more research is needed with particular attention to randomized, controlled studies comparing the efficacy of intra-articular injection of placebo, steroid, radionuclide, and the combination of the latter two.

One area that has received a great deal of attention is that of patient selection. A common theme in the literature is the inability to halt disease progression once a critical point in the clinical and radiographic course has been reached. In the past, this had been assessed through largely subjective means employing plain film interpretation in conjunction with standard scoring systems. Some authors have advocated the use of the blood pool images on a bone scan to detect active synovitis or Doppler ultrasound to assess synovial thickness and vascularity (4,5,7,21,41,52,53).

Much of the recent research centers on the use of gadolinium-enhanced MRI for this purpose. Some investigators have shown a significant decrease in both the synovial thickness and degree of synovial enhancement following radiosynovectomy. Other groups have found a direct correlation between synovial blood vessel density and fractional area, with rate of dynamic MRI enhancement, and demonstrated the ability to accurately distinguish active synovitis (hypervascular) from fibrotic pannus (54,55). As it has been postulated that the best results with radiosynovectomy are obtained in patients with a relatively thin, actively inflamed, and hypervascular synovium, perhaps screening should be attempted and results monitored with dynamic gadolinium-enhanced MRI. If MRI is not readily available or not felt to be cost-effective, bone scan or Doppler ultrasound can be utilized instead.

At this time, no consistently reproducible quantitative method of disease assessment has been agreed upon. It is crucial to the future of this procedure that an effective means of screening patients and selecting those most likely to benefit be established.

Despite the need for more research, radiosynovectomy has been utilized now for approximately 30 to 40 years as an alternative to surgical removal of the synovium and repeated intra-articular injections of steroids. Initial concerns over possible chromosomal damage and carcinogenic potential seem largely unfounded in light of the elimination of ^{98}Au, advent of newer agents, and use of careful technique resulting in minimal extra-articular leakage. To date, no cancers related to the treatment have been detected. Given the decreased morbidity and potential savings in terms of hospital stay and cost, radiosynovectomy appears to be here to stay.

REFERENCES

1. Ansell BM, Crook A, Mallard JR, Bywaters EGL. Evaluation of intra-articular colloidal Gold 198Au in the treatment of persistent knee effusions. Ann Rheum Dis 1963; 22:435–439.
2. Delbarre F, Menkes CJ, Aignan M, Roucayrol JC, Ingrand J, Sanchez A. Une nouvelle preparation radioactive pour la synoviorthese: le rhenium 186 colloidal. Avantages par rapport au colloide d'or 198. Nouv Presse Med 1973; 2:1372.
3. Taylor WJ, Corkill MM, Rajapaske CNA. A retrospective review of yttrium-90 synovectomy in the treatment of knee arthritis. Br J Rheumatol 1997; 36:1100–1105.
4. Kraft O, Kasparek R, Stepien A. Reradiosynoviothesis of the knee. Cancer Biother Radiopharm 2005; 20:356–362.
5. Rau H, Franke C, Goretzki G, Spitz J. Radionsynoviorthesis in osteoarthritis and other disorders with concomitant synovitis in comparison to rheumatoid arthritis. Cancer Biother Radiopharm 2005; 20:349–355.
6. Kresnick E, Mikosch P, Gallowitsch HJ, et al. Clinical outcome of radiosynoviorthesis: a meta-analysis including 2190 joints. Nucl Med Commun 2002; 23:683–688.
7. Gencoglu EA, Aras A, Kucuk O, et al. Comparison of Tc-99m HIG and three-phase Tc-99m MDP bone scintigraphy for evaluating the efficacy of yttrium-90 silicate radionuclide synovectomy. Clin Nucl Med 2003; 28:277–285.
8. Topp JR, Cross EG, Fam AG. Treatment of persistent knee effusions with intra-articular gold. Can Med Assoc J 1975; 112:1085–1089.
9. Menkes CJ, Algnan M, Galmiche B, et al. Le traitement des rhumatismes par les synoviortheses. Rhumatol 1972; 2(suppl 1):61–80.
10. Kampen WU, Voth M, Pinkert J, Krause A. Therapeutic status of radiosynoviorthesis of the knee with yttrium [90Y} colloid in rheumatoid arthritis and related indications. Rheumatology 2006; 46:16–24.
11. EANM Procedure Guidelines for Radionsynovectomy. Eur J Nucl Med 2003; 30:BP12–BP16.
12. Siegel ME, Siegel HJ, Luck JV Jr. Radiosynovectomy's clinical applications and cost effectiveness: a review. Sem Nuc Med 1997; 26:354–371.
13. Grove F. Radiosynovectomy: Clinical Review. World Congress of Nuclear Medicine and Biology, Sydney, Australia, 1995.
14. Alonso-Ruiz A, Perez-Ruiz F, Calabozo M, et al. Efficacy of radiosynovectomy of the knee in rheumatoid arthritis: evaluation with magnetic resonance imaging. Clin Rheumatol 1998; 17:277–281.
15. Heuft-Dorenbosch LLJ, de Vet HCW, van der Linden S. Yttrium radiosynoviorthesis in the treatment of knee arthritis in rheumatoid arthritis: a systematic review. Ann Rheum Dis 2000; 59:583–586.
16. Gobel D, Gratz S, von Rothkirch T, Becker W, Willert HG. Radiosynoviorthesis with rhenium-186 in rheumatoid arthritis: a prospective study of three treatment regimens. Rhematol Int 1997; 17:105–108.
17. Van der Zant FM, Jahangier ZN, Moolenburgh JD, Swen AA, Boer RO, Jacobs JWG. Clinical effect of radiation synovectomy of the upper extremity joints: a randomized, double-blind, placebo-controlled study. Eur J Nucl Med Mol Imaging 2007; 34:212–218.
18. Stucki G, Bozzone P, Treuer E, Wassmer P, Felder M. Efficacy and safety of radiation synovectomy with yttrium-90: a retrospective long-term analysis of 164 applications in 82 patients. Br J Rheumatol 1993; 32:383–386.
19. Gumpel JM, Roles NC. A controlled trial of intra-articular radiocolloids versus surgical synovectomy in persistent synovitis. Lancet 1975; 1:488–489.

20. Winfield J, Gumpel JM. An evaluation of repeat intraarticular injections of yttrium-90 colloids in persistent synovitis of the knee. Ann Rheum Dis 1979; 38:145–147.
21. Fischer M, Modder G. Radionuclide therapy of inflammatory joint diseases. Nucl Med Commun 2002; 23:829–831.
22. Bridgman JF, Bruckner F, Eisen V, Tucker A, Bleehen NM. Irradiation of the synovium in the treatment of rheumatoid arthritis. Q J Nuc Med 1973; 42:357–367.
23. Lee P. The efficacy and safety of radiosynovectomy. J Rheum 1982; 9:165–168.
24. Pelosi E, Pellerito R, Clerico P, Migliaretti G, Rosso U, Bisi G. Characterization of the ideal candidate for knee radiosynoviorthesis treatment in patients with rheumatoid arthritis. Nucl Med Commun 2004; 25:603–608.
25. Gratz S, Gobel D, Behr TM, Herrmann A, Becker W. Correlation between radiation dose, synovial thickness, and efficacy of radiosynoviorthesis. J Rheum 1999; 26:1242–1249.
26. Sledge CB, Zuckerman JD, Shortkroff S, et al. Synovectomy of the rheumatoid knee using intra-articular injection of dysprosium-165-ferric hydroxide macroaggregates. J Bone and Joint Surg 1987; 69A:970–975.
27. Nuss R, Kilcoyne F, Rivard GE, Murphy J. Late clinical, plain x-ray and magnetic resonance imaging findings in haemophilic joints treated with radiosyonviortheis. Haemophilia 2000; 6:658–663.
28. Nuss R, Kilcoyne F, Geraghty S, et al. MRI findings in haemophilic joints treated with radiosynoviorthesis with development of an MRI scale of joint damage. Haemophilia 2000; 6:162–169.
29. Van Soesbergen RM, Hoefnagel CA, Marcuse HR, Dijkstra PF, Bernelot Moens HJ. Radiosynoviorthesis of the knee: a doubleblind trial of 1 versus 5 mCi Gold-198. Clin Rheum 1988; 7:224–230.
30. Silva M, Luck JV Jr. ME Siegel. P32 chromic phosphate radiosynovectomy for chronic haemophilic synovitis. Haemophilia 2001; 7(suppl 2):40–49.
31. Ahlberg A, Pettersson H. Synoviorthesis with radioactive gold in hemophiliacs; clinical and radiological follow-up. Acta Orthop Scand 1979; 50:513–517.
32. Siegel HJ, Luck JV Jr, Siegel ME, Quines C, Anderson E. Hemarthrosis and synovitis associated with hemophilia: clinical use of P-32 chromic phosphate synoviorthesis for treatment. Radiology 1994; 190:257–261.
33. van Kasteren MEE, Novakova IRO, Th Boerbooms AM, Lemmens JAM. Long term follow up of radiosynovectomy with yttrium-90 silicate in haemophilic haemarthosis. Ann Rheum Dis 1993; 52:548–550.
34. Manco-Johnson MJ, Nuss R, Lear J, et al. 32P Radiosynoviorthesis in children with hemophilia. J Pediatr Hematol Oncol 2002; 24:534–539.
35. Heim M, Goshen E, Amit Y, Martinowitz U. Synoviorthesis with radioactive yttrium in haemophilia: Israel experience. Haemophilia 2001; 7(suppl 2):36–39.
36. Rodriguez-Merchan EC, Jiminez-Yuste V, Villlar A, Quintana M, Lopez-Cabarcos C, Hernandez-Navarro F. Yttrium-90 synoviorthesis for chronic haemophilic synovitis: Madrid experience. Haemophilia 2001; 7(suppl 2):34–35.
37. Fernandez-Palazzi F, Caviglia H. On the safety of synoviorthesis in haemophilia. Haemophilia 2001; 7(suppl 2):50–53.
38. Grmek M, Milcinski M, Fettich J, Benedik-Dolnicar M, Brecelj J. Radiosynoviorthesis for treatment of hemophilic hemarthrosis-Slovenian experience. Cancer Biother Radiopharm 2005; 20:338–343.
39. Rodriguez-Merchan EC. Radionuclide synovectomy (radiosynoviorthesis) in hemophilia: a very efficient and single procedure. Semin Thromb Hemost 2003; 29:97–100.
40. Kat S, Kutz R, Elbracht T, Weseloh G, Kuwert T. Radiosynovectomy in pigmented villonodular synovitis. Nuklearmedizin 2000; 39:209–213.
41. Fransen MJ, Boerbooms AM, Karthaus RP, Buijs WC, van de Putte LB. Treatment of pigmented villonodular synovitis of the knee with yttrium-90 silicate: prospective evaluations by arthroscopy, histology, and 99mTc pertechnetate uptake measurements. Ann Rheum Dis 1989; 48:1007–1013.
42. O'Sullivan MM, Yates DB, Pritchard MH. Yttrium 90 synovectomy-a new treatment for pigmented villonodular synovitis. Br J Rheum 1987; 26:71–72.
43. Visser ED, Veth RPH, Pruszcynski M, Wobbes T, van de Putte LBA. Diffuse and localized pigmented villonodular synovitis: evaluation of treatment of 38 patients. Arch Orthop Trauma Surg 1999; 119:401–404.
44. Wiss DA. Recurrent villonodular synovitis of the knee: successful treatment with yttrium-90. Clin Orthop 1999; 169:139–144.
45. Will R, Laing B, Edelman J, Surveyor I. Comparison of two yttrium-90 regimens in inflammatory and osteoarthropathies. Ann Rheum Dis 1992; 51:262–265.
46. Johnson LS, Yanch JC. Absorbed dose profiles for radionuclides of frequent use in radiation synovectomy. Arth and Rheum 1991; 34:1521–1530.
47. Tandon P, Malpani BL, Venkatesh M, Bhatt BC. Estimation of radiation dose at various depths for commonly used radionuclides in radiosynoviorthesis in a tissue equivalent material. Med Phys 2006; 33:2744–2750.
48. Chandler ST, Sattar MA. Does distribution and retention of Y90 within the joint influence the clinical outcome of radioactive synovectomy? Nucl Med Commun 1998; 19:1109.

49. Houtman PM, Bruyn GAW, Griep EN, Jansen TL, Hoving J. Does the distribution of Y90 within the knee joint influence the clinical outcome of radiosynovectomy? Nucl Med Commun 1999; 20:589–590.
50. Rivard GE, Mgirard, Belanger R, Jutras M, Guay JP, Marton D. Synoviorthesis with colloidal P32 chromic phosphate for the treatment of hemophilic arthropathy. J Bone and Joint Surg 1994; 76A:482–488.
51. Peters W, Lee P. Radation necrosis overlying the ankle joint after injection with yttrium-90. Ann Plast Surg 1994; 32:542–543.
52. Rodriguez-Merchan EC, Wiedel JD. General principles and indications of synoviorthesis (medial synovectomy) in haemophilia. Haemophilia 2001; 7(suppl 2):6–10.
53. Turkmen C, Zulflkar B, Taser O, et al. Radiosynovectomy in hemophilic synovitis: correlation of therapeutic response and blood-pool changes. Cancer Biother Radiopharm 2005; 20:363–370.
54. Gaffney K, Cookson J, Blades S, Coumbe A, Blake D. Quantitative assessment of the rheumatoid synovial microvascular bed by gadolinium-DTPA enhanced magnetic resonance imaging. Ann Rheum Dis 1998; 57:152–157.
55. Konig H, Sieper J, Wolf KJ. Rheumatoid arthritis: evaluation of hypervascular and fibrous pannus with dynamic MR imaging enhanced with Gd-DTPA. Radiology 1990; 176:473–477.

17 Aspiration of Calcific Tendonitis

Caroline Parlier-Cuau, Marc Wybier, and Jean-Denis Laredo
Department of Bone and Joint Radiology, Lariboisière Hôpital, Assistance Publique–Hopîtaux de Paris, Paris, France

INTRODUCTION

Calcific tendinitis results from the deposition of calcium hydroxyapatite crystals in or around tendons mostly in periarticular locations. The shoulder is the most frequent location of calcific tendinitis. However, calcific tendinitis may be encountered in a large variety of anatomical locations such as the hip, wrist, foot, and cervical spine. Some patients have a single joint involved, whereas many others have bilateral involvement of the shoulders or involvement of multiple different joints (calcium hydroxyapatite deposition disease).

Apatite deposits may be symptomatic. In such cases, pain may be related to the inflammation process caused by the presence of the calcification (1). The aim of needle aspiration in calcified deposits (NACD) is to remove a significant part of the calcification to decrease the patient's pain. Partial removal of the calcification by the needle aspiration is usually followed by a spontaneous resorption of the remaining calcium (2). This is probably facilitated by the opening of the calcium-containing cavity at the time of needle aspiration (3–5). NACD is mostly indicated in calcified tendinitis of the shoulder; however, it may be occasionally performed in other anatomic locations.

INDICATIONS

The procedure has three goals: evacuation of a maximum of calcium, opening of the calcium-containing cavity, and fragmentation of the residual calcific deposits in order to facilitate resorption during the following weeks, with reduction of inflammation secondary to the presence and migration of residual calcific deposits by in situ injection of corticosteroids.

The most important clinical selection criterion for NACD is exacerbation of pain at night (6,7). Conversely, absence of pain at night, pain caused by a specific kind of motion within a given arc, and limitation of a specific motion suggest that pain is related to an impingement syndrome rather than to the calcific deposit. Prior studies have shown that, in most cases, no cuff tear is associated with calcified tendinitis (5,8,9). Prior to NACD, a careful radiologic evaluation is necessary. The structure of the calcific deposits and their relation to the rotator cuff tendons and subacromial bursa must be determined (1,7). Other features to evaluate are number, size, density, contours, and homogeneity of the calcific deposits and their tendency to evacuate into the subacromial bursa (4,5). The absence of changes on successive X-rays suggests an indication for needle aspiration. The calcification must be larger than 5 mm in diameter. The composition of the deposit will determine the success or failure of the aspiration. Faint milky calcifications with fuzzy contours are usually liquid and easily aspirated (Fig. 1), whereas very dense calcifications with clearly defined margins are often very hard and cannot be aspirated. Irregular striated calcifications are usually located within tendon fibers (Fig. 2). They usually correspond to degenerative tendinitis and cannot be aspirated.

TECHNIQUE

The technique described here is performed under fluoroscopic guidance and involves the shoulder, which is the most frequent site of tendinous calcific deposits. Recently, ultrasonography (US) has also been shown to detect and localize rotator cuff calcifications reliably (10). NACD under US guidance has also been reported by two different teams (3,10–12).

A direct anteroposterior approach under fluoroscopic guidance is used. The patient is placed in a supine position on the radiographic table. The X-ray beam is centered vertically to

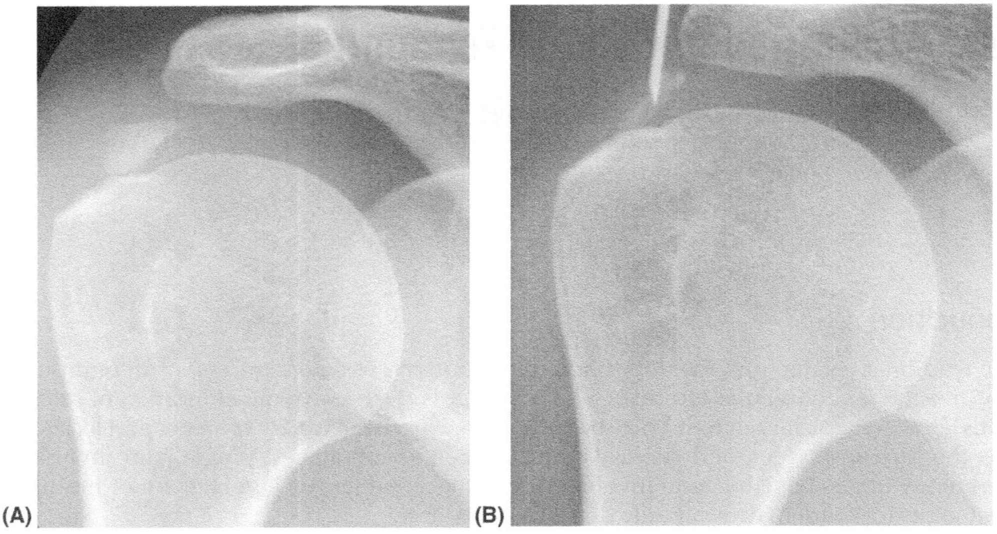

FIGURE 1 (**A**) Supraspinatus calcific deposit with faint and milky appearances and fuzzy contours (usually liquid and easily aspirated). (**B**) After needle aspiration the plain radiograph shows hardly any residual deposit.

the shoulder or slightly tilted if this allows better separation of the calcium deposit from the underlying bone on the image intensifier screen. Arm position is chosen according to the location of the calcification within the rotator cuff. Aseptic conditions are mandatory. The skin and superficial planes are anesthetized with 1% lidocaine. A 19-gauge needle is vertically advanced under fluoroscopic guidance to the center of the calcification, following a direction parallel to the X-ray beam (Figs. 3 and 4A). During the entire approach, the needle appears on the image

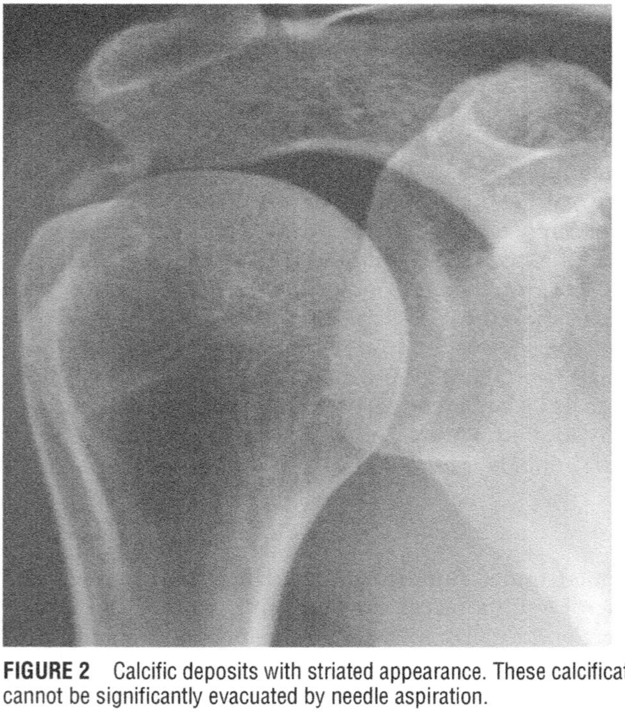

FIGURE 2 Calcific deposits with striated appearance. These calcifications are usually located within tendon fibers and cannot be significantly evacuated by needle aspiration.

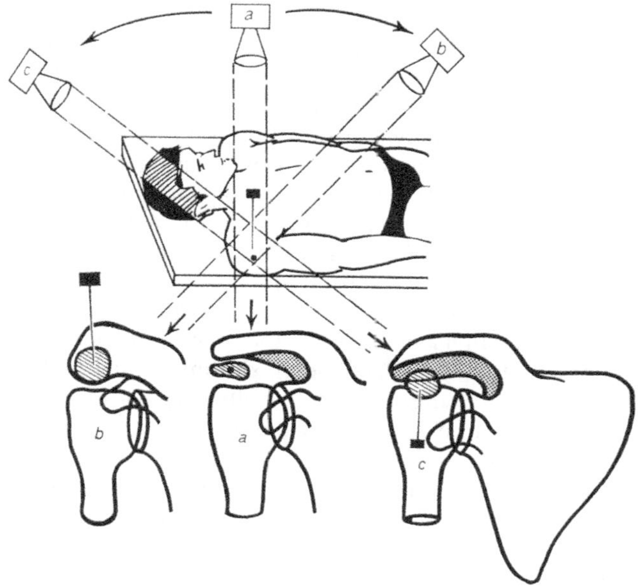

FIGURE 3 Fluoroscopic procedure to check that the needle tip is within the calcific deposits. (a) When the X-ray beam is directed along the needle axis, the needle appears as a dot in the center of the calcification. (b, c) The X-ray beam is then successively tilted in maximal cephalad and caudad directions. If correctly placed, the needle tip will remain within the calcification. *Source*: From Ref. 13.

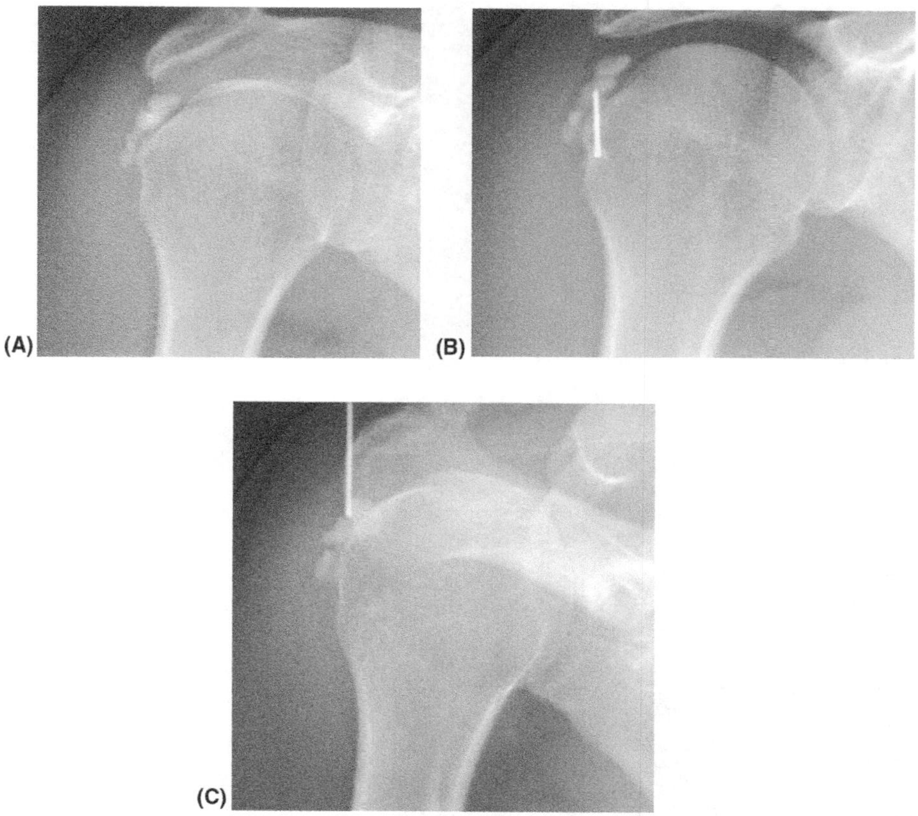

FIGURE 4 Anteroposterior radiograghs of the shoulder. (**A**) Supraspinatus calcific deposits. The needle is advanced parallel to the X-ray beam and appears on the screen as a single point in the center of the calcification. (**B, C**) The X-ray beam is successively tilted in maximal cephalad and caudad directions. The needle tip has to always be within the calcification.

intensifier screen as a single point in the center of the calcification (Figs. 3 and 4A). At any time of the procedure, the X-ray beam can be successively tilted cranially and caudally to confirm on the screen that the needle is actually within the calcification (Figs. 3B, C and 4B, C). A firm sensation is obtained when the calcification is reached. Calcium aspiration is then performed using a syringe containing first lidocaine and then sterile water or saline solution (1–2 mm), doing a succession of propulsions and suctions with the syringe piston. Aspirated calcium appears in the syringe as a white, cloudy return (Fig. 5). This procedure is repeated until maximal aspiration of calcium has been obtained (Figs. 1 and 6). In large and lobulated calcific deposits, insertion of two needles may be necessary (Fig. 7). The amount of calcium aspirated at the end of the procedure is variable and always incomplete (10–80%) (Figs. 1 and 6). In some cases, the calcification has a hard consistency and no calcium can be aspirated. However, grinding of the calcific deposit with the needle may in itself accelerate the process of spontaneous resorption (2,4,5,14) and is as important as aspiration. Once the maximum of calcium has been aspirated, 2 to 3-mL of prednisolone acetate (50–75 mg) is injected in situ (14). Radiographs are obtained at the end of the procedure to document evacuation of the deposits. In the great majority of cases, the procedure is well tolerated and painless.

Keeping the shoulder at rest for five or six days is recommended. One-third of patients have a painful reaction rarely lasting more than two to four days. This is managed with intermittent application of ice and prescription of pain medication and nonsteroid anti-inflammatory agents (14). This painful crisis is usually accompanied by an almost complete resorption of the remaining calcification.

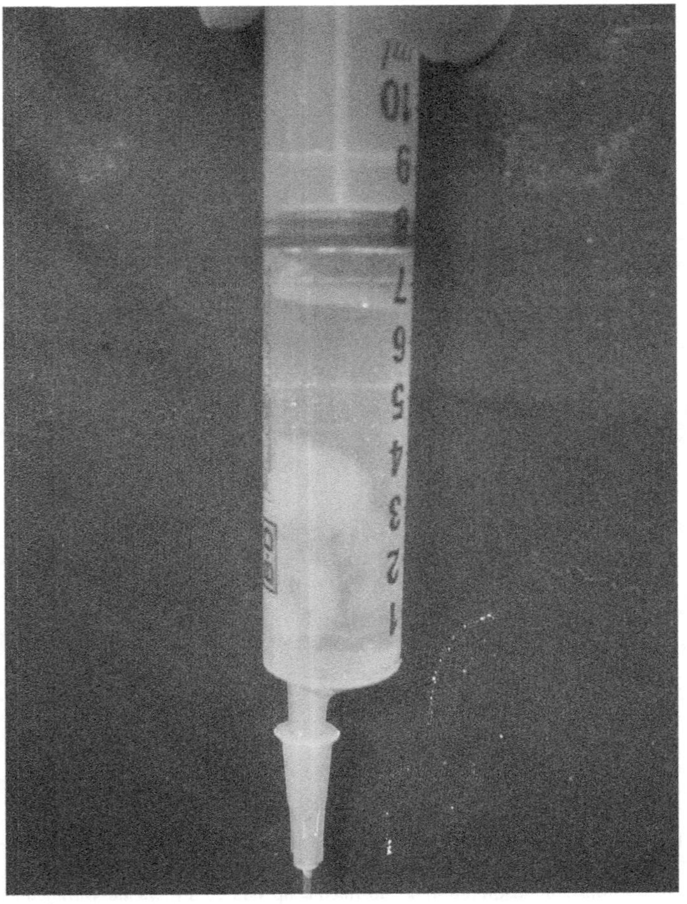

FIGURE 5 Aspirated calcium appears in the syringe as a white, cloudy return.

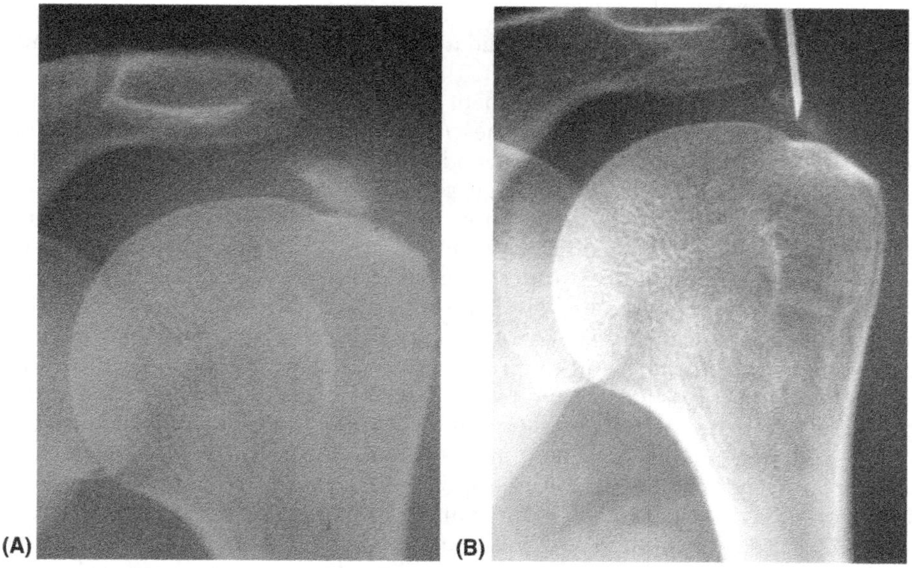

FIGURE 6 Supraspinatus calcific deposits before (**A**) and after (**B**) needle aspiration: the calcification is fragmented and faint.

RESULTS AND DISCUSSION

Although aspiration of calcific deposits is conventionally done under fluoroscopy, they can also be performed quickly and efficiently with US. Usually, rotator cuff calcifications appear under US as bright echogenic images with posterior acoustic shadowing (3.11). In this case, it is easy to localize the calcific depots; then US offers a nonionizing imaging technique and a valuable tool for guiding aspiration. Its realtime capabilities permit continuous monitoring of the needle position relative to the target and to surrounding structures. In few cases, calcifications appear

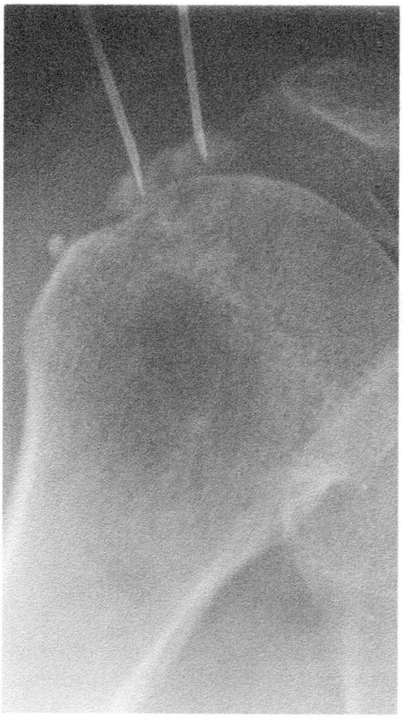

FIGURE 7 Large and lobulated calcific deposit of the supraspinatus. Two needles have been necessary to obtain a maximal aspiration of calcium.

without posterior acoustic shadowing (3), and US guidance could be more difficult than fluoroscopic guidance. Anyway US and fluoroscopic guidance both could be used, but they both mandate a proper training.

According to several studies, good and excellent results are achieved in 61% (5) to 74% (11) of cases. Comfort and Arafiles, in 1978, first described this technique and reported nine cases followed for an average of nine years (2). Good-to-excellent results were obtained by needle irrigation and aspiration, and follow-up on plain radiographs showed no residual deposits. In 1989, Normandin et al. reported 69 cases of calcific tendinitis treated by needle aspiration (5). The clinical results were good in 60.9% of cases at 11 to 45 months (mean, 24) of follow-up. Aspiration of large amounts of calcium and secondary resorption of residual deposits on plain radiographs in the following weeks were significantly associated with good results. By contrast, in cases of hard stone-like consistency of the calcification or in cases of calcification encrusted within the tendon fibers, no significant calcium was usually aspirated, and the result of the procedure was poor. In 1994, Pfister and Gerber reported on 212 patients with calcific tendinitis of the shoulder treated with needle aspiration (15,16). At five years of follow-up, 60% of the patients were free of pain, 34% had a marked pain relief, and 6% were unchanged. These results, however, were not compared to the spontaneous outcome of shoulder calcific tendinitis, which is often favorable. In 1996, Farin et al. reported 61 patients with rotator cuff calcifications treated by needle aspiration and lavage under US guidance (11). Clinical results were good in 74% of cases and moderate or poor in 26%. Clinical results were compared to the changes in the calcification on plain radiographs at one-year follow-up. In cases with a good result, the calcification decreased in size in 86% of cases, whereas no change was seen in 14% of cases. In cases with a moderate or poor result, the calcification decreased in size in 37% of cases, whereas no change was seen in 63% of cases (11).

Surgical excision of calcifications of the rotator cuff gives prompt, complete, and permanent relief of symptoms (17). At present, surgical excision of calcifications is mostly performed through arthroscopy (18,19). There are, however, many objections to the use of surgery as the primary procedure in rotator cuff calcifications mainly owing to a prolonged period of disability and potential complications of surgery, especially the risk of secondary reflex sympathetic dystrophy. Therefore, NACD should always be attempted first because it is a minimally invasive technique providing good results in at least two-thirds of cases (2,5,11,15). In our experience, there are no technical failures resulting from difficulty in locating the calcification with the needle, under fluoroscopic guidance. As in arthroscopic treatment of calcific tendinitis (18), it seems not essential with NACD to remove the deposit completely. If the size of calcification is unchanged after the procedure, however, it is important to obtain a decreased density of the calcification. The good results in such cases can be explained by the opening of the calcium-containing cavity and creation of a communication between the cavity and surrounding vascularized tissues (Fig. 8). This may cause local inflammation and hyperhemia, which in turn accelerates the process of resorption.

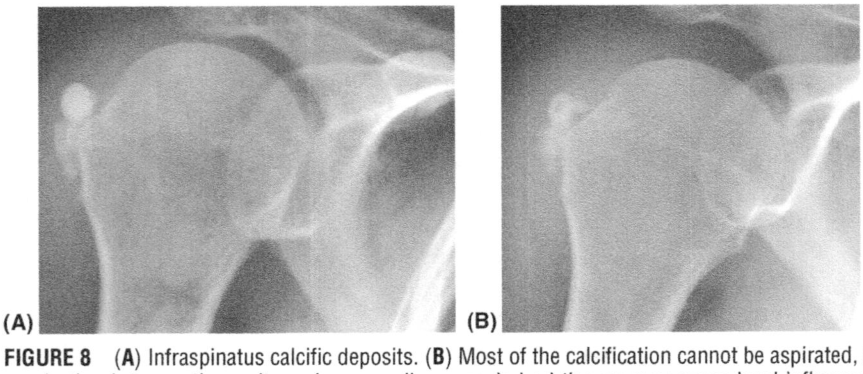

FIGURE 8 (**A**) Infraspinatus calcific deposits. (**B**) Most of the calcification cannot be aspirated, but creation of a communication between the cavity and surrounding vascularized tissues may cause local inflammation and hyperhemia, which accelerates the process of resorption.

Extracorporal shock-wave therapy (ESWT) in the treatment of shoulder calcific tendinitis has been recently described (20–25). It may offer a new and safer additional nonoperative treatment of chronic calcific tendinitis of the shoulder. Good and excellent results were obtained in 55% (24) to 64% (21,22) of cases. Charrin and Noel reported a series of 32 patients treated with ESWT (24). Improvement in pain was noted in 55.1% of the patients after 24 weeks. Sixty-two percent of the patients were at significantly improved at 12-week-follow-up in another series of 21 patients (23). In a series of 50 patients treated with ESWT, Rompe et al. had 60% good and excellent results after one year and 64% good and excellent results after two years. The results have been significantly improved by the use of exact focusing of ESWT (20,24) and high-energy shock waves (21–23). All these reports, however, are noncontrolled studies and have an open format. Ebenbichler et al. have conducted a randomized, double-blind study in patients with symptomatic calcific tendonitis (26). Results showed that ESWT helps resolve calcifications and is associated with short-term (six weeks) clinical improvement. But at nine months, the difference between the two groups was no longer significant.

CONCLUSION

In cases of calcific tendinitis, needle aspiration of tendinous calcific deposits is a well-tolerated conservative procedure that should be attempted after failure of medical treatment in chronically painful shoulders associated with rotator cuff deposits. Excellent and good results vary from 61% to 74% of patients. In these patients, dramatic and durable improvement is obtained without surgery. Therefore surgery should be reserved to failures of NACD. ESWT for calcific tendonitis appears to be an effective therapy in 55% to 64% of the cases, but additional controlled studies are necessary to confirm the value of this technique.

REFERENCES

1. Bosworth BM. Calcium deposits in the shoulder and subacromial bursitis. A survey of 122 shoulders. JAMA 1941; 116:2477.
2. Comfort TH, Arafiles RP. Barbotage of the shoulder with image-intensified fluoroscopic control of needle placement for calcific tendinitis. Clin Orthop 1978; 135:171.
3. Farin PU, Jaroma H, Soimakallio S. Rotator cuff calcifications: treatment with US-guided technique. Radiology 1995; 195:841.
4. Laredo JD, Bellaiche L, Hamze B, Naouri JF, Bondeville JM, Tubiana JM. Current status of musculoskeletal interventional radiology. Rad Clin North Am 1994; 32:377.
5. Normandin C, Seban E, Laredo JD, et al. Aspiration of tendinous calcific deposits. In: Bard M, Laredo JD, eds. Interventional Radiology in Bone and Joints. Wien, New York: Springer-Verlag, 1988:285.
6. Simon WH. Soft tissue disorders of the shoulder: frozen shoulder, calcific tendinitis and bicipital tendinitis. Orthop Clin North Am 1975; 6:521–539.
7. Welfing J. Les calcifications de l'épaule. Diagnostic clinique. Rev Rhum 1964; 1:265.
8. Jim YF, Hsu HC, Chang CY, Wu JJ, Chang T. Coexistence of calcific tendinitis and rotator cuff tear: an arthrographic study. Skeletal Radiol 1993; 22:183.
9. Matsen MLA, Kilcoyne RF, Davies PK, Sickler ME. US evaluation of the rotator cuff. Radiology 1985; 157:205.
10. Farin PU. Consistency of rotator cuff calcifications. Observations on plain radiography, sonography, computed tomography, and at needle treatment. Invest Radiol 1996; 31:300.
11. Farin P, Rasanen H, Jaroma H, Harju A. Rotator cuff calcifications: treatment with US-guided percutaneous needle aspiration and lavage. Skeletal Radiol 1996; 25:551.
12. Aina R, Cardinal E, Bureau NJ, Aubin B, Brassard P. Calcific shoulder tendinitis: treatment with modified US-guided fine-needle technique. Radiology 2001; 221(2):455–461.
13. Normandin C, Seban E, Laredo JD, et al. Aspiration of tendinous calcific deposits. In: Bard M, Laredo JD, eds. Interventional Radiology in Bone and Joints. Wien, New York: Springer-Verlag, 1988:265.
14. Moutounet J, Chevrot A, Godefroy D, et al. Ponction infiltration radioguidée dans le traitement des périarthrites calcifiantes rebelles d'épaules. J Radiol 1984; 65:569.
15. Pfister J, Gerber H. Treatment of calcific humero-scapular periarthropathy using needle irrigation of the shoulder: retrospective study. Z Orthop Ihre Grenzgeb 1994; 132:300.
16. Pfister J, Gerber H. Chronic calcifying tendinitis of the shoulder-therapy by percutaneous needle aspiration and lavage: a prospective study of 62 shoulders. Clin Rheumatol 1997; 16:169–274.
17. Mc Laughlin HL. The selection of calcium deposits for operation: the technique and results of operation. Surg Clin North Am 1963; 43:1501.

18. Ark JW, Flock TJ, Flatov EL, Bigliani LU. Arthroscopic treatment of calcific tendinitis of the shoulder. Arthroscopy 1992; 2:183.
19. Jerosch J, Strauss JM, Schmiel S. Arthroscopic treatment of calcific tendinitis of the shoulder. J Shoulder Elbow Surg 1998; 7(1):30–37.
20. Haake M, Deike B, Thon A, Schmitt J. Exact focusing of extracorporeal shock wave therapy for calcifying tendinopathy. Clin Orthop 2002; 397:323–331.
21. Rompe JD, Burger R, Hopf C, Eysel P. Shoulder function after extracorporal shock wave therapy for calcific tendinitis. J Shoulder Elbow Surg 1998; 7:505–509.
22. Rompe JD, Zoellner J, Nafe B. Shock wave therapy versus conventional surgery in the treatment of calcifying tendinitis of the shoulder. Clin Orthop 2001; 387:72–78.
23. Loew M, Daecke W, Kusnierczak D, Rahmanzadeh M, Ewerbeck V. Shock-wave therapy is effective for chronic calcifying tendinitis of the shoulder. J Bone Joint Surg Br 1999; 81(5):863–867.
24. Charrin JE, Noel ER. Shockwave therapy under ultrasonographic guidance in rotator cuff calcific tendinitis. Joint Bone Spine 2001; 68(3):241–244.
25. Wang CJ, Ko JY, Chen HS. Treatment of calcifying tendinitis of the shoulder with shock wave therapy. Clin Orthop 2001; 387:83–89.
26. Ebenbichler GR, Erdogmus CB, Resch KL, et al. Ultrasound therapy for calcific tendinitis of the shoulder. N Engl J Med 1999; 340(20):1533–1538.

18 Percutaneous Steroid Treatment for Plantar Fasciitis

Phoebe A. Kaplan and Robert G. Dussault
Musculoskeletal Radiology, Laënnec Radiology, Montreal, and Reso-Concorde MRI, Laval, Quebec, Canada

INTRODUCTION

"Painful heel" syndrome is a common condition, estimated to afflict nearly 10% of the adult population over a lifetime. There are many different possible causes of heel pain, including plantar fasciitis, calcaneal stress fracture, avascular necrosis of the calcaneus, calcaneal enthesophyte (especially if there is associated plantar heel pad atrophy), subcalcaneal rheumatoid nodules, subcalcaneal adventitial bursitis, nerve entrapment in the tarsal tunnel, infection, and tumor. The most common cause of heel pain syndrome is plantar fasciitis, which is an inflammatory and degenerative response in and around the plantar fascia.

ANATOMY AND FUNCTION

The plantar fascia, also known as the plantar aponeurosis, is made of multiple layers of collagen and elastic fibers, which form a structure essentially identical to a tendon. It allows great elasticity during weight bearing, is the most important structure in the foot for dynamic longitudinal arch support, assists in the push-off phase of gait, and absorbs forces from the mid-tarsal joints (1,2).

The plantar fascia originates on the undersurface of the medial tuberosity of the calcaneus (Fig. 1). It proceeds distally on the plantar aspect of the foot to attach to the flexor tendon sheaths and to the bases of the proximal phalanges. There are actually three different cords of the plantar fascia, the thick central cord, and the thinner medial and lateral cords with a thin aponeurosis connecting the cords. Only the thick central cord tends to become abnormal and symptomatic, and discussions in this chapter, which refer to the plantar fascia, are referring specifically to the central cord of the fascia. The upper and lower margins of the fascia are parallel and smooth, and are thickest at its origin (not more than 4 mm thick in the dorsoplantar dimension). The fascia gradually decreases in thickness from its origin to its insertion. The flexor digitorum brevis muscle is located immediately adjacent to the central cord of the fascia on its dorsal aspect.

ABNORMALITIES OF THE PLANTAR FASCIA

The plantar fascia may rupture completely (usually from acute trauma or repeated steroid injections), be affected by inflammation, partial tears, and degeneration (plantar fasciitis), or develop fibromatosis (benign, proliferative masses arising from the fascia). Only plantar fasciitis will be discussed here because it is the condition that may be treated with steroid injections.

Plantar fasciitis is an inflammation of the plantar fascia and the perifascial structures. Both local and systemic factors may result in this inflammatory response. Local factors are mechanical in nature and occur from overuse. Repetitive trauma is common in athletes, particularly runners, in people who stand for long periods, or those with certain foot deformities that place an increased stress on the fascia. The chronic, repetitive trauma causes microtears of fibers in the fascia, particularly at its origin on the calcaneus, that are accompanied by an inflammatory response. Chronic inflammation results in collagen degeneration and necrosis, angiofibroblastic hyperplasia, chondroid metaplasia, and matrix calcification (1). Systemic factors, particularly the seronegative spondyloarthropathies and, to a lesser extent, rheumatoid arthritis may cause inflammation about the origin of the plantar fascia that is identical to that caused by chronic, low-grade trauma and lead to plantar fasciitis.

(A)

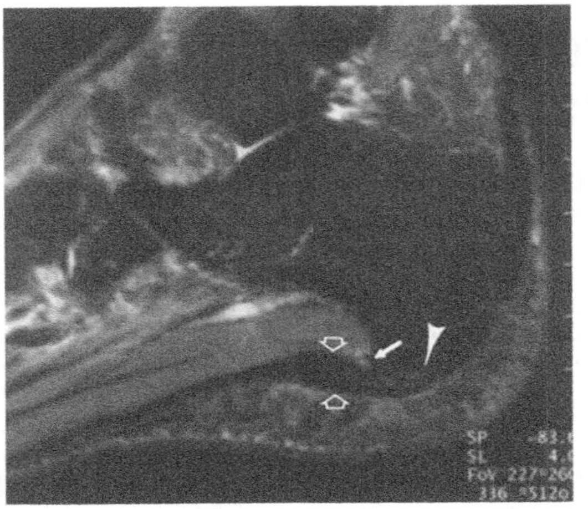

(B)

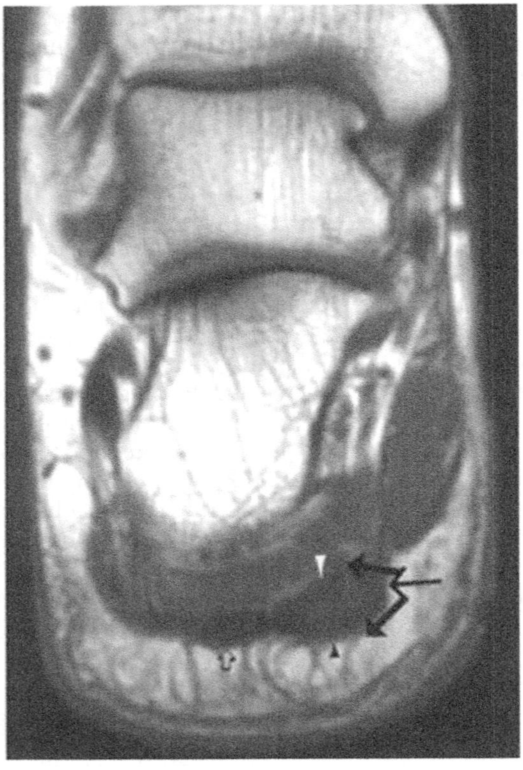

FIGURE 1 Plantar fascia anatomy in a patient with plantar fasciitis. (**A**) Sagittal short tau inversion recovery image through the medial side of the hindfoot demonstrates the central cord of the plantar fascia attaching to the medial tuberosity of the calcaneus, just plantar to a calcaneal enthesophyte (*solid arrow*). There are changes of plantar fasciitis, consisting of a thickened fascia (*open arrows*) at its site of origin, high signal within the fascia, high signal plantar to it (perifasciitis), and in the tuberosity of the calcaneus (*arrowhead*). The fascia decreases in thickness as it progresses distally. (**B**) Coronal T1-weighted image through the proximal portion of the fascia demonstrates the thickening and high signal of the central cord of the fascia (*arrowheads*). Compare this to the low signal, normal lateral cord of the fascia adjacent to it (*open arrow*). Also, note the flexor digitorum brevis muscle on the dorsal aspect of the fascia and the subcutaneous fat plantar to it. The arrows depict the trajectory the needle will take to reach the fascia (*central arrow*), and then walked off on the dorsal (*upper arrow*) and plantar (*lower arrow*) surfaces.

This cycle of chronic inflammation leading to degeneration results in abnormal thickening of the proximal plantar fascia, inflammatory fluid on either or both sides of the fascia (perifascial edema), bone marrow edema in the adjacent calcaneal tuberosity, or calcaneal bone erosions.

Although plantar calcaneal spurs (enthesophytes) are frequently present in people with heel pain, they usually are not the source of symptoms. It is particularly important to be aware that it is not the plantar fascia that causes the enthesophyte. Rather, it is the attachment of the flexor digitorum brevis and abductor hallucis muscles to the medial tuberosity that results in repetitive traction that leads to periostitis and spur formation. The plantar fascia is located plantar to a calcaneal spur that often is coincidentally present.

The clinical symptoms caused by plantar fasciitis include activity-related pain beneath the medial calcaneus, often worse with the first step taken and provoked by stretching the plantar fascia. Although the clinical diagnosis is usually straightforward, other causes of heel pain syndrome may cause an identical symptomatology. The clinical course for most individuals with plantar fasciitis consists of resolution of pain within a few weeks to several months. During this time, the management consists of stretching exercises, cushions placed beneath the calcaneus, and oral anti-inflammatory medications.

If there is no response to conservative therapy in six to nine months, there is a low likelihood of recovery after that without more aggressive measures being taken. Recalcitrant plantar heel pain should probably have documentation that the diagnosis is indeed plantar fasciitis rather than some other cause of similar heel symptoms. This may be done with magnetic resonance imaging or ultrasound. More aggressive therapy includes steroid injections or, ultimately, surgical release of the fascia, if the injections are unsuccessful in eliminating pain.

TECHNIQUE FOR STEROID INJECTIONS OF THE PLANTAR FASCIA

The first step in performing image-guided steroid injections for plantar fasciitis is to convince the physician taking care of the patient that image-guided plantar fascia injections are superior to blind injections. Thus, the ability to refer to radiologists who are skilled at doing the fascial injections has value. It is a common procedure in our practice.

Image-guided plantar fascia injections can be done with the aid of virtually any of the imaging modalities; however, ultrasound and fluoroscopic guidance are the two most practical techniques. Ultrasound allows for direct visualization of the fascia and permits the diagnosis of plantar fasciitis to be made as well as demonstrating the medication infiltrating around the fascia (3). Fluoroscopy is readily available and easy to operate. Fluoroscopy-guided fascia injections will be described in detail here, but the same principles may be applied to other forms of image guidance.

Goal

Pain relief from diffuse distribution of local anesthetic and steroids to *both* sides of an abnormal plantar fascia.

Patient Preparation

No preparation is necessary. However, if there is a history of contrast material allergy, premedication would be advised.

Materials

Alcohol and poviodine are utilized for skin cleaning; ethyl chloride topical anesthetic skin refrigerant; 1% lidocaine for local anesthesia; 1.5 in. 25-gauge needle; nonionic contrast material (2–3 mL, Omnipaque 300) in a 3 mL syringe with a short connecting tubing attached; and celestone, 1 mL (6 mg) mixed with 1 mL of 0.5% bupivacaine in a 3 mL syringe.

Fluoroscopic Technique

The patient is positioned on the fluoroscopy table so that the medial side of the foot is facing upwards and exposed for the procedure, whereas the lateral side is against the table. The foot should be in an anatomically lateral position, and the plantar aspect of the calcaneus should be identified with fluoroscopy. The skin is marked just anterior to the plantar aspect of the calcaneal tuberosity, or just anterior and plantar to a calcaneal spur, if present (Fig. 2). The plantar surface of the fascia can be appreciated with fluoroscopy because it is outlined by fat, but the dorsal surface is not evident because it blends with the density of the adjacent muscle.

The skin site is sprayed with ethyl chloride (a topical anesthetic skin refrigerant) for about five seconds to numb the skin before it is entered with a 25-gauge needle and lidocaine is infiltrated. Placing even the smallest needle anywhere into the foot can be exquisitely painful, and that is why we make the extra effort to numb it with ethyl chloride prior to putting the needle into the skin. Once the needle is through the skin, pain is minimal and the syringe with anesthetic is removed from the needle.

The 25-gauge needle is advanced using intermittent fluoroscopy to make certain that it is properly directed. There is a subtle difference in density of the tissues that can be detected with

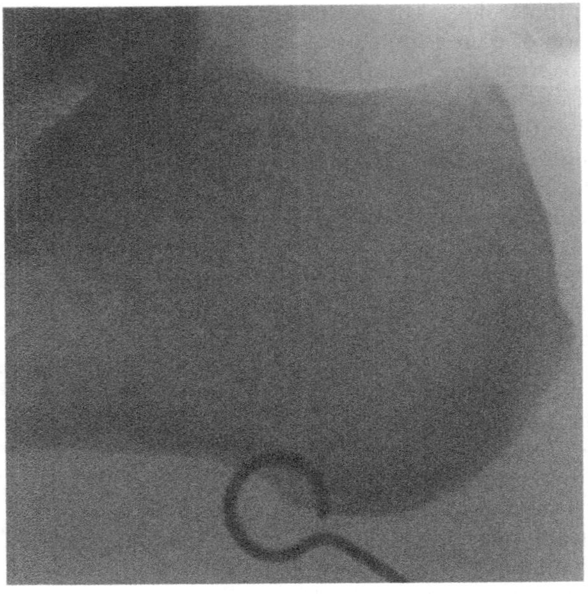

FIGURE 2 Marking the skin site for injection of the plantar fascia. A fluoroscopic spot film shows the foot in a lateral position. The marker is placed on the medial aspect of the hindfoot, just anterior to the tuberosity of the calcaneus.

experience. The plantar fascia is somewhat tough and firm as compared to the subcutaneous fat. The proper depth of the needle can be determined by appreciating this change in density of the tissues. Alternatively, the needle can be advanced until the medial side of the calcaneus is touched with the needle, and this can orient one as to the proper depth, because the fascia starts on the medial aspect of the calcaneus. Unfortunately, the latter method places the needle somewhat posterior to the ideal needle position for infiltrating around the proximal fascia. Once the needle is at the proper depth, using either technique, the needle can be walked off of the fascia on either its plantar or dorsal surface for about 5 mm. Again, a subtle tactile difference is encountered when the needle slides off the fascia and into the adjacent muscle (dorsal) or fat (plantar). A small amount of contrast material is injected through the connecting tubing to demonstrate the needle location. If properly positioned, contrast will flow linearly along the surface of the fascia (Fig. 3). If the contrast pools around the needle tip, the needle needs to be repositioned.

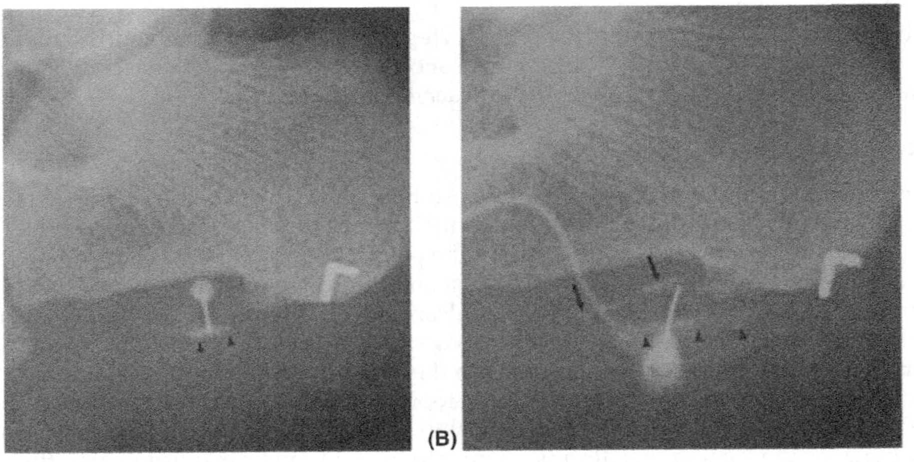

(A) (B)

FIGURE 3 Proper needle placement for fascia injection. (**A**) Fluoroscopic spot film shows the needle has been walked off the fascia in a plantar direction and contrast material runs in a linear fashion along the fascia (*arrowheads*). (**B**) The needle was repositioned dorsal to the fascia and linear contrast outlines that side of the fascia (*arrows*). The fascia is positioned between the two lines of contrast material. The arrowheads indicate the contrast on the plantar aspect of the fascia. Steroid and anesthetic deposited in these regions will bathe both surfaces of the inflamed fascia.

When injecting the dorsal surface of the fascia, contrast may infiltrate the muscle that lies immediately adjacent to it. This is not unusual, nor is it a problem, so long as the contrast material is also seen tracking on the surface of the fascia. Half of the steroid/anesthetic mixture (1 mL) is injected onto the surface of the proximal fascia.

After the first injection, the same needle is partially withdrawn and again placed on the surface of the fascia. It is then walked off the opposite side of the fascia and contrast material injected to show the linear distribution along the fascia (Fig. 3). The other half of the anesthetic/steroid solution is injected and the needle removed.

If the fascia is severely degenerated and partially torn, it may be soft enough that it can be infiltrated directly. Some people inject directly into the abnormal portion of the fascia. Others believe that if this occurs, the needle should be readjusted appropriately, because direct infiltration of steroids into the fascia may predispose it to rupturing. If the fascia is not so severely affected, it is very difficult to inject when the needle is located within the fascia. This is a helpful sign that allows one to adjust the needle tip into a more desirable position.

COMPLICATIONS OF PLANTAR FASCIA STEROID INJECTIONS

The complications that may occur from a steroid injection for plantar fasciitis are of little consequence and are rare. Potential risks include pain, infection, hematoma, allergic reaction to contrast material, fat pad atrophy from repeated injections, and rupture of the fascia. A 10% rate of plantar fascia rupture after corticosteroid injection has been reported (4); however, that has not been our experience.We have done hundreds of these injections without any occurrences of rupture. Perhaps this is because we do not directly inject steroids interstitially into the fascia, which predisposes the fascia to a higher rate of rupture than does infiltrating around the fascia. Regardless, rupture of the plantar fascia is not such a disaster if one remembers that the ultimate treatment for plantar fasciitis, when all other therapies fail, is a fascial release (fasciotomy), which is nothing more than a surgically created rupture of the fascia.

EFFICACY OF PLANTAR FASCIA INJECTIONS

Clinical improvement after blind steroid injections for plantar fasciitis has been reported in 35% to 70% of people (5,6). The highest rate of pain relief was in a group where each patient routinely received five to six injections. There are few reports of response to image-guided injections. One study that utilized bone scans to guide the injections in 15 patients had an 80% response rate (7). A report of ultrasound-guided injections (3) on the heels of five patients showed four with a positive response (80%). We reviewed a group of 28 of our patients who had magnetic resonance imaging proof of plantar fasciitis and recalcitrant and long-standing pain. These patients were injected using fluoroscopic guidance. Follow-up at one year or more after a single injection showed 79% with continued pain relief (unpublished data). Image guidance seems to increase the response rate as well as decrease the number of injections necessary because the steroid is being deposited at the proper location initially.

REFERENCES

1. Theodorou DJ, Theodorou SJ, Farooki S, Kakitsubata Y, Resnick D. Disorders of the plantar aponeurosis: a spectrum of MR imaging findings. AJR 2001; 176:97–104.
2. Spouge AR. Miscellaneous conditions of the foot and ankle. In: Spouge A, Pope TL, eds. Practical MRI of the Foot and Ankle. Boca Raton, FL: CRC Press LLC, 2000:226–228.
3. Kane D, Greaney T, Bresnihan B, Gibney R, FitzGerald O. Ultrasound guided injection of recalcitrant plantar fasciitis. Ann Rheum Dis 1998; 57:383–384.
4. Acevido JI, Beskin JL. Complications of plantar fascia rupture associated with steroid injection. Foot Ankle Int 1998; 19:91–97.
5. Wolgin M, Cook C, Graham C, Mauldin D. Conservative treatment of plantar heel pain: long-term follow-up. Foot Ankle Int 1994; 15:97–102.
6. Bao-Xing C, You-Lin Z. Treatment of painful heel combined with method of traditional Chinese medicine and western medicine. Chin Med J 1985; 98:471–474.
7. Dasgupta B, Bowles J. Scintigraphic localization of steroid injection site in plantar fasciitis. Lancet 1995; 346:1400–1401.

19 Needle Localization of Musculoskeletal Lesions

William B. Morrison
Department of Radiology, Thomas Jefferson University Hospital, Philadelphia, Pennsylvania, U.S.A.

INTRODUCTION

Needle localization, predominantly using hookwire needles, has been used for preoperative localization of breast lesions and pulmonary nodules (1–6). This technique is advantageous because the needle can be precisely positioned using imaging guidance. The hook at the end of the wire (Fig. 1) immobilizes it so that the patient can be transported to the surgical suite for excision of the lesion. This technique is also useful for musculoskeletal lesions (7). In this chapter, we will describe the technique, review previously reported musculoskeletal applications for this procedure, and discuss potential indications.

TECHNIQUE

Although needle localization of musculoskeletal lesions has been reported using only computed tomography (CT) guidance, which is the method discussed here, the technique should be applicable using whatever modality best visualizes the lesion. If ultrasound or magnetic resonance imaging (MRI) is chosen, specialized needles are commercially available to facilitate visualization. As with any biopsy, but particularly in this situation, consultation with the surgeon is essential to determine the optimal skin entry site that will be used for the surgical approach.

Using CT, the patient is initially scanned with 5 to 10 mm sections through the area of the lesion. Marker needles are placed on the skin, and repeat 3 mm sections are acquired at the level of the lesion (Fig. 2A) in order to plan the skin entry site and to measure the distance and angulation from the lesion to this point. The skin entry site is marked with an indelible marker, and this site and surrounding area is widely prepped. The hookwire needle selected is of a length determined by the depth required to reach the lesion, with an additional 5 to 10 cm external to the skin margin. Local anesthetic is administered subcutaneously. The hookwire (with the hook covered by the needle sheath) is then advanced (Fig. 2B) to the desired location next to or within the lesion based on the angle and depth determined using the preliminary scan. Repeat 3 mm CT sections through the area (Fig. 2C) document needle positioning. At this point, the needle can be repositioned if necessary. Once optimal positioning is confirmed, the hook is deployed by pulling back the sheath of the needle while holding the inner wire in place (Fig. 2D). Follow-up CT confirms final placement and serves as a surgical guide. The wire extending from the skin is coiled loosely and packed in sterile fashion with gauze and tape (Fig. 2E). The patient is then transferred to the surgical suite for excision of the lesion.

INDICATIONS

Osseous Lesions

Preoperative wire localization of osteoid osteoma has been described (8,9). However, these lesions can be ablated percutaneously with image guidance (10), making surgery less likely to be needed. Similarly, the majority of osseous lesions can be safely and effectively accessed and biopsied percutaneously using image guidance. Despite these developments, there may be situations in which operative excision is preferred. These include lesions subject to insufficient sampling via the percutaneous route, such as lesions with varying aggressiveness in different areas, such as chondroid or lipomatous lesions, or lesions suspected to be such that preservation of tissue architecture is

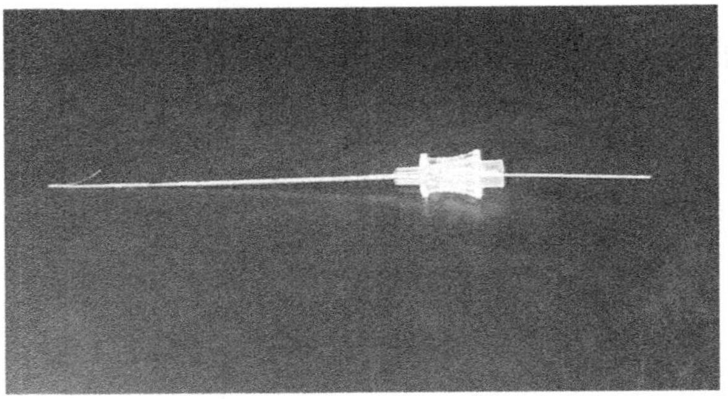

FIGURE 1 Hookwire needle with tip deployed.

required to make the diagnosis. In these settings, preoperative localization could be useful to mark a small lesion or to mark an area within a lesion suspected of having more aggressive characteristics. Other osseous lesions can be difficult or unsafe to sample using a percutaneous route, such as highly vascular lesions in noncompressible areas or sclerotic rib lesions. In particular, lesions in small or rounded bones, especially those near vital structures may prompt surgical rather than percutaneous management. Uncooperative or combative patients may also be easier to manage using needle localization. Finally, the preference of the surgeon, the patient, or the patient's family may drive the decision to perform surgery rather than percutaneous biopsy or ablation. The last category includes pediatric patients with tumors (11).

Soft-Tissue Lesions

Soft-tissue lesions may also be referred for preoperative localization. Again, the preference of the referring physician may drive the management decision. Nevertheless, the technique should be considered for some soft-tissue lesions. These include hard lesions within the soft tissues (such as a calcified mass) that would merely be deflected away from a biopsy needle, or lesions precariously close to vital structures that would be more safely approached surgically. Also, in patients with distorted anatomy or scar tissue surrounding the area of interest, a localizer needle can provide the surgeon with a helpful guide. In these situations, CT-guided hookwire needle localization of the lesion before surgical biopsy may be the best option for the patient.

RISKS/LIMITATIONS

Some issues and limitations regarding this technique should be recognized and discussed with the patient. First, the lesion must be readily visualized on the modality used for localization. For example, a rib lesion may be obvious on MRI, but have only subtle findings on CT; this coupled with rib motion during respiration as well as volume averaging effects from the oblique course of the ribs may complicate correct localization of the lesion. Localization of the wrong rib is a potential complication. However, this technique offers an improved chance of localizing a subtle rib lesion compared to blind surgical exploration.

Second, despite the hook at the tip of the wire, it is possible for the wire to migrate farther into the tissue or to be pulled out. When localizing an osseous lesion, the needle tip should be positioned such that it is abutting against the bone, preventing migration. When localizing lesions of bone with this technique, migration should theoretically be much less likely than in localization procedures involving loose soft tissues such as the breast or lung. In rare situations in which the needle tip must be positioned within or next to a lesion in the soft tissues of the musculoskeletal system, again, planning is necessary to assure that the needle course would not endanger any sensitive structures if it were to unintentionally advance. Also, after placement, the wire protruding from the skin can be bent with a hemostat, so that it lies flush against the skin and cannot advance further. In our experience there has been no migration of the needle after placement. After proper positioning of the wire and deployment of the hook, it should be carefully coiled and packed loosely on the skin to prevent dislodgement.

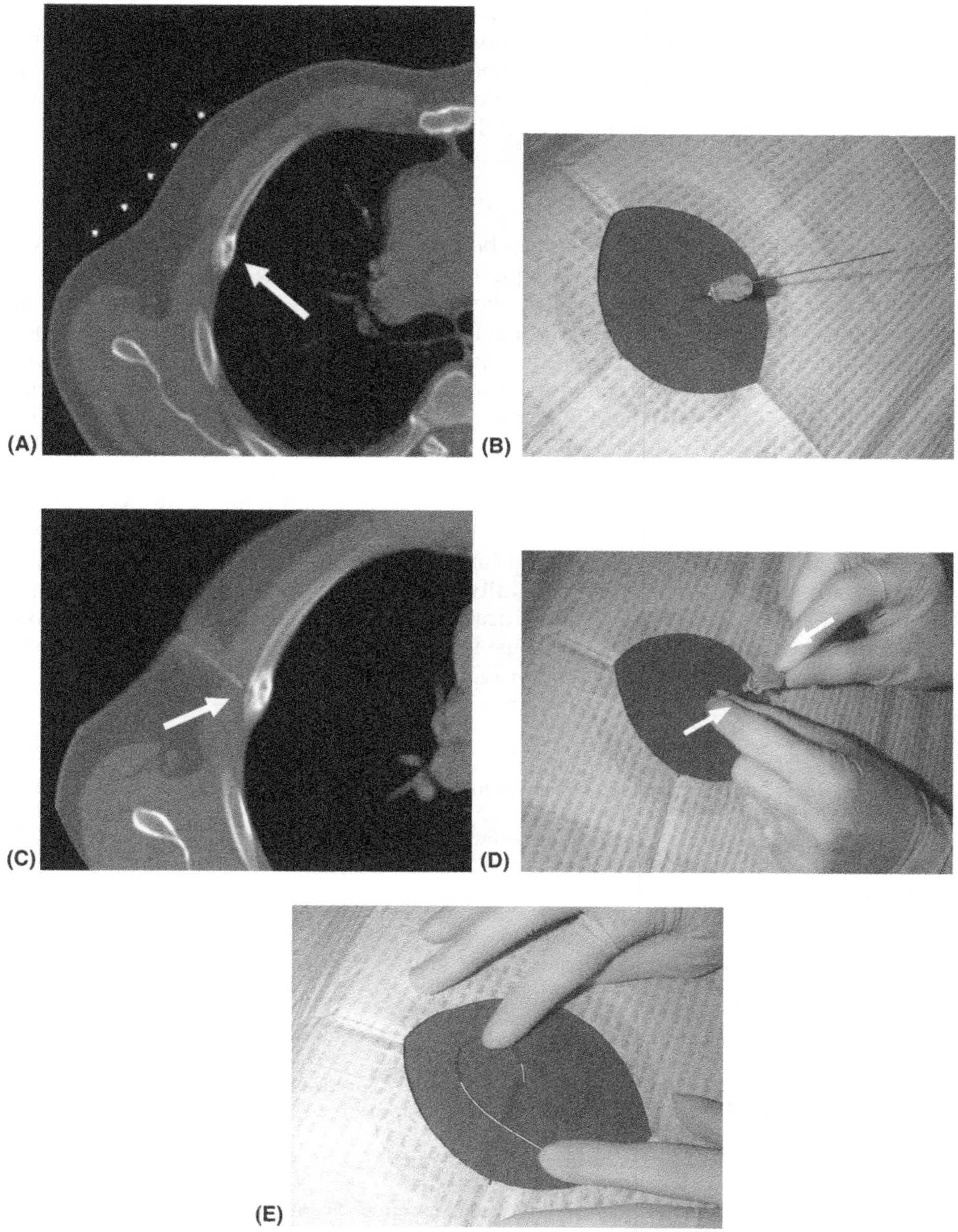

FIGURE 2 Technique for placement and deployment of a hookwire needle. (**A**) Scout axial computed tomography (CT) image showing a small, sclerotic lesion (*arrow*) in the sixth rib laterally (later proven to represent metastatic adenocarcinoma). Preprocedure images are used to plan the skin entry site as well as the angle and depth to the lesion from that point. (**B**) The skin entry site is marked; the surrounding area is prepped and draped. Hookwire needle (with the hook retracted) is positioned at the planned angle and depth. (**C**) Repeat CT verifies optimal placement of needle tip (*arrow*) against lesion. (**D**) Hookwire is deployed by holding the inner wire in place while withdrawing the outer needle (motion of hands indicated by *arrows*). (**E**) After repeat CT documents final location, wire protruding from skin is loosely packed in sterile fashion.

Surgical complications such as wire transection or loss are minimized by effective communication with the referring surgeon. The surgeon should be made aware of the type of needle used, and the length, and precise location of the needle course, tip, and hook relative to adjacent structures. We routinely print a sheet of images of the procedure including that of the final needle placement, which we send to the operating room along with the patient. Surgeons ordering this procedure are generally experienced at using hookwire localization of other organ systems; however, the radiologist performing the localization should discuss the technique with the surgeon to assure that the surgeon is familiar with the procedure.

SUMMARY

The vast majority of musculoskeletal lesions can be biopsied or ablated using image-guided percutaneous techniques. However, this technique may rarely be excessively difficult or may subject the patient to unacceptably high risk-to-benefit ratio. These situations are likely to include combinations of the following factors: small lesion size, lesions in small or curved bones (e.g., rib lesions); lesions with overlying sclerotic bone or thick cortex, with no soft-tissue extension or access to a drill; lesions close to vital structures; heterogeneous lesions subject to errors in sampling (e.g., chondroid lesions); hard (e.g., calcified) lesions in soft tissue, which would be pushed away by the biopsy needle; and lesions with significant bleeding risk (e.g., vascular lesions in noncompressible areas). Other lesions already planned for open surgical biopsy carry greater risk or less chance for success because of small lesion size, contiguity with major blood vessels or nerves, and surrounding scar tissue or distorted anatomy; in these situations, preoperative hookwire needle localization of the lesion can provide a reliable guide for the surgeon. This could potentially decrease operative morbidity and increase success rate of obtaining a diagnostic sample. Regarding the risk of complications, a general rule is that preoperative needle localization procedures should not be considered any less involved than percutaneous biopsies; the same preprocedure evaluation and preparation is essential.

REFERENCES

1. Kopans DB, DeLuca S. A modified needle-hookwire technique to simplify preoperative localization of occult breast lesions. Radiology 1980; 134:781.
2. Meyer JE, Kopans DB. Preoperative roentgenographically guided percutaneous localization of occult breast lesions: three-year experience with 180 patients and description of a method. Arch Surg 1982; 117:65–68.
3. Shah RM, Spirn PW, Salazar AM, et al. Localization of peripheral pulmonary nodules for thoracoscopic excision: value of CT-guided wire placement. AJR 1993; 161:279–283.
4. Kanazawa S, Ando A, Yasui K, et al. Localization of pulmonary nodules for thoracoscopic resection: experience with a system using a short hookwire and suture. AJR 1998; 170:332–334.
5. Sawhney R, McCowin MJ, Wall SD, Block MI. Fluoroscopically guided placement of the Kopans hookwire for lung nodule localization prior to thoracoscopic wedge resection. J Vasc Interv Radiol 1999; 10:1133–1134.
6. Thaete FL, Peterson MS, Plunkett MB, Ferson PF, Keenan RJ, Landreneau RJ. Computed tomography-guided wire localization of pulmonary lesions before thoracoscopic resection: results in 101 cases. J Thoracic Imaging 1999; 14:90–98.
7. Morrison WB, Sanders TG, Parsons TW, Penrod BJ. Preoperative CT-guided hookwire needle localization of musculoskeletal lesions. AJR 2001; 176:1531–1533.
8. Steinberg GG, Coumas JM, Breen T. Preoperative localization of osteoid osteoma: a new technique that uses CT. AJR 1990; 155:883–885.
9. Magre GR, Menendez LR. Preoperative CT localization and marking of osteoid osteoma: description of a new technique. J Comput Assist Tomogr 1996; 20:526–529.
10. Rosenthal DI, Springfield DS, Gebhardt MC, Rosenberg AE, Mankin HJ. Osteoid osteoma: percutaneous radiofrequency ablation. Radiology 1995; 197:451–454.
11. Hardaway BW, Hoffer FA, Rao BN. Needle localization of small pediatric tumors for surgical biopsy. Pediatr Radiol 2000; 30:318–322.

20 Discography

Bruno Grignon
Guilloz Department of Radiology, University Hospital of Nancy, Nancy, France

INTRODUCTION

Discography was first performed in 1948 by the Swedish radiologist K. Lindblom, who injected red lead contrast into cadaveric intervertebral discs (IVDs), to diagnose discal degenerative changes (1). At the beginning of the 21st century, discography is sometimes used as a diagnostic technique, mainly when computed tomography (CT), magnetic resonance (MR), or myelography is equivocal. It is followed, when possible, by CT examination (CT discography). Presently, however, discography is most often carried out as the first step of an operative procedure such as chemonucleolysis, intradiscal radiofrequency or laser therapy, or intradiscal injection of steroids or anesthetics.

ANATOMY OF THE INTERVERTEBRAL DISC

The IVD is interposed between the bodies of adjacent vertebrae and constitutes the main intervertebral joint (Fig. 1) (2–5). IVDs are cartilaginous synarthroses or symphyses, designed for weight bearing. The IVD is composed of two parts with more or less indistinct limits: a peripheral portion, the annulus fibrosus (AF), and a central space, the nucleus pulposus (NP). The AF is a ring structure, which is inserted by collagen fibers, also called Sharpey's fibers, into the adjacent vertebral bodies. It is dense and composed of concentric lamellae of fibrocartilage. The fibers that form each lamella run obliquely from one vertebra to another, and run at a right angle to those of adjacent lamellae (Fig. 2). This arrangement of fibers, at a right angle between two adjacent lamellae, explains the striation that is observed on discographic studies when the contrast medium is wrongly injected into the AF instead of the NP. The fibrous bundles are vertical in the periphery and oblique towards the center, providing a strong link between two adjacent vertebrae. However, it must be mentioned that at the lumbar level, the AF is thinner posteriorly than anteriorly and laterally.

The AF is made up of collagen fibers and fibroblasts and contains little water and proteoglycans. The NP is the central core of the IVD, situated between the cartilagenous plates of the vertebrae and surrounded by the AF (Fig. 1). At the lumbar level, it is located more posteriorly than anteriorly. It is of gelatinous consistency, and contains a large amount of water (about 90% for a young nucleus), proteoglycans (which act on the water content, and are the target of chymopapain), and a few collagen fibers that are scattered randomly (Fig. 2). After the age of 20, the center of the NP is usually penetrated by collagen fibers, which has been described as the intranuclear cleft (6). Due to its high water content, the IVD acts both as a shock absorber against axial forces and as a fluid ball bearing during flexion, extension, and rotation. Unfortunately, the NP becomes progressively dehydrated with age and undergoes a gradual fibrous transformation. Tears may appear in the AF. The IVD is poorly supplied by blood vessels, which can be seen only in its periphery. The disc is partly maintained by the posterior and anterior longitudinal ligaments, which run, respectively, on the posterior and anterior aspects of the vertebral bodies, from the occipital bone to, respectively, the coccyx and the sacral bone. The posterior longitudinal ligament (PLL) is attached to the IVD and is often considered as contributing to the prevention of posterior herniation of the disc. However, this ligament is narrower and weaker than the anterior longitudinal ligament. In addition, the lower the lumbar vertebral level, the narrower the PLL. Posteriorly to the IVD, the vertebral canal contains the epidural space with the initial segment of the spinal nerves with their dural sheat crossing over at each level.

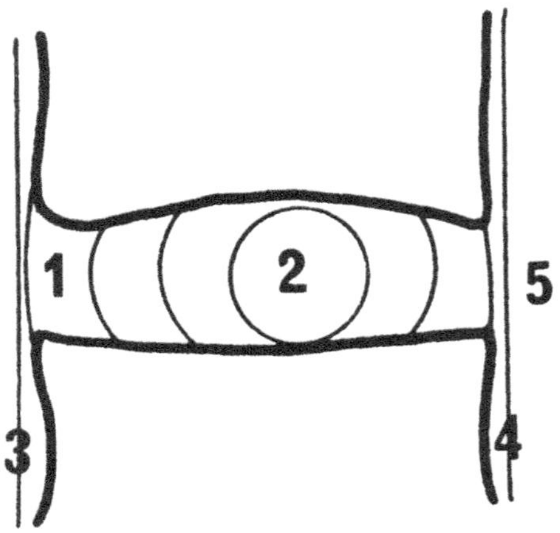

FIGURE 1 Anatomy of the disc. Line drawing showing the intervertebral disc in the sagittal plane. The annulus fibrosus (*1*) constitutes a fibrous peripheral ring, whereas the nucleus pulposus (*2*) is the central core of the disc and is the injection target in discography. Anterior longitudinal ligament (*3*). Posterior longitudinal ligament (*4*). Vertebral canal (*5*).

TERMINOLOGY OF DEGENERATIVE DISC DISORDERS

- It must be kept in mind that, from one language to another, degenerative disc disorder terminology can be very confusing. Several words are commonly used to describe an identical degenerative discal feature, and the same terms may have a different meaning in another

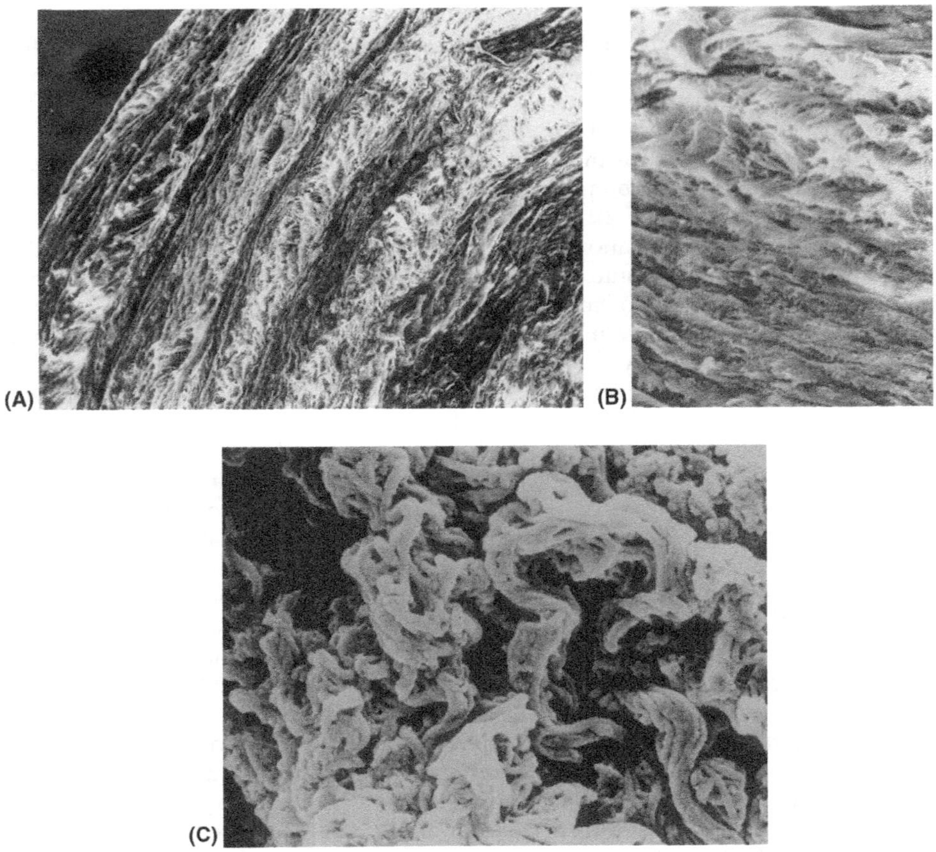

FIGURE 2 Microscopic anatomy of the disc. Annulus fibrosus (AF) (**A, B**). Scanning electron microscopy demonstrating the characteristic arrangement of the collagen fibers in the concentric lamellae, the disposition at a right angle of the fibers of two adjacent lamellae (A: × 8000), and the tight structure of the fibers in two adjacent lamellae (B: × 12,000). Nucleus pulposus (**C**). Scanning electron microscopy (× 12,000) showing the disposition of the collagen fibers that are scattered randomly, contrasting with those of the AF.

language, or even in the everyday use of the same language (7–13). Despite some attempts, no official standardization has been obtained (12,13).

- The term *"disc degeneration"* is a general term, which means that the water content of the disc is decreased or lost. Consequently, the disc height is reduced, and there is a diffuse, circumferential, symmetric, annular bulge. Disc signal intensity is reduced on T2-weighted MR images. Degenerated discs often have tears or fissures in the AF, which may be seen at discography (Fig. 3).
- Milette proposed a morphologic nomenclature based on the assessment of the disc contour (13,14). The degree of the disc extension beyond the interspace or DEBIT may be described as follows. Four categories of disc contour can be seen: *normal* (no disc extension), *bulging* disc, *protruded*, and *extruded* disc. The term *"bulge"* refers to a circumferential and symmetric DEBIT. Asymmetric or focal DEBIT is designated by the term *"protrusion."* When the DEBIT is focal and obvious or when there is no more connection between the disc material and the parent disc, it is referred to as an *"extrusion"* (13,14). The distinction between disc protrusion and disc extrusion has been discussed. The term *"herniation"* has been proposed to designate both lesions (15). However, this distinction has a clinical relevance, because disc protrusion as well as bulging disc can be asymptomatic, whereas disc extrusions are uncommon in asymptomatic patients (13). Moreover, the term *"herniation"* initially referred to a focal extension of the NP coming out of a disc space, whereas the true term was *"herniation of the NP."* This is too restrictive a definition, because it has been shown that the herniated material may be nuclear, annular, or endplate fragments (13).

INDICATIONS OF DISCOGRAPHY

- Discography can be useful as a diagnostic technique or as the first step of therapeutic procedures.
- As a diagnostic technique, discography is mainly used as an imaging method, but it can also provide a volumetric, a manometric, and a pain provocation evaluation of the disc (16), which will be further detailed.

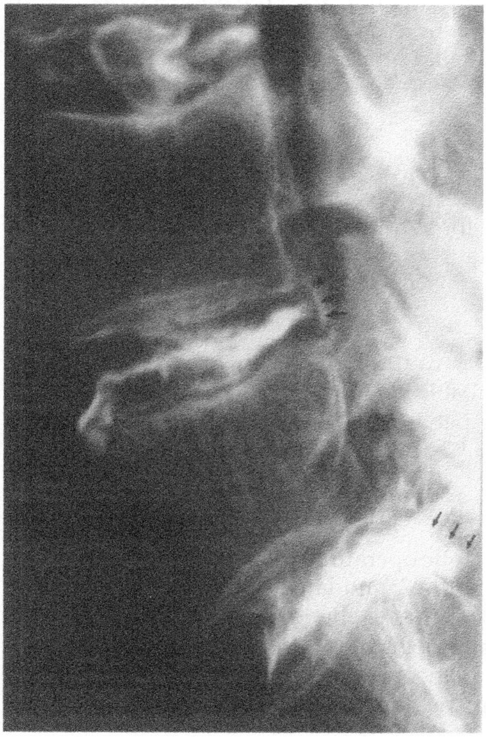

FIGURE 3 Discographies performed at the L3–L4, L4–L5, and L5–S1 discal levels of a cadaveric spine. At the L3–L4 level: mildly degenerated disc. The nucleus pulposus (NP) is of large size, with a light center and pseudopolyps and communicates with a posterior radial tear and a thinner anterior tear in the annulus fibrosus (AF). The posterior discal border is normal at this level (*arrows*). At the L4–L5 level: degenerated disc with a flattened NP, several annular tears, and a bulge easily seen on the posterior border (*arrows*). At the L5–S1 level: large posterior protrusion opacified from the NP by a radial tear in the AF, with caudal migration, and mild epidural extravasation of contrast agent.

- However, because it is an invasive technique, it should be reserved for patients with symptoms severe enough to require invasive therapy or surgery and when noninvasive diagnostic techniques have failed to reach a diagnosis, in particular, in cases of nerve root pain with negative or equivocal MR images, CT, or myelogram, or when disc disease is shown at multiple levels (9,17).
- An uncommon indication is cases of low back pain where surgery is discussed. Goal of discopathy in such cases is to prove the discal origin of the back pain (9,18).
- Most often, discography is performed as the first step of an operative procedure such as chemonucleolysis, nucleotomy, intradiscal radiofrequency, or laser therapy, or sometimes an intradiscal injection of cortisone or anesthetic (9).

GENERAL RULES OF TECHNIQUE

Preparation

It is crucial to first interview the patient and review the physician's request and all the patient's medical and imaging records. Location, type, and nature of the pain must be clearly studied, as well as the patient's present and past history, and, in particular, a history of prior surgery. A physical examination is required, and the patient's psychological state must also be taken into account. Conversely, the patient has to be informed about discography, including the details of the technique, the risks, and the complications of the procedure.

Premedication and Neuroleptic Analgesia

Recent coagulation tests must be checked. If there is an allergy, the patient should have an allergological assessment (19,20). Conscious sedation is routinely used during discography to allow a comfortable procedure for the patient with the opportunity to reproduce the patient's pain.

Specific Case of the Discography Carried Out as the First Step of Chemonucleolysis

In such cases, attention must be paid to the possibility of a recent myelography. Because it has been shown that intrathecal injection of chymopapain may induce fatal hemorrhaging in rabbits and dogs (21), a minimum delay of 24 hours (21) or even seven days (22,23) is required between myelogram and chemonucleosis in order to allow the dural leak to seal. In addition, specific allergology tests should be performed.

Prediscography Computed Tomography

A prediscography CT in a prone position is not routine. However, it can be helpful to determine the location of the colon in this position, and to look for abnormalities of organs and vessels in order to choose the best path for the needle (8).

Plain Radiographs

Anteroposterior (AP) and lateral plain radiographs as well as "coned down" sagittal and AP views centered on the selected IVD level must be available before the discography. At the L5–S1 level, it is necessary to have a plain radiograph of the pelvis in order to study the position of the L5–S1 disc with regard to the transverse processes of L5 and the sacral bone.

X-Ray Room

Discography is carried out in the X-ray room under aseptic operating room conditions, including sterile drapes, gloves, and surgical gowns. Because both AP and lateral views of the disc must be available during the procedure in order to control the needle position, a portable C-arm image intensifier unit is the best device to carry out a discography. However, if such equipment is not available, it is also possible to use a single plane fluoroscopy table and an auxiliary X-ray tube, or even to rotate the patient to obtain biplane fluoroscopic control.

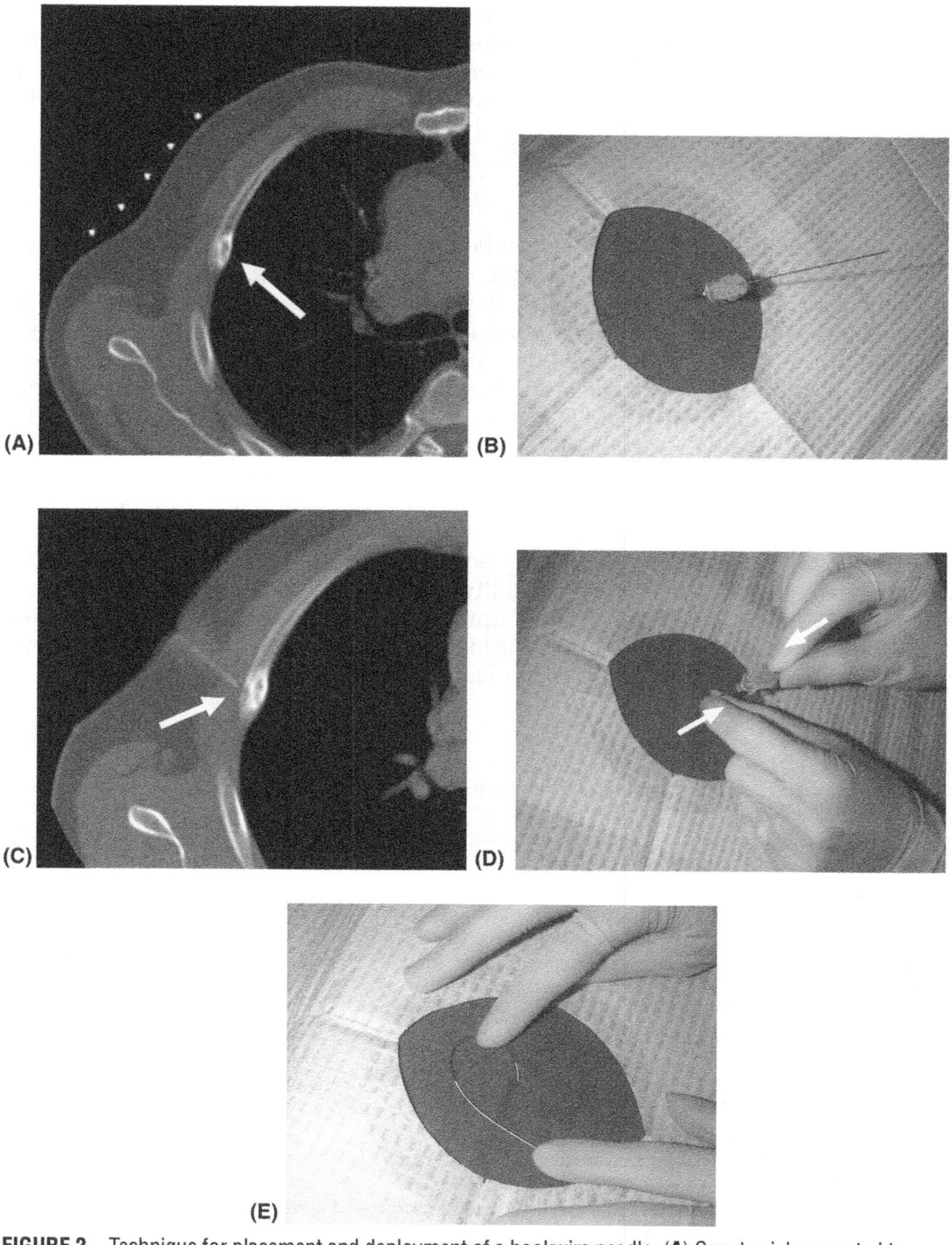

FIGURE 2 Technique for placement and deployment of a hookwire needle. (**A**) Scout axial computed tomography (CT) image showing a small, sclerotic lesion (*arrow*) in the sixth rib laterally (later proven to represent metastatic adenocarcinoma). Preprocedure images are used to plan the skin entry site as well as the angle and depth to the lesion from that point. (**B**) The skin entry site is marked; the surrounding area is prepped and draped. Hookwire needle (with the hook retracted) is positioned at the planned angle and depth. (**C**) Repeat CT verifies optimal placement of needle tip (*arrow*) against lesion. (**D**) Hookwire is deployed by holding the inner wire in place while withdrawing the outer needle (motion of hands indicated by *arrows*). (**E**) After repeat CT documents final location, wire protruding from skin is loosely packed in sterile fashion.

Surgical complications such as wire transection or loss are minimized by effective communication with the referring surgeon. The surgeon should be made aware of the type of needle used, and the length, and precise location of the needle course, tip, and hook relative to adjacent structures. We routinely print a sheet of images of the procedure including that of the final needle placement, which we send to the operating room along with the patient. Surgeons ordering this procedure are generally experienced at using hookwire localization of other organ systems; however, the radiologist performing the localization should discuss the technique with the surgeon to assure that the surgeon is familiar with the procedure.

SUMMARY

The vast majority of musculoskeletal lesions can be biopsied or ablated using image-guided percutaneous techniques. However, this technique may rarely be excessively difficult or may subject the patient to unacceptably high risk-to-benefit ratio. These situations are likely to include combinations of the following factors: small lesion size, lesions in small or curved bones (e.g., rib lesions); lesions with overlying sclerotic bone or thick cortex, with no soft-tissue extension or access to a drill; lesions close to vital structures; heterogeneous lesions subject to errors in sampling (e.g., chondroid lesions); hard (e.g., calcified) lesions in soft tissue, which would be pushed away by the biopsy needle; and lesions with significant bleeding risk (e.g., vascular lesions in noncompressible areas). Other lesions already planned for open surgical biopsy carry greater risk or less chance for success because of small lesion size, contiguity with major blood vessels or nerves, and surrounding scar tissue or distorted anatomy; in these situations, preoperative hookwire needle localization of the lesion can provide a reliable guide for the surgeon. This could potentially decrease operative morbidity and increase success rate of obtaining a diagnostic sample. Regarding the risk of complications, a general rule is that preoperative needle localization procedures should not be considered any less involved than percutaneous biopsies; the same preprocedure evaluation and preparation is essential.

REFERENCES

1. Kopans DB, DeLuca S. A modified needle-hookwire technique to simplify preoperative localization of occult breast lesions. Radiology 1980; 134:781.
2. Meyer JE, Kopans DB. Preoperative roentgenographically guided percutaneous localization of occult breast lesions: three-year experience with 180 patients and description of a method. Arch Surg 1982; 117:65–68.
3. Shah RM, Spirn PW, Salazar AM, et al. Localization of peripheral pulmonary nodules for thoracoscopic excision: value of CT-guided wire placement. AJR 1993; 161:279–283.
4. Kanazawa S, Ando A, Yasui K, et al. Localization of pulmonary nodules for thoracoscopic resection: experience with a system using a short hookwire and suture. AJR 1998; 170:332–334.
5. Sawhney R, McCowin MJ, Wall SD, Block MI. Fluoroscopically guided placement of the Kopans hookwire for lung nodule localization prior to thoracoscopic wedge resection. J Vasc Interv Radiol 1999; 10:1133–1134.
6. Thaete FL, Peterson MS, Plunkett MB, Ferson PF, Keenan RJ, Landreneau RJ. Computed tomography-guided wire localization of pulmonary lesions before thoracoscopic resection: results in 101 cases. J Thoracic Imaging 1999; 14:90–98.
7. Morrison WB, Sanders TG, Parsons TW, Penrod BJ. Preoperative CT-guided hookwire needle localization of musculoskeletal lesions. AJR 2001; 176:1531–1533.
8. Steinberg GG, Coumas JM, Breen T. Preoperative localization of osteoid osteoma: a new technique that uses CT. AJR 1990; 155:883–885.
9. Magre GR, Menendez LR. Preoperative CT localization and marking of osteoid osteoma: description of a new technique. J Comput Assist Tomogr 1996; 20:526–529.
10. Rosenthal DI, Springfield DS, Gebhardt MC, Rosenberg AE, Mankin HJ. Osteoid osteoma: percutaneous radiofrequency ablation. Radiology 1995; 197:451–454.
11. Hardaway BW, Hoffer FA, Rao BN. Needle localization of small pediatric tumors for surgical biopsy. Pediatr Radiol 2000; 30:318–322.

20 Discography

Bruno Grignon
Guilloz Department of Radiology, University Hospital of Nancy, Nancy, France

INTRODUCTION

Discography was first performed in 1948 by the Swedish radiologist K. Lindblom, who injected red lead contrast into cadaveric intervertebral discs (IVDs), to diagnose discal degenerative changes (1). At the beginning of the 21st century, discography is sometimes used as a diagnostic technique, mainly when computed tomography (CT), magnetic resonance (MR), or myelography is equivocal. It is followed, when possible, by CT examination (CT discography). Presently, however, discography is most often carried out as the first step of an operative procedure such as chemonucleolysis, intradiscal radiofrequency or laser therapy, or intradiscal injection of steroids or anesthetics.

ANATOMY OF THE INTERVERTEBRAL DISC

The IVD is interposed between the bodies of adjacent vertebrae and constitutes the main intervertebral joint (Fig. 1) (2–5). IVDs are cartilaginous synarthroses or symphyses, designed for weight bearing. The IVD is composed of two parts with more or less indistinct limits: a peripheral portion, the annulus fibrosus (AF), and a central space, the nucleus pulposus (NP). The AF is a ring structure, which is inserted by collagen fibers, also called Sharpey's fibers, into the adjacent vertebral bodies. It is dense and composed of concentric lamellae of fibrocartilage. The fibers that form each lamella run obliquely from one vertebra to another, and run at a right angle to those of adjacent lamellae (Fig. 2). This arrangement of fibers, at a right angle between two adjacent lamellae, explains the striation that is observed on discographic studies when the contrast medium is wrongly injected into the AF instead of the NP. The fibrous bundles are vertical in the periphery and oblique towards the center, providing a strong link between two adjacent vertebrae. However, it must be mentioned that at the lumbar level, the AF is thinner posteriorly than anteriorly and laterally.

The AF is made up of collagen fibers and fibroblasts and contains little water and proteoglycans. The NP is the central core of the IVD, situated between the cartilagenous plates of the vertebrae and surrounded by the AF (Fig. 1). At the lumbar level, it is located more posteriorly than anteriorly. It is of gelatinous consistency, and contains a large amount of water (about 90% for a young nucleus), proteoglycans (which act on the water content, and are the target of chymopapain), and a few collagen fibers that are scattered randomly (Fig. 2). After the age of 20, the center of the NP is usually penetrated by collagen fibers, which has been described as the intranuclear cleft (6). Due to its high water content, the IVD acts both as a shock absorber against axial forces and as a fluid ball bearing during flexion, extension, and rotation. Unfortunately, the NP becomes progressively dehydrated with age and undergoes a gradual fibrous transformation. Tears may appear in the AF. The IVD is poorly supplied by blood vessels, which can be seen only in its periphery. The disc is partly maintained by the posterior and anterior longitudinal ligaments, which run, respectively, on the posterior and anterior aspects of the vertebral bodies, from the occipital bone to, respectively, the coccyx and the sacral bone. The posterior longitudinal ligament (PLL) is attached to the IVD and is often considered as contributing to the prevention of posterior herniation of the disc. However, this ligament is narrower and weaker than the anterior longitudinal ligament. In addition, the lower the lumbar vertebral level, the narrower the PLL. Posteriorly to the IVD, the vertebral canal contains the epidural space with the initial segment of the spinal nerves with their dural sheat crossing over at each level.

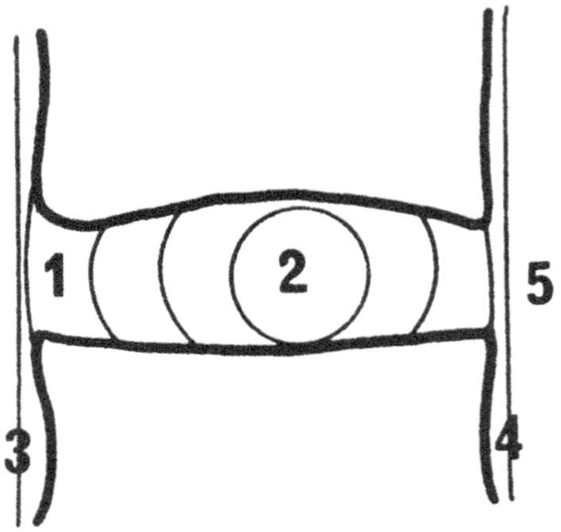

FIGURE 1 Anatomy of the disc. Line drawing showing the intervertebral disc in the sagittal plane. The annulus fibrosus (*1*) constitutes a fibrous peripheral ring, whereas the nucleus pulposus (*2*) is the central core of the disc and is the injection target in discography. Anterior longitudinal ligament (*3*). Posterior longitudinal ligament (*4*). Vertebral canal (*5*).

TERMINOLOGY OF DEGENERATIVE DISC DISORDERS

- It must be kept in mind that, from one language to another, degenerative disc disorder terminology can be very confusing. Several words are commonly used to describe an identical degenerative discal feature, and the same terms may have a different meaning in another

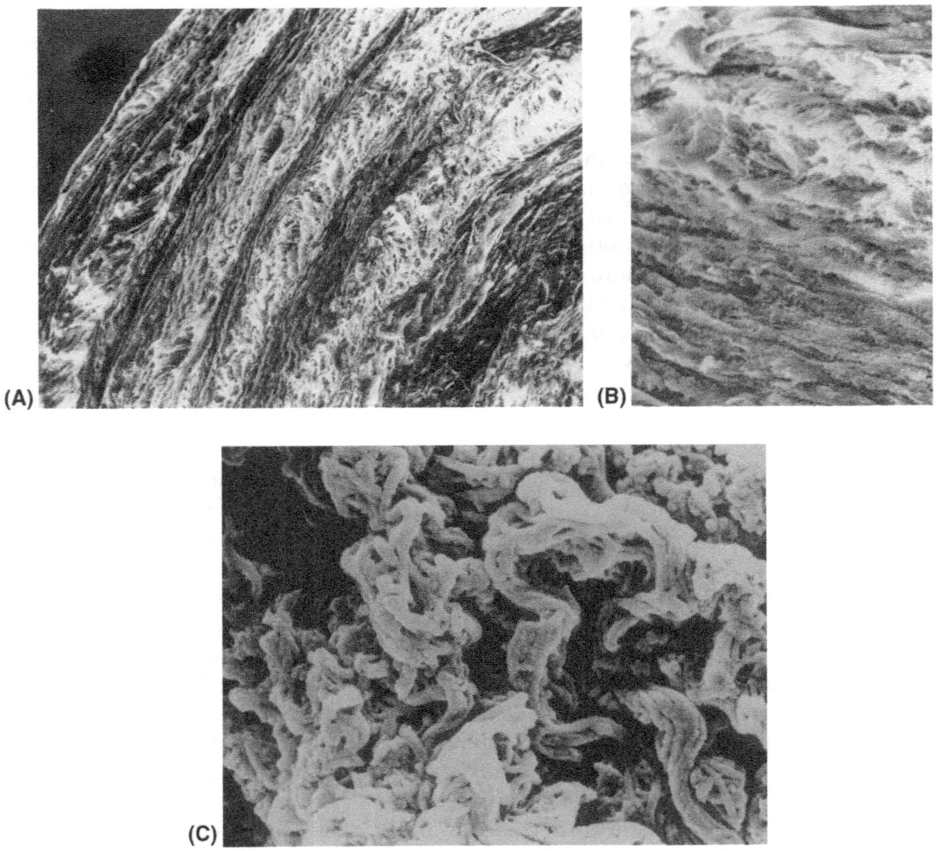

FIGURE 2 Microscopic anatomy of the disc. Annulus fibrosus (AF) (**A, B**). Scanning electron microscopy demonstrating the characteristic arrangement of the collagen fibers in the concentric lamellae, the disposition at a right angle of the fibers of two adjacent lamellae (A: × 8000), and the tight structure of the fibers in two adjacent lamellae (B: × 12,000). Nucleus pulposus (**C**). Scanning electron microscopy (× 12,000) showing the disposition of the collagen fibers that are scattered randomly, contrasting with those of the AF.

The Standard Discography Tray

It should include the following:

- For local anesthesia: 10 mL syringe, 1% or 2% lidocaine (Xylocaïne®), a 25-gauge (G) needle for the superficial planes, and, according to some authors, a spinal 9-cm 22-G needle for the deep planes (18).
- Needles for the discography include a 15-cm 22-G and a 9-cm 18-G needle (L4–L5 level and upper levels) or a 20-cm 22-G and a 15-cm 18-G (L5–S1 level and corpulent patients).
- A 2, 2.5, or 3 mL syringe for the contrast agent, a nonionic intrathecally well-tolerated contrast media, and sterile water.
- Semiradiopaque ruler and indelible ink marker.

Patient Monitoring

An electrocardiogram monitor and a pulse oximeter allow monitoring of the patient.

Approach

Four approaches are theoretically possible to reach the disc space: the posterior, posterolateral (Erlacher), lateral extradural, and true lateral approaches (Fig. 4). Posterior and posterolateral approaches are usually contraindicated because nervous structures are on the needle path (24,25). With a true lateral approach, the colon may be punctured before the disc, with the risk of discal infection (22).

Therefore, a lateral extradural approach must be used (22,25–27), which is termed lateral as opposed to the posterolateral intradural of Erlacher (22,25). Using this approach, the site of insertion of the needle in the skin is 8 to 12 cm from the midline of the spine, with an angulation of 40° to 60° to the sagittal plane. The angle and direction of the needle depend on the IVD level and patient's morphology (Fig. 5). Because there is no relation between the side of approach and the side of the pain, theoretically either a right- or left-side approach may be used. However, for the lumbar spine, some authors recommend injecting on the right side because of the risk of wounding the artery of Adamkiewickz, which is four times more frequently located on the left than on the right (9).

Position of Patient

Discography may be performed with the patient in a strict lateral position, a prone position, or an oblique prone position. Each technique has both advantages and disadvantages, which will be further detailed.

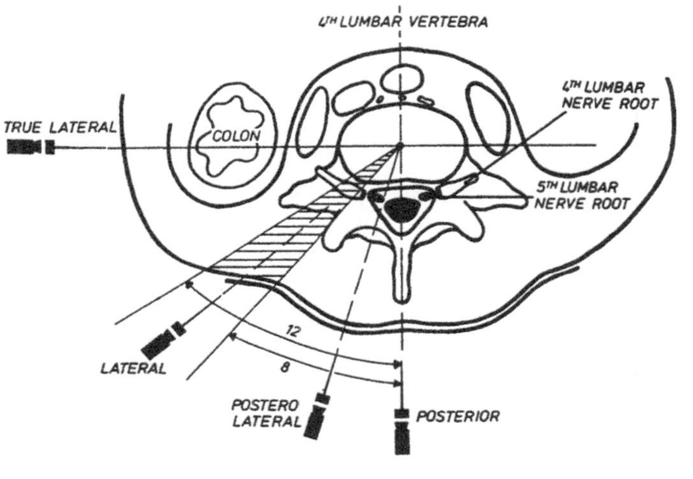

FIGURE 4 Line drawing in the axial plane showing the various approaches of the nucleus pulposus. The "lateral" extradural approach alone is to be used. *Source*: From Ref. 16.

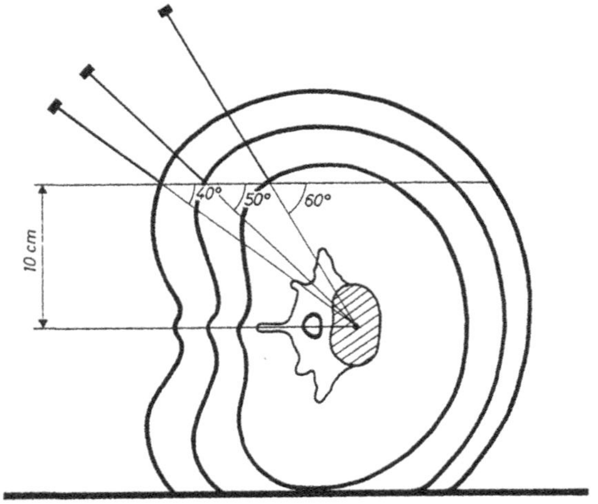

FIGURE 5 Line drawing showing the influence of the patient's build on the distance of the needle insertion point from the midline and on the angulation of the needle. *Source*: From Ref. 16.

TECHNIQUE OF LUMBAR DISCOGRAPHY

Lumbar Discography with Lateral Positioning

The main advantage of the lateral position is a good view of bony obstacles, which allows an easy approach to the disc space, with a direction parallel to the vertebral end plates, and a rapid control of the placement of the needles. However, this position requires a clear three-dimensional view of the anatomy, and an exact lateral positioning of the patient (22).

Patient Positioning

The patient is placed in a lateral decubitus position, on his left side, on the radiolucent table. The hips and knees are flexed at 60° to 90° to reduce lumbar lordosis. The knees, ankles, and the left arm are protected by foam cushions. A radiolucent block or a folded sheet is placed beneath the flank to compensate for the thoracolumbar curve due to lateral decubitus (Fig. 6). The head is elevated. Perfect lateral positioning of the patient is essential. This position must be checked by fluoroscopy. It is crucial to have a strictly orthogonal orientation of the radiographic tube to the X-ray table (22).

Needle Insertion Point

This point lies about 7 to 11 cm lateral to the midline of the spine, just above the iliac crest, found by palpation. The distance between the midline and the needle insertion point is theoretically 7 cm at the upper lumbar spine and 11 cm at the lower lumbar spine, but depends in fact on the patient's build (Fig. 5). A semiradiopaque ruler is used to simulate the path of the needle on the fluoroscopic screen to determine the craniocaudal angle of the needle and the level of its site of

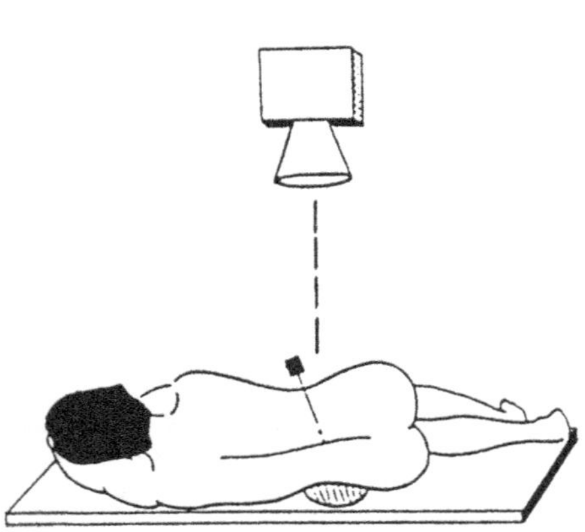

FIGURE 6 Lumbar discography with lateral positioning of the patient. The patient is placed on their left side. During the first part of the procedure, the X-ray beam must be strictly vertical and perpendicular to the table. *Source*: From Ref. 16.

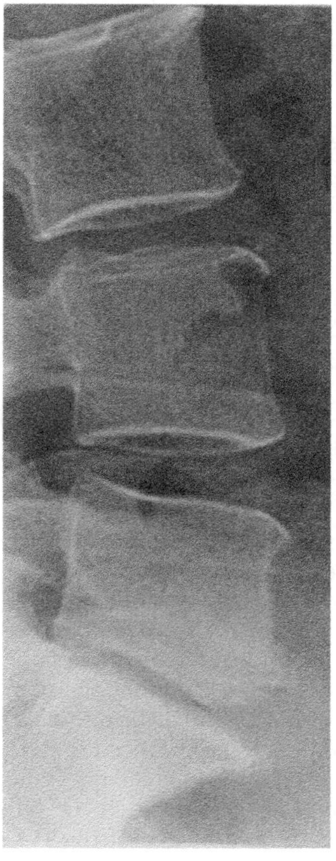

FIGURE 7 Lumbar discography with lateral positioning of the patient. Lateral plain radiograph showing the radiopaque semiruler placed on the side of the patient and tilted as desired to choose the craniocaudal angle of the needle.

insertion (Fig. 7). This ruler is placed on the side of the patient and it is tilted as desired so that it projects between the needle insertion point and the posterior margin of the disc. The transverse processes must not be in the path of the needle, which runs as parallel as possible to the vertebral end plates. This angle is then marked on the skin (22).

Disinfection

The skin is prepared. A sterile surgical field is placed.

Local Anesthesia

The superficial planes the needle will go through are anesthetized with a 25-G needle and 3 to 5 mL of lidocaine. Deeper planes may be anesthetized with a spinal 9-cm 22-G needle (18).

Needle Approach

The needle is inserted at an angle of 40° to 60° to the sagittal plane (Fig. 5). Its advancement must be controlled on the fluoroscopic screen. The needle can be first advanced to the facet joint. After contact, the needle is withdrawn a little and then advanced in a more sagittal direction (Fig. 8). Bony obstacles may be encountered and identified with the image amplifier. At the level of the pedicles, bony impingement is most often due to the transverse process, and at the level of the lower half of the vertebral body, to the facet joint. It is crucial to keep a clear three-dimensional image of the lumbar region in one's mind (22). Problems may also arise from the displacement of the patients, who often rotate their pelvis away from the needle. It may be useful to check the truly lateral position of the patient with the image intensifier (22). As soon as the needle has passed the line joining the transverse processes, the X-ray beam can be tilted to demonstrate the disc interspace (Fig. 9). The needle is then advanced toward the disc.

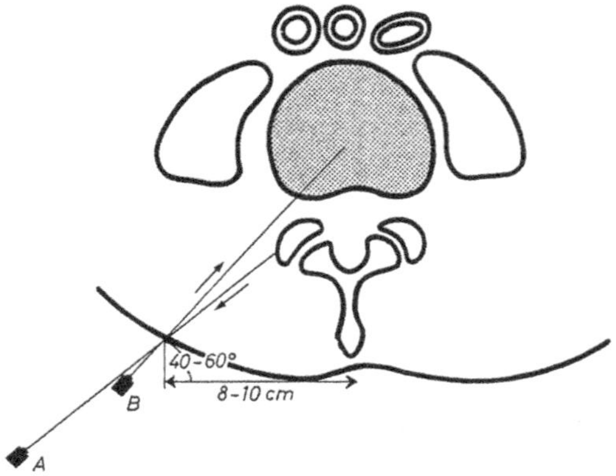

FIGURE 8 Lumbar discography with lateral positioning of the patient. The course of the needle, which is first advanced to the facet joint, then withdrawn a few centimeters and more sagittally oriented. *Source*: From Ref. 16.

When the contact with the AF is felt, the tip of the needle should be exactly on the vertical line passing through the posterior borders of the vertebral bodies (Fig. 10) (22). If the tip of the needle is anterior to this line, the approach is too lateral and too sagittal, and the needle will pass anterolateral to the NP. The angle of the needle insertion must be changed to a more coronal direction. If the tip of the needle is posterior to the line, the approach is too medial and coronal (Fig. 10). Once the needle is correctly placed, the patient is rotated to an oblique position and the X-ray beam is tilted to the same angle as the needle in order to check the position of the needle tip, seen as a metallic dot, in relation to the disc width (22). The ideal needle position is at the junction between the medial and middle-third of the disc width (Fig. 11). In that position, the NP can be reached without bending the needle. The stylet of the 18-G needle can be removed, and a 22-G needle can be inserted into the 18-G needle. The thin needle is then advanced toward the center of the NP, where a sensation of sudden decreased resistance can be perceived. If the needle is lateral to this point, a mild bending of the 22-G needle is needed to reach the NP. During this phase of the procedure, proper needle placement at the mid-height of the IVD should also be obtained. An AP view is performed to check the correct position of the tip of the 22-G needle, which is ideally located on the midline and equidistant from the two vertebral end plates. On a lateral view, the tip of the thin needle is ideally located at the junction between the posterior and middle-third of the disc (Fig. 12). AP and lateral plain radiographs should be obtained before disc injection.

Lumbar Discography with Prone Oblique Positioning

The main advantage of this position is easy three-dimensional control of the needle position, because the puncture point projects as a simple dot on the fluoroscopic screen. It is also the

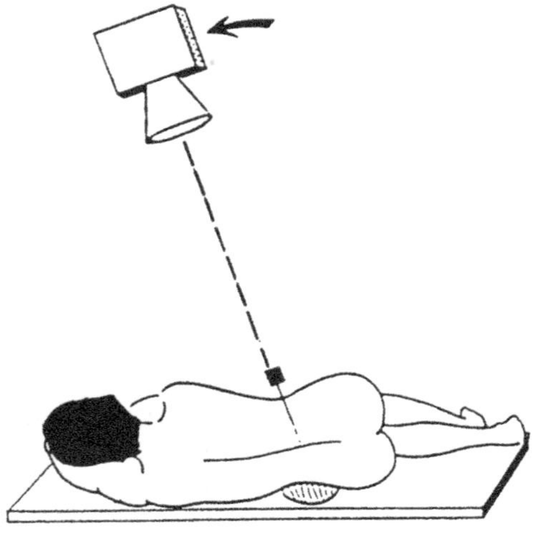

FIGURE 9 Lumbar discography with lateral positioning of the patient. The X-ray beam is tilted in order to profile the discal space exactly as soon as the needle has passed the line joining the transverse processes. *Source*: From Ref. 16.

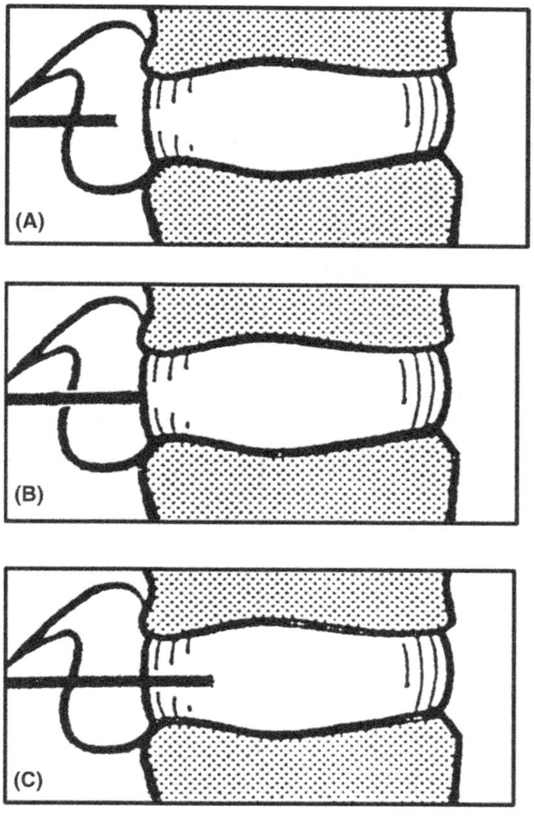

FIGURE 10 Lumbar discography with lateral positioning of the patient. Line drawing showing the placement of the 18-G needle on a lateral fluoroscopic view. In (**A**), the needle placement is too medial, and the angulation of the needle must be corrected in a more sagittal direction. In (**B**), the placement is proper: the tip of the needle projects on the line of the posterior border of the vertebral bodies. In (**C**), the needle placement is too lateral; the needle must be inserted in a more coronal direction. *Source*: From Ref. 16.

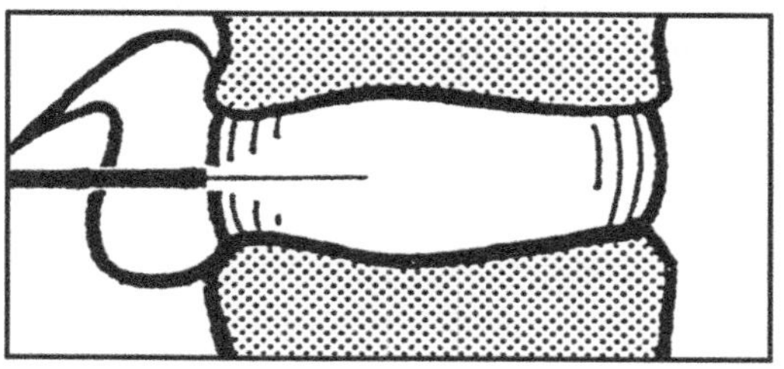

FIGURE 11 Lumbar discography with lateral positioning of the patient. Line drawing showing the right position for the needle when the patient is placed in an oblique position. The tip of the needle must appear at the junction between the internal and middle-third of the width and at midheight of the disc.

easiest technique of approach especially for the less-experienced physician. However, with this technique, it is somewhat more difficult to approach disc space in a direction parallel to the vertebral end plates (18,22).

Patient Positioning

The patient is placed in an oblique prone position. The patient's right side is up, and a radiolucent block is put under his left side. The right hip and knee are flexed. A small cushion is placed under the right knee for his comfort. The X-ray beam is tilted to profile the disc space (Fig. 13). The patient is then rotated to obtain a "triangle of puncture" delineated at the L4–L5 level, by the right superior facet of L5, the lower end plate of L4, and the iliac crest (Fig. 14). Disinfection and local anesthesia are then performed as previously described.

Needle Approach

The 18-G needle is inserted following the angle of inclination of the X-ray tube, so that its distal tip projects in the center of the triangle of puncture. The advancement of the needle is

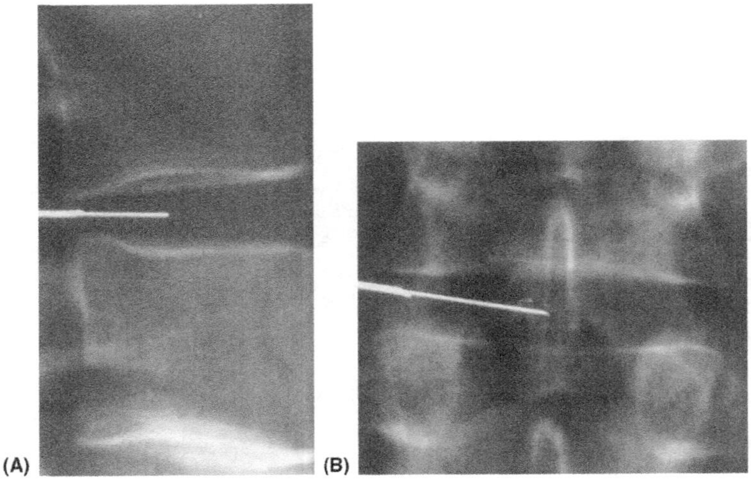

FIGURE 12 Lateral (**A**) and anteroposterior (**B**) plain focussed radiographs showing the proper placement of the needles. The tip of the 22-G needle is projected at the union of the posterior third and middle third of the disc on the lateral view and on the spine midline.

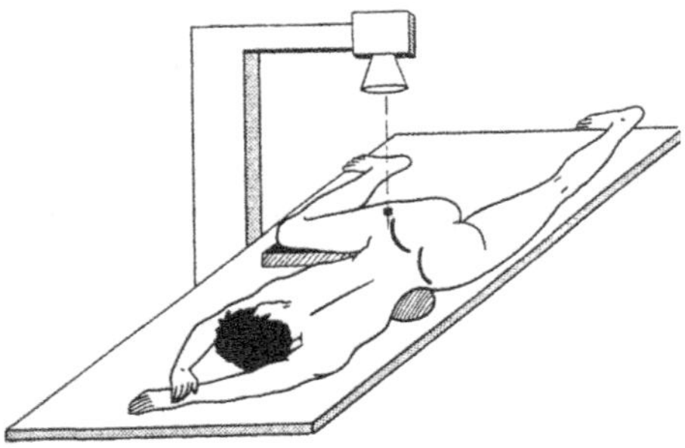

FIGURE 13 Lumbar discography in prone oblique position at the L4–L5 level. Positioning of the patient. The X-ray beam is tilted to profile the disc space in order to obtain the triangle of puncture. *Source*: From Ref. 16.

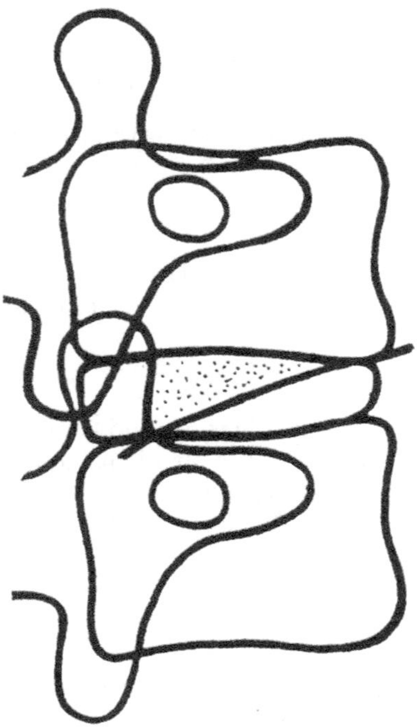

FIGURE 14 Lumbar discography in prone oblique position at the L4–L5 level. Line drawing demonstrating the "triangle of puncture," as seen on oblique roentgenogram. The triangle is bordered by the inferior end plate of the L4 vertebra, the superior facet joint of L5 posteriorly, and the iliac crest anteriorly.

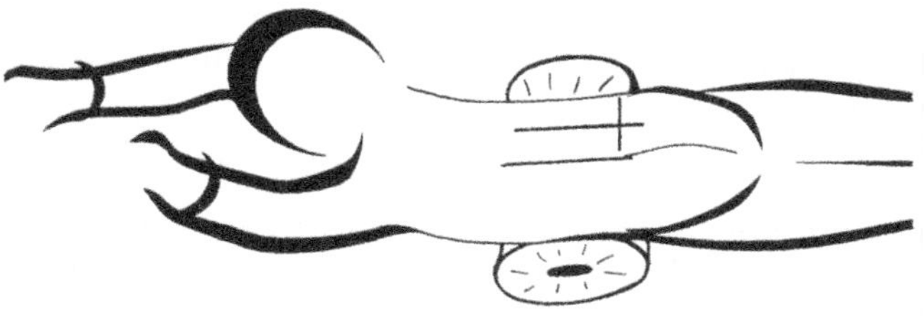

FIGURE 15 Lumbar discography in prone position. Line drawing showing the positioning of the patient, placed prone on the fluoroscopic table with a cushion under the abdomen. The spine midline, an 8 cm paramedian lateral line to the midline, and the desired discal level are marked on the skin with indelible marker.

then controlled fluoroscopically. When the needle has reached the AF, its stylet is removed, and the 22-G needle is inserted into it. The patient is then placed at a strict lateral position, and the X-ray beam is tilted in order to profile the disc space. The 22-G needle is then carefully advanced until its distal tip reaches the junction between the posterior-third and the middle-third. The position of this needle should also be checked on an AP projection. It must be projected in the center of the IVD. If this is not the case, the needle must be repositioned.

Lumbar Discography with Prone Positioning

The greatest advantage of the prone position is that it is the most stable and provides the best immobility of the patient (9). It is also easier for the physician to orient the needle toward the disc space (9). However, this approach is particularly difficult for the L5–S1 level.

Patient Positioning

The patient is placed in a prone position. A cushion is put under his abdomen in order to reduce his lumbar lordosis (Fig. 15). The patient's knees and feet are placed on a small cushion. The knees are semiflexed.

Skin Markers

A vertical paramedian line at 8 cm (9–10 cm in obese patients) from the midline of the back of the patient and the desired disc level defined with fluoroscopy are marked on the skin. At the L3–L4 and L4–L5 intervertebral levels, the site of skin puncture is at the intersection between these two lines. For the L5–S1 level, the needle has to be inserted higher, roughly midway between L3–L4 and L4–L5 (9).

Needle Approach

For the L3–L4 and L4–L5 levels, the needle is approached at an angle of 30° to 60° to the sagittal plane. Needle advancement is controlled on the X-ray intensifier screen. At the L5–S1 level, an additional caudal angle must be given. Moreover, at this level, it is almost always necessary to bend the tip of the 22-G needle to reach the disc center.

Discography at the L5–S1 Level

Anatomical Specificities

This approach of the disc space is more difficult at the L5–S1 level because of the presence of the iliac crest (Fig. 16), a reduced intersomatic space, the caudal orientation of the disc plane at this

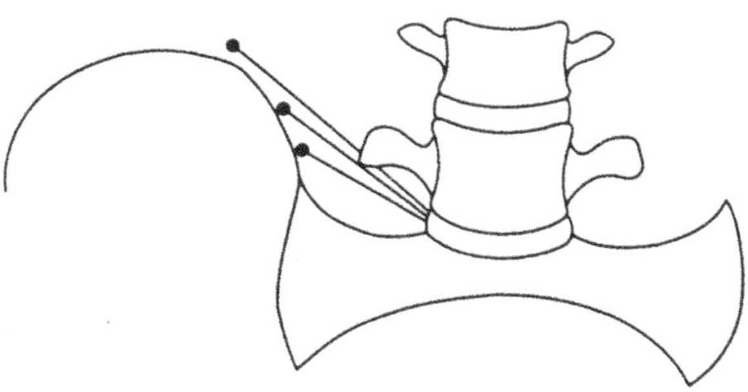

FIGURE 16 Puncture at the L5–S1 level. Line drawing showing the pathway of the needle in the frontal plane with regard to the iliac crest. The more lateral the site of insertion of the needle, the higher it will be, and the more obliquely downward the pathway of the needle will be. *Source*: From Ref. 16.

level (9), and an increased width of the neural arch (22). The space between the transverse process of L5 and the sacral bone is often narrow especially when the L5–S1 disc is deeply embedded. The build of the patient may make the procedure even more difficult.

Patient Positioning

Lateral positioning of the patient is recommended for this level, particularly when the L5–S1 disc is deep relative to the pelvis, because the radiolucent block placed under the flank of the patient is more efficient in this position than in the prone oblique position. The latter position and the prone position are sometimes possible when the iliac crest is not overly high, with the knowledge that the needle approach cannot be strictly parallel to the vertebral end plates (18,22).

Needle Insertion Point

The side with the maximum distance between the sacral wing and the L5 transverse process on the plane radiograph is selected for skin puncture (22). The patient is positioned lateral on the opposite side. It is important to place a radiolucent block beneath his flank in order to open both the IVD and the space between the sacral wing and the transverse process, as well as to lower the iliac crest. A marker is placed at 8-cm to 10-cm lateral to the midline, which indicates the point of puncture. As described for the L4–L5 level, a ruler simulating the path of the needle is tilted as required in order to project between the puncture point and the posterior border of the L5–S1 disc. The approach must be as parallel as possible to the disc space. A line representing this approach is marked on the patient's skin.

Needle Approach

The 15-cm 18-G needle is inserted at an angle of 40° to 60° to the sagittal plane as at the L4–L5 level, but with an additional caudal orientation. However, because the neural arch is wider at the L5–S1 level than at the level above, it is often necessary to bend the 22-G needle in order to reach the center of the disc. The distal part of the needle must be carefully curved in such a manner that its bevel is placed on the convex aspect of the curve. The required degree of curvature can be estimated on the oblique view, as illustrated by Figure 17. The curved 22-G is then advanced in the 18-G needle and then rotated, so that its concavity is medial and cephalad (22).

DISCOGRAPHY

Needle Placement

- The contrast agent is injected once the proper needle placement has been checked on both AP and lateral views (Fig. 18). The contrast media must flow away from the tip of the needle and cross the discal midline as soon as the injection begins (Fig. 19).

Annulograms

In case of injection into the AF, the contrast media may appear to be in a correct location, central into the disc space on one radiographic view, and will be lateral and not central into the disc

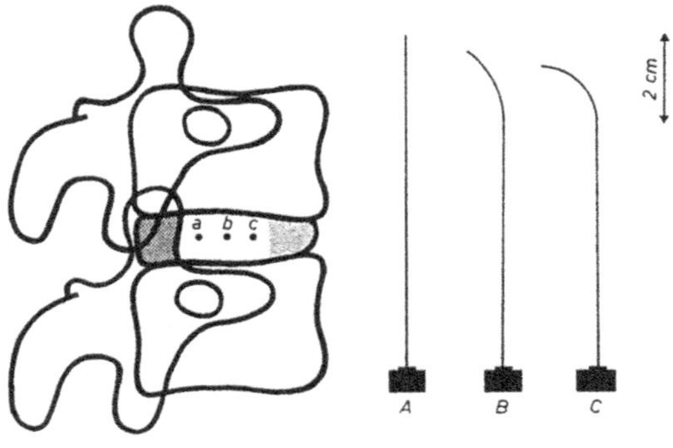

FIGURE 17 Degree of curvature of the 22-G needle according to the position of the 18-G needle on an oblique view. *Source*: From Ref. 16.

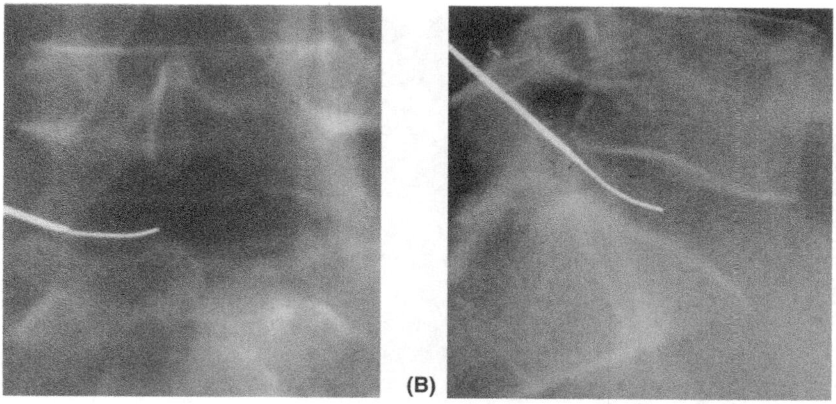

(A) (B)

FIGURE 18 Anteroposterior (**A**) and lateral (**B**) plain-focussed radiographs showing the proper placement of the needle at the L5–S1 level.

space on the orthogonal view (18,20). A striated appearance of the contrast media can also be observed, due to the arrangement of the AF fibers (Fig. 20).

Virtual Space

Quinnel and Stockdale have also showed that the contrast medium may be injected in a virtual space, between the NP and the inner AF. In such cases, the contrast media appears as a crescent, displacing the true nuclear cavity (28).

Vascular Opacification

In case of chemonucleolysis, absence of vascular opacification should be confirmed at fluoroscopy before chymopapain injection because this finding contraindicates enzyme injection (22,29–31).

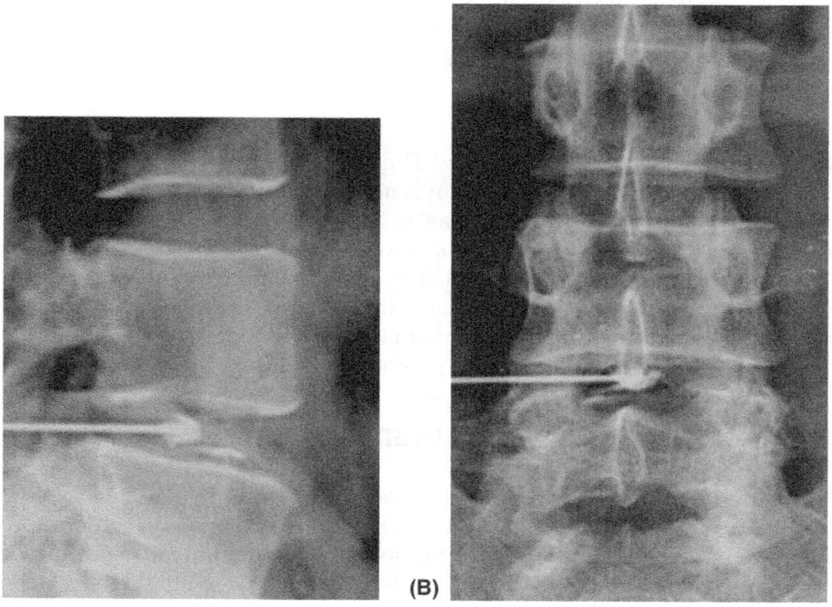

(A) (B)

FIGURE 19 Anteroposterior (AP) (**A**) and lateral (**B**) plain radiographs during the beginning of the injection of contrast agent at the L4–L5 level, using a single-needle technique. Dense opacification in the center of the disc on both AP and lateral views as well as the flowing away of contrast media from the tip of the needle, together ensure the right positioning of the needle.

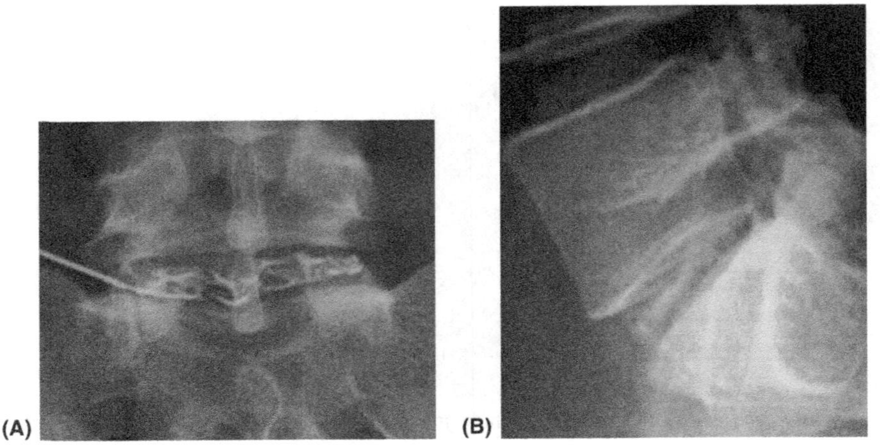

FIGURE 20 Annulography. Anteroposterior (**A**) and lateral (**B**) plain radiographs show a rectangular opacification in the disc which remains anterior. A striation of the opacification can be seen.

Communication Between Subarachnoid Space and Disc

A communication space between the discal cavity and the subarachnoid space may rarely be shown either by discography or myelography (Fig. 21). Due to the extreme toxicity of intrathecal injection of chymopapain into the subarachnoidian space (22,32), chemonucleolysis is absolutely contraindicated in such cases (8).

Amount of Contrast Media

The amount of contrast media injected depends on the indication of the discography. When the discography is carried out as a diagnostic procedure, a volume of 1.5 to 2 mL is injected in order to obtain a good opacification of the disc herniation and to reproduce the patient's pain (provocative test). However, when discography is the first step of a chemonucleosis, a smaller amount of contrast material (0.5–1 mL) is injected in order to reduce the inhibitory action of the contrast media on chymopapain activity and the reflux of the enzyme along the needle (21–23,32).

Provocation Pain Test

Severe pain may be induced by injecting into the disc. The provocation test is positive if it reproduces the patient's pain spontaneously (33). This pain is not to be confused with the pain due to discal distension. A number of authors have emphasized the value of discography as both an imaging and a clinical diagnostic method (9,16,17,33). However, the value of the provocation test has also been questioned, particularly in the cervical spine (34–37). Provocation discography may be of use in clinically difficult cases, in particular, when noninvasive diagnostic techniques are equivocal. Some authors recommend that the patient not be warned of the time of injection, to better observe his reaction, especially his facial expression, which may be videotaped (9).

POST-DISCOGRAM RADIOGRAPHS AND COMPUTED TOMOGRAPHY

- AP and lateral plain radiographs should be performed shortly after injection to avoid contrast media resorption. A number of techniques such as digital substraction tomography in coronal and sagittal planes, and craniocaudal or caudocranial tangential radiographic views have been recommended by certain authors in corpulent patients, or to obtain more detailed information (9,38,39).
- At present, CT discography is strongly recommended (Fig. 22) (40). CT discography can provide findings up to two hours after injection (9). Thin slices spiral acquisition allows axial, sagittal, and coronal reformations.

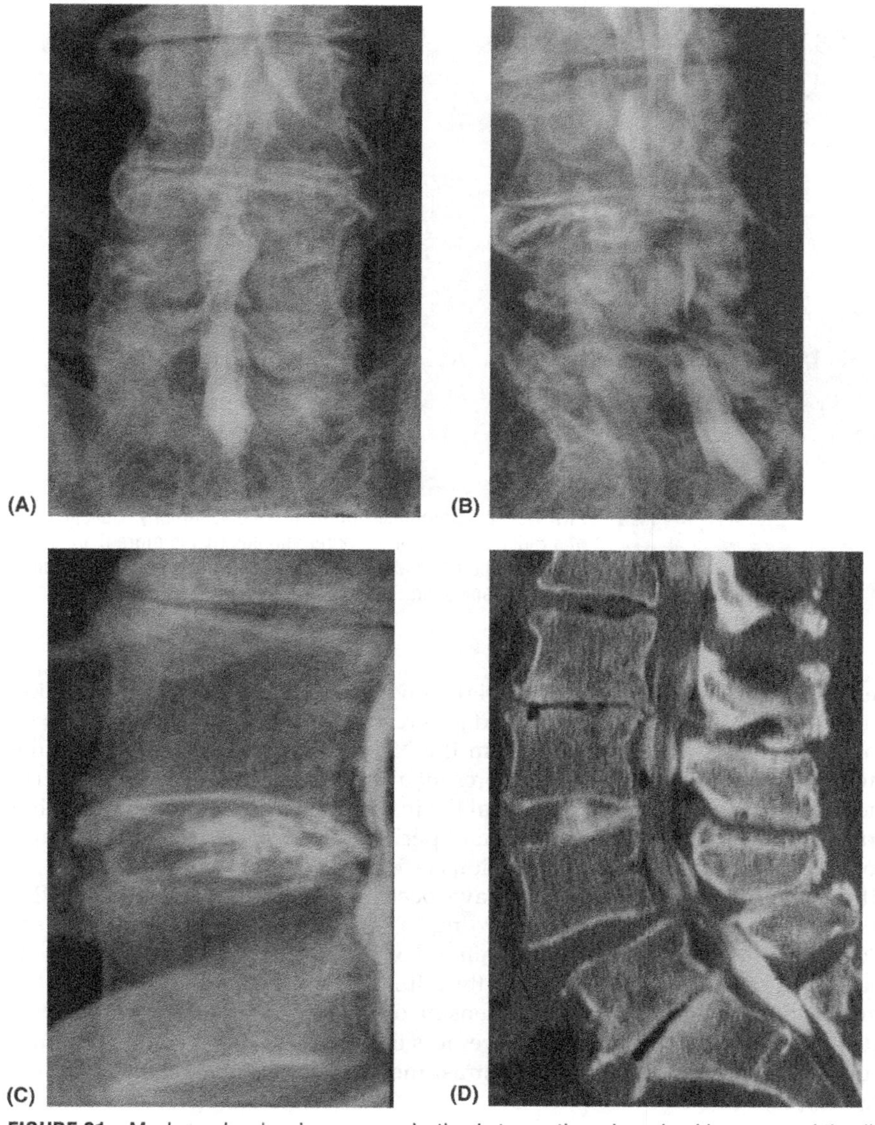

FIGURE 21 Myelography showing communication between the subarachnoid spaces and the disc at the L3–L4 level. Anteroposterior (**A**), oblique (**B**), lateral (**C**) plain radiographs. Sagittal computed tomography reformation (**D**). *Source*: Courtesy of Dr. X. Zhu.

NORMAL AND ABNORMAL DISCOGRAMS

Normal Pattern

Only the NP is opacified. A number of normal shapes of the NP can be encountered (Fig. 23) (8,9,18). The NP may be round, oval, bilocular, triangular, H-shaped, or rectangular, and small or large in size. A lateral position has also been described (8), but likely corresponds to the vitual space of Quinnel and Stockdale (28), as previously mentioned. Light centers and pseudopolyps have also been showed (8). It may sometimes be difficult to differentiate mild degeneration from a variation of normal (8).

Degenerative Disc Disease

When the disc is degenerated, the pressure becomes lower in the NP, the disc height is narrowed and tears may be seen in the annulus (Fig. 3). Three main kinds of tears—concentric,

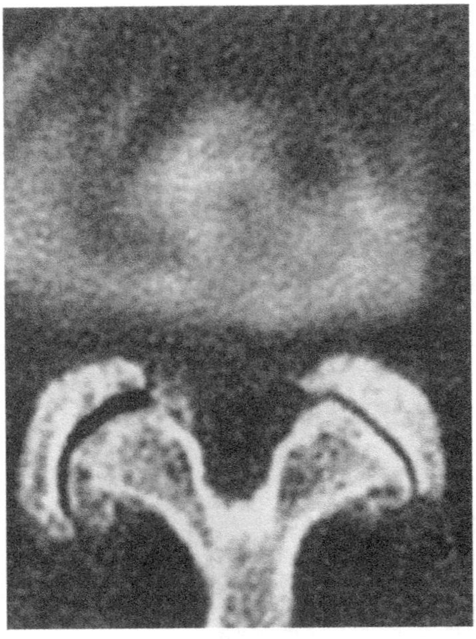

FIGURE 22 Computed tomography discography. Opacification of a radial tear and of the outer annulus that is more than 30° of the disc circumference: grade 4 lesion, according to the modified Dallas classification.

radial, and transverse—have been described (9,28,41–44). Concentric tears are due to delamination of the annulus lamellae. They are usually considered as a result of the aging process and to be not clinically significant. Radial tears originate from the NP and extend radially toward the periphery of the disc. They are theoretically always present in case of disc herniation. Transverse tears are due to the avulsion of the peripheral fibers at the insertion of the annulus onto the ring apophysis. A fourth kind of tear described as small and peripheral has also been detected, and probably corresponds to the initial stage of the previous lesions (45).

Some classifications of these tear patterns have been proposed. The modified Dallas classification (46) is the most often used (9). According to this classification, five grades are defined. In grade 0, the contrast media remains within the NP. Grade 1 is defined by the extension of contrast media in a radial fissure of the inner-third limited to the annulus. In grade 2, the contrast reaches the middle-third of the annulus. Extension of the contrast medium in the outer-third of the annulus, focally or radially, which involves less than 30° of the disc's circumference, corresponds to a grade 3 lesion. In grade 4, the contrast media occupies more than 30° of the outer annulus (Fig. 22) (46).

Disc Herniation

When disc herniation is present in a mildly degenerated disc, the annular tear and the discal protrusion are clearly opacified by contrast media. In contrast, if the discal degeneration is severe, it may sometimes be difficult to see it because of the presence of multiple fissures (18). In such cases, epidural leakage of contrast medium is very common and is not a contraindication to chymopapain injection (32).

After Discal Surgery

Most often, in cases of recurrent radiculopathy after discal surgery, CT and especially MR images are used for differentiating recurrent disc lesions from fibrous scars (47). However, in about 10% of the cases, MR may be equivocal, and CT discography is then indicated because it has been shown that sensitivity and specificity of CT discography are superior to MR (47–49). Clarisse proposed a double-contrast technique to perform disco-CT to differentiate a true recurrent protrusion from a residual cavity–containing liquid. The gas will fill the cavity in the latter case, and not in the former (50).

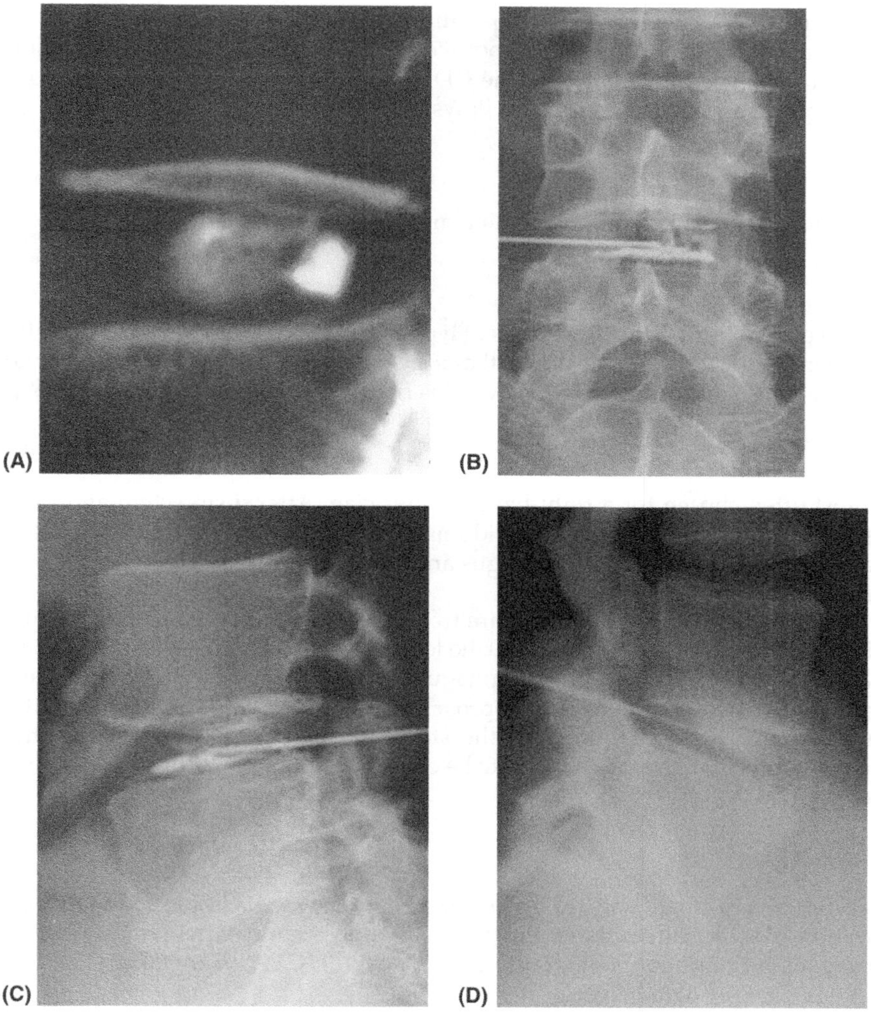

FIGURE 23 Different shapes of nucleus pulposus (NP). Bilocular NP (**A**), horizontal H-shaped NP (**B**, **C**), slightly rectangular NP (**D**).

COMPLICATIONS

It must be kept in mind that discography is an invasive technique. The most frequent complication of discography is disc infection. The rate of disc infection after discography has been reported in a range of 0.1% to 2.7% (9). As a result, some authors recommend the prophalactic use of antibiotics (51). The risk could be lower when the double-needle technique is used (52). Other infectious complications such as meningitis, arachnoiditis, subdural or epidural abscesses have also been reported. Intrathecal and retroperitoneal hemorrhages have rarely been observed (26). Accidental puncture of the retroperitoneal structures such as the ureter and the kidney is possible but exceptional (53). Nausea, convulsions, headache, and increased pain can sometimes be encountered (9,17,26,39,54). Benign reactions to contrast media such as urticaria may also be observed (49). Disk herniation after discography has been reported (55), but other studies have shown that repeated discography does not damage the disc (56,57).

CERVICAL DISCOGRAPHY

Cervical discography (CD) was first described in 1957 (58). The usefulness of CD remains controversial. In particular, the value of the pain provocation test (to confirm the discal origin of a cervicobrachial pain), as described above, is frequently questioned (9,34,35,59,60). Moreover, injection of chymopapain at the cervical level has not received Food and Drug Administration or European approval (61). The indications for the CD are identical to those for the lumbar spine. The specific technical aspects of CD are as follows.

Material

A single-needle technique with a 22-G or a 23-G spinal needle is usually used (9,59).

Patient Positioning

The patient is placed supine on the radiolucent table. Hyperextension of the neck is obtained by placing cushions under his lower neck and upper thoracic spine (Fig. 24) (9). Alternatively, the patient may be placed in an oblique prone position, with his right side elevated with cushions (59).

Needle Approach

The right side is most often chosen for a right-handed physician. After disinfection and local anesthesia in the medial border of the sternocleidomastoid muscle, the needle is inserted between the large vessels laterally and the esophagus and trachea medially, which are pushed away by the fingers (9).

Great care must be taken during the procedure to avoid puncturing the carotid. Specific techniques to avoid this problem have been proposed: no local anesthesia because of the proximity of cephalic vessels; displacing the trachea, the esophagus, and the carotid medially (together) away from the path of the needle; more lateral approach; and laterally and posteriorly to the jugulocarotid vessels at the posterior border of the sternocleidomastoid muscle. As at the lumbar level, proper placement of the needle tip is to be checked with biplanar fluoroscopy and AP and lateral radiographs.

Injection of Contrast Agent

Because the cervical NP is very small, only 0.5 mL or less of contrast media is injected in the NP. Although it is questioned by certain authors, the provocation test can be attempted (9,58,59). As with the lumbar level, postinjection radiographs and, if possible, CT scan are taken (Fig. 25).

Complications

Specific complications may be observed (62). The major complication is puncturing the spinal cord, if the tip of the needle is placed too posteriorly, Tetraplegia, due to intracanalar discal

FIGURE 24 Cervical discography. Line drawing showing the position of the patient, placed supine on the radiolucent table. A cushion under the neck allows hyperextension of the cervical spine. The needle must be inserted between the carotid sheath laterally and the trachea and esophagus medially.

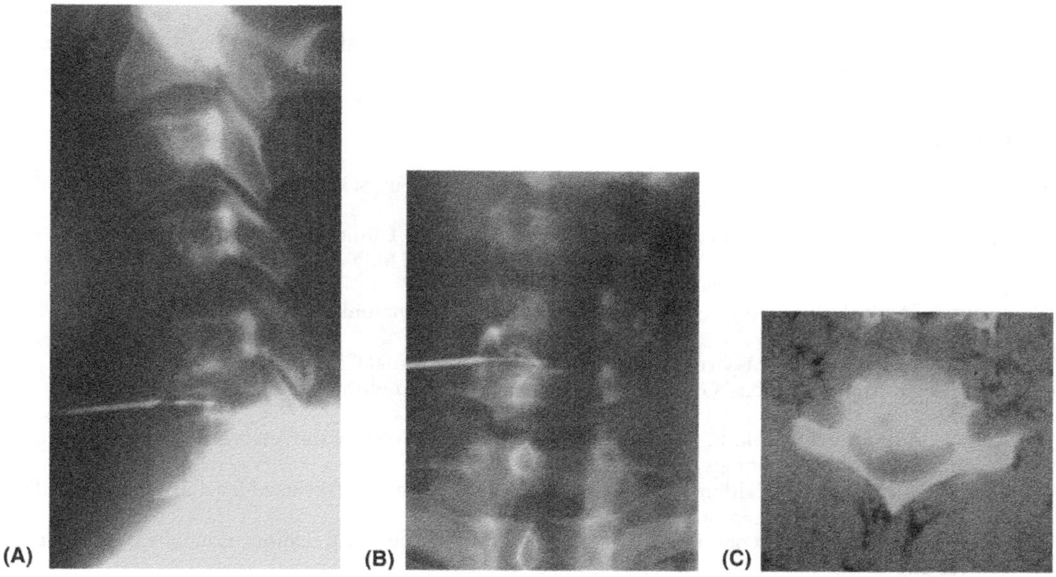

FIGURE 25 Cervical discography (**A**, **B**) and computed tomography discogram at the C5–C6 level (**C**) showing a large discal hernia near the right intervertebral foramen. *Source*: Courtesy of Pr. J. Roland.

herniation, has been reported (62). Checking for pulsation is the safest method to avoid inadvertent puncture of the carotid. But if it occurs, compression is applied immediately. The risk of disc-space infection is greater in the cervical discs than in the lumbar ones. This may be due to the presence of hair follicles in the skin, lack of utilization of the double needle technique, contact of the needle tip with the fingers (because anesthesia and needle approach are more difficult at the location of the cervical discs than at the lumbar spine), or inadvertent puncturing of the esophagus (62).

THORACIC DISCOGRAPHY

Thoracic discography is rarely performed because disc herniations are far less common in this spine segment because of the reduced mobility of the thoracic spine. However, some authors carry out this technique in Scheuerman's disease to choose the correct therapeutic procedure (63,64).

The patient is placed prone on the CT table. The procedure is identical to that described above concerning the technique for discography of the lumbar spine in the prone position. However, first, CT axial images must be obtained in order to determine the paramedian line to be marked on the skin of the patient as well as the desired discal level. Great care must be taken during the procedure to avoid lesions in the lungs and the spinal cord. A 22-G needle and 0.5 to 1 mL of contrast agent are usually used (9).

MAGNETIC RESONANCE DISCOGRAPHY

Recently, it has been shown that gadolinium (gadopentetate dimeglumine) can be used as an intradiscal contrast agent, in a new imaging technique, using MR imaging termed "MR discography" (65). A prospective study was carried out on 13 patients (42 disc levels). After intradiscal injection of gadolinium and water-soluble iodinated contrast, AP and lateral radiographs, CT scans, and T1-weighted MR were performed. The results were interpreted in blinded fashion by six physicians. Interscan interpretation variability was studied for MR discography and CT discography by statistical tests, and a high correlation was found between them (65).

In conclusion, despite some controversies, discography and CT discography are still important diagnostic and therapeutic procedures. However, these are invasive techniques, which must be restricted for use in selected indications.

REFERENCES

1. Lindblom K. Diagnostic puncture of intervertebral disc in sciatica. Acta Orthop Scand 1948; 17:213–229.
2. Moore KL, Dalley AF. Joints of the vertebral column. In: Clinically Oriented Anatomy. 4th ed. Philadelphia, Baltimore, New York, London, Buenos Aires, Hong Kong, Sydney, Tokyo: Lippincott, Williams, and Wilkins, 1992:450–455.
3. Testut L. Articulations de la colonne vertébrale. Traité d'anatomie humaine. 4th ed. Paris: Doin, 1899:410–412.
4. Rabischong P, Louis R, Massarre C. The intervertebral disc. Anat Clin 1978; 1:55–64.
5. Nazarian S. Anatomical basis of intervertebral disc puncture with chemonucleosis. Anat Clin 1985; 7:23–32.
6. Aguila LA, Piraino DW, Modic MT, Add M. The intranuclear cleft of the intervertebral disk Magnetic Resonance Imaging. Radiology 1985; 155:155–158.
7. Fardon DF, Pinkerton S, Balderston R, Garfin S, Nasca R, Salib R. Terms used for diagnosis by English speaking spine surgeons. Spine 1993; 18:274–277.
8. Fabris G, Lavaroni A, Leonardi M. Discography. In: Discography. Udine: Edizioni del centauro, 1990:11–21.
9. Tehranzadeh J. Discography 2000. Radiol Clin North Am 1998; 36:463–495.
10. Laredo JD, Bard M. Saillies discales. In: Scanner du rachis dégénératif. Paris, Barcelona, Milan, Mexico: Masson, 1988:17–34.
11. Irstam L. Point of view. Spine 1995; 20:1264.
12. Brant-Zawadzki MN, Jensen MC, Obuchowski N, Ross JS, Modic MT. Interobserver and intraobserver variability in interpretation of lumbar disc abnormalities. Spine 1995; 20:1257–1264.
13. Milette PC. Classification, diagnostic imaging and imaging characterization of a lumbar herniated disk. Imaging of low back pain. Radiol Clin North Am 2000; 38: 1267–1292.
14. Milette PC. The proper terminology for reporting lumbar intervertebral disk disorders. Am J Neuroradiol 1997; 18:1859–1866.
15. Czervionke LF. Lumbar intervertebral disc disease. Neuroimaging Clin North Am 1993; 3:465–485.
16. Bernard TN Jr. Don't discard discography. Radiology 1987; 162:285.
17. Guyer RD, Ohnmeiss DD. Contemporary concepts in spine care: lumbar discography. Spine 1995:2048–2059.
18. Champsaur P, Parlier-Cuau C, Hamze B, Laredo JD. Discographie, discoscanner et arthrographies lombaires et thoraciques. In: Arthrographie, arthroscanner et arthro- IRM. Membre supérieur et tronc. Paris: Masson, 2000:13–29.
19. Laroche D, Aimone-Gastin I, Dubois F, et al. Mechanisms of severe immediate reactions to iodinated contrast media. Radiology 1998; 209:183–190.
20. Clément O. L'allergie aux produits de contraste existe! Implications pratiques. J Radiol 2001; 82:963.
21. McCulloch JA. Chemonucleolysis: experience with 2,000 cases. Clin Orthop 1980; 146: 35–128.
22. Laredo JD, Busson J, Wybier M, Bard M. Technique of lumbar chemonucleolysis. In: Interventional Radiology in Bone and Joint. Wien, New York: Springer-Verlag, 1988: 101–122.
23. Benoit M, Deburge A, Heripret J, Busson J, Rigot J, Cauchoix J. Treatment of lumbar disc herniation by chymopapain chemonucleolysis: a report on 120 patients. Spine 1982; 7:613–617.
24. Keck C. Discography, technique and interpretation. Arch Surg 1960; 80:580–585.
25. Erlacher P. Nucleography. J Bone Joint Surg (Br) 1952; 34:204–210.
26. McCulloch JA, Waddell G. Lateral lumbar discography. Br J Radiol 1978; 51:498–502.
27. Laredo JD, Bellaiche L, Hamze B, Naouri JF, Bondeville JM, Tubiana JM. Current status of musculoskeletal interventional radiology. Radiol Clin North Am 1994; 32:337–398.
28. Quinnell RC. An investigation of artefacts in lumbar discography. Br J Radiol 1980; 53: 831–839.
29. Ray CD. Danger of intravenous injection during chemonucleosis. J Neurosurg 1984; 60: 1327.
30. Crock HV. Chymopapain. Lancet 1986; 2:1159.
31. Crock HV. Discography and vertebral venography before chemonucleolysis. J Neurosurg 1984; 60: 149–150.
32. McCulloch JA. Chemonucleolysis. J Bone J Surg (Br) 1977; 59:45–52.
33. Colhoun E, McCall IW, Williams L, Pullicino VNC. Provocation discography as a guide to planning operations on the spine. J Bone J Surg (Br) 1988; 70:267–271.
34. Holt EP. Fallacy of cervical discography. JAMA 1964; 188:799–801.
35. Travera J. Is discography a useful diagnosis procedure? J Can Assoc Radiol 1967; 18:294–295.
36. Bogduck N, Modic M. Controversy lumbar discography. Spine 1996; 21:402–404.
37. Shapiro R. Lumbar discography: an outdated procedure. J Neurosurg 1986; 64:686–691.

38. Hecht ST, Greenspan A. Digital subtraction lumbar discography. J Spinal Disord 1993; 6:68–70.
39. Massie WK, Stevens DB. A critical evaluation of discography. J Bone J Surg (Am) 1967; 49:1243–1244.
40. Videman T, Malmivaara T, Moonet V. The value of axial view in assessing discograms: an experimental study with cadavers. Spine 1987; 12:299–304.
41. Resnick D. Degenerative disease of the spine. In: Diagnosis of Bone and Joint Disorders. Philadelphia, London, Toronto, Montreal, Sydney, Tokyo: WB Saunders Company, 2002:1382–1475.
42. Coventry MB, Ghormley RK, Kernohan JW. The intervertebral disc: its microscopic anatomy and pathology. Part I. Anatomy, development and physiology. J Bone Joint Surg (Br) 1945; 105-A:105–112.
43. Coventry MB, Ghormley RK, Kernohan JW. The intervertebral disc: its microscopic anatomy and pathology. Part II. Changes in the intervertebral disc concomitant with age. J Bone Joint Surg (Br) 1945; 27-A:233–247.
44. Coventry MB, Ghormley RK, Kernohan JW. The intervertebral disc: its microscopic anatomy and pathology. Part III. Pathological changes in the intervertebral disc. Bone Joint Surg 1945; 27-A: 460–474.
45. Osti OL, Vernon-Roberts B, Fraser RD. Annulus tears and intervertebral degeneration. An experimental study using an animal model. Spine 1990; 15:762–767.
46. Sachs BL, Vanharanta H, Spivey MA. Dallas discography description: a new classification of CT/discography in low back disorders. Spine 1987; 12:287–294.
47. Hodge JC, Ghelman B, Schneider R. Computed tomography (CT), discography, and CT myelography. J Spinal Disord 1994; 7:470–477.
48. Bernard TN Jr. Lumbar discography followed by computed tomography. Spine 1990; 15:361–369.
49. Bernard TN Jr. Using computed tomography/discography and enhanced magnetic resonance imaging to distinguish between scar tissue and recurrent lumbar disc herniation. Spine 1990; 19:2826–2832.
50. Clarisse J, Pruvo JP, Leclerc X, et al. Etude radio-anatomique des fentes discales et des espaces épiduraux lombaires. Rachis 1990; 2:111–126.
51. Osti OL, Fraser RD, Vernon-Roberts B. Discitis after discography: the role of prophylactic antibiotics. J Bone J Surg (Br) 1990; 72:271–274.
52. Fraser RD, Osti OL, Vernon-Roberts B. Discitis after discography. J Bone J Surg 1987; 69:31–35.
53. Krause D, Drape JL, Maitrot D, Woerly B, Tongio J. CT: anatomic aspects. In: CT and MRI of disk herniations. Berlin, Heidelberg, New York, London, Paris, Tokyo, Hong Kong, Barcelona: Springer-Verlag, 1990:79–80.
54. Tallroth K, Soini J, Antti-Poika L. Premedication and short-term complications in Iohexol discography. Ann Chir Gynaecol 1991; 80:49–53.
55. Grubb SA, Lipscomb HJ, Guilford WB. The relative value of lumbar roentgenograms, methrizamide myelography and discography in the assessment of patients with chronic low back pain. Spine 1987; 12:282–285.
56. Flanagan NM, Chung BU. Roentgenographic changes in 188 patients 10–20 after discography and chemonucleosis. Spine 1986; 11:444–448.
57. Johnson RG. Does discography injure normal discs? An analysis of repeat discograms. Spine 1989; 14:424–426.
58. Smith GW, Nichols P. The technic of cervical discography. Radiology 1957; 68:718–720.
59. Morvan G, Mathieu P. Discographie, discoscanner et arthrographies cervicales. In: Arthrographie, arthroscanner et arthro-IRM. Membre superieur et tronc. Paris: Masson, 2000:1–12.
60. Kikuchi S, McNab I, Moreau P. Localisation of the level of symptomatic disc degeneration. J Bone Joint Surg (Br) 1981; 63:272–277.
61. Krause D, Tongio J, Drape L. Nucléolyse cervicale. Rev Im Med 1991; 3:485–496.
62. Connor PM, Darben BV. Cervical discography complications and clinical efficacy. Spine 1993; 18:744–747.
63. Bradford DS, Ahmed KB, Moe JH. The surgical management of patients with Scheuermann's disease. J Bone Joint Surg (Am) 1980; 62:705–712.
64. Winters RB, Schellhas KP. Painful adult thoracic Scheuermann's disease: diagnosis by discography and treatment by combined arthrodesis. Am J Orthop 1996; 25:783–786.
65. Huang TS, Zucherman JF, Hsu KY, Saphiro M, Lentz D, Gartland J. Gadopentetate dimeglumine as an intradiscal contrast agent. Spine 2002; 27:839–843.

21 A Presentation of Percutaneous Treatments of Nerve Root Pain Due to Lumbar Disc Herniation

Jean-Denis Laredo, Laurence Bellaïche, and Nathalie Azoulay
Department of Bone and Joint Radiology, Lariboisière Hôpital, Assistance Publique–Hopîtaux de Paris, Paris, France

INTRODUCTION

Percutaneous, minimally invasive treatments of lumbar disc herniation have been a field of research for more than 40 years, because there is a huge need for such treatments. Some preliminary remarks are necessary in the evaluation of the results provided by the different existing methods. It is first necessary to differentiate, in the cohorts of patients included in trial, those with predominant leg pain from those with predominant or exclusive back pain. The pathophysiology of back pain is much more complicated than that of nerve root pain, and trials that include patients with either type of complaint are altogether invalid. The second remark concerns the format of the study. As a first step, open studies may be useful to explore a new treatment. However, only randomized studies are able to establish the definitive value of a treatment.

To our knowledge, chymopapain chemonucleolysis (CC) is the sole percutaneous treatment of nerve root pain due to lumbar disc herniation, whose efficiency has been controlled against placebo (1,2) and surgery (3).

In the following chapter we will review several percutaneous treatments of nerve root pain due to disc herniation. All of them provided success rates of approximately 70% to 80% in open studies. However, none of them except CC had a proven efficiency in controlled studies against placebo or surgery. Such studies are mandatory to evaluate the efficiency of these new treatments.

These percutaneous treatments may be classified into two categories, depending on whether their mechanism of action is predominantly a decompression of the nucleus pulposus (NP) through mechanical NP piece removal [automated percutaneous discectomy (APD), Nucleotomy using Dekompressor®] or NP vaporization (intradiscal laser therapy) or radiofrequency ablation and coagulation (coblation nucleoplasty) or predominantly a drug injection acting through a chemical effect (CC, ozone therapy, and ethanol injection).

It may be expected that decompression techniques will work only if the depression in the NP communicates with the herniated disc compartment. Therefore, decompression techniques may not work in migrated herniated discs, for example, where this communication is less likely to operate. On the contrary, drugs injected in the NP under pressure are more likely to reach the herniated disc compartment and to have a direct action on the herniated disc material, even in cases of migrated disc herniation in a relatively separated pressure compartment.

PERCUTANEOUS INTRADISCAL DRUG INJECTION TECHNIQUES

Chymopapain Chemonucleolysis

Chymopapain is no longer available for chemonucleolysis since Abbott laboratories decided to stop its commercialization in 2001, probably because this drug was not providing enough profit. Previously, CC was mainly used in Europe.

Chymopapain is a proteolytic enzyme extracted from latex papaya, which breaks down peptide links between aminoacids of the proteoglycans of the NP (4). Depolymerization of proteoglycans causes reduction in the osmotic pressure and progressive dehydration of the NP

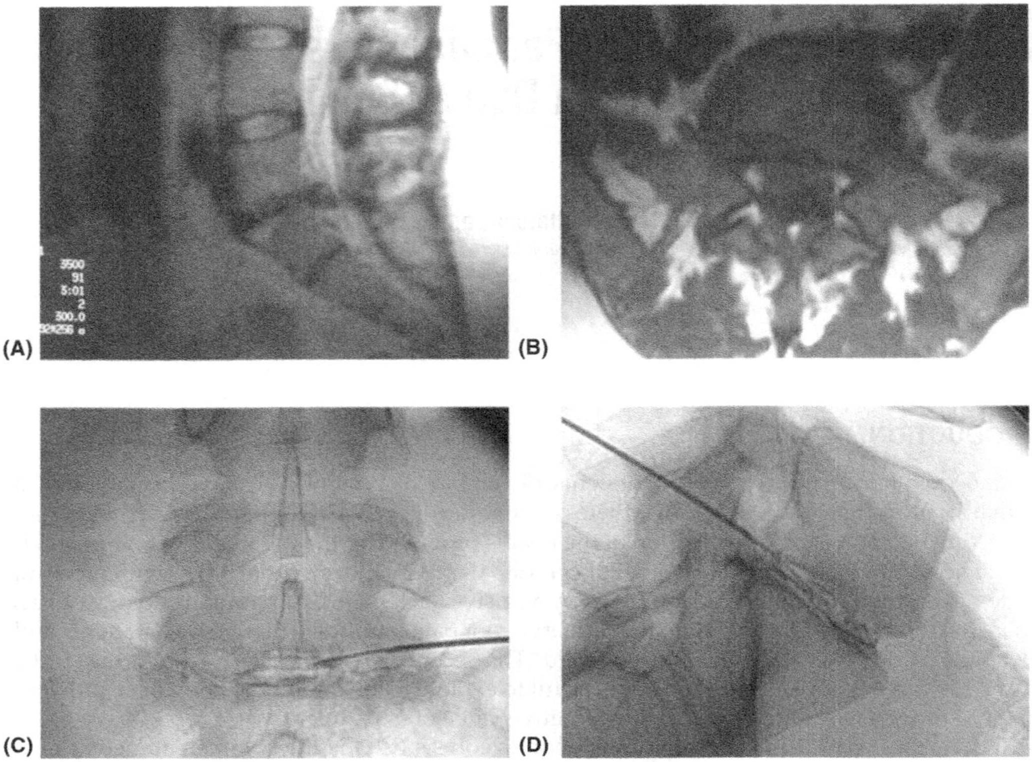

FIGURE 1 Large disc herniation on magnetic resonance imaging (**A, B**) treated by chymopapain chemonucleolysis after discography (**C, D**).

(5) while the annulus is preserved (6). CC is the reference percutaneous treatment of nerve root pain due to lumbar herniated discs, because its efficiency has been controlled in randomized trials where CC provided a success rate almost twice as great as that with placebo (1,2) and similar or slightly different from that of surgical discectomy (3).

CC has been accused of causing serious complications including neurological deficits and anaphylactic shocks. However, neurological complications are the consequence of a wrong technique with subarachnoid injection. An anaphylactic shock occurs in approximately 0.5% of cases and even less with the use of cutaneous tests (prick test), which allow the detection of the most highly sensitive patients (7).

CC provided good results even in cases of large disc herniations (Fig. 1) and foraminal disc herniations, and in the elderly. Local contraindications included a clearly sequestered disc fragment, mostly calcified disc herniations, and stenosis of the central or lateral spinal canal not predominantly due to disc herniation.

Ozone Therapy (Intradiscal and Foraminal Injections of O_2–O_3 Mixture)

This technique is used mostly in Italy and Germany. The mechanism of action of O_2–O_3 mixture could be a direct effect on mucopolysaccharides with dehydration and shrinkage of the disc (Fig. 2). Other possible actions include an oxydative stress reduction of the cell-mediated process with inhibition of proteinase release, increase in immunosuppressor cytokines, and inhibition of inflammation and pain mediators (8).

Four milliliters of the O_2–O_3 mixture at a concentration of 27 to 30 mcg/mL are injected into the NP and 8 mL in the periradicular space through the same needle. Andreula et al. obtained 70.3% good and excellent results (Mac Nab criteria) in their experience on 300 patients (9).

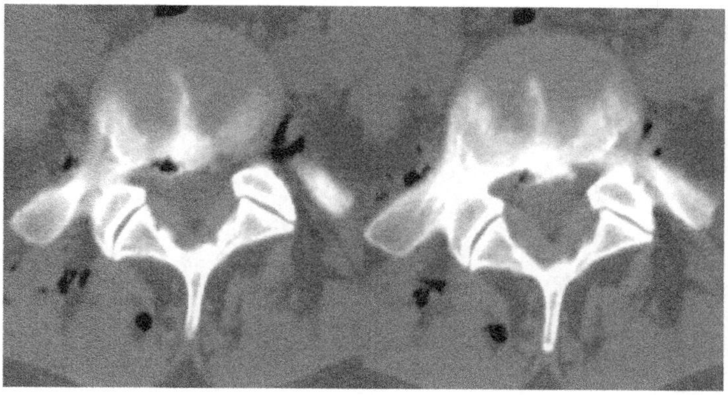

FIGURE 2 Ozone therapy.

Intradiscal Injection of Alcohol

Intradiscal injection of ethanol (absolute alcohol, 0.4 mL per disc) has been proposed by French authors in the treatment of lumbar disc herniations (10). They obtained 97.6% good results in 118 patients without complications (10), despite the toxicity of alcohol for nerves.

PERCUTANEOUS DISC DECOMPRESSION TECHNIQUES

Intradiscal Laser Therapy

Laser therapy is another method to obtain vaporization of the NP. Nd-Yag 1064 or 1318 nm is the most frequently used (11,12), using an energy of 15 W with 0.5 or 1 second pulse mode, with 4 to 10 seconds interval, delivering a total energy of 1500 to 2000 Joules (Figs. 3 and 4). The presence of gas in the disc herniation is associated with a good result according to Gangi et al. (12). Positioning of the laser fiber too close to an endplate may result in endplate burning responsible for bone edema and postoperative pain. This technique provided 74% to 80% good results (Mac Nab criteria) in open studies (11–14).

Coblation Nucleoplasty

Coblation (ablation + coagulation) nucleoplasty is a technique of vaporization of the NP by the application of high-voltage radiofrequency energy. Using the ablation mode, the wand is advanced into the NP. Then, in the coagulation mode, the wand is withdrawn. This technology uses low temperatures (50–70°) and causes minimal damage to the surrounding tissue. Disc volume is reduced by 10% to 20% at the end of the procedure (15,16). Coblation nucleoplasty provided 80% good results in an open study (17,18).

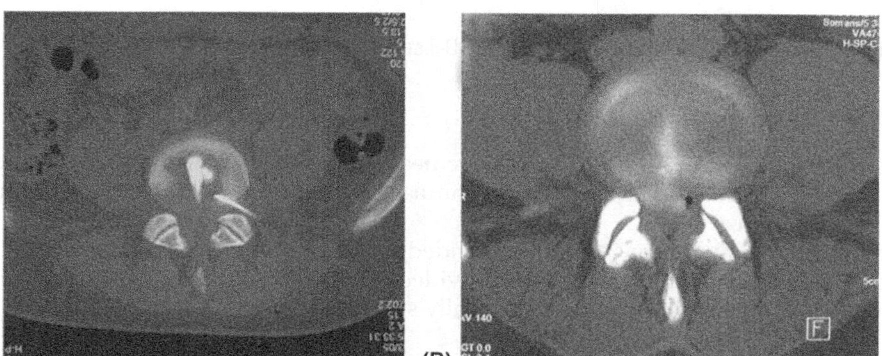

FIGURE 3 Laser therapy of herniated disc. (**A**) CT-discography at the beginning of the procedure. (**B**) Gas bubble in the disc herniation at the end of the procedure (different patient).

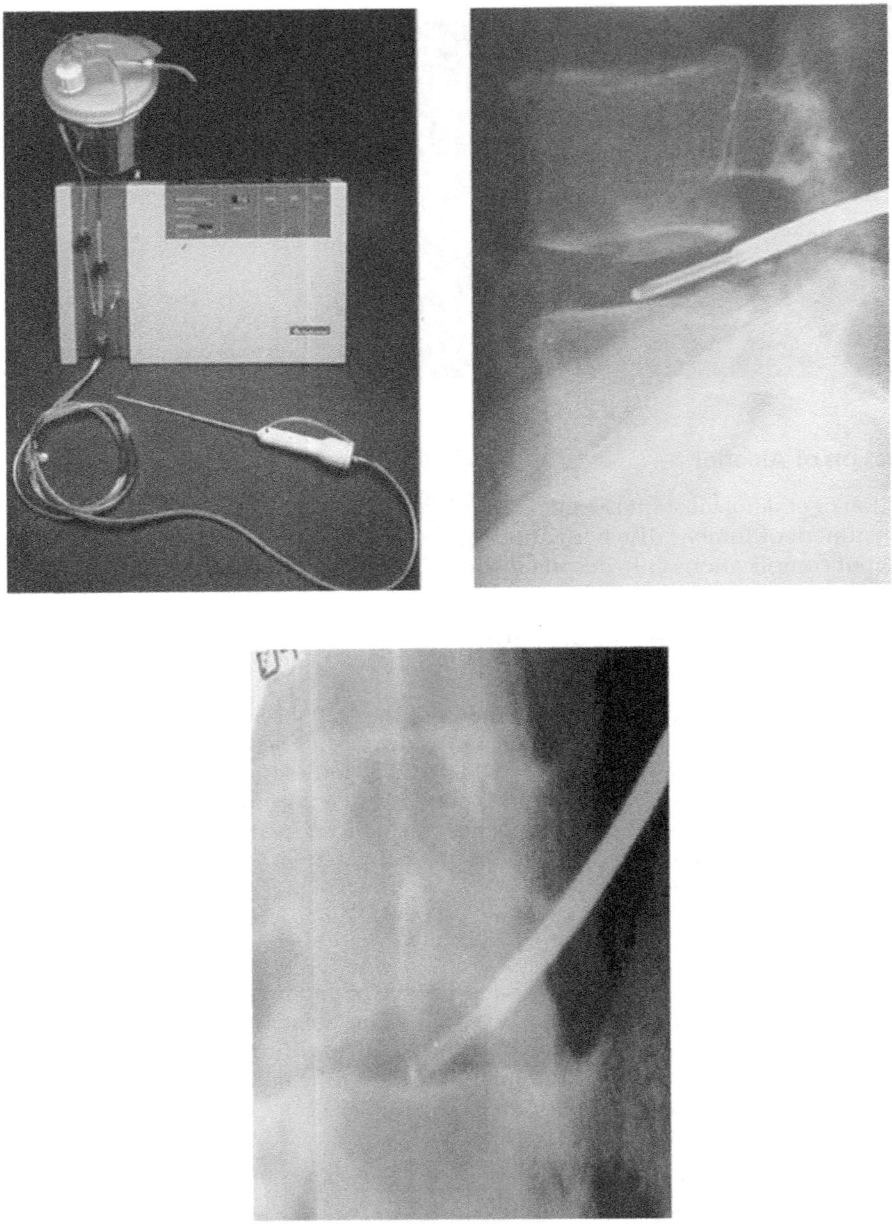

FIGURE 4 Material for automated percutaneous nucleotomy (Delmed®, Irvine, California, U.S.A.).

Automated Percutaneous Discectomy

APD, proposed by Onik et al. (19), consists in piecemeal removal of the NP using a reciprocating suction cutter introduced through a 2.8-mm cannula. This technique was indicated in small, contained disc herniations.

In a randomized trial versus CC, APD provided a 44% success rate after six months of follow-up and 37% after one year, whereas CC provided 61% and 66% success rates at the same follow-up ranges (20). The difference was statistically significant.

Percutaneous Discectomy with a 1.5-mm Probe

A new type of percutaneous discectomy using a 1.5-mm probe connected to a disposable motor (Dekompressor, Stryker®), which mechanically aspirates the nucleus, has been proposed (Fig. 5)

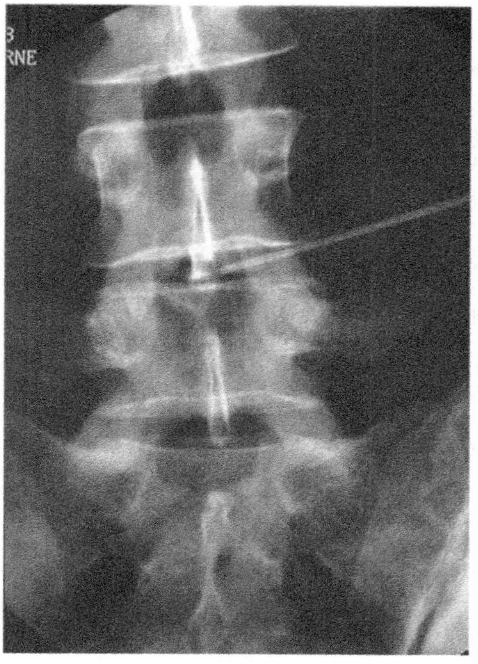

FIGURE 5 Nucleotomy with the Dekompressor (Stryker®).

(21,22). Advantages of this technique are the small caliber of the probe, its simplicity of use (small portable instrument), and its high tolerance for an ambulatory procedure, provided the operative technique is accurate. However, again, controlled trials remain to be done with this technique also.

CONCLUSION

Several techniques are currently available for the percutaneous treatment of nerve root pain due to lumbar disc herniations, which may be classified into injection techniques and decompression techniques. Only CC was found to have a proven efficiency in controlled trials, but it is no longer available. Other treatments have only been tested in open trials and there is a need for controlled trials.

Based on their alleged mechanism of action, indications for decompression techniques should be restricted to disc herniations clearly communicating with the parent disc at imaging.

REFERENCES

1. Fraser RD. Chymopapain for the treatment of intervertebral disc herniation. A preliminary report of a double-blind study. Spine 1982; 7:608–612.
2. Javid MJ, Nordby EJ, Ford LT, et al. Safety and efficacy of chymopapain (chymodiactin) in herniated nucleus pulposus with sciatica. Results of a randomized, double-blind study. JAMA 1983; 249:2489–2494.
3. Javid MJ. Chemonucleolysis versus laminectomy. A cohort comparison of effectiveness and charges. Spine 1995; 20:2016–2022.
4. Garvin P, Jennings R, Smith L. Chymopapain: a pharmacologic and toxicologic evaluation in experimental animals. Clin Orthop 1965; 42:204–223.
5. Gibson M, Buckley J, Mulholland R, Worthington B. The changes in the intervertebral disc after chemonucleolysis demonstrated by magnetic resonance imaging. J Bone Joint Surg 1986; 68(B):719–723.
6. Suguro T, Oegema R, Bradford D. The effects of chymopapain on prolapsed human intervertebral disc. A clinical and correlative histochemical study. Clin Orthop 1986; 213:223–231.
7. Dabezies E, Beck C, Shoji H. Chymopapain in perspective. Clin Orthop 1986; 206:10–14.
8. Andreula C, Muto M, Leonardi M. Interventional spinal procedures. Eur J Radiology 2004; 50:112–119.

9. Andreula CF, Simonetti L, De Santis F, Agati R, Ricci R, Leonardi M. Minimally invasive oxygen–ozone therapy for lumbar disc herniation. AJNR Am J Neuroradiol 2003; 24:994–1000.
10. Riquelme C, Musacchio M, Mont'Alverne F, Tournade A. Chemonucleolysis of lumbar disc herniation with ethanol. J Neuroradiol 2001; 28:219–229.
11. Choy DSJ. Percutaneous laser disc decompression (PLDD): twelve years experience with 752 procedures in 518 patients. J Clin Laser Med Surg 1998; 16:325–331.
12. Gangi A, Dietemann JL, Ide C, Brunner P, Klinkert A, Warter JM. Percutaneous laser disk decompression under CT and fluoroscopic guidance: indications, technique and clinical experience. Radiographics 1996; 16:89–96.
13. Gronemeyer DH, Buschkamp H, Braun M, Schirp S, Weinsheimer-Gevargez A. Image guided percutaneous laser disk decompression for herniated lumbar disks: a 4-year follow-up in 200 patients. J Clin Laser Med Surg 2003; 21:131–138.
14. Ohnmeiss DD, Guyer RD, Hochschuler SH. Laser disc decompression: the importance of proper patient selection. Spine 1994; 19:2054–2059.
15. Chen YC, Lee SH, Chen D. Intradiscal pressure study of percutaneous disc decompression with nucleoplasty in human cadavers. Spine 2003; 28:661–665.
16. Chen YC, Lee SH, Saenz Y, Lehman NL. Histologic findings of disc, end plate and neural elements after coblation of nucleus pulposus: an experimental study. Spine J 2003; 3:466–470.
17. Sharps L. Percutaneous disc decompression using nucleoplasty. Pain Physician 2002; 5(2):121–126.
18. Vijay S et al. Percutaneous disc decompression using Coblation® (Nucleoplasty™) in the treatment of chronic discogenic pain. Pain Physician 2002; 5(3):250–259.
19. Onik G, Helms CA, Ginsburg L, Hoaglund FT, Morris J. Percutaneous lumbar diskectomy using a new aspiration probe. Am J Radiol 1985; 144:1137–1140.
20. Revel M, Payan C, Vallee C, et al. Automated percutaneous lumbar discectomy versus chemonucleolysis in the treatment of sciatica. A randomized multicenter trial. Spine 2004; 18:1–7.
21. Alo KM, Wright RE, Sutcliffe J. Percutaneous lumbar discectomy: clinical response in an initial cohort of fifty consecutive patients with chronic radicular pain. Pain Practice 2004; 4:19–29.
22. Amoretti N, Huchot F, Flory P, Brunner P, Chevallier P, Bruneton JN. Percutaneous nucleotomy: preliminary communication on a decompression probe (Dekompressor) in percutaneous discectomy: ten case reports. J Clin Imaging 2005; 29:98–101.

22 Percutaneous Laser and Radiofrequency Disc Decompression

Afshin Gangi, Xavier Buy, Stephane Guth, and Jean Pierre Imbert
Department of Radiology B, University Hospital of Strasbourg, Strasbourg, France

INTRODUCTION

The long-term outcome, the complications, and the suboptimal results that may accompany open disc surgery have led to the early development of other treatment techniques that would avoid a surgical approach through the spinal canal and an extensive disc ablation. Percutaneous removal of nucleus pulposus has been performed with a variety of chemical and mechanical techniques for the past several years. Several percutaneous techniques have been used, including chemonucleolysis, percutaneous lumbar discectomy, laser disc decompression, and radiofrequency nucleoplasty (RFN) (1–8).

These techniques consist of removing all or part of the nucleus pulposus to induce more rapid healing of the pathological lumbar disc. These procedures have a reported success rate of 70% to 80%, but each procedure has its limitations. Chymopapain chemonucleolysis has been widely used to treat leg pain due to herniated discs; however, it is no longer available in the United States and Europe. Percutaneous nucleotomy is now widely used, because it is much less invasive than surgical discectomy. Automated percutaneous lumbar discectomy is in limited use today due to a combination of unfavorable clinical results. Two other minimally invasive therapies are very promising in the treatment of disc herniations: percutaneous laser disc decompression (PLDD) and RFN.

PLDD and RFN are minimally invasive techniques that can be performed in about 25% of patients needing surgical intervention for lumbar disc herniation with leg pain. PLDD andRFN have the same goal through slightly different means and are designed to remove a small portion of nucleus pulposus. PLDD has been performed since 1984 and seems to be a safe and effective modality for the treatment of herniated intervertebral disc. In 1984, PLDD was pioneered by Choy et al. (9), and the first case of PLDD in a patient with symptomatic disc herniation was performed in September 1988 with Nd: YAG (1318 nm) laser (10–13).We will review both techniques with their indications, advantages, and limitations.

PRINCIPLE OF PERCUTANEOUS LASER DISC DECOMPRESSION

The aim of PLDD is to vaporize a small portion of nucleus pulposus, thereby reducing the volume and pressure of the affected disc. The advantage of this percutaneous technique is the reduction of volume and pressure of the pathological disc without the damage of other spinal structures. In this procedure, the laser energy is transmitted through a thin optical fiber into the disc. The intact intervertebral disc can be compared to a closed hydraulic space surrounded by the annulus fibrosus and the end plates. The ablation of a relatively small volume of the nucleus results in an important reduction of intradiscal pressure, thus inducing reduction of disc herniation (11,14–17).

In the case of herniated intervertebral disc, the reduction of intervertebral disc pressure changes the gradient between the nucleus pulposus and the tissue surrounding the herniation, with the result that the herniation tends to be "aspirated" back into the disc. After laser irradiation, intradiscal pressures decrease and the nucleus pulposus is gradually replaced by cartilaginous fibrous tissue (Fig. 1) (12). Therefore sequestrated disc fragment is a contraindication to PLDD. An experimental study has proven that during the PLDD, no temperature change is detected in the neural foramina and spinal canal (12). PLDD is a minimally invasive procedure available on an outpatient basis.

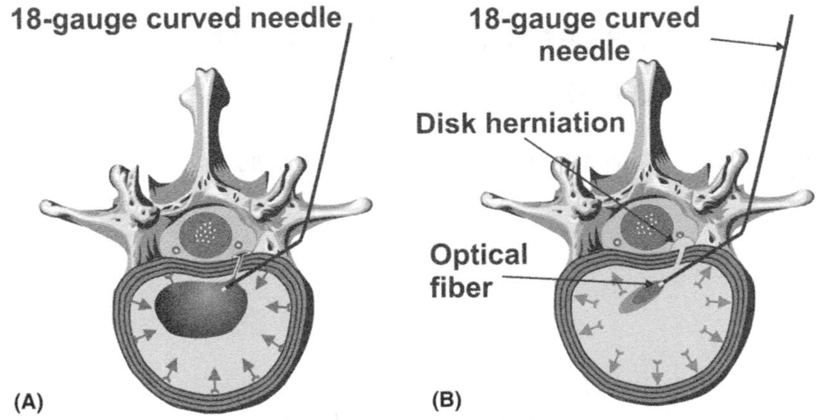

FIGURE 1 Diagram representing the percutaneous laser disc decompression technique. (**A**) The ablation of a relatively small volume of the nucleus results in an important reduction of intradiscal pressure. (**B**) Depending to the disc size, 1200 to 1500 J of energy is delivered to the nucleus. The vaporization is monitored by computed tomography scan. The fiber position can be modified during the procedure.

The reduction of intradiscal pressure does not seem to be the only effect of laser. Denervation of the annulus fibrosus and ligaments of the spinal column may be another. Heating the disc with the laser can lead to the destruction of nerve transmitter production sites (L-glutamate, substance P, peptides, and quinine).

PRINCIPLE OF RADIOFREQUENCY NUCLEOPLASTY

The aim of percutaneous RFN like PLDD is to reduce the volume of the pathological disc with ablation and coagulation of a small portion of the nucleus (18). The ablation of a relatively small volume of the nucleus results in an important reduction of intradiscal pressure, thus inducing the reduction of disc herniation. Percutaneous disc decompression or Nucleoplasty using the Perc-DLE™ SpineWand™ relies on Coblation® technology for ablating and coagulating soft tissue, combining both approaches for partial disc removal. Coblation technology removes tissue by using a low-energy bipolar radiofrequency (RF) wave to create an ionic plasma field from sodium atoms within the nucleus (Fig. 2). This low-temperature plasma field removes tissue from the treatment area via a molecular dissociation process that converts the tissue into

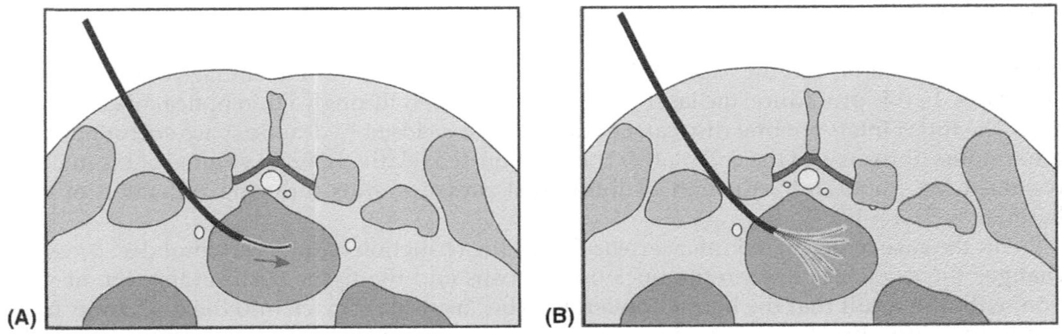

FIGURE 2 Diagram representing the radiofrequency nucleoplasty procedure. (**A**) After the first channel is created, the SpineWand™ is rotated clockwise. (**B**) As the SpineWand device is curved, each rotation changes its direction and creates a new channel inside the nucleus. Approximately 6 to 12 channels are created in total, depending on the desired amount of tissue reduction.

gases that exit the treatment site. Unlike other RF systems that are temperature driven, it does not rely on heat energy to remove tissue; so thermal damage and tissue necrosis is avoided. On withdrawal of the SpineWand, the channels are thermally treated, producing a zone of thermal coagulation.

ADVANTAGES OF PERCUTANEOUS TECHNIQUES

The advantage of these percutaneous techniques is to reduce volume and pressure of the pathological disc without damage to other spinal structures. These minimally invasive techniques avoid the drawbacks of classical surgery:

- No significant soft-tissue injury.
- No risk of fibrosis.
- No extensive hospitalization (outpatient basis).
- No general anesthesia; PLDDand RFN are performed under local anesthesia.
- Minimal recovery time of six weeks or less.
- No scar.
- Lower cost (25–30% of the surgical treatment cost).

INDICATIONS

Both techniques have the same indications and contraindications. Patient selection is crucial for treatment effectiveness.

The indications are contained disc herniations determined by computed tomography (CT) scan or magnetic resonance imaging (MRI), with positive and consistent neurologic findings (leg pain of greater intensity than back pain, positive straight-leg–raising test, decreased sensation, normal motor response, and tendon reflex) with failure of six weeks of conservative therapy (12,13,18–20).

CONTRAINDICATIONS

The contraindications are nerve paralysis, hemorrhagic diathesis, spondylolisthesis, spinal stenosis, previous surgery at the same level, significant psychological disorders, significant narrowing of disc space, workplace injuries with monetary gain, and local infection of cutaneous or subcutaneous or muscular layers.

MATERIALS

For Percutaneous Laser Disc Decompression

- The laser energy could be provided by many lasers, for example, KTP, Ho: YAG, Nd: YAG (1064 nm), and diode laser (805 nm) (7,8,11,13).
- Once the 18-gauge needle is in appropriate position in the disc, the stylet is removed, and the optical fiber is inserted coaxially.
- With CT and/or fluoroscopic guidance (13,21–26).

For Radiofrequency Nucleoplasty

- The material for Nucleoplasty consists of a needle, a bipolar RF probe, the SpineWand, and a RF generator (ArthroCare™).
- Once the 17-gauge needle is in appropriate position in the disc, the stylet is removed, and the SpineWand RF probe is placed through it in a coaxial fashion
- The RF bipolar probe is the ArthroCare SpineWand, which uses Coblation to remove tissue and to create small channels within the disc. To achieve coagulation, the device uses a higher energy bipolar RF mode (27,28).

TECHNIQUE

Disc Puncture

The patient is placed in prone position. A standard extrapedicular posterior–lateral approach is used. In order to open up the posterior aspect of the disc space, rolls are positioned under the abdomen to place the lumbar spine in a semiflexed position (particularly helpful for L5–S1 level).

For fluoroscopic guidance, the patient can be positioned in prone or lateral decubitus. We prefer the prone technique because the patient is in a more stable position. A C-arm is used for the guidance. The patient's feet are rested on a pillow. Under fluoroscopy, the desired disc levels are marked. A paramedian line 8 to 10 cm lateral and parallel to the midline is drawn depending on the patient corpulence. The site of insertion of the needle is the point of crossing the paramedian line with the level of the disc (lateral fluoroscopy). However, the site of insertion is higher for L5–S1 level. After the skin is washed and antiseptic solution is applied three times, a 22-gauge spinal needle is used for local anesthesia. The pathway of the needle is anesthetized, and the tip of the needle should touch the articular process. After local anesthesia, a direct or coaxial technique is used to reach the disc. Under lateral fluoroscopy, the needle is first positioned in contact with the articular process to avoid the nerve root and then rolled up the articular process to reach the disc. The needle must remain parallel to and midway between the two endplates on fluoroscopic control (29).

The "Scotty dog" technique requires an oblique projection. The C-arm (or the patient) is rotated until the superior articulating process (ear of the "Scotty dog") is centered midway between the anterior and posterior aspects of the vertebral body. The superior end plates of the same vertebral body should superimpose. For L5–S1 level, significant caudal angulation is required for optimal visualization (29).

For CT-guidance, the CT gantry is tilted if necessary to be parallel to the intervertebral disc plane. The entry point and the pathway are determined precisely by CT, avoiding the nerve root and visceral structures. At the L5/S1 level, curved needles can be necessary. Once the entry point is determined, a lateral fluoroscopy view is obtained at the desired disc level. In this way, the operator can visualize the pathway and the correct angulation of the needle toward the intervertebral disc in real time (13). In order to confirm contained disc herniation, or if any doubt persists, a discography can be performed just before the procedure.

PERCUTANEOUS LASER DISC DECOMPRESSION PROCEDURE

The optical fiber is checked. Before the optical fiber is placed, it is inserted in an 18-gauge needle mounted by a side-arm fitting to measure the appropriate length of the fiber. The distal part of the optical fiber should extend 5 mm beyond the needle tip. The proper length of the fiber is indicated with sterile strip to avoid excessive advancement of the fiber. After removal of the stylet of the 18-gauge needle, the optical fiber is inserted into the disc.

When satisfactory needle position is obtained, the laser procedure begins (Fig. 3A and B). The laser is turned on to produce 15 to 20 W with 0.5 to 1 second pulses and 4 to 10 second intervals, depending on patient comfort. Recommended laser doses for PLDD range from 1200 to 1500 J for L1–L2, L2–L3, L3–L4, and L5–S1 levels and 1200 to 2000 J maximum for L4–L5 (biggest disc). Using this technique with these short exposure times, there is no heating of the adjacent bone structures. A CT scan is performed every 200 J at the disc level to visualize the vaporized area (Fig. 3C). The patient must be able to communicate and respond to pain during the entire procedure. General anesthesia is therefore absolutely contraindicated. If pain occurs, the intervals between pulses are increased and aspiration is applied to reduce pressure within the disc (13).

RADIOFREQUENCY NUCLEOPLASTY PROCEDURE

A 17-gauge needle is percutaneously inserted. The tip of the 17-gauge needle must reach the posterior part of the nucleus pulposus. The tip of the needle is placed at the nuclear/annular junction (Fig. 4A), so that when the device is introduced through the needle, the active tip may pass directly into the nucleus.

Tissue removal creates channels through the nucleus to the opposite (anterior) side of the disc, where a depth stopper prevents ablation into the annulus. Prior to the activation of the

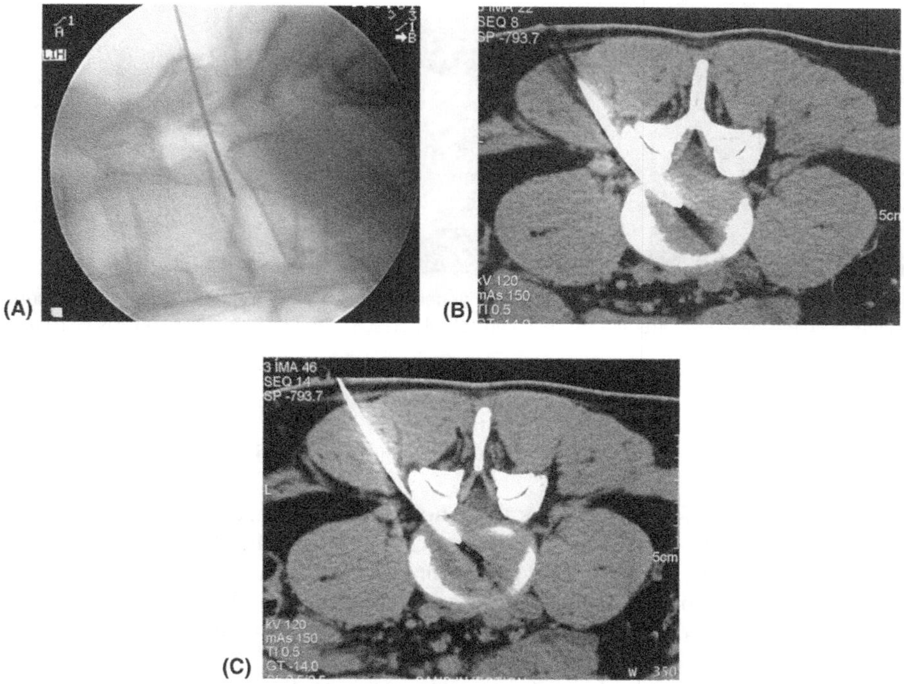

FIGURE 3 Percutaneous laser disc decompression. L4–L5 disc herniation. (**A**) The 18-gauge needle is positioned in the nucleus under fluoroscopic guidance. (**B**) Computed tomography control before vaporization. (**C**) Vaporization of the nucleus.

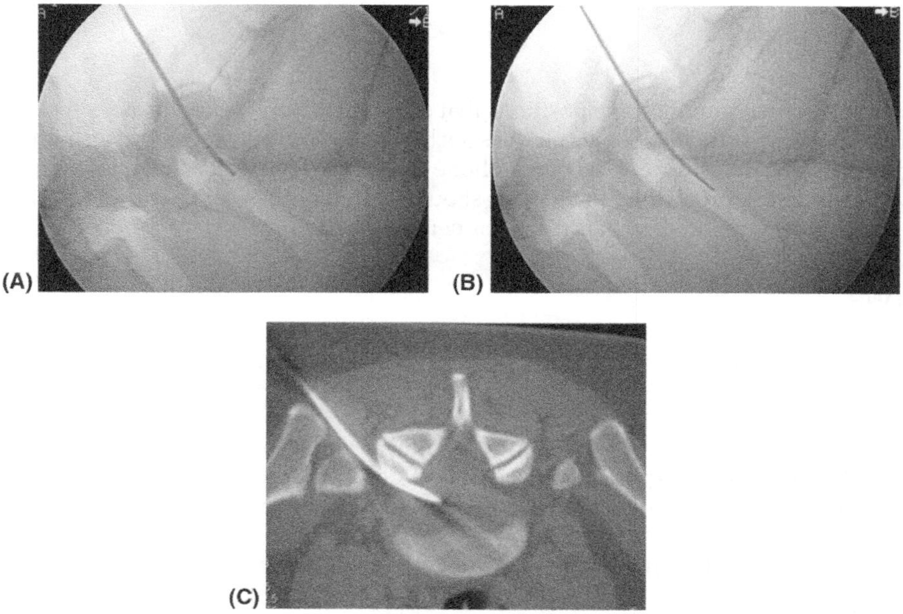

FIGURE 4 Percutaneous radiofrequency nucleoplasty in a posterolateral L5–S1 disc herniation. (**A**) The tip of the 17-gauge needle is placed under fluoroscopy at the nuclear/annular junction. (**B**) Tissue removal creates channels through the nucleus to the opposite (anterior) side of the disc. (**C**) Needle position (the SpineWand™) in the nucleus, computed tomography control.

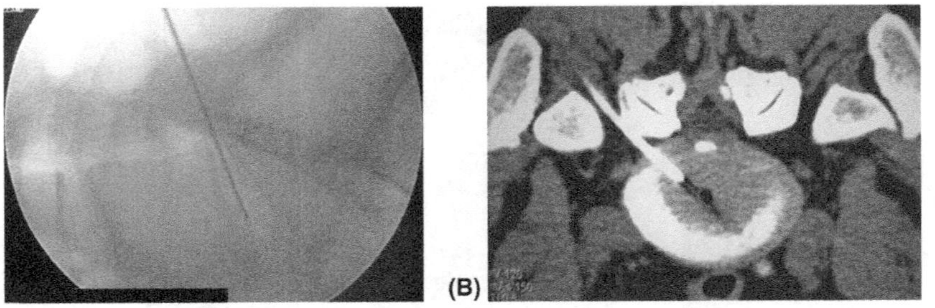

FIGURE 5 L5–S1 disc herniation. Radiofrequency nucleoplasty under computed tomography (CT) and fluoroscopy guidance. (**A**) 17-gauge needle placement under fluoroscopy control. (**B**) Tissue removed by SpineWand™ is visualized by CT scan: gas filling the disc.

device, fluoroscocopic and/or CT confirmation of the placement of the device and the extent of the channel is advisable.

Patient must be monitored for pain during the whole intervention, and the needle has to be repositioned if radicular pain occurs. In order to confirm contained disc herniation or if any doubt persists, a discography can be performed just before nucleoplasty.

While monitoring the patient, the SpineWand is advanced into the disc under continuous fluoroscopy control. As the SpineWand is advanced, the Coblation plasma mode is activated, so that tissue along the path of the device is removed (Fig. 4B and C) (30).

After stopping at a predetermined depth (prior to reaching the anterior annulus), the SpineWand is slowly withdrawn, in coagulation mode, to the starting position. During withdrawal, RF heating at the tip of the device causes the adjacent tissue to coagulate, with the result that the ablation channel is thermally treated. Sufficient thermal energy is generated to denature nerve fibers adjacent to the channel within the nucleus pulposus.

After the first channel is created, the SpineWand is rotated clockwise. As the SpineWand device is curved, each rotation changes its direction and creates a new channel inside the nucleus. Approximately 6 to 12 channels are created in total, depending on the desired amount of tissue reduction (Fig. 5).

Follow-Up

For two weeks after the intervention, some positions that could induce hyperkyphosis, as well as athletic activities should be restricted. Some activities such as bathroom privileges are permitted. Follow-up begins the same day and consists of telephone calls twice a week for the following six weeks. Three weeks after the procedure, the patient should undergo physical therapy to stretch the trunk area. At six weeks, an office visit is made for neurologic examination.

COMPLICATIONS

- The major complication is septic diskitis. To avoid this complication, severe sterility during the intervention is mandatory.
- Back pain.
- Thermal aseptic diskitis is another complication, leading to severe backache for three to six months. Unlike PLDD, there is no potential risk of thermic aseptic diskitis with this technique. RFN does not rely on heat for tissue removal, and does not introduce excessive heat to cause such tissue damage in the disc.
- Recurrence of disc herniation or free-fragment evacuation with recurrence of leg pain.

RESULTS

Percutaneous Laser Disc Decompression

PLDD has been performed in our institution since 1991. MacNab criteria were used to grade the response to treatment. The overall success rate was 76%, with 55.6% good and 20.2% fair.

In four cases, the PLDD was performed at two levels. After 6 to 12 months, a reduction of disc herniation was observed with CT or MRI. These cases were evaluated a second time with a mean follow-up of 53 months; these longterm results had a 71% success rate. These data provide encouraging information substantiating the validity of percutaneous laser nucleotomy for contained lumbar disc herniation.

Radiofrequency Nucleoplasty

Since 2002, RFN procedure has been used to treat 35 patients in our institution. Mean age was 45 years. The longest follow-up was 15 months; the average followup was six months. MacNab criteria were used to grade the response to treatment. The overall success rate was 81% according to MacNab's criteria with 50% good and 31% fair. This study is still running; the number of cases must be increased and further comparative studies are necessary to confirm these first data. A recent peer-reviewed article found that percutaneous disc Nucleoplasty successfully treated 79% of patients, based on up to a year of follow-up evaluation, and had a success rate of 82% when patients with prior surgery were excluded (30).

CONCLUSION

This minimally invasive technique avoids the drawbacks of classical surgery and has overcome some of the limitations of previous and currently existing percutaneous disc decompression. The success rates of PLDD and RFN are similar. Early success rates of RFN are comparable to or slightly better than PLDD. However, this difference can be explained by a better selection of patients and a better technique due to experienced operators.

RFN seems to be more valuable than PLDD as there is no risk of thermal diskitis. There is no thermal injury to the disc [high temperature systems (greater than 100°C) including laser remove tissue by exploding molecules with extreme temperatures; however, remaining tissue can be severely burned or charred. This is particularly of concern in the disc where there are no blood vessels to allow necrotic tissue to be resorbed into the body.] RFN does not rely on heat for tissue removal, and does not introduce excessive heat to cause such tissue damage in the disc. The RFN procedure is faster and less complicated than PLDD. However, the price of the needle electrode is higher than an optical fiber.

The three most critical elements to successful PLDD and RFN are proper patient selection, correct needle placement, and nucleus vaporization, and/or ablation. Even if further random comparative studies are necessary to confirm these data, PLDD and RFN are valuable, minimally invasive alternatives to conventional surgery for treating disc herniation.

REFERENCES

1. MaNab I, McCulloch JA, Weiner DS. Chemonucleolysis. Can J Surg 1971; 14:280.
2. Benoist M, Bonneville JF, Lassale B, et al. A randomized double-blind study to compare low-dose with standard-dose chymopapain in the treatment of herniated lumbar intervertebral discs. Spine 1993; 18:28–34.
3. Kambin P, Sampson S. Posterolateral percutaneous suction-excision of herniated lumbar intervertebral discs. Clin Orthop 1986; 207:37–43.
4. Hijikata S. Percutaneous nucleotomy. A new concept technique and 12 years' experience. Clin Orthop 1989; 238:9–23.
5. Onik G, Helms CA, Ginsberg L, Hoaglund FT,Morris J. Percutaneous lumbar diskectomy using a new aspiration probe: porcine and cadaver model. Radiology 1985; 155:251–252.
6. Kambin P. Percutaneous posterolateral discectomy. Clin Orthop 1985; 145:145–154.
7. Liebler WA. Percutaneous laser disc nucleotomy. Clin Orthop 1995; 310:58–66.
8. Maroon JC. Current concepts in minimally invasive discectomy. Neurosurgery 2002; 51:137–145.
9. Choy DS, Case RB, Fielding W, Hughes J, Liebler W, Ascher P. Percutaneous laser nucleolysis of lumbar disks [letter]. N Engl J Med 1987; 317:771–772.
10. Choy DS. Intervertebral disc pressure as function of fluid volume infused. In: HH Sherk, ed. Laser Discectomy. Philadelphia: Hanley & Belfus, 1993:11–15.
11. Choy DS, Altman P. Fall of intradiscal pressure with laser ablation. In: Sherk HH, ed. Laser Discectomy. Philadelphia: Hanley & Belfus, 1993:23–29.
12. Choy DS, Ascher PW, Saddekni S, et al. Percutaneous laser disc decompression. A new therapeutic modality. Spine 1992; 17:949–956.

13. Gangi A, Dietemann JL, Ide C, Brunner P, Klinkert A, Warter JM. Percutaneous laser disk decompression under CT and fluoroscopic guidance: indications, technique, and clinical experience. Radiographics 1996; 16:89–96.
14. Choy DS, Altman PA, Case RB, Trokel SL. Laser radiation at various wavelengths for decompression of intervertebral disk. Experimental observations on human autopsy specimens. Clin Orthop 1991; 267:245–250.
15. Rhodes A, Black J, Lane GJ, Lee SJ, Sherk HH. Clinical use of the 2.1-μm Holmium: YAG laser and percutaneous lumbar discectomy. In: HH Sherk, ed. Laser Discectomy. Philadelphia: Hanley & Belfus, 199:49–54.
16. Prodoehl JA, Lane GJ, Black J, Lee SJ, Cummings RS, Sherk HH. The effect of lasers on intervertebral disc pressures. In: HH Sherk, ed. Laser Discectomy. Philadelphia: Hanley & Belfus, 1993:17–21.
17. Hellinger J, Linke R, Heller H. A biophysical explanation for Nd: YAG percutaneous laser disk decompression success. J Clin Laser Med Surg 2001; 19:235–238.
18. Chen YC, Lee SH, Chen D. Intradiscal pressure study of percutaneous disc decompression with nucleoplasty in human cadavers. Spine 2003; 28:661–665.
19. Choy DS. Response of extruded intervertebral herniated discs to percutaneous laser disc decompression. J Clin Laser Med Surg 2001; 19:15–20.
20. Gevargez A, Groenemeyer DW, Czerwinski F. CT-guided percutaneous laser disc decompression with Ceralas D, a diode laser with 980-nm wavelength and 200-microm fiber optics. Eur Radiol 2000; 10:1239–1241.
21. Schreiber A, Suezawa Y, Leu HJ. Does percutaneous nucleotomy with discoscopy replace conventional discectomy? Clin Orth Rel Res 1989; 238:35–42.
22. Grönemeyer DHW, Seibel RMM. Combination of CT and fluoroscopy. In: Seibel RMM, Grönemeyer DHW, eds. Interventional Computed Tomography. Boston: Blackwell Scientific Publications, 1990:313–314.
23. Gangi A, Kastler B, Dietemann JL. Percutaneous vertebroplasty guided by a combination of CT and fluoroscopy. AJNR 1994; 15:83–86.
24. Damascelli B, Marchiano A, Spreafico C, et al. CT and fluoroscopy. Toward a dual unit. J Interven Radiol 1992; 7:91–96.
25. Costello P, Gaa J. Clinical assessment of an interventional CT table. Radiology 1993; 189:284–285.
26. Seibel RMM, Grönemeyer DHW, Sorensen RAL. Percutaneous nucleotomy with CT and fluoroscopic guidance. JVIR 1992; 3:571–576.
27. Dworkin GE. Advanced concepts in interventional spine care. J Am Osteopath Assoc 2002; 102:8–11.
28. Welch WC, Gerszten PC, McGrath P. Intradiscal electrothermy: indications, techniques, and clinical results. Clin Neurosurg 2001; 48:219–25.
29. Tehranzadeh J. Discography 2000. Radiol Clin North Am 1998; 36:463–495.
30. Sharps L, Isaacs Z. Percutaneous disc decompression using nucleoplasty. Pain Physician 2002; 5:121–126.

23 Complications of Percutaneous Vertebroplasty and Their Prevention

Jean-Denis Laredo and Bassam Hamzé
Department of Bone and Joint Radiology, Lariboisière Hôpital, Assistance Publique–Hopîtaux de Paris, Paris, France

INTRODUCTION

Vertebroplasty consists in the percutaneous injection of polymethylmethacrylate (PMMA) into vertebral collapses (VC) in order to obtain pain relief and mechanical strengthening of the vertebral body. It was first proposed by Galibert et al. in the treatment of malignant and aggressive vertebral hemangiomas (1). It is also now extensively used in osteoporotic VC, especially in the United States. It is an efficient treatment that is not free of complications. The Food and Drug Administration (FDA) has recently issued an alert about possible complications following the use of PMMA to treat osteoporotic compression fracture. Reports to the FDA concerned soft-tissue damage and nerve root compression related to the leakage of bone cement. In addition, the FDA mentioned the lack of prospective, randomized, controlled trials to characterize long-term safety and effectiveness of vertebroplasty and kyphoplasty. This paper will review the different types of cement extravasation, complications due to cement extravasation, local and general reactions not due to cement extravasation, and the prevention of other complications.

CEMENT LEAKAGE

Cement extravasation is a very frequent occurrence in vertebroplasty. It has been reported to occur in 38% (2) to 72.5% of cases (3) in malignant collapses, and in 30% (4), 59.5% (5), and 65% of the cases (6) in osteoporotic VC. Fortunately, it is well tolerated in the large majority of patients. However, cement extravasation is also the main source of clinical complications. Cement may leak into a large variety of anatomical compartments including needle track and the prevertebral soft tissue in 6% (7) to 52.5% (3) of the cases, and the spinal canal in up to 37.5% of the cases (3). It may also leak into the intervertebral disk in 5% (7) to 25% (8), prevertebral veins in 5% (3) to 16.6% (9), and epidural veins in 16.5% of the cases (7), with metameric artery, inferior vena cava (2,10), aorta and lungs also being reported.

Cement leakage in the prevertebral soft tissue is almost always asymptomatic (Fig. 1) except in two cases of transitory femoral neuropathy related to PMMA leakage into the psoas muscle (3,9). It may be secondary to preexistent cortical destruction or cortical breakthrough at time of biopsy preceding the vertebroplasty. Due to the round shape of the vertebral body, a needle may pass through the anterior cortex even though it appears to be within the vertebral body, both on anteroposterior (AP) and lateral views.

Cement leakage into the spinal canal is frequent when destruction of the posterior cortex of the vertebral body is present. In patients in whom a soft-tissue mass has developed in the anterior part of the spinal canal, such as malignant diseases and aggressive hemangiomas, the cement commonly fills the intraspinal soft-tissue mass (Fig. 2). Cement leakage into the spinal canal is usually well tolerated if there is enough residual space for the thecal sac.

Cement leakage into the foramen (Fig. 3) is less frequent because the transpedicular approach is preferred to the classical posterolateral approach, which crosses the foramen. Both foraminal and spinal canal cement leakage may be due to breakthrough of the medial or inferior cortex of the vertebral pedicle at the time of approach.

Intervertebral disk leakage is a frequent occurrence (Fig. 4) especially in cases of severe VC. Peh et al. reported 35% disk leakage in a series of severe VC and found that the leakage onset was not dependant on the shape of the VC (gibbus deformity, vertebra plana, and

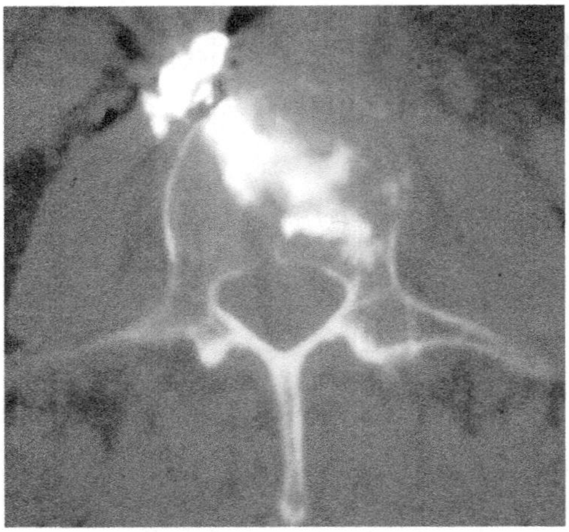

FIGURE 1 Metastatic disease: cement extravasation in the prevertebral soft tissue.

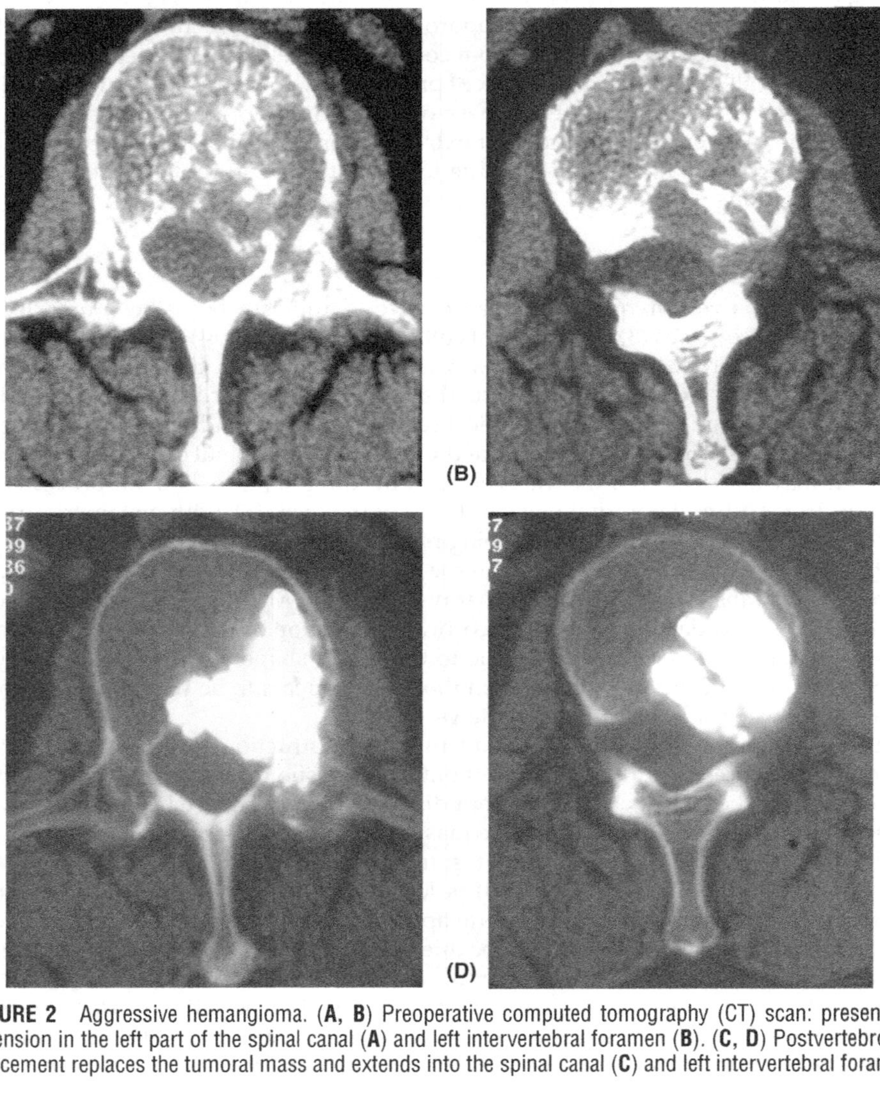

FIGURE 2 Aggressive hemangioma. (**A, B**) Preoperative computed tomography (CT) scan: presence of a tumoral extension in the left part of the spinal canal (**A**) and left intervertebral foramen (**B**). (**C, D**) Postvertebroplasty CT scan: the cement replaces the tumoral mass and extends into the spinal canal (**C**) and left intervertebral foramen (**D**).

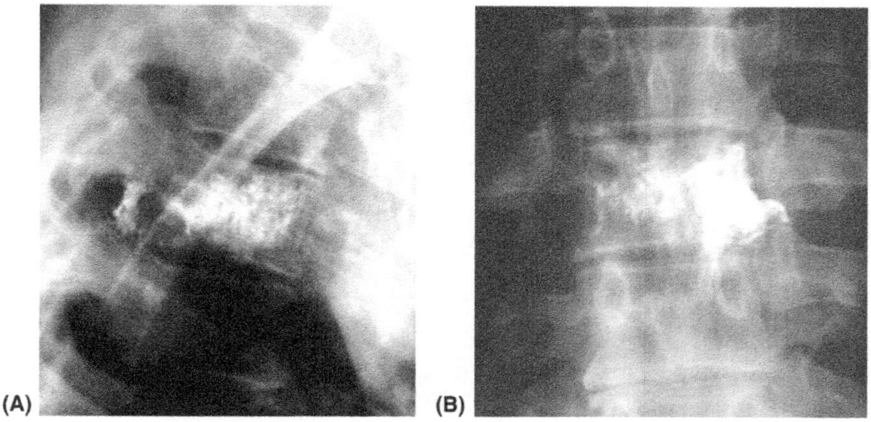

FIGURE 3 Fifty-year-old man with a very painful metastatic involvement of T5 with slight vertebral collapse due to a cutaneous basocellular carcinoma of the face: lateral (**A**) and anteroposterior (**B**) radiographs after vertebroplasty showing filling of T5 vertebra with polymethylmethacrylate. There is cement leakage into the left T5–T6 foramen. Despite the cement leakage, the patient experienced no nerve root pain after the procedure.

H-shape) (10). Intervertebral disk leakage is asymptomatic, but may have mechanical long-term consequences on adjacent vertebrae (11).

Venous cement leakage is very frequent (Figs. 5–7). It is not serious on its own but carries a risk of pulmonary embolism. Venous leakage may extend into the inferior vena cava without clinical symptoms (2,9). Padovani et al. reported a case of pulmonary embolism with a favorable outcome (12). Scroop et al. reported a case of paradoxical cerebral arterial embolization of PMMA together with pulmonary embolism of PMMA in a 78-year-old woman after multilevel intraoperative vertebroplasty for spinal fixation surgery (13). Multiple pulmonary emboli of PMMA precipitated pulmonary hypertension and right-to-left shunting into the venous circulation through a patent foramen ovale (13). This occurred because of failure to recognize venous migration of cement during the procedure. Aebli et al. in an experimental study on vertebroplasty in sheep, found in one sheep a string of cement wrapped into an ovoid shape approximately 3 cm long, in a main pulmonary artery (14).

Finally, arterial leakage may occur in highly vascularized lesions, with high intralesional pressure at time of cement injection, which results in reflux of cement into the metameric artery and even the aorta (Fig. 8).

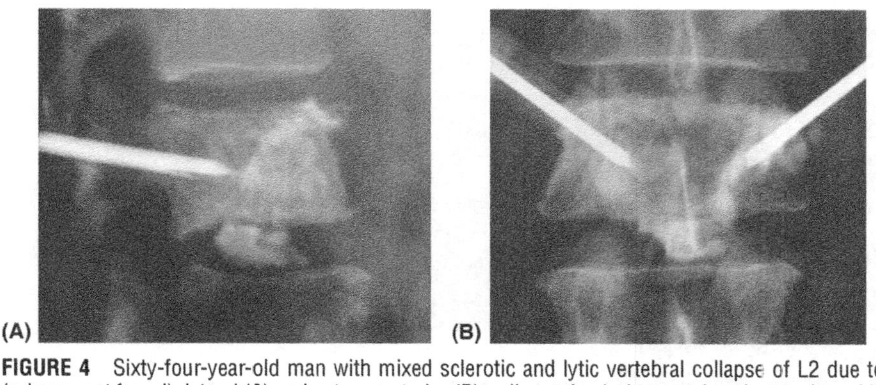

FIGURE 4 Sixty-four-year-old man with mixed sclerotic and lytic vertebral collapse of L2 due to metastatic disease (primary not found): lateral (**A**) and anteroposterior (**B**) radiographs during vertebroplasty using a bilateral transpedicular approach. There is cement leakage into the L2–L3 intervertebral disk. Due to the mostly sclerotic nature of the lesion, cement injection required high pressure, which may explain cement leakage into the intervertebral disk.

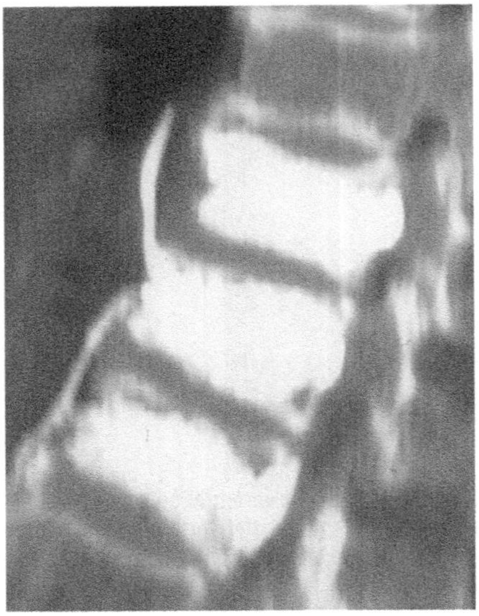

FIGURE 5 Multiple myeloma. Vertebroplasty at three vertebral levels: postvertebroplasty computed tomography-reformatted image showing cement leakage into the prevertebral veins.

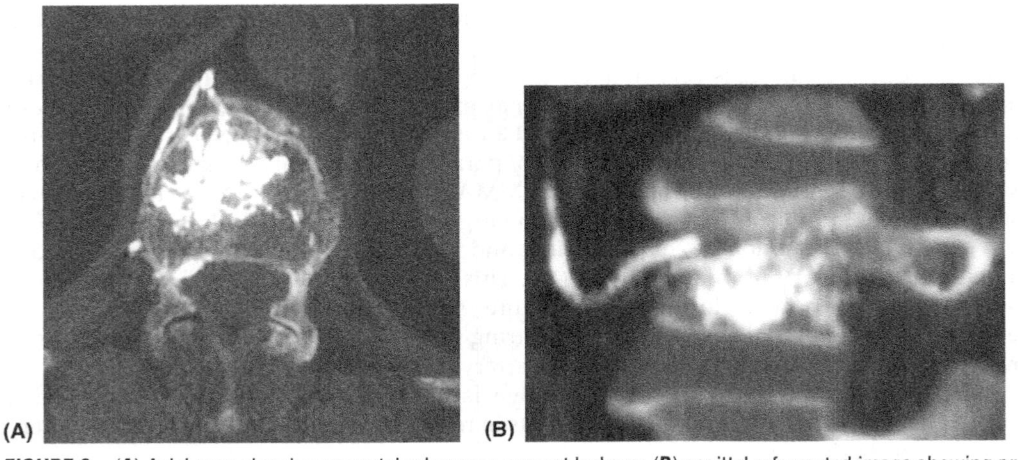

FIGURE 6 (**A**) Axial scan showing prevertebral venous cement leakage; (**B**) sagittal reformated image showing prevertebral venous cement leakage.

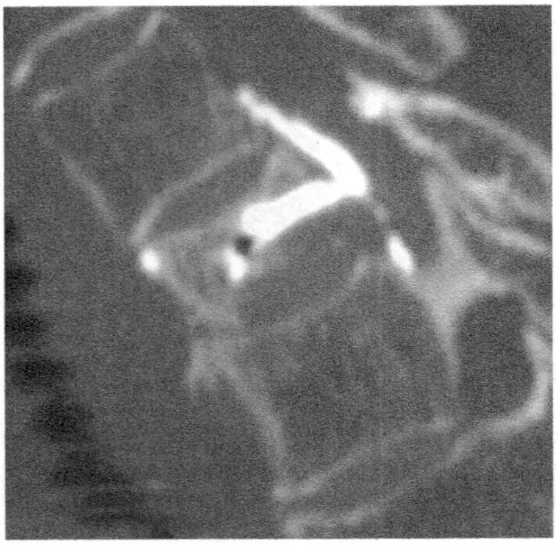

FIGURE 7 Metastatic disease of T4: postvertebroplasty computed tomography scan showing epidural veins cement leakage.

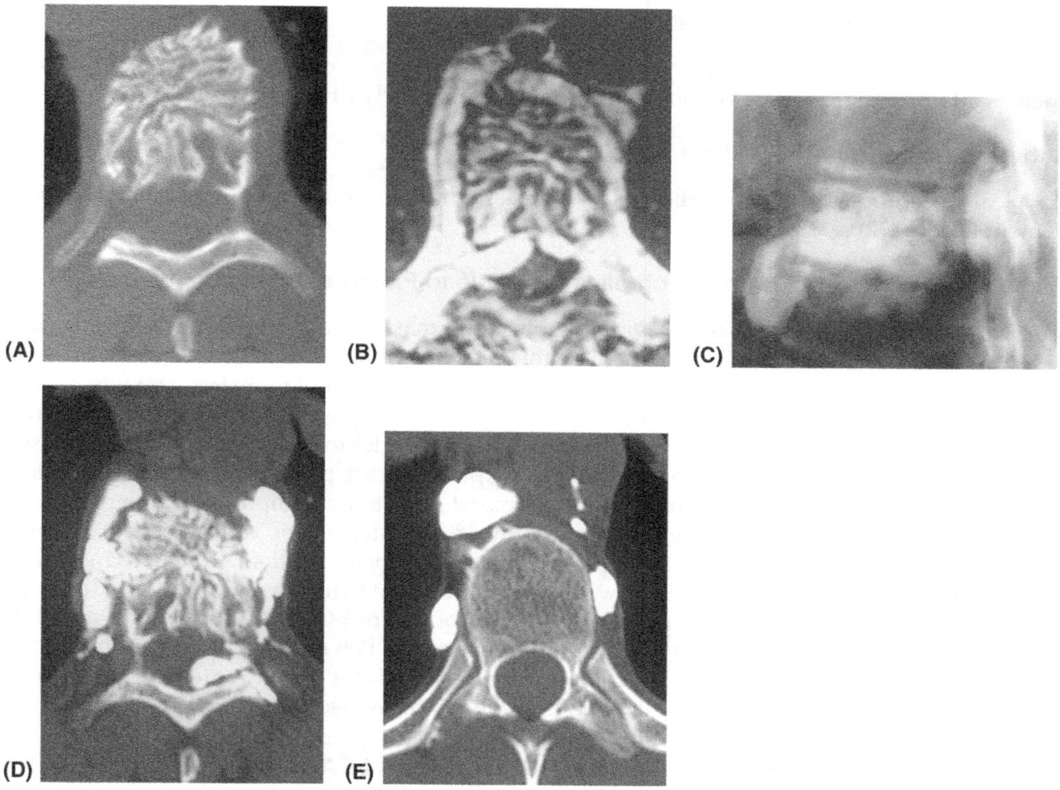

FIGURE 8 Metastatic leiomyosarcoma of T9. (**A**) Preoperative transverse computed tomography (CT) scan showing pseudoangiomatous appearance with intravertebral parallel bone channels. (**B**) Preoperative axial magnetic resonance (MR) postgadolinium T1-weighted MR scan showing intense enhancement of the intravertebral bony channels prevertebral and epidural soft-tissue masses. (**C**) Postvertebroplasty lateral radiograph showing cement extravasation in the prevertebral soft tissue and posteriorly into the spinal canal. (**D**) Postoperative transverse scan showing extravasation of cement into the prevertebral soft tissue and posterior aspect of the spinal canal through the intravertebral body channels. (**E**) Axial CT scan showing extravasation of cement in the prevertebral soft tissue and a small drop of cement into the aorta.

CLINICAL COMPLICATIONS DUE TO PMMA LEAKAGE

Neurologic Complications

Paraplegia due to spinal cord compression by PMMA leakage is one of the most dreaded complications of vertebroplasty, but it is fortunately uncommon, occurring in 1 (0.3%) out of 274 cases collected by Chiras and Deramond between 1992 and 1996 (7). This case occurred in metastatic disease and the neurologic deficit partially recovered after surgical decompression. In fact, the risk of nerve root compression in the foramen seems much higher, occurring in 2% (15) to 8% (2,3) of the cases. PMMA leakage in a narrow foramen is apparently less well tolerated than in the spinal canal. In the series reported by Cotten et al., spinal canal leakage was well tolerated in all 15 cases while two out of eight cases of foraminal cement leakage were associated with a radiculopathy (3). Foraminal leakage of PMMA may be due to a number of causes including pre-existing cortical destruction, inadequate needle puncture hole with inadvertent cortical perforation, inadequate needle approach passing through the neural foramina, reflux of cement along the needle track in cases of posterolateral needle approach, and cement leakage into the foraminal veins (3). Some cases of induced radiculopathy are managed by systemic corticosteroids and nerve root block, while surgical decompression is needed in other cases (2,3). As mentioned above, a case of paradoxical cerebral arterial embolization of PMMA together with pulmonary embolism of PMMA has been reported in a patient with right-to-left shunting through a patent foramen ovale (13).

Pulmonary Embolism of Polymethylmethacrylate

As already mentioned, pulmonary embolism of PMMA may occur (12,13), especially where there is a failure to recognize venous migration of cement early during the procedure.

Other Reactions

Cervical vertebroplasty may be followed by transient difficulties in swallowing related to focal cement leakage (2).

Factors Influencing the Rate of Complications Due to PMMA Extravasation

Cortical destruction, presence of an epidural soft tissue mass, highly vascularized lesions, and severe VC are factors that are likely to increase the rate of complications. However, some authors found that complications of vertebroplasty for vertebral tumors are not more frequent when there is destruction of the posterior cortex of the vertebral body (2,7) or epidural tumoral mass (2). However, the complication rate is probably much higher in vertebroplasty for malignant diseases than osteoporotic collapse. In a historical study using primarily the posterolateral (extrapedicular) approach to the vertebral body, Chiras and Deramond reported a complication rate of 10% in metastatic disease, 2.5% in hemangiomas, and only 1% in osteoporotic collapse (7,16). Several authors have considered that severe collapse of the vertebral body (as defined by a reduction of two-thirds or more of the normal height) was a contraindication to the procedure (2,17). However O'Brien et al. used a bipedicular approach in six cases of severe VC, with the needle inserted in the far lateral aspect of the vertebral body (18). More recently, Peh et al. reported vertebroplasty in 48 severe osteoporotic VCs without complications and did not find it more difficult than in less advanced collapses (10). It is not known, however, if this is also true in malignant cases.

LOCAL AND GENERAL REACTIONS NOT DUE TO PMMA LEAKAGE

Local Reactions

Local pain may increase within hours or days following the procedure. It usually lasts less than 72 hours and may depend on the amount of cement injected (2,7).

In the author's knowledge, only one case of secondary infection has been reported in an immunocompromised patient (16). This case required surgery.

Complications Due to Methylmethacrylate Exposure

Acute exposure of rhesus monkey to extremely high levels of methylmethacrylate (MMA) vapor can cause liver necrosis, pulmonary edema, and pulmonary emphysema (19). Occupational exposure of medical personnel is well below the levels necessary to elicit these toxic effects (20). McLaughlin et al. found occupational exposure of medical personnel to MMA to be at an acceptable level during hip arthroplasty in which the amount of PMMA used (20 mL) is the same as for typical vertebroplasty (21). Samples collected during vertebroplasty yielded MMA vapor levels of approximately five parts per million (ppm) per one hour procedure while the recommended maximum exposure is 100 ppm over the course of an eight-hour workday (20). However, less dramatic effects may occur with small doses because MMA is known to be a potent allergenic sensitizer (22) and a potential pulmonary toxin (23). Occupational exposure to MMA may cause nausea, lack of appetite, chronic cough, bronchospasm, and asthma (20,24). A case of acute reversible bronchospasm during a vertebroplasty in a technologist, who had a history of asthmatic attacks but had never before been exposed to PMMA, has been reported recently (25). Use of a closed malaxer to mix the MMA fluid and the rMMA powder is therefore useful to reduce the exposure of medical personnel who are sensitive to MMA vapor. There is insufficient information acquired to date on whether exposure of personnel to MMA vapor carries the risk of carcinogenicity, decreased fertility, or adverse effects on the fetus (20).

Risk of Collapse of Adjacent Vertebrae After Vertebroplasty

A difficult question to answer is whether vertebroplasty increases the risk of collapse of adjacent vertebrae. There is no prospective randomized study in the literature comparing the incidence of new vertebral fractures in patients with osteoporotic VCs either treated with vertebroplasty or managed conservatively.

In a series of 177 osteoporotic patients treated with vertebroplasty and reviewed retrospectively after two years or more, 22 (12.4%) developed a total of 36 new vertebral body fractures (26). In another small series of 25 patients with osteoporosis who had a total of 34 levels treated with vertebroplasty, 13 (52%) developed at least one new vertebral fracture at an average follow-up of 48 months (range 12–84 months) (27). However, these figures should be compared to the natural history of osteoporotic VC. Lindsay et al. evaluated the risk of new vertebral fractures within one year following a vertebral fracture in patients with osteoporosis enrolled in four large three-year prospective clinical trials (28); they found an incidence of 19.2% new vertebral fractures in the year following the fracture (21). Therefore, there is no evidence that the overall incidence of new vertebral fractures is increased following vertebroplasty.

However, more precisely, the incidence of new vertebral fractures in adjacent vertebrae may be increased by vertebroplasty, which increases stiffness in the vertebral segment. Grados et al. found that the relative risk of fracture of a vertebral segment adjacent to a treated vertebral body was 2.27 (27). Sixty-seven percent of the new VCs involved vertebrae adjacent to the previously treated vertebral level in the study by Uppin et al. and interestingly, 67% of the new vertebral fractures occurred within 30 days after treatment of the initial fracture (26).

However, nothing can be considered as established concerning the increased risk of collapse of adjacent vertebrae since it has been postulated that bone loss may occur in vertebral bodies adjacent to an original fracture (29) and it is a common observation that osteoporotic VC involve several adjacent vertebrae in a short period of time. Therefore, prospective randomized studies are needed to resolve this issue.

General Reactions

Systemic reactions may occur during vertebroplasty in the absence of cement leakage. Vasconcelos et al. reported one case of sudden decrease in blood pressure shortly after beginning PMMA injection with only minimal leak in a lateral paravertebral vein (9). Weill et al. reported the death of a patient through pulmonary embolism in which no cement was found on the chest radiograph (2). Systemic reactions occurring after PMMA injection during hip arthroplasty have been well documented (30,31). Pathogenic mechanisms advocated to account for these general reactions include pulmonary embolism caused by tissue debris or bone marrow expelled from the medullary cavity with fat embolism, a neurogenic reflex, and direct toxic or vasodilatating effects of the bone cement, especially the cement monomer.

Release of MMA monomer in the general circulation is a potential cause of adverse general reaction after cement injection. MMA monomer was constantly found in blood samples taken from the radial artery and the inferior vena cava during hip replacement (30). However, although a decrease in blood pressure was sometimes associated with the peak value of the monomer, no statistical relationship was found between the monomer concentration and the hypotensive event. Absence of systemic allergic manifestation such as hives, erythema or bronchodilatation makes an allergic reaction unlikely.

Fat embolism is a well-known complication observed during cemented and noncemented hip and knee arthroplasties, after fracture of the long bones, and during intramedullary nailing (14,32). Although some authors have mentioned fat embolism as a potential complication of vertebroplasty (33,34), there is no report of a general complication which can be attributed to fat embolism.

Aebli et al. showed in an experimental study in six sheep that vertebroplasty (mean 5.3 mL of PMMA) provoked fat embolism and resulted in a two-phase decrease in arterial blood pressure (14). A first very rapid (2–5 seconds) decrease in heart rate and increase in venous pressure followed by a fall in arterial pressure was attributed to a reflex mechanism, and a second fall in arterial pressure beginning at 18±2 seconds was a consequence of fat emboli passing through the heart and getting trapped in the lungs. Histology confirmed the presence of intravascular fat globules and bone marrow cells in lung tissue (14). The response of a reflex mechanism to PMMA injection during the first rapid fall in arterial pressure is supported by the work of

Rudigier and Ritter (35) who pressurized the medullary space of the tibia in rabbits with the femoral vein and muscles of the ipsilateral leg ligated, thus preventing emboli from reaching the general circulation, and observed a decrease in arterial blood pressure within two seconds after the pressure was applied. These authors hypothesized that this reflex hypotension results from stimulation of the sensory nerve endings by the MMA monomer. The hypothesis was supported by the findings of Ahmed et al. (36) and Antonacci et al. (37) that vertebral bodies are equipped with sensory nerve endings that are thought to have a vasoregulatory function.

However, it is not established that it is the MMA monomer that is responsible for the release of this reflex reaction because Breed (38) obtained a fall in blood pressure with a time course similar to that resulting from bone cement injection when he injected wax into the medullary space instead of bone cement (34). Another argument against the direct role of cement components is the work by Kaufmann et al. (39) who found no clinically significant association between the use of PMMA in percutaneous vertebroplasty (3–4 mL per level) and cardiovascular, or pulmonary dysfunction (arterial blood pressure, heart rate, and oxygen saturation).

Finally, although generalized complications due to PMMA injections have been well documented in the case of hip arthroplasty, where a much larger amount of cement is injected, there are no strong arguments to support the responsibility of PMMA injections in reported general reactions not due to PMMA embolism in the case of percutaneous vertebroplasty. Rather, there are experimental arguments that favor the role of a very rapid reflex reaction to intramedullary injection and a second reaction due to fat embolism.

PREVENTION OF COMPLICATIONS

General Conditions

Safety of vertebroplasty is a multifactorial issue. Some preliminary conditions such as correct patient selection, good three-dimensional (3-D) understanding of vertebral bone, nerve, and vascular anatomy, high-quality fluoroscopy preferably of the biplane type, and patient positioning are critical points. We have no personal experience of computed tomography (CT) guidance with C-arm scanning, but, in our opinion, if the operator has a good 3-D understanding of spinal anatomy, the combination of AP and lateral high-quality fluoroscopy provides an evaluation of needle placement as accurate as CT-scan guidance. In addition, at time of cement injection, biplane fluoroscopy allows optimal and permanent visualization of cement progression without moving the patient and interrupting the injection as required with CT-scan guidance. Use of local anesthesia and conscious sedation probably allows earlier detection of any changes in the clinical condition of the patient who can indicate if new, unacceptable pain develops (10,16). In cases of spinal tumor with destruction of the posterior vertebral body wall, Sarzier and Evans have proposed to combine vertebroplasty with intrathecal injection of contrast medium to monitor early encroachment of the thecal sac and spinal cord or nerve roots due to PMMA leakage or posterior displacement of the tumor mass (40).

Needle Approach and Placement

1. A beveled needle is preferred, which is easier to anchor into the cortical bone and may be reoriented at any time of the needle advancement by turning the bevel.
2. In any case, a pedicular route (transpedicular, parapedicular, or suprapedicular routes) is preferable to the posterolateral route, which crosses the neural foramen and may subject the nerve root to injury, especially in cases of cement extravasation along the needle track (Fig. 9).
3. A bilateral transpedicular approach is preferred to a unilateral one in cases of increased risk of cement leakage such as severe VC and large central disk herniation (Fig. 10). Bilateral approach enables the placement of each needle tip in the lateral part of the vertebral body (18), avoiding the central part where disk and venous leakages are more likely to occur. Less pressure is needed when injecting the cement.
4. Typically, for a bilateral transpedicular approach at the thoracolumbar junction, the point of skin incision projects 5 mm cranial and lateral to the superior and lateral edge of the vertebral body (Fig. 10A) on an AP-fluoroscopic view. Skin incision should be a little more lateral (8–10 mm) for the unilateral transpedicular approach (Fig. 10B).

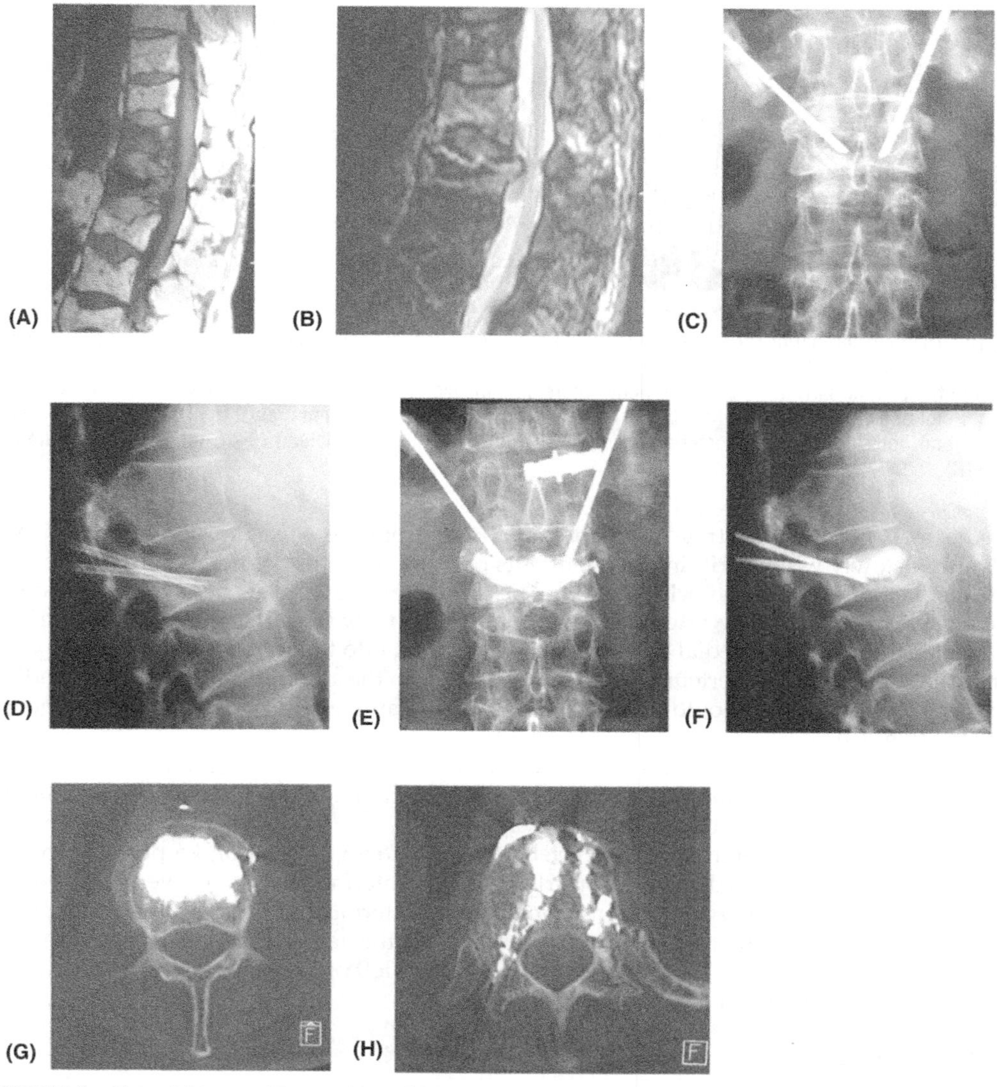

FIGURE 9 Sixty-eight-year-old man with multiple myeloma. (**A**) Sagittal T1-weighted magnetic resonance (MR) image showing vertebral fracture with signal intensity changes of T12 and L1 and chronic collapse of T11, L2, L3, and L4. An area of signal void is seen into L1 vertebral body. (**B**) Fat.Sat.Fast Spin Echo image showing high signal intensity of L1 vertebral collapse with an anterior and central area of signal void and diffuse high signal intensity of T12. (**C, D**) Anteroposterior (**C**) and lateral (**D**) radiographs showing intravertebral needle placement through a bilateral transpedicular approach during vertebroplasty. Needle tips are placed at the posterior edge of the signal void area to allow filling of the intravertebral cleft with cement. (**E, F**) Anteroposterior (**E**) and lateral (**F**) radiographs during vertebroplasty showing complete filling of the intravertebral cleft with cement without significant extravertebral cement leakage. (**G**) Transverse computed tomography (CT) image showing complete filling of the L1 intravertebral cleft. There is a small anterior venous cement leakage. (**H**) Transverse CT image through the pedicles of L1 showing cement reflux along the needle track.

5. If possible, needle tip is brought below a nonfractured endplate, which minimizes the risk of disk leakage. When the VC is not too severe, placement of the needle tip is avoided in the mid-height of the vertebral body where venous drainage is abundant.
6. A correct site for needle penetration of the neural arch at the beginning of the procedure is critical in order to make a single path in the vertebra, which minimizes the risk of cement extravasation along a previous needle track. Several attempts are often necessary to find the optimal site for bone panetration and to achieve an accurate needle placement. Very often, the needle tip slides over the irregular cortex of the neural arch. Turning the bevel of the

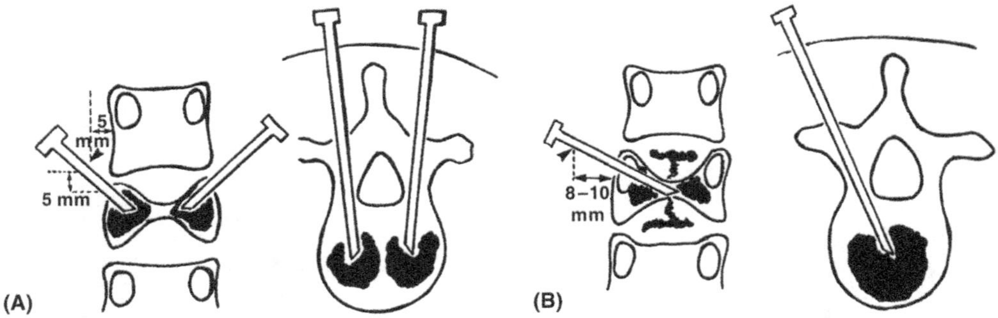

FIGURE 10 Diagram comparing the bilateral (**A**) and unilateral (**B**) transpedicular vertebral cement filling. Intervertebral disk and venous cement leakages are more likely to occur with the unilateral technique, because the needle tip has to be placed in the central part of the vertebral body where compression is more severe and intraosseous veins more abundant. Point of skin puncture for the bilateral approach is 5 mm lateral and 5 mm cranial to the superolateral edge of the vertebral body on an anteroposterior radiograph (**A**) and slightly more lateral (8–10 mm) for the unilateral approach (**B**).

needle may help to penetrate the cortex at the appropriate site. The needle should be advanced slowly while verifying frequently its position on both AP and lateral views.

7. The medial border of the pedicle should not be crossed on the AP view before it has crossed the posterior cortex of the vertebral body on the lateral view (Fig. 11), unless there is a risk of breakthrough of this medial cortex and reflux cement into the spinal canal (Fig. 12).
8. Once the needle is in the vertebral body, making a hole in the anterior cortex with the needle tip or a biopsy cannula should be avoided because it can create a potential site of cement leakage (Fig. 13).
9. The round shape of the vertebral body should be kept in mind. Therefore, a needle may appear to be contained within the vertebral body on both the AP and lateral views while it has actually crossed the anterior cortex (Fig. 13).
10. In severe VCs with separation of the vertebral body into an anterior and a posterior part, the needle tip is advanced very anteriorly into the anterior part of the vertebral body (Fig. 14). The vertebral body will fill from anterior to posterior (10). In cases of vacuum phenomenon or fracture cleft or gap, it is important to place the needle tip in or very close to the vacuum area to be able to fill it with cement. Filling the cleft or gap with cement is probably critical to obtain pain relief and stabilization.

Cement Application

Optimal visualization of the cement and permanent fluoroscopic control at time of application are crucial to the achievement of vertebroplasty. Early recognition of a cement leak enables the

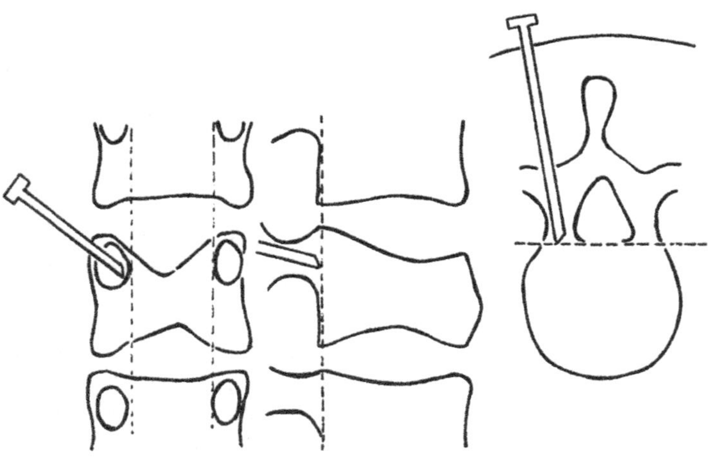

FIGURE 11 Diagram showing that the needle tip should not cross the medial border of the pedicle on the anteroposterior view (*left*) before it has reached the posterior cortex of the vertebral body on the lateral view (*center*).

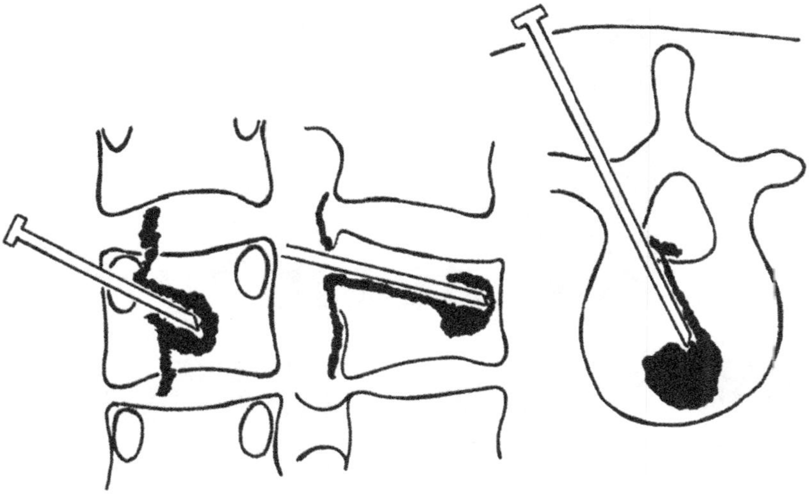

FIGURE 12 Diagram showing an example of wrong needle track: needle placement is apparently satisfactory on the anteroposterior (*left*) and lateral (*center*) views. However, the medial cortex of the pedicle has been broken at the time of approach thus creating a path for cement leakage into the spinal canal.

injection to be stopped and prevents clinical complications. The following precautions are helpful for early detection of cement extravasation:

1. High-quality biplane (or C-arm) fluoroscopy as well as perfect C-arm positioning is essential.
2. Optimal opacification of the cement requires at least 30% barium sulfate or equivalent by weight (41) or addition of tungsten or tantalum powder (11,17). The greater the cement opacity is, the faster the operator will notice a cement leakage.
3. Refrigeration of the cement gives a longer working time.
4. It is important to use the recommended proportions of MMA polymer (powder) and monomer (fluid), respectively. It must be kept in mind that increasing the proportion of fluid (monomer) accelerates cement hardening. Once the polymer and the monomer are mixed, the mixture is stirred at a rate of one turn per second. A more rapid movement accelerates cement hardening.

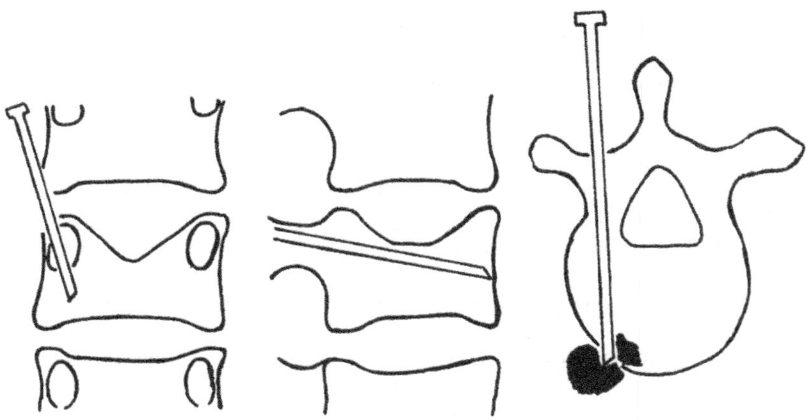

FIGURE 13 Diagram showing an example of wrong needle placement: the needle tip is apparently contained within the vertebral body both on the anteroposterior (*left*) and lateral (*center*) views. However, the needle tip is actually partly in the prevertebral space due to the round shape of the vertebral body (*right*). This is more likely to occur in the case of a too-sagittal needle path.

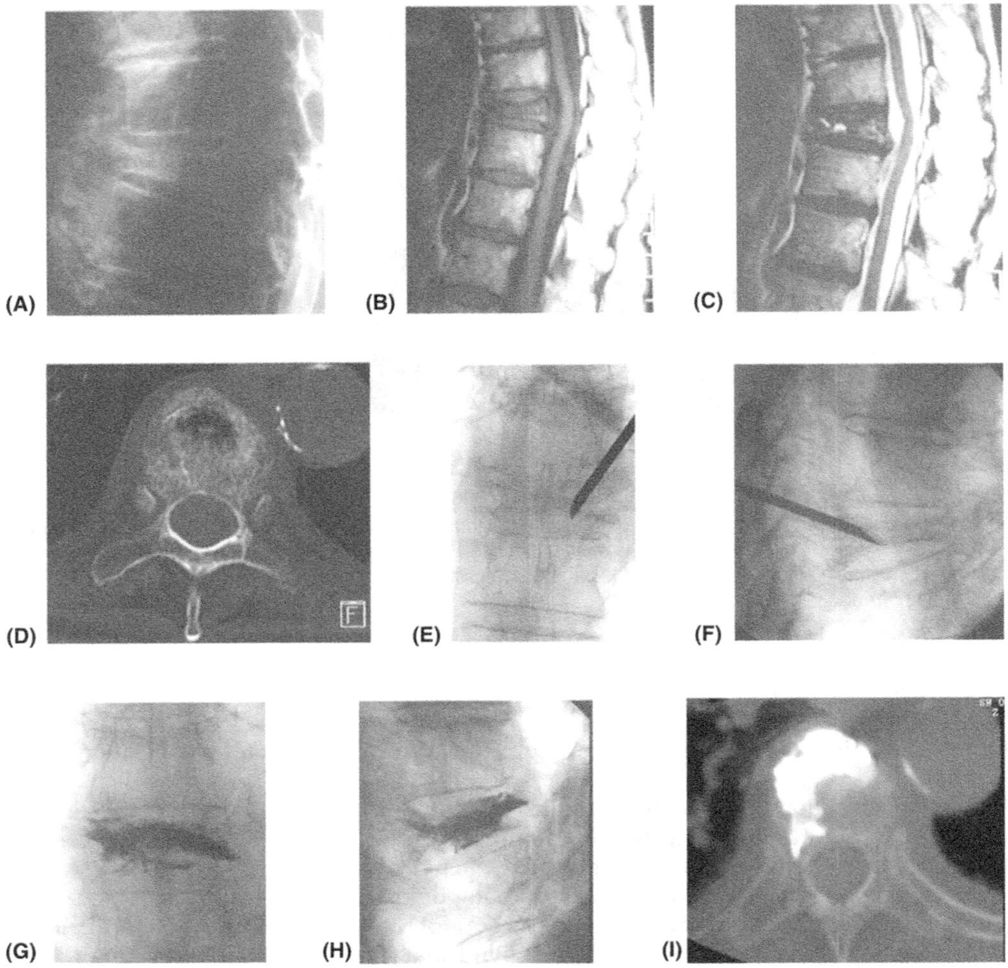

FIGURE 14 Seventy-one-year-old man with L1 osteoporotic fracture with high level of pain despite high doses of morphinics. (**A**) Lateral radiographs showing marked wedge fracture of L1. (**B**) Sagittal magnetic resonance T1-weighted image showing low signal intensity in L1 vertebral fracture. (**C**) Sagittal T2-weighted image showing intravertebral high fluid-like, signal intensity in L1 vertebral fracture. (**D**) Transverse computed tomography (CT) image through L1 vertebral fracture showing anterior vacuum phenomenon in L1 vertebral fracture. (**E, F**) Anteroposterior (**E**) and lateral (**F**) radiographs of L1 showing placement of the needle into the central part of L1 just posterior to the area of vacuum phenomenon. (**G, H**) Anteroposterior (**G**) and lateral (**H**) radiographs during vertebroplasty after cement injection showing cement filling of the L1 intravertebral cleft. There is some cement leakage in L1–L2 intervertebral disk and in the anterior prevertebral space. (**I**) Postvertebroplasty CT image showing cement distribution into the right part of L1 and replacing the vacuum phenomenon.

5. The cement is injected when it has a tooth-paste consistency. To avoid the problem of an overly liquid cement, Wong and Mathis perform a "drip test" in which the operator waits until the cement balls up at the end of the needle and no longer flows downward from the tip (41).
6. Use of a screwing cement injection syringe also permits injection of a more viscous cement and more time for injection.
7. In the presence of a vascular bone, allowing 30 to 60 seconds or more for a small amount of PMMA from the first injection to set in the vertebral body before injecting all the PMMA, is helpful (Lawler GJ, personal communication to Peh and Gilula) (42).
8. If viscous cement clogs in the needle during the application process, the needle stylet can be used as a plunger to push the cement into the vertebral body with a high degree of control (41).

9. In cases of PMMA leakage, the cement injection is stopped, the needle tip is displaced, and the hardening of the first PMMA injected is awaited, so that it can serve as a plug (personal communication L. Gilula). This is easier to perform with cements that harden relatively slowly, as compared to plain PMMA such as Simplex®. Keeping the PMMA monomer in the freezer up to the last moment and maintaining it in an ice bath between use also allows more time for injection. The room temperature must be taken into account. High room temperature will result in accelerated cement hardening.
10. One should not try to obtain the perfect image (complete vertebral body filling) and it must be kept in mind that it is not necessary to completely fill the vertebral body. Clinical result is not closely dependent on the amount of cement injected (3). In addition, Belkoff et al. have shown that 2 to 4 mL of cement is sufficient to restore strength and stiffness of the vertebral body (43).

Intraosseous Venography

Intraosseous venography is advocated by some authors (4,42) in order to identify potential sites of cement leakage along fracture lines, especially into the spinal canal and intervertebral foramen, and dangerous venous runoffs such as the junction between the basivertebral venous plexus and the anterior epidural venous plexus, in cases of highly vascularized tumors or suboptimal needle placement in the basivertebral venous plexus, or in direct connection with a paravertebral vein. This may lead to needle repositioning, thus avoiding potential complications. According to Peh and Gilula, the operator performing venography should verify in cases of early venous drainage that the contrast material passes first into bone trabeculae before leaving the bone through draining veins (42). Direct filling of a venous structure should lead to readjustment of needle position, typically by advancing the needle more anteriorly into the vertebral body (42). In addition, McGraw et al. have shown that intraosseous venography has a high predictive value in determining whether cement will cross the midline and is therefore helpful in deciding if unipedicular needle placement will suffice (44). Venograms that demonstrated a bilateral marrow blush predicted flow of PMMA across the midline to adequately fill the contralateral hemivertebra in 95% of the cases while a unilateral marrow blush predicted the necessity of a second puncture 97% of the time (44). Intraosseous venography also predicted PMMA entering vertebral end plates or cortical defects in all cases and venous structures in 29% of the cases. Another advantage of intraosseous venography in cases of fracture cleft in the vertebral body is to confirm that contrast material actually fills the cleft to eliminate the patient's presenting pain (42).

Some authors however do not believe that intraosseous venography can accurately predict PMMA cement flow because of the difference in viscosity and flow pattern between the two materials (11). McGraw et al. did not demonstrate that intraosseous venography improves safety (44). In addition, many authors routinely place two needles transpedicularly because it decreases the risk of cement leakage by avoiding needle placement in the center of the vertebral body where disk and venous leakages are more likely to occur.

Gaughen et al. found no difference in the rate of cement extravasation in patients who did not undergo venography prior to vertebroplasty and concluded that venography does not significantly improve the effectiveness or safety of percutaneous vertebroplasty performed by qualified experienced operators (5). For example, Jensen et al. reported occurrences of asymptomatic pulmonary embolism in patients who underwent antecedent venography (4). In addition, intraosseous venography may interfere with cement visualization during its application if the contrast pools in a necrotic cavity, especially in cases of vertebral osteonecrosis (39) or malignant tumors in which there is no washout (11,41). Therefore, Peh and Gilula recommend using a small amount (0.5–2 mL) of low concentration contrast medium and repeated flushing of the venography needle with sterile saline to overcome the problems of excessive staining of the tumor or tissue and filling of the fracture cleft (42).

Finally, the value of antecedent venography is very controversial. We believe that it should be performed in each case where a highly vascularized lesion such as arteriovenous malformations or tumors with arteriovenous shunting is suspected or when disc leakage is likely to occur, especially in the case of adjacent intradiscal vacuum phenomenon. Otherwise, according to Do, venography may not augment the safety of vertebroplasty when it is performed by experienced operators but may guide less experienced operators, especially with the delineation of the venous anatomy around the vertebral body, which may improve early detection of cement extravasation (45).

KYPHOPLASTY

Kyphoplasty is a variant of vertebroplasty, which aims to restore to some extent the height of vertebral compression fractures (46). It includes the introduction of a cannula into the vertebral body followed by the insertion of an inflatable bone tamp designed to reduce the vertebral body fracture while creating a cavity to be filled with bone cement (46). The cement is injected into the low-pressure preformed cavity at a more viscous phase than with classical vertebroplasty thus reducing the risk of extravasation. In a series of 70 kyphoplasties performed on 30 patients mostly suffering from osteoporotic VCs, Lieberman reported no clinical complication and 6 (8.5%) cement leakages involving the epidural space (one case, 1.4%), the disk space (two cases, 2.8%), or the prevertebral soft tissue (three cases, 4.2%) (46). More data are necessary, especially studies comparing kyphoplasty to vertebroplasty, to conclude that kyphoplasty actually reduces the complication rate.

CONCLUSION

Cement extravasation is a very frequent occurrence in vertebroplasty and is well tolerated in the large majority of cases. However, it is also the main source of complications. Prevention of complications is a multifactorial issue involving all the steps of the procedure including preparation, needle approach and placement, and cement application. Optimal visualization of a well-opacified cement and permanent fluoroscopic control at time of cement application are especially critical. Conscious sedation allows early detection of any clinical change in the condition of the patient.

REFERENCES

1. Galibert P, Deramond H, Rosat P, Le Gars D. Note préliminaire sur le traitement des angiomes vertébraux par vertébroplastie acrylique percutanée (preliminary report on the percutanenous vertebroplasty with acrylic cement as a treatment of vertebral angioma). Neurochirurgie 1987; 33:166–168.
2. Weill A, Chiras J, Simon JM, Rose M, Sola-Martinez T, Enkaoua E. Spinal metastases: indications for and results of percutaneous injection of acrylic surgical cement. Radiology 1996; 199:241–247.
3. Cotten A, Dewatre F, Cortet B, et al. Percutaneous vertebroplasty for osteolytic metastases and myeloma: effects of the percentage of lesion filling and the leakage of methyl methacrylate at clinical follow-up. Radiology 1996; 200:525–530.
4. Jensen ME, Avery JE, Mathis JM, Kallmes DF, Cloft HJ, Dio JE. Percutaneous polymethylmethacrylate vertebroplasty in the treatment of osteoporotic vertebral compression fractures: technical aspects. AJNR Am J Neuroradiol 1997; 18:1897–1904.
5. Gaughen J, Jensen ME, Schweickert PA, Kaufmann TJ, Marx WF, Kallmes DF. Relevance of antecedent venography in percutaneous vertebroplasty for the treatment of osteoporotic compression fractures. AJNR 2002; 23:594–600.
6. Cortet B, Cotten A, Boutry N, et al. Percutaneous vertebroplasty in the treatment of osteoporotic vertebral compression fractures: an open prospective study. J Rheumatol 1999; 26:2222–2228.
7. Chiras J, Deramond H. Complications des vertébroplasties. In: Saillant G, Laville C, eds. Echecs et Complications de la Chirurgie du Rachis: Chirurgie de Reprise. Montpellier, France: Sauramps Medical, 1995:149–153.
8. Cyteval C, Baron Sarrabère MP, Roux JO, et al. Acute osteoporotic vertebral collapse: open study on percutaneous injection of acrylic surgical cement in 20 patients. AJR 1999; 173:1685–1690.
9. Vasconcelos C, Gailloud PH, Martin JB, Murphy KJ. Transient arterial hypotension induced by polymethylmethacrylate injection during percutaneous vertebroplasty [letter]. J Vasc Interv Radiol 2001; 12:1001–1002.
10. Peh WCG, Gilula LA, Peck DD. Percutaneous vertebroplasty for severe osteoporotic vertebral body compression fractures. Radiology 2002; 223:121–126.
11. Deramond H, Depriester C, Galibert P, et al. Percutaneous vertebroplasty with polymethylmethacrylate: technique, indications, and results. Radiol Clin North Am 1998; 36: 533–546.

12. Padovani B, Kasriel O, Brunner P, Peretti P. Pulmonary embolism caused by acrylic cement: a rare complication of percutaneous vertebroplasty. AJNR Am J Neuroradiol 1999; 20:375–377.
13. Scroop R, Eskridge J, Britz GW. Paradoxical cerebral arterial embolization of cement during intraoperative vertebroplasty: case report. AJNR 2002; 23:868–870.
14. Aebli N, Krebs J, Dabis G, Walton M, Williams MJA, Theis JC. Fat embolism and acute hypotension during vertebroplasty. Spine 2002; 27:460–466.
15. Barr JD, Barr MS, Lemley TJ, McCann RM. Percutaneous vertebroplasty for pain relief and spinal stabilization. Spine 2000; 25:923–928.
16. Chiras J, Depriester C, Weill A, Sola Martinez M-T, Deramond H. Vertébroplasties percutanées, Technique et indications. J Neuroradiol 1997; 24:45–59.
17. Cotten A, Boutry N, Cortet B, et al. Percutaneous vertebroplasty: state of the art. Radiographics 1998; 18:311–320.
18. O'Brien JP, Sims JT, Evans AJ. Vertebroplasty in patients with severe vertebral compression fractures: a technical report. AJNR 2000; 21:1555–1558.
19. Kessler MJ, Kupper JL, Brown RJ. Accidental methyl methacrylate inhalation toxicity in a rhesus monkey (Macaca mulata). Lab Animal Scl 1977; 27:388–390.
20. Cloft HF, Easton DN, Jensen ME, et al. Exposure of medical personnel to methylmethacrylate vapor during percutaneous vertebroplasty. AJNR 1999; 20:352–353.
21. McLaughlin RE, Barkalow JA, Allen MS. Pulmonary toxicity of methylmethacrylate vapors: an environmental study. Arch Environ Health 1979; 34:336–338.
22. Wheater RIL. Hazard of methyl methacrylate to operating room personnel. JAMA 1976; 235:3652.
23. Marez T, Edmé JL, Boulenguez C, Shirali P, Hauguenoer JM. Bronchial symptoms and respiratory functions in workers exposed to methylmethacrylate. Br J Ind Med 1993; 50:894–897.
24. Wong HY, Vidovich MI. Acute bronchospasm associated with polymethylmethacrylate cement. Anesthesiology 1997; 87:696–698.
25. Kirby BS, Doyle A, Gilula LA. Acute bronchospasm due to exposure to polymethylmethacrylate vapors during percutaneous vertebroplasty. AJR 2003; 180:543–544.
26. Uppin AA, Hirsch JA, Centenera LV, Pfiefer BA, Pazianos AG, Choi IS. Occurrence of new vertebral body fracture after percutaneous vertebroplasty in patients with osteoporosis. Radiology 2003; 226:119–126.
27. Grados F, Depriester C, Cayrolle G, Hardy N, Deramond H, Fardellone P. Long-term observations of vertebral osteoporotic fractures treated by percutaneous vertebroplasty. Rheumatology (Oxford) 2000; 39:1410–1414.
28. Lindsay R, Silverman S, Cooper C, et al. Risk of new vertebral fracture in the year following a fracture. JAMA 2001; 285:320–323.
29. Wasnich U. Vertebral fracture epidemiology. Bone 1996; 18:S179–S183.
30. Philipps H, Cole PV, Lettin AW. Cardiovascular effects of implanted bone cement. Br Med J 1971; 3:460–461.
31. Peebles DJ, Ellis RH, Stride SDK, Simpson BRJ. Cardiovascular effects of methylmethacrylate cement. Br Med J 1972; 1:349–351.
32. Ereth MH, Weber JG, Abel MD, et al. Cemented versus noncemented total hip arthroplasty: embolism, hemodynamics, and intrapulmonary shunting. Mayo Clin Proc 1992; 67:1066–1074.
33. Bai B, Jazrawi LM, Kummer FJ, et al. The use of an injectable, biodegradable calcium phosphate bone substitute for the prophylactic augmentation of osteoporotic vertebrae and the management of vertebral compression fractures. Spine 1999; 24:1521–1526.
34. Gangi A, Kastler BA, Dietemann JL. Percutaneous vertebroplasty guided by a combination of CT and fluoroscopy. AJNR Am J Neuroradiol 1994; 15:83–86.
35. Rudigier JF, Ritter G. Pathogenesis of circulatory reactions triggered by nervous reflexes during the implantation of bone cements. Res Exp Med (Berl) 1983; 183:77–94.
36. Ahmed M, Bjurholm A, Kreicbergs A, et al. Neuropeptide Y, tyrosine hydroxylase, and vasoactive intestinal polypeptide-immunoreactive nerve fibers in the vertebral bodies, discs, dura mater, and spinal ligaments of the rat lumbar spine. Spine 1993; 18:268–273.
37. Antonacci MD, Mody DR, Heggeness MH. Innervation of the human vertebral body: a histologic study. J Spinal Disord 1998; 11:526–531.
38. Breed AL. Experimental production of vascular hypotension, and bone marrow and fat embolism with methylmethacrylate cement: traumatic hypertension of bone. Clin Orthop 1974; 0:227–244.
39. Kaufmann TJ, Jensen ME, Ford G, Gill LL, Marx WF, Kallmes DF. Cardiovascular effects of polymethylmethacrylate use in percutaneous vertebroplasty. AJNR 2002; 23: 601–604.
40. Sarzier JS, Evans AJ. Intrathecal injection of contrast medium to prevent polymethylmethacrylate leakage during percutaneous vertebroplasty. AJNR 2003; 24:1001–1002.
41. Wong W, Mathis J. Is intraosseous venography a significant safety measure in performance of vertebroplasty. J Vasc Interv Radiol 2002; 13:137–138.
42. Peh WCG, Gilula LA. Additional value of a modified method of intraosseous venography during percutaneous vertebroplasty. AJR 2003; 180:87–91.

43. Belkoff SM, Mathis JM, Erbe EM, et al. Biomechanical evaluation of a new bone cement for use in vertebroplasty (see comments). Spine 2000; 25:1061–1064.
44. McGraw JK, Heatwole EV, Strnad BT, Silber JS, Patzilk SB, Boorstein JM. Predictive value of intraosseous venography before percutaneous vertebroplasty. J Vas Interv Radiol 2002; 13:149–153.
45. Do HM. Intraosseous venography during percutaneous vertebroplasty: is it needed? AJNR 2002; 23:508–509.
46. Lieberman IH, Dudeney S, Reinhardt M.-K, Bell G. Initial outcome and efficacy of "kyphoplasty" in the treatment of painful osteoporotic vertebral compression fractures. Spine 2001; 26:1631–1638.

24 Kyphoplasty: A Minimally Invasive Treatment for Painful Vertebral Compression Fractures

Jeff S. Silber
North Shore/LIJ Health System, New Hyde Park, New York, U.S.A.

Alexander R. Vaccaro
Department of Orthopedics-Spine Division, Rothman Institute, Thomas Jefferson University, Philadelphia, Pennsylvania, U.S.A.

INTRODUCTION

Osteoporosis is a systemic disease characterized by low bone mass or quantity and microarchitectural deterioration of bone tissue, leading to enhanced bone fragility and a consequent increase in fracture risk, often after minimal or no trauma. "Osteopenia" is a term usually applied to the radiographic findings of reduced bone density. Both these terms are used with increasing frequency as the U.S. population continues to age.

EPIDEMIOLOGY

More than 24 million Americans are considered osteoporotic and approximately 500,000 to 700,000 people annually sustain a vertebral body compression fracture. Approximately 85% of these fractures are due to primary osteoporosis and the rest are from secondary osteoporosis and malignancies affecting the vertebral bodies. Furthermore, 25% of all postmenopausal women experience a vertebral body fracture, and out of those, 84% have pain that impacts negatively on the quality of their life (1).

Once a patient sustains a vertebral compression fracture, depending on its location, a viscous cycle may begin with the potential for subsequent contiguous vertebral fractures leading to progressive sagittal deformity and a significant change in spinal biomechanics. The presence of one compression fracture increases the risk of other vertebral compression fractures by a factor of five due to the increased force transmitted to vertebrae above or below.

The consequences of sustaining a vertebral body compression fracture dramatically impacts the quality of life for that individual, and therefore indirectly on the patient's family and caregivers. The presence of severe pain affects the patient's activities of daily living, which may lead to decreased mobility, a decreased appetite, depression, and, for some, a bedridden existence. Other less common symptoms associated with an acute vertebral compression facture include a transient ileus, urinary retention, and, rarely, symptomatic spinal cord compression.

The economic cost can be tremendous. There are approximately 150,000 hospital admissions annually of those who suffer from a vertebral body compression fracture. The average cost for the hospital stay in 1995 was about $12,300 and the average stay approximately eight days (2). Once discharged from the hospital, these patients are often noted to have a prolonged disability that can last up to three months or more. In fact, there is documented increased long-term morbidity and mortality in patients suffering from a vertebral body compression fracture. The overall five-year survival is approximately 61% in patients who sustain a vertebral body compression fracture as opposed to 76% in the general population in age-matched controls (3). From a pulmonary standpoint, it has been reported that there is a 9% loss in predicted forced vital capacity with each vertebral fracture (4).

PREVENTION TREATMENT

Prevention of osteoporotic compression fractures is the mainstay of treatment. Traditional preventative interventions are geared to decrease the progression of osteoporosis as a whole. These measures include periodic bone density determinations (i.e., a Dual Energy X-ray Absorptiometry scans), a regular exercise program, vitamin D and calcium supplementation, and hormonal (estrogen) replacement. Recently other medications have become available for clinical use including Salmon Calcitonin, Alendronate Sodium (Fosamax), a bisphosphonate that acts by decreasing osteoclastic activity, and Raloxifene Hydrochloride (Evista) known as a specific estrogen receptor modulator (SERM) . Clinical studies have shown as much as a 50% reduction in osteoporosis-related fractures including vertebral body compression fractures in postmenopausal women on these medications (5,6). Fosamax promotes bone formation and decreases resorption by preventing the differentiation of precursor stromal cells into osteoclasts. Evista acts as an estrogen receptor stimulant and through feedback mechanisms decreases osteoclastic activity.

NONOPERATIVE TREATMENT

Due to the usually poor vertebral bone quality in osteoporotic patients, conventional spinal instrumentation application is often fraught with instrumentation migration, pullout, or lack of rigid stability. Due to this fact, the vast majority of patients are treated nonoperatively even in the presence of a progressive spinal deformity. Traditional nonoperative treatment algorithms include (*i*) observation, as most fractures heal within 6 to 12 weeks with relief of the acute pain; (*ii*) bed rest, which initially makes the patient somewhat comfortable but when prolonged increases the rate of bone loss as well as decreases muscle mass and tone; (*iii*) pain medication, which may disturb the patient's sensorium and increase the risk of a fall; (*iv*) bracing, which may stabilize the fracture but contributes to the risk of bone loss and is often poorly tolerated by the elderly patient; (*v*) physical therapy, which decreases the rate of bone loss; and (*vi*) behavioral modifications with advice on lifting, bending, and protecting the spine from injury by minimizing forces that may contribute to future fractures.

OPERATIVE TREATMENT: KYPHOPLASTY

Although the majority of patients respond to nonoperative intervention, a small subset of patients develop worsening or chronic symptoms that affect their overall level of function. This fact has led to new advances in minimally invasive treatment methods for those patients who continue to have severe pain, discomfort, and disability following failure of nonoperative treatment for an osteoporotic compression fracture. The ideal minimally invasive intervention should be designed to not only treat the present pain and discomfort from the compressed vertebrae but also change the natural history of future compression of adjacent vertebrae from altered spinal biomechanics. This is the basis of a minimally invasive procedure known as kyphoplasty. This procedure involves the insertion of a bone tamp, that is, "pump," via a small cortical window within the vertebral body, allowing low-pressure injection of bone cement, that is, polymethylmethacrylate (PMMA) into a void created within an elevated compression fracture (Fig. 1). This process is designed to restore vertebral body height and provide immediate fracture stability. The desired restoration of spinal alignment and enhanced vertebral stability will theoretically render proper biomechanics, thereby reducing pain from altered sagittal alignment and potentially decreasing the incidence of future compression fractures. Vertebroplasty in contrast is defined as a minimally invasive procedure involving the high-pressure injection of bone cement (PMMA) through both pedicles or through other bony passageways into a vertebral body fracture to restore structural stability and therefore provide relief of pain. The difference between vertebroplasty and kyphoplasty is that kyphoplasty delivers a cement substance in a low pressure environment due to the void created by a balloon tamp intended to elevate the vertebral end plate to improve sagittal plane alignment.

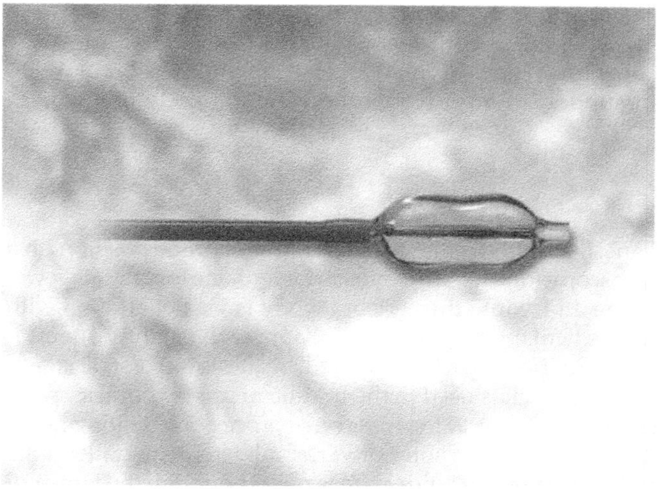

FIGURE 1 A photograph of the inflatable bone tamp (pump) in the inflated position.

INDICATIONS FOR KYPHOPLASTY

Indications for kyphoplasty include continued, unrelenting pain (more than four to six weeks) following an identified vertebral compression fracture. Identifying the appropriate pain generator in the presence of multiple compression fractures of differing ages can be a challenging endeavor. This requires a careful physical examination and scrutiny of selected imaging modalities. The use of magnetic resonance imaging, computed tomography scanning, and nuclear medicine scanning offers additional information to plain radiography in determining the acuteness of a fracture or fractures (Fig. 2). Some authors have recommended the use of vertebroplasty and kyphoplasty in fractures less than four to six weeks of age. Intervention in this setting is

FIGURE 2 Post-kyphoplasty lateral plain radiograph of the thoracolumbar spine in an elderly female with multiple compression fractures. The painful vertebral compression fracture (pain generator) was properly identified preoperatively to be at the T12 level by physical examination, plain radiography, and magnetic resonance imaging.

predicated on the failure to improve, following two to three weeks of conservative measures in the setting of intense unrelenting pain. Other indications for kyphoplasty include prophylactic augmentation of noncompressed osteopenic levels adjacent to an instrumented fusion in osteoporotic patients, and anterior column support of an osteoporotic vertebral body following a posterior decompression.

KYPHOPLASTY TECHNIQUE

After induction of general or regional anesthesia, the patient is placed in the prone position on a radiolucent table. Biplanar (C-arm) fluoroscopy (one for anteroposterior and one for lateral views) is utilized throughout the entire procedure. Entry into the vertebral body is via penetration with a Jamshidi needle either through a transpedicular or extrapedicular approach (slightly superior and lateral to the pedicle). The transpedicular approach is favored in the lower thoracic and lumbar spine whereas the extrapedicular approach is used in the middle to upper thoracic spine. A posterolateral entry into the vertebral body anterior to the transverse process and exiting nerve root is another approach that allows direct access to the anterior vertebral body, especially in the lumbar region. Once the vertebral body is penetrated, a guide wire is placed through the Jamshidi needle into the posterior portion of the vertebral body. A blunt cannula is then placed over the guide wire followed by a working cannula. The inner cannula is removed, which allows placement of a handheld drill into the vertebral body. The path created by the drill is directed toward the anterior half of the vertebral body allowing delivery of the inflatable bone tamp (Fig. 3). The proper position of the inflatable bone tamp (balloon) is identified by radiographic markers by anteroposterior and lateral fluoroscopy. The balloon should reside in the anterior half of the vertebral body. The balloon is then inflated, elevating the end plates in an attempt to restore vertebral body height. Inflation should be discontinued if the balloon comes in contact with any cortical surface or if the maximum volume of the balloon is reached depending on the size of the balloon (15 mm in the thoracic spine and 20 mm in the lumbar spine). The balloon is filled with a radiographic contrast medium to evaluate the size of balloon expansion. The contralateral balloon can now be placed allowing for symmetric elevation of the vertebral end plates. Once both bone tamps are in place, an alternating slow filling of each balloon is undertaken until the pressure manometer indicates a pressure of approximately 150 psi (maximum 220 psi).

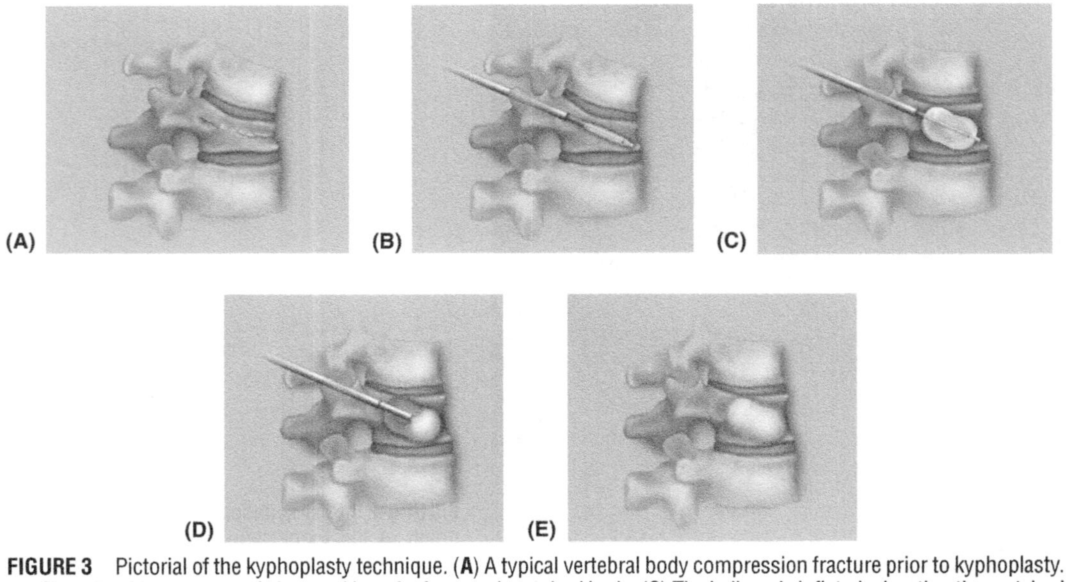

FIGURE 3 Pictorial of the kyphoplasty technique. (**A**) A typical vertebral body compression fracture prior to kyphoplasty. (**B**) The inflatable bone tamp is inserted into the fractured vertebral body. (**C**) The balloon is inflated, elevating the vertebral end plates attempting to restore vertebral height. (**D**) The balloon is deflated and withdrawn leaving a void in the vertebral body. Polymethylmethacrylate (PMMA) is then inserted through the cannula. (**E**) The void is filled with PMMA. The vertebral body is now elevated and stabilized. *Source*: Courtesy of Kyphon, Inc., Santa Clara, California, U.S.A.

The end point of fracture reduction occurs when (*i*) adequate reduction of the compression fracture is accomplished, (*ii*) pressure readings approach 220 psi, (*iii*) cortical proximity of the bone tamp occurs based on orthogonal fluoroscopic images, and (*iv*) maximum inflation volumes as read from the syringe barrel connected to the inflation device (4 cc for 15 mm length and 6 cc for 20 mm length tamps) occurs. Factors that ultimately determine the final pressure and volume of the balloon tamp depend on the volume and density of the vertebral body (trabecular microarchitecture), as well as the age (healing rate) of the compression fracture. Higher pressures are generally needed for more dense bone as well as for subacute fractures whereas low pressures are commonly experienced in less dense bone and acute fractures. The surgeon must be vigilant in detecting any evidence of a cortical breech or leakage of any contrast material from the balloon prior to the insertion of cement. Once the desired reduction is obtained or maximal pressure is achieved, the bone tamp is deflated and removed on only one side to prevent the occasional loss of reduction in more acute fractures. PMMA, which at present (2001) is not approved for use in the spine by the Food and Drug Administration, is then mixed with sterile barium and placed into a 20 cc syringe. The viscosity of the cement should be on the thicker side to prevent inadvertent leakage through any cortical defect or via a venous conduit. The volume of cement used normally exceeds the volume of balloon inflation by 1 cc. Under low-pressure injection force, the PMMA is delivered through the cannula into the created void under continuous fluoroscopic monitoring. Cement extravasation may occur if the PMMA is too runny and/or a cortical margin is violated. If cement extravasation is noted, injection is immediately halted and time should be given for further cement hardening before further cement injection is commenced. Although PMMA leakage is common, significant embolization of the lungs and significant canal intrusion are rare. Once cement injection is completed, and approximately two to three minutes pass, the cannulas are removed and the PMMA is allowed to further harden. A final anteroposterior and lateral image is obtained to document cement location and end plate elevation (Fig. 4).

Contraindications for performing kyphoplasty include (*i*) cases where there is significant vertebral body compression (i.e., the vertebral body height is smaller than the sagittal pedicle height), making the procedure difficult to perform; (*ii*) when an obvious defect is present in the middle column (posterior cortex) that may lead to inadvertent cement extravasation into the spinal canal; and (*iii*) an uncorrectable coagulopathy. A relative contraindication for this procedure is seen in the patient with multiple levels of compression where the painful level cannot be identified.

PAIN RELIEF

Although there are multiple theories as to why both vertebroplasty and kyphoplasty relieve pain, the precise reasons are still unknown. Theories proposed include the following: (*i*) the exothermic polymerization of PMMA may result in thermal damage of nociceptive nerve endings during cement hardening; (*ii*) the hardening of PMMA increases the structural stability of the vertebral body decreasing micromotion at the fracture sites; and (*iii*) the release of toxic substances (cement monomers) may destroy or denature nociceptor pain fibers leading to the reduction of pain. Additionally, the procedure kyphoplasty is also designed to relieve pain by reducing fracture alignment through vertebral height restoration and therefore restoring proper spinal biomechanics. Studies at present have demonstrated no correlation between PMMA volume injected and the degree of pain relief.

OUTCOMES

The outcomes following kyphoplasty are limited at this time due to the recent introduction of this new procedure to the general population. The results of vertebroplasty from North America and Europe to date have been extremely encouraging.

Vertebroplasty

Jensen et al. in 1997 presented 29 patients (47 compression fractures) who underwent vertebroplasty after failing conservative treatment (7). Twenty-six patients out of 29 (90%) had reduction

(A) (B) (C)

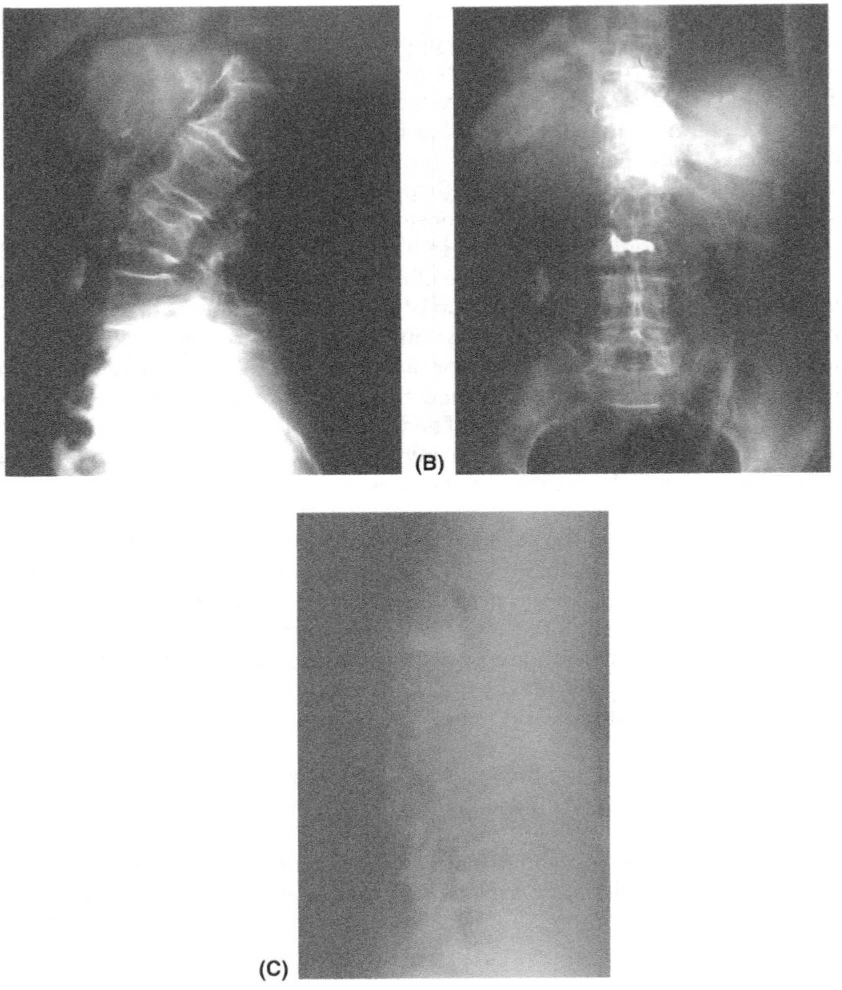

FIGURE 4 (**A**) Preoperative lateral plain radiograph showing an osteoporotic vertebral compression fracture at the T12 level. (**B, C**) Postoperative anteroposterior and lateral plain radiographs following kyphoplasty. Reinforcement of the vertebral body with polymethylmethacrylate with some restoration of vertebral height is noted.

of pain within 24 hours of the procedure. Barr et al. reported on 47 patients (84 compression fractures) who underwent vertebroplasty three weeks to three months after the onset of symptoms. Twenty-four of 47 (63%) patients had complete pain relief, 12 (32%) had moderate pain relief, and only 1 (5%) had no pain relief. The most contemporary study is from Evans et al. who reported a multicenter study that evaluated 467 cases of vertebroplasty with a follow-up three-part subjective scale system that included (*i*) pain relief (a 10-point scale), (*ii*) activities of daily living, and (*iii*) abnormal ambulation. Evaluation of preoperative to postoperative status showed a mean drop in the pain scale of 9 out of 10 to 3.2 out of 10. In regard to activities of daily living, only 9% of patients reported having no significant problem preoperatively, which increased to 67% postoperatively. Preoperatively 98% of patients had an abnormal gait-pattern ambulation compared to 28% postoperatively (8).

Kyphoplasty

Lieberman et al. recently reported a prospective analysis involving 30 consecutive patients (70 vertebral body compression fractures) who were managed with kyphoplasty (the inflatable bone tamp) [Lieberman IH, Dudeney S, Reinhart MK, Bell G. Initial outcome and efficacy of "kyphoplasty" in the treatment of painful osteoporotic compression fractures. Spine (manuscript submitted)]. Twenty-four patients had an osteoporotic compression fracture and six had a compression fracture secondary to multiple myeloma. All had failed nonoperative treatment ranging from 1 to 13 months. Seventy percent of the fractures showed some or total vertebral height restoration postoperatively measured on lateral plain radiography. The average height restoration was reported as 35% (range 0–100%) measured by the equation: (height regained/ height lost) × 100. The height lost was estimated by measuring the above adjacent uncompressed vertebrae for the prefracture height minus the height of the preoperative collapsed vertebral fracture height. In patients in whom vertebral height could be manipulated (21 patients), a 47% increase in vertebral height was observed. The patients showed a significant improvement in six of eight short form 36 scales including bodily pain, physical function, role physical, vitality, mental health, and social functioning. General health and role emotional scales did not show significant improvement. Complications included PMMA extravasation at six levels (8.6%), which entered the epidural space in one case, the disc space in two cases, and the paraspinal tissues in three cases with no long-term sequelae. Lane et al. and Yuan et al. reported a retrospective multicenter study involving 30 consecutive kyphoplasty procedures (52 fractures in 26 patients) (9,10). The compression fractures were between 2 and 16 weeks old. Twenty-three of 25 (92%) patients reported significant pain relief within hours following the procedure and all have remained independently mobile at 3- to 10-month follow-up. The average restoration of vertebral body height was 45% in the anterior vertebral body (10 patients), 71% in the middle of the vertebral body (13 patients), and 54% in the posterior vertebral body (8 patients). The average reduction in kyphosis was 17 degrees (6 patients) and no complications were reported. The authors concluded that kyphoplasty can be safely used for the treatment of vertebral compression fractures, with significant reduction in pain, disability, and deformity. At present, an ongoing industry-sponsored study has gathered data on 1121 compression fractures in 696 patients. Ninety percent of the patients in this study have experienced significant pain relief and 90% of patients were able to return to their normal activities of daily living. A complication rate of less than 2% was found and none were related to the balloon used in the procedure.

SUMMARY

Painful vertebral compression fractures are alarmingly common and are associated with a significant morbidity to both the patient and society. Prevention is the key in light of the new increased awareness of this systemic disorder. This has accelerated development of various medications including the bisphosphonates and the SERMs. Based on early clinical results, the minimally invasive percutaneous techniques, including vertebroplasty and kyphoplasty, have both shown excellent pain relief and improved function. Each procedure serves an important role in the spectrum of painful osteoporotic compression fractures. The results of early studies have demonstrated that kyphoplasty is a well-tolerated procedure indicated for the surgical treatment of painful osteoporotic vertebral compression fractures with a documented low complication rate. The findings of low rates of PMMA (cement) extravasation seen with kyphoplasty, compared with published rates for vertebroplasty, support the theory of injection of higher viscosity cement under low pressure into a created void cavity. This appears to be safer in terms of cement extravasation than the injection of cement under higher pressure into an unreduced cancellous portion of the vertebral body as that which occurs with vertebroplasty (11). Furthermore, the reduction of the vertebral body height via the inflatable bone tamp, with restoration of sagittal alignment and proper spinal biomechanics, may protect adjacent vulnerable vertebral levels from fracturing through minimization of excessive forces at these levels.

Current and future research regarding these procedures will hopefully answer certain questions as to the utility of a uniportal versus a biportal approach as well as the proposal of future bone resorbing and replaceable cements, which cure in an isothermic setting. Recent reports have shown that a uniportal approach may be just as efficacious as a biportal approach with less-attendant morbidity. The future use of bioactive corals, cements, and ceramics along with the delivery of growth factors will hopefully lessen the small but potential adverse consequence of the long-term sequelae of PMMA within a vertebral body. Additionally, the use of a biabsorbable balloon that allows the injection of cement within itself, followed by immediate resorption of its outer surface to allow cement interdigitation will further decrease the potential morbidity of cement extravasation. The development of intravertebral cavitating implants will allow the placement of intravertebral implants that in and of themselves will function to stabilize the vertebral segment as well as to function as a carrier for biologically active compounds.

The future of kyphoplasty may prove to be successful in the surgical treatment of elderly patients suffering from acute or subacute osteoporotic vertebral body compression fractures in order to enable early mobilization and prevent or decrease the high morbidity and mortality rates experienced in this group of patients. Long-term clinical studies are ongoing and are needed to determine the efficacy and outcome of this revolutionary procedure. At present, it appears to have enormous potential and promise.

REFERENCES

1. Cooper C, Atkinson EJ, O'Fallon WM, Melton LJ III. Incidence of clinically diagnosed vertebral fractures: a population-based study in Rochester Minnesota, 1985–1989. J Bone Mineral Res 1992; 7:221–227.
2. Riggs BL, Melton LJ III. The worldwide problem of osteoporosis: insights afforded by epidemiology. Bone 1995; 17:505–511.
3. Cooper C, Kinson EJ, Atkinson J, Jacobsen SJ. Population based study of survival after osteoporotic fractures. Am J Epidemiol 1993; 137:1001–1005.
4. Leech JA, Dulberg C, Kellie S, Pattie L, Gray J. Relationship of lung function to severity of osteoporosis in women. Am Rev Respir Dis 1990; 141:68–71.
5. Ettinger B, Black DM, Mitlak BH, et al. Reduction of vertebral fracture risk in postmenopausal women with osteoporosis treated with raloxifene: results from a 3-year randomized clinical trial. JAMA 1999; 282:637–645.
6. Black DM, Cummings SR, Karpf DB, et al. Randomized trial of effect of alendronate on risk of fracture in women with existing vertebral fractures. Lancet 1996; 348:1535–1541.
7. Jensen ME, Evans AJ, Mathis JM, Kallmes DF, Cloft HJ, Dion JE. Percutaneous polymethylmethacralate vertebroplasty in the treatment of osteoporotic vertebral body compression fractures: technical aspects. Am J Neurorad 1997; 18:1897–1904.
8. Evans AJ, Jensen ME, Negin GA, et al. Pain reduction and improvement in functional mobility after vertebroplasty for compression fractures. Presented at the 15th Annual Meeting of the North American Spine Society, New Orleans, Louisiana, Oct 25–28, 2000:60.
9. Lane JM, FpGirardi, Khan SN, et al. Preliminary outcomes of the first 226 consecutive kyphoplasties for the fixation of painful osteoporotic vertebral compression fractures. J Bone Mineral Res 2000; 15(suppl 1):198.
10. Yuan HA, Garfin SR, Lieberman IH, Reiley MA, Yonemura K. Early clinical outcomes with kyphoplasty, the minimally-invasive reduction and fixation of painful osteoporotic vertebral body compression fracture (VCF). Presented at the 16th Annual Meeting Section on Disorders of the Spine and Peripheral Nerves, Palm Springs, California, Feb 1 23–26, 2000:66.
11. Cotton A, Dewatre F, Cortet B, et al. Percutaneous vertebroplasty of osteolytic metastases and myeloma: effects of the percentage of lesion filling and the leakage of methylmethacralate at clinical follow-up. Radiology 1996; 200:525–530.

25 Percutaneous Acetabuloplasty

Xavier Demondion and Anne Cotten
Service de Radiologie Ostéoarticulaire, Hôpital R. Salengro, Lille, France

INTRODUCTION

Patients with malignant osteolytic lesions of the acetabular roof often experience severe pain and functional disability. Surgical procedures are the treatment of choice for such lesions but may be contraindicated in patients in poor health, with extensive local disease, or who have multiple lesions (1,2). Radiation therapy gives partial or complete pain relief in more than 90% of the patients. Unfortunately, radiotherapy results in minimal and delayed bone strengthening, which does not allow patients with lytic lesions of the acetabular roof to ambulate. Moreover, bone reconstruction is sometimes preceded by transitory osteoporosis, which further increases the risk of pathological fracture. The technique of percutaneous vertebroplasty (3) has been adapted to osteolytic metastases and myeloma of the acetabular roof to provide pain relief as well as bone strengthening (4,5). This procedure consists of the injection, under fluoroscopic guidance, of methyl methacrylate into lytic lesions of the acetabulum.

INDICATIONS

This technique is a palliative procedure and consequently should be offered only to patients who are unable to tolerate surgery. The major indication is osteolytic metastases or myelomatous lesions that involve the weight-bearing part of the acetabulum (the acetabular roof), are painful, produce disability, or threaten acetabular strength. However, the bone destruction must be not too extensive because of the subphysiologic buttressing effect of cement. This procedure may be performed prior to radiation therapy, which complements its ameliorative action on pain, following radiation therapy that failed to relieve pain, or after local recurrence of treated disease.

PROCEDURE

Radiography and computed tomography (CT) scan should be performed a few days prior the procedure to assess the precise location, extent of cortical destruction, and involvement of the articular surface. Cement injection is performed under conscious sedation with the use of additional local anesthesia, because acute worsening of pain may occur during cement injection. A posterior, posterolateral, or lateral approach is chosen. The choice is dependent on the relative involvement of the posterior part of the acetabular roof while ensuring that the needle is perpendicular to the cortical bone to facilitate insertion. Under (biplane) fluoroscopic guidance, a 10-gauge needle is inserted into the lesion. Then methyl methacrylate polymer, with the consistency of paste, is injected through the needle until resistance is met (Fig. 1). This initial injection allows an injection of 4 to 8 mL (average, 7 mL) of bone cement (4,5). The incomplete distribution of bone cement may necessitate one or more additional focal injections (Fig. 2). Injection is stopped if leak of bone cement into the joint space or soft tissue occurs. The procedure takes one to two hours (mean, 1.5 hour) to perform. CT scanning is performed following cement injection to more precisely assess the extent of lesion filling and to better depict subtle soft-tissue leakage.

CLINICAL RESULTS

Pain relief is the main clinical result of this procedure and should be experienced by three hours to four days (mean, 24 hours) in most patients (87.5% of our cases) (4,5). Rapid improvement in

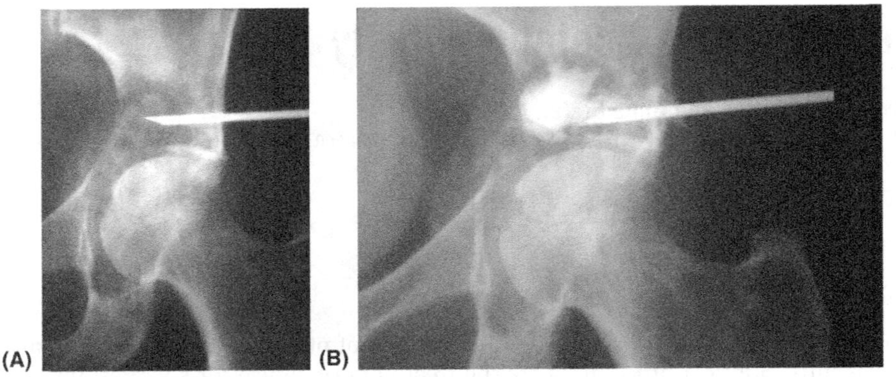

FIGURE 1 Acetabular osteoplasty in a patient with metastasis. Bone cement fills the acetabular roof. However, there is a risk of acetabular protrusion due to the presence of an osteolysis of the quadrilateral surface, which cannot be filled by the cement.

ambulation (within one to four days) is the secondary advantage of this technique. This improvement was seen in all our patients and even those who were previously bedridden (4). It is very likely that pain relief provided by cement injection plays a fundamental role in improvement in walking, but filling in of lytic lesions by cement (although usually incomplete) also plays a role in better distributing mechanical forces. It is important to stress that improved walking and pain relief do not seem directly proportional to the quality of lesion filling. In some cases, despite what appeared to be quite insufficient filling of the lesion, excellent clinical and functional results were obtained (4). The best results occur in patients with more limited osteolysis.

Because of the rapid clinical response (pain and ambulation) to cementoplasty, the hospitalization time is short (mean, four days), especially important in patients with a limited life expectancy. This procedure has to be complemented by radiation therapy, which further prevents tumor growth, decreases pain, and aids walking while preserving the mechanical integrity of the bone cement. At short-term and mid-term follow-up, clinical improvement seems to be maintained unless the lytic lesion progresses or extends further into bone or soft tissue. It may be difficult to differentiate clinical improvement related to cementoplasty from that caused by radiation therapy.

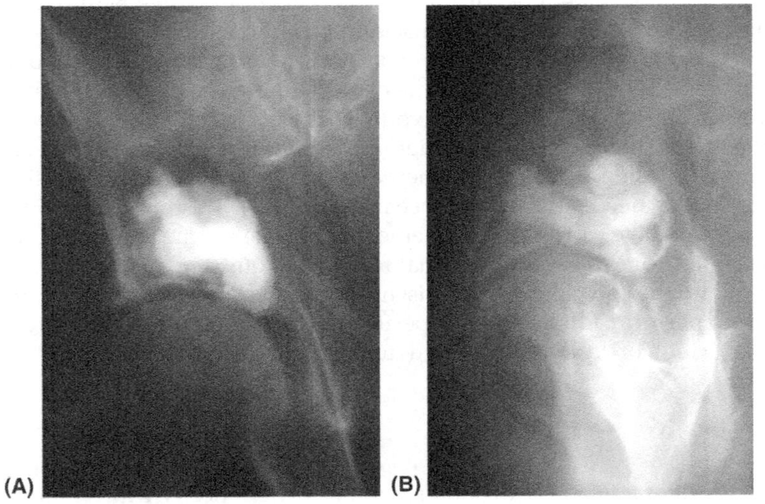

FIGURE 2 Patient with myeloma. Satisfactory filling of the acetabular osteolysis on the anteroposterior (**A**) and lateral (**B**) views.

SIDE EFFECTS AND COMPLICATIONS

Side effects are usually transitory. They consist of worsening of pain or fever as a systemic inflammatory reaction to the heat of acrylic polymerization. To minimize these side effects, nonsteroidal anti-inflammatory drugs may be administrated for two to four days following the procedure. We also have observed one case of postprocedural renal insufficiency in a patient with myeloma, possibly related to his primary disease.

The principal risk to periarticular percutaneous procedures is leak of bone cement into the joint space. Upon fluoroscopic detection of such a leak, the injection should be stopped immediately with the hip joint mobilized to flatten the articular cement prior to complete solidification. In our experience, this leak was associated with a striking transitory upsurge in pain, which, surprisingly, did not prevent eventual pain relief and improvement in walking (4). However, a case of rapid chondrolysis of the hip, appearing after the intra-articular leakage of cement and requiring hip replacement, has been reported (6). This event suggests a direct chondrolytic effect of the acrylic cement.

The principal complication at follow-up is the traumatic acetabular protrusion resulting from acetabular fossa tumoral osteolysis despite radiation therapy.

CONTRAINDICATIONS

This interventional procedure is contraindicated in patients with coagulation disorders because of the large diameter of the needle. Cortical destruction of the acetabular roof increases the risk of intra-articular leak of bone cement, but is not an absolute contraindication of the technique. However, acetabular fossa osteolysis potentiating acetabular protrusion may limit the benefits of cementoplasty.

ADVANCED ETHANOL INJECTION

When osteolysis does not involve the weight-bearing part of the acetabulum, or when the acetabular bone destruction is very extensive, adequate bone strengthening cannot be obtained and the sole treatment goal is pain relief. Therefore, ethanol may be used in place of methyl methacrylate (5,7) . Concomitant ethanol and methylmethacrylate injections may be performed if both weight-bearing and nonweight-bearing parts of the acetabulum are involved or extensive soft-tissue involvement is present.

CONCLUSION

Percutaneous osteoplasty is an attractive interventional procedure allowing rapid and marked pain relief and walking improvement in most cases. However, this a palliative procedure that should be offered only to patients who cannot undergo surgery, and in association with radiotherapy.

REFERENCES

1. Harrington KD. The management of acetabular insufficiency secondary to metastatic malignant disease. J Bone Joint Surg Am 1981; 63:653–664.
2. Murray JA, Parrish FF. Surgical management of secondary neoplastic fractures about the hip. Orthop Clin North Am 1974; 5:887–901.
3. Deramond H, Darrasson R, Galibert P. Percutaneous vertebroplasty with acrylic cement in the treatment of aggressive spinal angiomas. Rachis 1989; 1:143–153.
4. Cotten A, Deprez X, Migaud H, Chabanne B, Duquesnoy B, Chastanet P. Malignant acetabular osteolyses: percutaneous injection of acrylic bone cement. Radiology 1995; 197:307–310.
5. Cotten A, Demondion X, Boutry N, et al. Therapeutic percutaneous injections in the treatment of malignant acetabular osteolyses. RadioGraphics 1999; 19:647–653.
6. Leclair A, Gangi A, Lacaze F, et al. Rapid chondrolysis after an intra-articular leak of bone cement in treatment of a benign acetabular subchondral cyst: an unusual complication of percutaneous injection of acrylic cement. Skeletal Radiol 2000; 29(5):275–278.
7. Gangi A, Dietemann JL, Schultz A, Mortazavi R, Jeung MY, Roy C. Interventional radiologic procedures with CT guidance in cancer pain management. Radiographics 1996; 16:1289–1304.

26 Percutaneous Radiofrequency Ablation of Osteoid Osteoma

Martin Torriani and Daniel I. Rosenthal
Division of Musculoskeletal Radiology, Department of Radiology, Massachusetts General Hospital, Boston, Massachusetts, U.S.A.

INTRODUCTION

Osteoid osteoma, a benign but painful bone tumor, may occur over a wide age range and in almost any bone, but is usually found in the lower extremities of children and young adults. The tumors are small, rarely exceeding 1.5 cm in diameter, appear to have little or no growth potential, and do not undergo malignant degeneration. Treatment is performed for pain control.

Some authorities advocate medical management of these patients (1), but surgical treatment is still widely practiced. Generous surgical margins increase the probability of cure, but resection of the weight-bearing bone can lead to a prolonged period of recovery and necessitate internal fixation and/or bone grafting.

During the past 10 years, we have performed almost 300 radiofrequency (RF) treatments for osteoid osteoma. We believe that RF ablation should be the primary treatment modality for those lesions for which it is suitable.

PATIENT SELECTION

Osteoid osteoma may arise in cortical bone, on the periosteal surface of bone, or less often in the bone marrow. All locations can be successfully treated by RF. The tumor may contain a variable amount of internal ossification, but in most cases at least the perimeter of the tumor is lucent. The lesions are often elongated in the direction of the long axis of the bone. Ninety percent of lesions are between 3 and 10 mm in longest diameter.

Thin-section computed tomography (CT) (less than 3.0 mm thick) is the preferred method for the demonstration of osteoid osteoma. Radioisotope bone scans consistently detect these lesions. The literature indicates that osteoid osteoma may occasionally fail to show uptake, but we have not seen such an example.

Diagnosis based upon imaging and clinical features is highly accurate; however, several different entities can be mistaken for osteoid osteoma, including benign cartilage tumors, hemangioma, eosinophilic granuloma, stress fractures, herniation pits (on the anterior surface of the femoral neck), abscess, osteoblastoma, and (extremely rarely) intracortical osteosarcoma.

We usually do not treat lesions of the spine or hands. Convalescence following operative treatment of tumors in these locations is shorter than that which follows lower extremity lesions. Most such tumors are in close proximity to nerves that control important functions, and therefore the benefits do not appear to justify the risks in these locations.

Tumors adjacent to joints can be treated. Although the articular cartilage is at risk for a focal injury, none of our patients has had joint symptoms following the procedure. Bleeding has not been a problem with this technique, despite the fact that most patients take significant amounts of nonsteroidal anti-inflammatory medications and presumably have prolonged bleeding times.

TECHNIQUE

We use general anesthesia for these procedures. Local anesthesia with or without conscious sedation is adequate for most bone biopsies. However, needle puncture of an osteoid osteoma is much more painful than biopsy of other lesions. When the needle enters the tumor, sudden marked tachycardia and tachypnea often occur, even under general anesthesia. The anesthetist should be aware of this phenomenon to prevent patient movement. An intravenous nonsteroidal anti-inflammatory drug such as ketorulac given approximately 20 minutes before completion helps with postprocedure pain.

A grounding plate is attached to the skin opposite to the site of entry. Tumor localization is performed as for a bone biopsy except that the scan thickness should not exceed 3 mm because of the small size of these lesions. Once the lesion has been demonstrated, the radiation exposure should be minimized by decreasing milliamperage (mA) to the lowest level that is adequate.

We use a drill to penetrate dense cortical bone (Bonopty Penetration Set, RADI Medical System AB, Uppsala, Sweden). This device produces a hole that admits a 16G (1.6 mm) biopsy needle. It is also possible to use a trephine needle or other bone biopsy tool. In either case, it is necessary to pause at frequent intervals while drilling through bone to clear the trephine or the threads of the drill. If this is not done, it becomes increasingly difficult to advance; indeed the biopsy device may become wedged into the bone, making it very difficult to withdraw.

For the relatively common lesions of the medial upper femur, the patient can be placed in the "frog-lateral" position once anesthetized (Fig. 1). When there is no safe direct access to the lesion a "transosseous" approach can be used, drilling through the entire thickness of the bone from the contralateral side (Fig. 2). When the drill comes into close proximity with the tumor (1 mm), but has not entered it, it is exchanged for the biopsy needle.

We most often use the Ostycut Bone Biopsy Needle (C. R. Bard Inc., Covington, Georgia, U.S.A.), because the sharp stylet exhibits little tendency to slip when used on the surface of the bone, and because it will fit through the cannula of the Bonopty drill. The external screw threads are helpful in advancing through the lesion, and it yields a biopsy sample of good quality. A 1.6 mm (16 G) needle is appropriate for the 1 mm outer diameter of the electrode.

We generally do not request "frozen section" analysis because tissue preservation is not optimal by this method, and the histological diagnosis rarely alters the treatment. If infection is a serious diagnostic possibility, a frozen section may be helpful, as thermocoagulation could potentially worsen the condition by producing a sequestrum.

After the biopsy is taken, the RF electrode is introduced and CT imaging is repeated to confirm the placement. We use electrodes with either 5 or 8 mm exposed tip. The electrode is connected to the RF generator (RFG-3C, Radionics, Burlington, Massachusetts, U.S.A.), and the tip temperature is slowly (1° every one to two seconds) raised to 90°C. The temperature is maintained

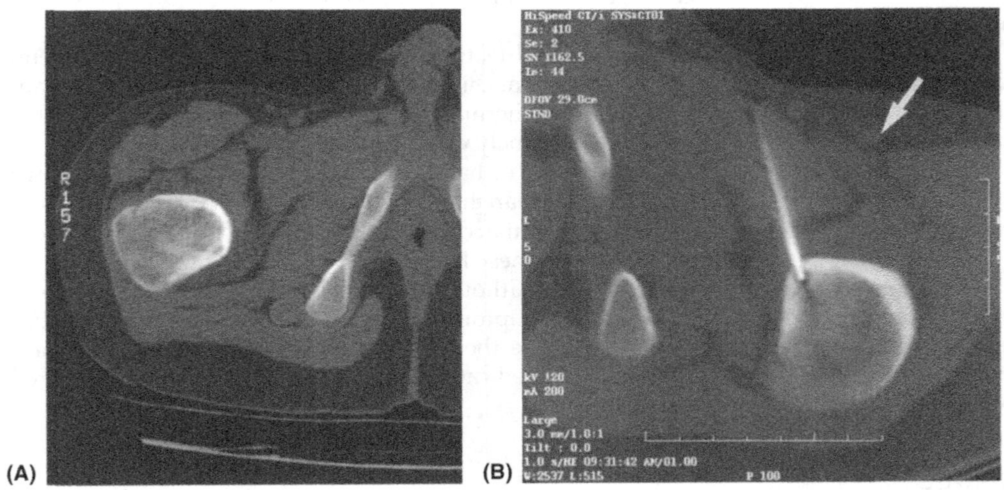

FIGURE 1 (**A**) Computed tomography scan of the left hip shows an osteoid osteoma in the posteromedial cortex of the proximal femur. (**B**) Images obtained with the patient in the "frog-lateral" position. This rotates the lesion anteriorly, allowing it to be safely approached medial to the femoral triangle (*arrow*).

(A)

(B)

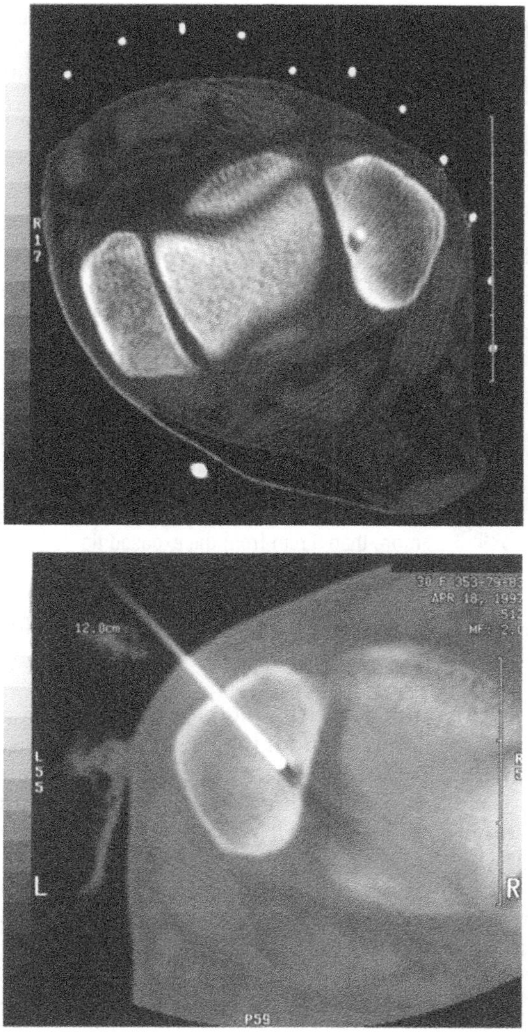

FIGURE 2 (**A**) Computed tomography scan through the lateral malleolus demonstrates a typical lytic lesion involving the media aspect of the fibula, adjacent to the talus. (**B**) An anterolateral transosseous approach has been used, to spare the ankle mortise.

for six minutes, after which the generator is turned off, and the procedure is complete. These parameters (90°C, six minutes) were selected because they consistently result in a spherical (5 mm electrode) or cylindrical (8 mm electrode) lesion with a radius of 5 to 6 mm. Therefore, the electrode placement must be such that no portion of the tumor is more than 5 to 6 mm away from its exposed tip (Fig. 3). If the tumor is large, or if electrode placement is off-center, more than one treatment must be performed to ensure that the entire tumor is treated (Fig. 4).

It is apparent that a minimum of 6 mm separation between the electrode and any vital structures is required to avoid tissue damage. For safety sake, we generally will not perform this procedure when the distance is less than 1 cm. The thermal effects do not reliably respect anatomical boundaries such as bone cortex.

In our experience, intralesional administration of anesthetics and postprocedure imaging is not helpful.

MANAGEMENT AFTER THE PROCEDURE

Pain in the recovery room is very variable. In those individuals who have pain, it is most intense almost immediately upon awakening. It may require narcotics for control, but once brought under control, it improves steadily and rapidly. Patients requiring narcotics in the recovery

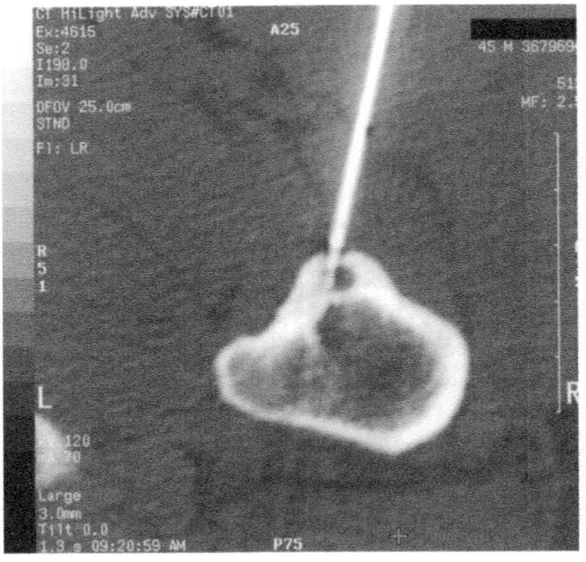

FIGURE 3 Computed tomography scan at the level of the lesser trochanter in a 45-year-old man. The electrode is seen within an osteoid osteoma of the periosteum. Although the electrode is positioned eccentrically, no portion of the lesion is more than 5 mm from the exposed tip.

room are discharged with a prescription for enough oral narcotics to last 24 hours, but seldom need more than one or two doses. Occasionally, a patient will complain of moderate-to-severe pain lasting for several days after the procedure.

The patient remains in the recovery room only until awake enough for discharge (one to two hours). The patient should be able to walk immediately without the use of casts, braces, or crutches, and should resume all daily activities within 24 hours. Physical therapy is not required. Teenaged and adult patients are advised to avoid prolonged running and/or jumping for three months. This is not required for younger children who are unlikely to be involved in serious training programs.

Approximately 50% of patients state that the tumor pain is gone within 24 hours. The remainder improves over the next several days. Symptoms are almost always gone within a week. Subsequent follow-up is minimal and performed at the discretion of the referring clinician.

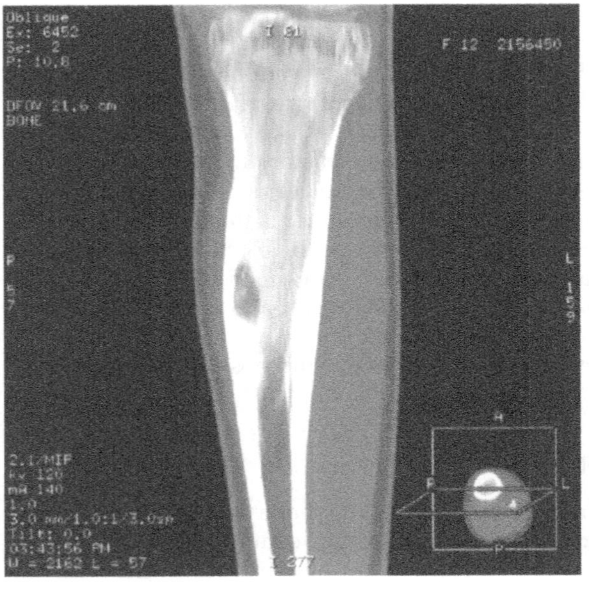

FIGURE 4 Coronal reformation of axial computed tomography (CT) images of the left tibia in a 12-year-old female demonstrates an elongated tumor in the longitudinal direction. Lesions of this shape frequently must be treated in more than one location to ensure complete coverage. Axial CT images (not shown) revealed a typical lytic lesion having a rounded cross section and showing internal ossification.

The radiographic appearance of the tumor changes very slowly after treatment. Within the first month, little or no change is apparent on radiographs. The X-ray examination may become normal after a few months in younger children, but it is not unusual for an asymptomatic density to persist at the tumor site for a year or more in adults.

OUTCOMES

In our experience, approximately 85% of patients have complete and permanent success from a single procedure (patient has no symptoms and takes no medications two years later). About 7% to 8% of patients undergo a second procedure because of recurrence, and another 7% to 8% describe some continuing symptoms, not significant enough to merit treatment. Recurrences after a year without symptoms are rare, and after two years exceedingly rare.

The probability of a diagnostic needle biopsy at the time of the procedure is about 70%. A nondiagnostic result is ignored if the patient is free of symptoms. Most recurrences are seen between three and six months following treatment.

Several published studies of osteoid osteoma treatment by RF ablation have shown 88% to 95% cure rates with follow-up periods ranging between 2 and 41 months (2–6). In all cases, a second RF treatment cured the recurrence (2,3,6,7). This success rate is similar to that observed following operative treatment (7).

COMPLICATIONS

We have seen no permanent procedure-related complications. A few patients have had somewhat prolonged postprocedural pain of unknown significance. It resolved spontaneously after approximately one week. In one case, an apparent cellulitis was treated with oral antibiotics and resolved. In another case, dusky blue skin discoloration developed approximately two weeks after treatment. It extended from the tibial treatment site distally to involve the entire leg and foot, and was present only when the leg was dependent. It was not accompanied by signs of skin atrophy and resolved spontaneously after several more weeks. Neither fractures nor major or delayed complications have been reported (2,3,6,7).

COMPARISON OF TECHNIQUES

We believe that CT-guided RF treatment is the most broadly applicable of the available minimally interventional methods. Excision, even with percutaneously placed hollow needles and drills, results in structural weakening and the risk of fracture (8). Large lesions exceed the size of most trephines, and necessitate several passes, adding to the hazards of the procedure.

The distribution of thermal necrosis with RF is more predictable than ethanol injection (9,10). Finally, interstitial laser photocoagulation through an optical fiber (11–13) yields similar results. However, the necessary equipment is more costly, and the purported advantage of using a fine optical fiber can only be accomplished by sacrificing the opportunity for biopsy.

REFERENCES

1. Kneisl JS, Simon MA. Medical management compared with operative treatment for osteoid-osteoma. J Bone Joint Surg Am 1992; 74:179–185.
2. Lindner NJ, Ozaki T, Roedl R, Gosheger G, Winkelmann W, Wortler K. Percutaneous radiofrequency ablation in osteoid osteoma. J Bone Joint Surg Br 2001; 83:391–396.
3. Woertler K, Vestring T, Boettner F, Winkelmann W, Heindel W, Lindner N. Osteoid osteoma: CT-guided percutaneous radiofrequency ablation and follow-up in 47 patients. J Vasc Interv Radiol 2001; 12: 717–722.
4. Barei DP, Moreau G, Scarborough MT, Neel MD. Percutaneous radiofrequency ablation of osteoid osteoma. Clin Orthop 2000; 373:115–124.
5. Rosenthal DI, Alexander A, Rosenberg AE, Springfield D. Ablation of osteoid osteomas with a percutaneously placed electrode: a new procedure. Radiology 1992; 183:29–33.
6. de Berg JC, Pattynama PM, Obermann WR, Bode PJ, Vielvoye GJ, Taminiau AH. Percutaneous computed-tomography-guided thermocoagulation for osteoid osteomas. Lancet 1995; 346:350–351.

7. Rosenthal DI, Hornicek FJ, Wolfe MW, Jennings LC, Gebhardt MC, Mankin HJ. Percutaneous radiofrequency coagulation of osteoid osteoma compared with operative treatment. J Bone Joint Surg Am 1998; 80:815–821.
8. Sans N, Galy-Fourcade D, Assoun J, et al. Osteoid osteoma: CT-guided percutaneous resection and follow-up in 38 patients. Radiology 1999; 212:687–692.
9. Adam G, Neuerburg J, Vorwerk D, Forst J, Gunther RW. Percutaneous treatment of osteoid osteomas: combination of drill biopsy and subsequent ethanol injection. Semin Musculoskelet Radiol 1997; 1:281–284.
10. Sanhaji L, Gharbaoui IS, Hassani RE, Chakir N, Jiddane M, Boukhrissi N. A new treatment of osteoid osteoma: percutaneous sclerosis with ethanol under scanner guidance. J Radiol 1996; 77:37–40.
11. Gangi A, Dietemann JL, Gasser B, et al. Interstitial laser photocoagulation of osteoid osteomas with use of CT guidance. Radiology 1997; 203:843–848.
12. Gangi A, Dietemann JL, Guth S, et al. Percutaneous laser photocoagulation of spinal osteoid osteomas under CT guidance. Am J Neuroradiol 1998; 19:1955–1958.
13. Gangi A, Dietemann JL, Gasser B, et al. Interventional radiology with laser in bone and joint. Radiol Clin North Am 1998; 36:547–557.

27 Interstitial Laser Photocoagulation of Osteoid Osteoma

Afshin Gangi, Lisa Wong, Jean Pierre Imbert, Xavier Buy, and Stephane Guth
Department of Radiology B, University Hospital of Strasbourg, Strasbourg, France

INTRODUCTION

Osteoid osteoma is a benign neoplasm of bone. Osteoid osteoma occurs more often in men. The age range is from 2 to 50 years, but 90% occur before the age of 25 (1,2). Osteoid osteoma produces local pain that is worse at night and improves dramatically with aspirin. The presence of nerve fibers within the tumor nidus and the stimulatory role of prostaglandins are thought to produce the pain. The characteristic findings of this tumor in clinical and radiologic examinations can lead to a high level of diagnostic confidence in many instances. The treatment of this tumor is achieved with the complete removal of the nidus. Conventional treatment is surgical or percutaneous excision of the nidus. Less invasive techniques such as computed tomography (CT)–guided core drill excision (3–9), radio-frequency ablation (10–17), alcohol injection (18), or interstitial laser photocoagulation (ILP) (19–22) have been used. ILP was a technique first described in 1983 (23). In vitro and experimental animal studies showed the production of a predictable size of tissue coagulation by the use of bare optical fibers with a low-power (typically 2–4 W) laser technique. A thin optical fiber is inserted percutaneously into the tumor, which is then coagulated and destroyed by direct heating. The ability to control the treated area with a high degree of precision, applicability in joints, and an excellent dose–response characteristic makes ILP a valuable treatment method for osteoid osteomas. This technique has been applied clinically to treat tumors of liver, pancreas, prostate, brain, breast, and lymph nodes (24–30).

THE PRINCIPLE OF INTERSTITIAL LASER PHOTOCOAGULATION OF OSTEOID OSTEOMAS

ILP consists of percutaneous insertion of optical fibers into the tumor. The tumor is coagulated and destroyed by direct heating. With a low-power laser technique, a well-defined coagulation of predictable size and shape can be obtained in the bone tissue. The extent of thermal necrosis has been shown to be dependent on the laser wavelength, the energy deposited, the power used (tip temperature), the thermal and optical properties of the target tissue, and the type of optical fiber used (23,31–33). Compared with the 1604-nm neodymium yttrium aluminum garnet (Nd: YAG) laser, the 805-nm diode laser was able to produce larger diameters of coagulation and charring at both higher and lower energies (31).

Experimental histopathologic examinations have evaluated the mean diameter of coagulation produced by 805-nm diode laser using a 400-μm polymer–clad fiber with a constant power of 2 W. The mean diameter of coagulation varies—3.5 mm with 200 J, 5 mm with 400 J, 6 mm with 600 J, 7.5 mm with 800 J, and 9 mm at 1000 J in femurs of pigs. The thermal data were significantly higher, with lesions of 16 mm in diameter for 1200 J (34). These results are consistent with those demonstrating that the transmission of heat within bone is sharply limited by blood flow, and that lethal temperatures cannot be sustained over great distances (33,35).

We speculated that cellular damage that would be too subtle to detect on histologic examination done immediately after injury could occur. This experimental work has shown that a reproducible area of coagulative necrosis is obtained around the fiber, with good correlation between the energy delivered and the lesion size, and with conservation of the biomechanical properties of the bone tissue in the treated area. The size of osteoid osteomas falls within the range that can effectively be coagulated by one or two fibers. Because the nidus of the osteoid osteoma can be precisely identified on CT scanning, we have performed ILP under CT guidance for treatment of this tumor in our institution since June 1993.

Advantage of Interstitial Laser Photocoagulation Compared to Surgical or Percutaneous Excision

- There is no weakening of bone structures through surgical removal. This is particularly important in weight-bearing bones and in children; there is no need for immobilization or osteosynthesis, and no limitation of activity.
- The technique is applicable in joints.
- It does not require extensive hospitalization. It is an ambulatory procedure.
- It does not require general anesthesia. The ILP can easily be performed under local anesthesia and neuroleptanalgesia in adults (general anesthesia is needed for children).
- It involves minimal recovery time.
- It does not leave any scar.
- It is a cost-effective procedure.

INDICATIONS AND CONTRAINDICATIONS

Patient selection is crucial for treatment effectiveness.

The indications are as follows:

- Osteoid osteomas determined by CT scan or magnetic resonance imaging and scintigraphy with positive and consistent clinical findings

The contraindications are as follows:

- Hemorrhagic diathesis
- Lesions too close to neurological structures

MATERIALS

- Laser—diode laser (Diomed, Cambridge, U.K.) for ILP
- 400-μm, precharred optical fiber, Y connector
- A drilling device—2-mm–diameter hand drill, 14-gauge Bonopty Penetration set (RADI Medical Systems Uppsala, Sweden), or 14-gauge bone Ostycut biopsy needle (Ostycut, Angiomed/Bard, Karlsruhe, Germany)

TECHNIQUE

A CT scan is performed to localize the tumor precisely. CT is used to measure the diameter of the nidus. The largest diameter of the nidus determines the energy that will be necessary to coagulate the tumor. For diameters larger than 10 mm, we use usually two fibers to ensure tumor destruction. The entry point and the pathway are determined by CT, avoiding nervous, vascular, and visceral structures. The penetration of the needle into the nidus is always extremely painful; therefore, ILP is performed under general anesthesia or regional block. The procedure is performed under strictly sterile conditions. The 18-gauge needle is guided safely under CT guidance. Fluoroscopy is used in conjunction with CT whenever drilling is necessary.

The tip of an 18-gauge needle must be placed into the central part of the nidus. Sometimes, bone drilling is required to reach the nidus, depending on perilesional hyperostosis (Fig. 1).

Subperiostal nidi or cortical nidi without major ossification are directly punctured with an 18-gauge spinal needle (Becton Dickinson, Rutherford, New Jersey, U.S.A.). In cases with mild ossification or small cortex surrounding the lesion, a 14-gauge bone biopsy needle is more adequate (Ostycut, Angiomed/Bard, Karlsruhe, Germany). In cases of dense ossification, or of dense cortical bone surrounding the lesion, drilling is necessary.

In these cases, we use a 2-mm diameter hand drill or a 14-gauge Bonopty Penetration set (Radi Medical Systems Uppsala, Sweden) to allow insertion of the 18-gauge needle.

The 18-gauge needle tip is inserted into the center of the nidus. Before the optical fiber is placed, it is inserted in an 18-gauge needle mounted by a sidearm fitting to measure the

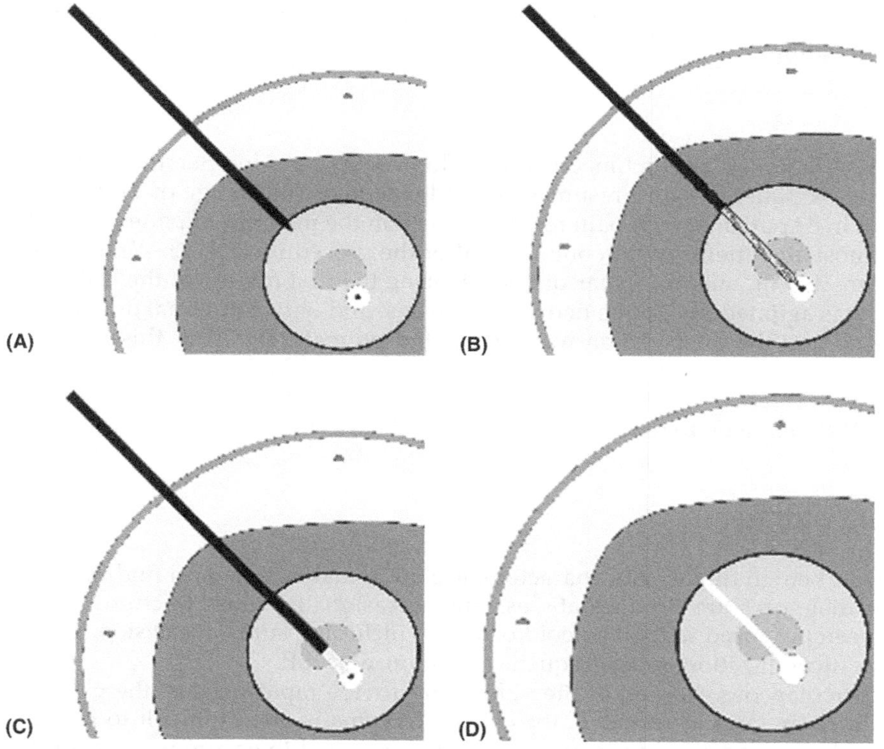

FIGURE 1 Diagram representing the technique of interstitial laser photocoagulation under computed tomography (CT) guidance. (**A**) Percutaneous trocar needle (14 gauge) placement under CT guidance. (**B**) Perforation of cortical bone with a coaxial drill. (**C**) Coaxial insertion of the 18-gauge needle with the optical fiber in the center of the nidus. (**D**) End of the procedure.

appropriate length of the fiber. The 400-μm precharred fiber is then inserted through the needle; the needle is withdrawn about 5 mm so that the tip of the bare fiber lies within the center of the tumor.

The diode laser (805 nm) is turned on in continuous wave mode, at a power of 2 W for 200 to 600 seconds, depending on the nidus size (energy delivered 400–1200 J). CT control scans are performed during the procedure to detect vaporization gas.

COMPLICATIONS

Complications of ILP are very rare. Possible complications of ILP are as follows:

- The major complication of ILP is septic osteitis. To avoid this complication, severe sterility during the intervention is mandatory.
- Hematoma.
- Reflex sympathetic dystrophy.
- Recurrence of osteoid osteoma.

In all 84 cases, no severe complications were encountered during the procedure. In one patient with an osteoid osteoma within the lunate bone, there was mild reflex sympathetic dystrophy of the wrist occurring one week after the procedure, and this consisted of burning pain, hyperalgesia, hyperesthesia, and vasomotor disturbances. A short course of high-dose prednisone and a beta-blocking agent was given in addition to physical therapy. There is, however, no correlation between the type and severity of the injury with the occurrence or course of this entity. There is evidence that regional analgesia, including sympathetic blockade, during the intervention provides pain relief and improves circulation of the injured part and helps prevent reflex sympathetic dystrophy (36,37). The patient's symptoms were completely relieved after two months.

RESULTS

From 1993 to 2001, CT-guided or combined CT- and fluoroscopic-guided percutaneous ILP were performed on 84 patients with presumed osteoid osteomas (mean age of 21.9 years). ILP was successful in 83 patients, with pain relief observed in the majority of patients within one day and in almost all patients within one week after the procedure (Fig. 2). We had only one unsuccessful treatment, with a 16-year-old boy. During the first minute of the ILP treatment, the patient was agitated due to the neuroleptic drugs, and only a minimal dose of 300 J could be delivered and the intervention was terminated prematurely. After this incident, only general anesthesia or regional block was used. Six patients had recurrence of pain, occurring from 6 weeks to 27 months after the first treatment. All these patients had a successful second ILP treatment to the residual or recurrent nidus and remained pain-free on recent follow-up.

CONCLUSION

Osteoid osteoma is a benign tumor with characteristic clinical and radiological findings. The major differential diagnoses are Brodie's abscess and, occasionally, stress fractures. Many techniques have been proposed to treat osteoid osteoma, including surgical excision, percutaneous extraction, alcoholization, radio-frequency ablation, and ILP.

Surgical or percutaneous excision of the nidus can provide rapid relief of the patient's symptoms (3–9). In open surgical resection, the osteoid osteoma is often difficult to identify intraoperatively, and an excessive amount of bone is often resected to ensure removal of the nidus. This weakens the cortical bone that is important for weight bearing, increases the risk of fracture, and imposes limitation of activity for up to three months after surgery in these patients. In addition, osteoid osteoma situated in a deep-seated, intra-articular or epiphyseal site may require wide excision or arthrotomy for its excision. Compared to open surgical resection, percutaneous extraction of nidus under CT guidance allows precise localization of tumor by CT with removal of less amount of bone and, hence, offers less risk than open surgery. However,

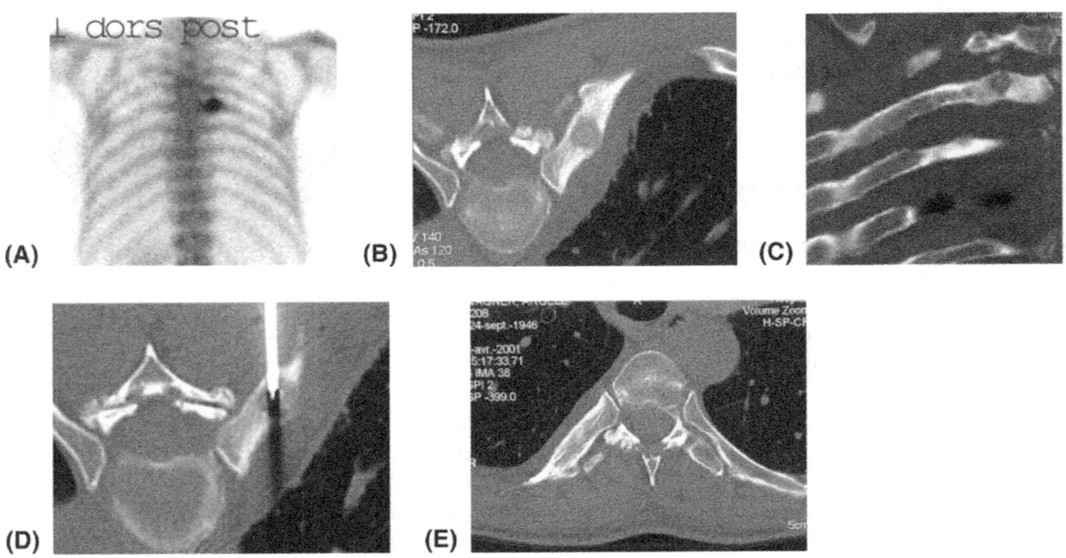

FIGURE 2 Osteoid osteoma of the posterior end of the rib. (**A**) Bone scan with hyperfixation of the rib. (**B,C**) The nidus is localized with computed tomography (CT) scan. Note the subpleural thickening. (**D**) Procedure performed under general anesthesia under CT guidance. (**E**) Control CT scan one year after the procedure.

(A) (B)
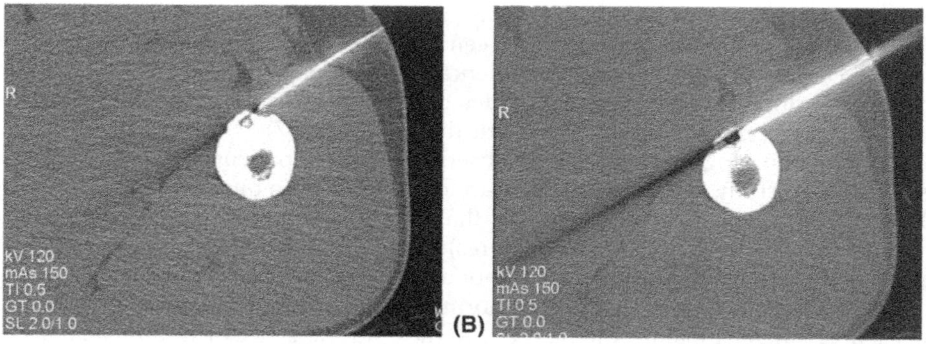

FIGURE 3 Osteoid osteoma of the femoral bone. (**A**) Typical subperiostal nidus. (**B**) Interstitial laser photocoagulation under computed tomography control.

the trephine needle used is often large, ranging from 7 to 10 mm internal diameter (8,9). The large size of the instrument may also incur risk of neurologic and vascular injury in some anatomic regions, particularly in children. Although smaller diameter drills and needles (3–4 mm) have been used, the small size of the core increases the risk of incomplete nidus removal and recurrences, and multiple passes are often required to complete the resection of the nidus (3–5). Although immediate weight bearing is possible when small drill or needle has been used (5), weight bearing is not allowed for four to six weeks to prevent risk of fracture when large-size needle or drill is used (8,9). Reported complications with percutaneous resection include skin burns at the entry site of instrument, soft-tissue hematoma, dysesthesia, superficial and deep infection including osteomyelitis, and fracture (6,8,9).

CT-guided percutaneous drilling with subsequent alcoholization has also been used (18), but this technique does not have precise control over the size and morphology of the tissue damage. In this respect, thermal ablation by radio-frequency ablation or ILP has the advantage of precise control over the size of the tissue damage achieved, with an excellent dose–response characteristic (Fig. 3) (11,34,38).

Since June 1993, ILP has been performed on patients with presumed osteoid osteomas in our institution. ILP was only performed in cases where the clinical and radiological features typical of osteoid osteomas existed in the patients. Because this tumor is completely benign, with no case of reported malignant transformation (1), and the size of the tumor is within the range of thermal coagulation by ILP, it is amenable to curable treatment by ILP. ILP has proven to be effective in 83 of 84 patients, with pain relief occurring within one week of the procedure in most patients. Although we had six recurrences in our series, with an average follow-up of 47.1 months (range 4–97 months), these recurrences were successfully treated with a second ILP. In four of the recurrences, the largest diameter of the nidus ranged from 10 to 24 mm, and the large tumor size may account for the recurrence (12). Two recurrences had occurred at six weeks and at four months after the initial ILP, and these were due to imprecise needle positioning within the tumor nidus leading to incomplete tumor coagulation. The other four recurrences occurred 12 to 27 months from the initial ILP. Two of these recurrences (children under 10-years-old), in fact, had previous surgical recurrences despite histological demonstrations of complete resection margins, and both were located within the tibial shafts. We postulate that these cases of recurrences with a pain-free interval of 12 months or more after the initial treatment may have triggered the regeneration of a new tumor at the same site. This may explain the two cases of repeated recurrences, one occurring after complete surgical resection and one occurring after ILP treatment, in similar locations.

Three out of the six recurrences occurred in an intra-articular location, two within the hip joint, and one within the zygoapophyseal joint. This is of a higher-than-expected proportion in view of the 32% of intra-articular tumors in our series. It seems that the intra-articular location may be predisposed to the recurrence, but the underlying mechanism is unclear. All these recurrences happened in patients under 16 years of age.

Reflex sympathetic dystrophy was the only major complication occurring in one of our patients. There is, however, no correlation between the type and severity of the injury and the occurrence or course of this entity. There is evidence that regional analgesia, including sympathetic blockade, during the intervention provides pain relief and improves circulation of the injured part and helps prevent reflex sympathetic dystrophy (36,37).

There was no neurological complication observed in all our patients. All 10 spinal osteoid osteomas were successfully treated, with no neurological deficit. In our previous publications, we had limited ILP treatment to tumor nidus that occurred at least 8 mm away from vital tissues (e.g., neurological and tendinous structures) to avoid tissue damage (39,40). This limitation, however, had been overcome with the use of epidural normal saline cool bath technique. This technique consisted of slow infusion of normal saline at room temperature at a rate of 70 mL/hr via a 22-gauge spinal needle placed into the epidural space or foramina adjacent to the tumor (Fig. 4). The maximum coagulation time is 10 minutes. With this technique, we were able to successfully treat four cases of spinal tumor nidus that were located less than 8 mm from adjacent neurological structures without any neurological complications or tumor recurrence (follow-up, 24–48 months).

There are many advantages in the ILP treatment of osteoid osteoma. The fiber used for delivering the laser energy is flexible and thin (400 μm), and this can be put into a small caliber needle (18 gauge) with minimal alteration to the existing structure (39–43). The thin fiber also

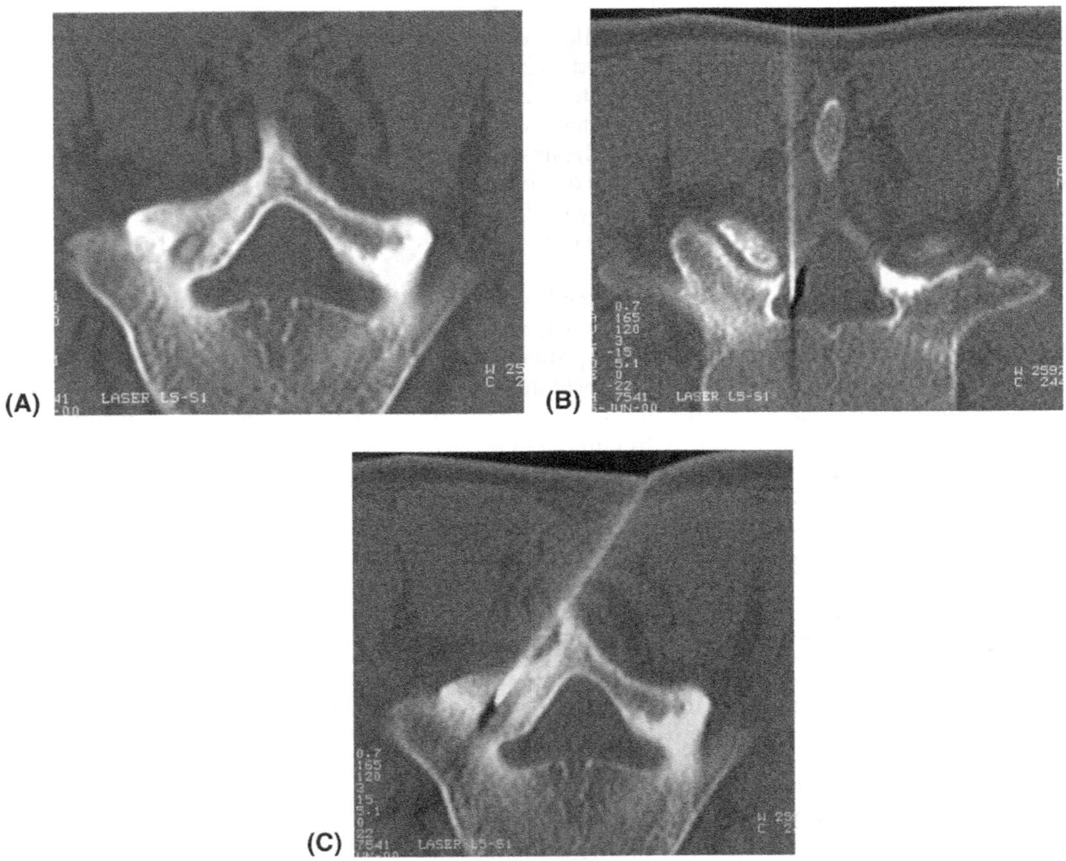

FIGURE 4 Osteoid osteoma of the spine with use of epidural normal saline cool bath technique. (**A**) Localization of the nidus. (**B**) A 22-gauge spinal needle placed into the epidural space close to the tumor. In difficult cases, a thermocouple could be used with the same technique for thermal monitoring to avoid any accidental coagulation. (**C**) Percutaneous interstitial laser photocoagulation of the tumor.

allows precise placement within tumors and is particularly useful in small, deep-seated osteoid osteomas. Because the fiber is a single-use fiber, there is reduced risk of contamination. The semiconductor laser is simple to use and does not require watercooling or three-phased electrical power. It is small, compact, and portable. With low power (typically 2 W), the laser source can produce a predictable size of necrosis in proportion to the energy delivered, which is much more precise than alcoholization. Most lesions are treated effectively by use of one fiber. For lesions larger than 15 mm in diameter, up to four fibers can be placed in the nidus and simultaneously fired with a 1 × 4–coupler fiber splitter for complete ablation. Nd:YAG or diode laser is usually available in the majority of medium- to large-sized hospitals, because it is also used by specialists of other disciplines. The cost of the procedure will be further reduced if the laser machine can be shared among specialists in the hospital. This technique is less expensive than surgical intervention. Most patients either are treated on an outpatient basis or require one night of hospitalization with ILP. Overall, this is more cost effective when compared to the costs incurred with several days of hospitalization usually required for surgical intervention. ILP has less associated morbidity compared to surgical resection or percutaneous resection, with maximum work interruption of two weeks and usually return to normal activities in one week for most of our patients.

The drawbacks of this technique are lack of routine histological verification and the use of quite expensive material. When biopsy was performed in surgical resection or radiofrequency ablation, the confirmation of osteoid osteoma is obtained in 57% to 79% of cases (5,6,9,11). We do not routinely obtain histological confirmation, because the small 18-gauge needle introduced does not allow sufficient material to be obtained to confirm the diagnosis. However, with a 14-gauge needle or in atypical cases, when the size of the nidus exceeds 2 cm, histological confirmation was obtained.

Thermal ablation of osteoid osteoma by percutaneously placed radio-frequency electrodes has also been described (10–17). It is a safe and effective treatment, with the failure rate of single radio-frequency treatment being 7.5% (13), and this is comparable to that of ILP in our series. This method has also been used for treatment of osteoid osteoma in the spine (15–17).

ILP is safe and effective for the treatment of osteoid osteomas (19–22,31). The small size of the needle required for accessing the tumor, no or short hospital stay, a quick postprocedural recovery, and the lack of significant complications are major advantages over open or percutaneous surgical resection. Recurrences can be treated easily and effectively with repeat ILP. We propose that all osteoid osteomas should be treated by thermal ablation techniques, radio-frequency ablation, or ILP, rather than surgical or percutaneous resection. Surgical resection should only be performed in the very rare cases of percutaneously inaccessible osteoid osteomas.

REFERENCES

1. Mirra JM, Picci P, Gold RH. Osseous tumors of intramedullary origin: osteoid osteoma. In: Mirra JM, Picci P, Gold RH, eds. Bone Tumors: Clinical, Radiologic and Pathologic Correlations. Philadelphia: Lea and Febiger, 1989:226–248.
2. Resnick D, Greenway GD. Tumors and tumor-like lesions of Bone. In: Resnick D, 2d ed. Bone and Joint Imaging. Philadelphia: Saunders, 1996:991–998.
3. Doyle T, King K. Percutaneous removal of osteoid osteomas using CT control. Clin Radiol 1989; 40:514–517.
4. Voto SJ, Cook AJ, Arrington G, Weiner DS, Ewing JW. Treatment of osteoid osteoma by computed tomography-guided excision in a pediatric patient. J Pediatr Orthop 1990; 10:510–513.
5. Mazoyer JF, Kohler R, Bossard D. Osteoid osteoma: CT-guided percutaneous treatment. Radiology 1991; 181:269–271.
6. Assoun J, Railhac JJ, Bonnevialle P, et al. Osteoid osteoma: percutaneous resection with CT guidance. Radiology 1993; 188:541–547.
7. Graham HK, Laverick M, Cosgrove A, Crone M. Minimally invasive surgery for osteoid osteoma of the proximal femur. J Bone Joint Surg 1993; 75-B:115–118.
8. Parlier-Cuau C, Nizard R, Champsaur P, Bassam H, Laredo JD. Percutaneous resection of osteoid osteomas. Semin Musculoskeletal Radiol 1997; 1:257–264.
9. Sans N, Galy-Fourcade D, Assouon J, et al. Osteoid osteoma: CT-guided percutaneous resection and follow-up in 38 patients. Radiology 1999; 212:687–692.

10. Rosenthal DI, Alexander A, Rosenberg AE, Springfield D. Ablation of osteoid osteomas with a percutaneously placed electrode: a new procedure. Radiology 1992; 183:29–33.
11. Rosenthal DI, Springfield DS, Gebhardt MC, Rosenberg AE, Mankin HJ. Osteoid osteoma: percutaneous radio-frequency ablation. Radiology 1995; 197:451–454.
12. De Berg JC, Pattynama PM, Obermann WR, Bode PJ, Vielvoye GJ, Taminiau AH. Percutaneous computed-tomography-guided thermocoagulation for osteoid osteomas. Lancet 1995; 346: 3590–3591.
13. Rosenthal DI. Percutaneous radiofrequency treatment of osteoid osteomas. Semin Musculoskeletal Radiol 1997; 1:265–272.
14. Rosenthal DI, Hornicek FJ, Torriani M, Gebhardt MC, Mankin HJ. Osteoid osteoma: percutaneous treatment with radiofrequency energy. Radiology 2003; 229:171–175.
15. Cové JA, Taminiau AH, Obermann WR, Vanderschueren GM. Osteoid osteoma of the spine treated with percutaneous computed tomography-guided thermocoagulation. Spine 2000; 25:1283–1286.
16. Osti OL, Sebben R. High-frequency radio-wave ablation of osteoid osteoma in the lumbar spine. Eur Spine J 1998; 7:422–425.
17. Blanco Sequeiros R, Hyvönen P, Blanco Sequeiros A, et al. MR imaging-guided laser ablation of osteoid osteomas with use of optical instrument guidance at 0.23 T. Eur Radiol 2003; 13: 2209–2314.
18. Adam G, Neuerburg J, Vorwerk D, Forst J, Gunther R. Percutaneous treatment of osteoid osteomas: combination of drill biopsy and subsequent ethanol injection. Semin Musculoskeletal Radiol 1997; 1:281–284.
19. Gangi A, Dietemann JL, Gasser B, et al. Interstitial laser photocoagulation of osteoid osteomas with use of CT guidance. Radiology 1997; 203:843–848.
20. Witt JD, Hall-Craggs MA, Ripley P, Cobb JP, Bown SG. Interstitial laser photocoagulation for the treatment of osteoid osteoma. Results of a prospective study. J Bone Joint Surg (Br) 2000; 82-B:1125–1128.
21. DeFriend DE, Smith SP, Hughes PM. Percutaneous laser photocoagulation of osteoid osteomas under CT guidance. Clin Radiol 2003; 58:222–226.
22. Sequeiros RB, Hyvonen P, Sequeiros AB, et al. MR imaging-guided laser ablation of osteoid osteomas with use of optical instrument guidance at 0.23 T. Eur Radiol 2003; 13:2309–2314.
23. Bown SG. Phototherapy of tumors. World J Surg 1983; 7:700–709.
24. Nolsoe CP, Torp-Petersen S, Burcharth F, et al. Interstitial hyperthermia of colorectal liver metastases with a US-guided Nd:YAG laser with a diffuser tip: a pilot clinical study. Radiology 1993; 187:333–373.
25. Amin Z, Donald JJ, Masters A, et al. Hepatic metastases: interstitial laser photocoagulation with real-time US monitoring and dynamic CT evaluation of treatment. Radiology 1993; 187:339–347.
26. Dachman AH, Smith MJ, Burris JA, VanDeMerwe W. Interstitial laser ablation in experimental models and clinical use. Semin Intervent Radiol 1993; 10:101–112.
27. Steger AC, Lees WR, Shorvon P, Walmsley K, Bown SG. Multiple-fibre lowpower interstitial laser hyperthermia: studies in the normal liver. Br J Surg 1992; 79:139–145.
28. Roux FX, Merienne L, Fallet-Bianco C, et al. Stereotaxic laser interstitial thermotherapy. A new alternative in the therapeutic management of some brain tumors. Neurochirurgie 1992; 38: 238–244.
29. Castro DJ, Lutkin RB, Saxton RE, et al. Metastatic head and neck malignancy treated using MRI guided interstitial laser phototherapy: an initial case report. Laryngoscope 1992; 102:26–32.
30. Goldstein NS. Laser therapy for small breast cancers. Am J Surg 2004; 187:149–150.
31. Amin Z, Buonaccorsi GA, Mills TN, Harries SA, Lees WR, Brown SG. Lower power interstitial laser photocoagulation in rat liver: importance of fiber type, laser wavelength, and tissue charring. Proc SPIE 1993; 1882:172–183.
32. Sherk HH. The use of lasers in orthopedic procedures. J Bone Joint Surg (Am) 1993; 75:768–776.
33. Lundskog J. Heat and bone tissue: an experimental investigation of the thermal properties of bone and threshold levels for thermal injury. Scand J Plast Reconstr Surg 1972; 9:1–80.
34. Gangi A, Gasser B, De Unamuno S, et al. New trends in interstitial laser photocoagulation of bones. Semin Musculoskeletal Radiol 1997; 1:331–337.
35. Tillotson CL, Rosenberg AE, Rosenthal DI. Controlled thermal injury of bone: report of a percutaneous technique using radiofrequency electrode and generator. Invest Radiol 1989; 24:888–892.
36. Bonica JJ. Causalgia and other reflex sympathetic dystrophies. In: Bonica JJ, ed. The Management of Pain. 2d ed. Philadelphia, PA: Lea & Febiger, 1990:220–243.
37. Ladd AL, De Haven KE, Thanik J, Patt RB, Feuerstein M. Reflex sympathetic imbalance: response to epidural blockade. Am J Sports Med 1989; 17:660–668.
38. Amin Z, Bown SG, Lees WR. Local treatment of colorectal liver metastases: a comparison of interstitial laser photocoagulation and percutaneous alcohol injection. Clin Radiol 1993; 48:166–171.
39. Gangi A, Dietemann JL, Gasser B, et al. Interventional radiology with laser in bone and joint. Radiol Clin North Am 1998; 36:547–557.

40. Gangi A, Dietemann JL, Guth S, et al. Percutaneous laser photocoagulation of spinal osteoid osteomas under CT guidance. AJNR 1998; 19:1955–1958.
41. DeFriend DE, Smith SP, Hughes PM. Percutaneous laser photocoagulation of osteoid osteomas under CT guidance. Clin Radiol 2003; 58:222–226.
42. Witt JD, Hall-Craggs MA, Ripley P, Cobb JP, Bown SG. Interstitial laser photocoagulation for the treatment of osteoid osteoma. J Bone Joint Surg Br 2000; 82:1125–1128.
43. Cool P, Williams DH, Pullicino V. Interstitial laser photocoagulation for the treatment of osteoid osteoma. J Bone Joint Surg Br 2001; 83:462.

28 Percutaneous Alcohol Injection for the Treatment of Metastatic Disease

Anne Cotten
Service de Radiologie Ostéoarticulaire, Hôpital R. Salengro, Lille, France

Luc Ceugnart
Service de Radiologie, Centre Oscar Lambret, Lille, France

INTRODUCTION

Patients with osteolytic metastases usually experience severe pain and functional disability. Surgery is usually not undertaken in most of these patients because of the multifocal nature of their disease. Moreover, surgery can be dangerous in debilitated patients, related to underlying disease or prior treatment. Therefore, the clinical benefit of surgery must be carefully weighed against the financial, social, and emotional costs and complications in patients with a fair life expectancy.

Radiation therapy, which may prevent tumor growth, usually results in partial or complete pain relief after the completion of the treatment. Most patients begin to experience some response 10 to 14 days after the start of the therapy. Unfortunately, some patients may demonstrate insufficient pain relief or local tumoral recurrence after the initial radiation therapy. Moreover, this treatment results in only minimal and often delayed (two to four months after the start of irradiation) bone strengthening, which may not allow patients with extensive lytic lesions of the weight-bearing part of bone (acetabulum and vertebra) to stand.

Percutaneous injection of ethanol is a palliative procedure that may be helpful in patients with metastatic disease because it may provide early and frequently striking pain relief, especially when the initial pain is considerable. Ethanol is an efficient sclerosing agent leading to a marked reduction in tumor vascularity and to tumor destruction. In contrast to cementoplasty, this procedure provides no or minimal bone strengthening.

INDICATIONS

Ethanol injection is indicated in patients with malignant osteolytic lesions and who are unable to tolerate surgery. This palliative procedure may be performed prior to or after complementary radiation or following local recurrence. It also has been suggested that ethanol injection can have value in treating bone metastases from thyroid carcinoma that fails to respond to radioiodine (1).

Because ethanol injection results in minimal and delayed bone strengthening, pain relief is the sole object of the procedure. Consequently, ethanol injection is indicated when osteolysis does not involve the weight-bearing part of a bone. Indeed, owing to the mechanical properties of bone cement, which hardens as polymerization occurs, injection of methylmethacrylate is preferred when osteolysis involves the weight-bearing part of a bone. If bone destruction is extensive, ethanol injection or methylmethacrylate injections may be performed together at both weight-bearing and nonweight-bearing parts of a bone.

Ethanol injection also results in tumor volume reduction, especially when extensive soft-tissue involvement is present and repeat ethanol injections have been peformed. Therefore, associated soft-tissue involvement is another excellent indication for ethanol injection.

TECHNIQUE

Radiography and computed tomography (CT) scan must be performed prior to alcohol injection to assess the location and extent of the lytic process, the presence of cortical destruction or fracture, and the presence of soft tissue involvement. Alcohol injection is performed under strict

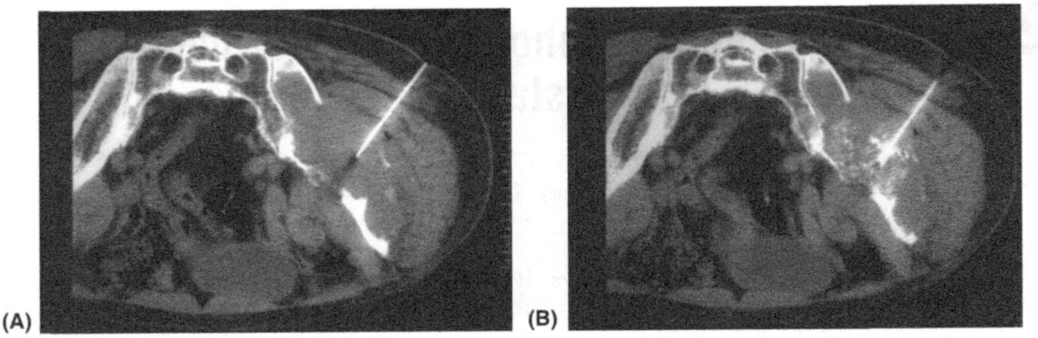

FIGURE 1 (A and B) Contrast injection into a metastasis of the ilium. Despite cortical osteolysis, no leak of contrast into the adjacent soft tissue is detected. Ethanol injection can be performed with caution.

sterile conditions in patients under conscious sedation or general anesthesia, because severe pain may occur during ethanol injection.

CT guidance is usually performed for both needle placement and injection monitoring to allow precise assessment of the bone and soft-tissue involvement (especially for the pelvic bones, the posterior arch of a vertebra, or the anterior part of a vertebral body with involvement of the adjacent soft tissue), but fluoroscopic guidance can be used in the appendicular skeleton, especially for femoral lesions (2,3). Ultrasound guidance has also been reported for skull and sternal metastases (1,4).

Percutaneous ethanol injection is technically easy to perform. A thin (20–22-gauge) needle is first inserted into the lesion, and a mixture of lidocaine (1%) and contrast material is injected first to assess the expected distribution of ethanol and to decrease the pain produced by the ethanol injection (Figs. 1 and 2). If leakage of contrast media is detected, especially into the joint space, the needle is repositioned and a second test injection is performed. If no contrast material is visualized, the needle has been positioned intravascularly and needs to be repositioned.

Next, a solution of 95% ethanol is injected slowly. Injection volume depends on lesion size and the diffusion of the contrast material. A volume of 1 to 4mL is usually sufficient for osteolytic lesions, but in patients with extensive soft-tissue involvement, the volume used can reach 20 to 30 mL. One or more injections may be performed at different parts of the lesion, or multiple lesions may be injected on the same day. Also, repeat injections performed over several weeks may further improve pain relief and decrease tumor size.

CONTRAINDICATIONS

There is no absolute contraindication for percutaneous ethanol injection. In the case of fracture or destruction of the articular cortex, the use of extreme caution during the procedure to prevent leakage into the adjacent joint space or soft tissue is strongly recommended. The thin needle

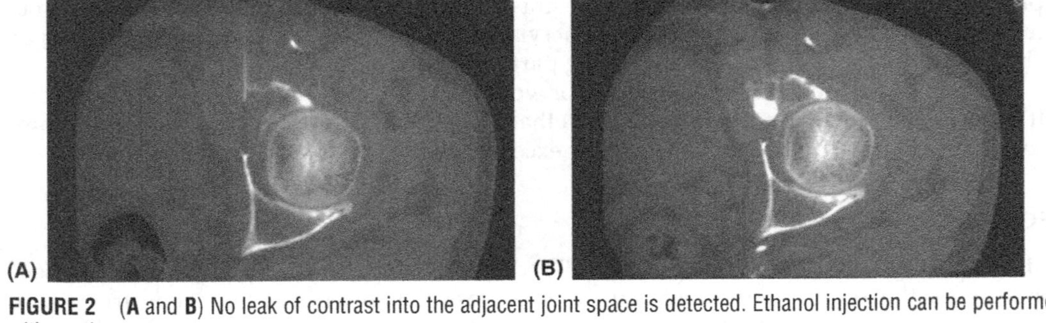

FIGURE 2 (A and B) No leak of contrast into the adjacent joint space is detected. Ethanol injection can be performed with caution.

used for the injection allows injections to be performed, albeit cautiously, even in patients with coagulation disorders.

SIDE EFFECTS AND COMPLICATIONS

A transitory worsening of pain may occur secondary to inflammatory reaction in the hours following injection (25% of cases in our experience) (2,4,5). This pain typically resolves spontaneously within two days following the procedure. To minimize this side effect, postprocedure nonsteroidal anti-inflammatory drugs may be administered. Few other complications of this procedure have been described in the literature. Skin necrosis requiring reconstruction has been reported, as well as gastrointestinal bleeding two days following ethanol injection (5). Other potential complications include vascular injury, nervous injury, infection, and leaks of ethanol. Intra-articular leaks may produce potential chondrolysis, and intramuscular leaks may lead to muscle necrosis. Neuropathic pain may also occur secondary to a leak of ethanol adjacent to a nerve.

CLINICAL EFFECT

The principal effect of percutaneous injection of ethanol is early (within 6 to 48 hours) and frequently striking pain relief. In our experience, significant pain relief was obtained in 30% to 40% of the patients after a single alcohol injection. In patients with no or only moderate pain relief following the first injection, a marked or a moderate pain relief was obtained after a second ethanol injection in 10% and 40% of the cases, respectively. This ethanol-induced pain relief is likely related to tumor necrosis and destruction of adjacent nerve endings. Owing to the rapid onset of clinical improvement, hospitalization time is short. This is an important feature in patients with a short life expectancy. Repeat ethanol injections may be performed, increasing pain relief and in some cases reducing tumor volume, especially in extensive lesions (1–3).

CONCLUSION

Percutaneous injection of ethanol is a palliative procedure that may be helpful in the treatment of bone metastases, because this procedure may provide early and frequently striking pain relief, tumor necrosis, and, in some cases, reduction in size of the lesion. This procedure is easy to perform and will be safe if a test with contrast media and a slow injection of ethanol is performed. However, it now contends with other therapeutic procedures such as percutaneous radiofrequence ablation.

REFERENCES

1. Nakada K, Kasai K, Watanabe Y, et al. Treatment of radioiodine-negative bone metastasis from papillary thyroid carcinoma with percutaneous ethanol injection therapy. Ann Nucl Med 1996; 10:441–444.
2. Cotten A, Demondion X, Boutry N, et al. Therapeutic percutaneous injections in the treatment of malignant acetabular osteolyses. RadioGraphics 1999; 19:647–653.
3. Gangi A, Dietemann J.-L, Schultz A, Mortazavi R, Jeung MY, Roy C. Interventional radiologic procedures with CT guidance in pain management. RadioGraphics 1996; 16:1289–1304.
4. Isaka T, Yoshimine T, Fujimoto K, et al. Direct ethanol injection for skull metastasis from hepatocellular carcinoma. The techniques and consequences of a therapeutic trial. Neurol Res 1998; 20:737–741.
5. Uflacker R, Paolini RM, Nobrega M. Ablation of tumor and inflammatory tissue with absolute ethanol. Acta Radiol Diagn (Stockh) 1986; 27:131–138.

29 Radiofrequency Ablation of Osseous Metastatic Disease

Damian E. Dupuy
The Warren Alpert Medical School of Brown University and Rhode Island Hospital, Providence, Rhode Island, U.S.A.

In the past three decades, many investigators have applied thermally mediated ablative procedures for the treatment of musculoskeletal neoplasms. Much of this preliminary work has been performed in animal models and small, limited clinical studies. These studies have shown the techniques of cryosurgery and radiofrequency (RF) to have some utility in specific clinical situations and specific tumors. To date, the most widely utilized techniques are cryosurgery in the operative setting of malignant or locally recurrent neoplasms such as giant cell tumor and percutaneous RF ablation (RFA) of osteoid osteomas and osseous metastatic disease using computed tomographic (CT) guidance. This chapter will give a brief overview of the problem of osseous metastatic disease, discuss the principles of RFA, and explain how I have applied the new minimally invasive technique of percutaneous CT-guided RFA as a treatment option for this group of patients.

Metastatic cancer is the commonest neoplasm involving the skeletal system (1). Of approximately 965,000 new cancer cases each year in the United States, approximately 30% to 70% develop skeletal metastases. Breast, lung, and prostate cancers account for more than 80% of these cases of metastatic bone disease. Bone metastases can cause significant morbidity from pain, pathologic fracture, and neural compression. Metastatic bone pain can be related to both mechanical and chemical factors. Mechanical factors are related to the local mass effect on the periosteum or adjacent neural structures that in turn causes local or radiating pain. Chemical factors are a result of hemorrhage from local osteoclastic activity that leads to a local release of bradykinin, prostaglandins, histamine, and substance P, which can irritate the endosteal nerves as well as local nerves (2).

Traditionally, primary treatment for symptomatic bone metastases has relied upon radiation therapy (3) with or without systemic chemotherapy or hormonal therapy. Newer systemic treatments with radionuclides and bisphosphonates have also shown some success (4–7). Surgical therapy is performed in certain instances where mechanical strengthening is necessary, such as an impending fracture. Many of these conventional therapies are unsuccessful in pain reduction, and patients may require significant doses of narcotics. Radiation therapy works by killing the local tumor and inflammatory cells that are responsible for causing the pain. Many prospective trials have been performed by studying the ability of external beam radiation therapy to palliate pain or control progression of osseous metastatic disease (8–15). For example, the Radiation Therapy Oncology Group study measured the patient's response to radiation therapy with a pain scale and narcotic requirement scale of 1 to 4 (8). The study showed a complete response rate of 54% and a partial response rate of 90%. Ninety-six percent of the patients had at least minimal relief within the first four weeks but only 50% had complete relief in four weeks. Yet there was a 30% relapse rate within the patients who survived at least 12 weeks. Patients in the study with lung cancer or with severe constant pain at the outset tended not to improve after radiation. Madsen reported a response rate of only 48% when measuring a patient's pain using a visual analog scale (9). Overall response rates vary according to the length of follow-up, type of radiation treatment, and the measurement of patient response. However, as pointed out by a review of all published reports by Ratanatharathorn et al., the relapse after initial response is frequent, the pain relief in all studies is poor, and the practices of radiation therapy need to be improved (16).

The whole premise of treating osseous metastatic disease with radiation therapy is to cause cell death in the tumor as well as the local inflammatory cells. With this fact in mind, percutaneous ablative therapies such as ethanol injection (17), vertebroplasty (18), and RFA (19) have been applied to this group of patients. The life expectancy of patients with osseous metastatic disease is limited, with an average median survival of between three and six months. Therefore, effective local therapies that can be done at a single outpatient sitting may be beneficial.

RFA was pioneered in 1920 by Harvey Cushing to create small lesions within the central nervous system. Since then, the technique has been refined such that precise control of lesion size is achieved by measuring the local temperature and impedance within the treated tissue. RFA's ability to create localized necrotic lesions is also why it is well established as the treatment of choice for many symptomatic cardiac arrhythmias (20). Today RFA is commonly used in the musculoskeletal system for treatment of intractable back pain due to failed-back syndrome and chronic back pain due to facet joint osteoarthritis (21,22). CT-guided RFA, as was discussed by Dr. Rosenthal in Chapter 26, has been shown to be a cost-effective surgical alternative in the treatment of osteoid osteomas (23) and in many institutions has replaced surgery. The potential advantages of RF versus other ablative methods in the treatment of bone metastases are multifold: cell death is immediate, lesion size can be accurately controlled, lesion temperature can be monitored, electrode placement can be achieved with a percutaneous image-guided procedure, and the procedure can be performed under local anesthesia and conscious sedation.

RFA is a technique whereby an alternating electrical current operating in the frequency of radio waves (460–480 kHz) is emitted from the tip of an electrode or needle placed directly into tissues. The alternating current causes the local ions to vibrate, producing heat and inducing cell death by coagulative necrosis. The cytotoxic temperature threshold is 50°C, but with RFA techniques, temperatures can exceed this and actually reach the boiling point of water. Until recently, a major limitation of RFA was the small lesion size created by this technique. Within the past few years, technical advances in RF systems have improved such that heat lesions larger than 5 cm in diameter can be created with a single treatment (24). There are currently three RF systems that are being utilized for treating tumors. Two of the systems (RITA Medical Systems Inc., Mountain View, California, U.S.; Radiotherapeutics-Boston Scientific, Natick, Massachusetts, U.S.) utilize a deployable array RF electrode that consists of 4 to 16 small wires that are deployed through a 15-gauge needle. The third RF system (Radionics Inc., Burlington, Massachusetts, U.S.) utilizes either single nonperfused or single and triple "cluster" perfused electrodes. The nonperfused conventional thermistor electrodes are of approximately 18-gauge and have a small 5 mm exposed tip. These smaller electrodes can ablate tissue with a treatment diameter of approximately 10 to 16 mm and are used for the ablation of osteoid osteomas, nervous tissue, or small tumors. The internally cooled RF electrode can increase the volume of induced coagulation necrosis to 4 to 7 cm as reported in liver (25), but may be greater in soft-tissue malignancies, due to the diminished cooling effect of the regional blood flow. Perfusion of the RF electrode tip reduces the local impedance, thus allowing a greater radius of RF energy deposition. The perfused treatment electrodes have a diameter of 18-gauge, but a thin, electrically insulated material coating increases its effective diameter to approximately 17-gauge. The electrode has a 1 to 3 cm uninsulated tip. Each needle contains an RF electrode with a thermocouple embedded at the tip to measure temperature. Circuitry incorporated within the generator and needle permits continuous measurement of tissue impedance. Two internal lumens extend to each electrode tip. One is used to deliver chilled perfusion fluid (sterile saline or water) to the electrode tip, whereas the second returns the perfusion effluent to a collection unit located external to the patient. A peristaltic pump is used to infuse the perfusion fluid into the cooling lumen of the electrode at 80 mL/min. A grounding pad or pads is/are placed on the patient. If a single RF treatment probe is used, at least two grounding pads (96 cm^2 surface area each) or the equivalent (minimum total surface area of approximately 200 cm^2) must be used. In the event that a cluster electrode is used, at least four grounding pads or their equivalent (approximately 400 cm^2 surface area) must be used. The pads must be placed horizontally to maximize the grounding effect and prevent thermal injury to the skin. The grounding pad(s) and electrode are connected to the RF generator. When the RF generator is activated, current flows between the conductive electrode tip and the grounding pad(s) or "dispersive electrode." The increase in the tissue temperature is proportional to the current density. Because the current density is highest near the conductive electrode tip, coagulation is induced in the tissue surrounding the treatment probe. When performing an RF treatment, the coagulation depth is controlled from the length of the uninsulated "active" electrode tip. The diameter of coagulation necrosis produced around the tip of the treatment electrode depends on the current, duration of treatment, and local tissue blood flow.

RFA may be applied in patients who have local pain. Our early results from a pilot study (26) in which we treated 18-metastatic bone tumors in 16 patients indicate that RFA reduced pain in a heterogeneous group of patients, some of whom could not have additional radiation therapy. By the visual analogue pain scale of 0 (*no pain*) to 10 (*worst*), the mean of average pain intensity was 6.5 prior to RFA and 4.62 at one week (P=0.039) and 4.64, at one month (P=0.036) after the procedure. Similarly the mean of the worst pain described by the patients was 8.5 before the procedure, and was 6.93 (P=0.005) and 5.64 (P=0.003) at one week and one month, respectively. No incidences of bleeding or infection at the site of procedure were identified, and the only side effects were swelling, local pain, and postablation syndrome, whereby flu-like symptoms occur for several days after the treatment. Patients with larger tumors involving the pelvis and sacroiliac joints did not have significant pain relief (Fig. 1). However, patients with smaller tumors and tumors involving the chest wall had significant pain relief (Fig. 2). Callstrom et al. (27) published

(A) (B)

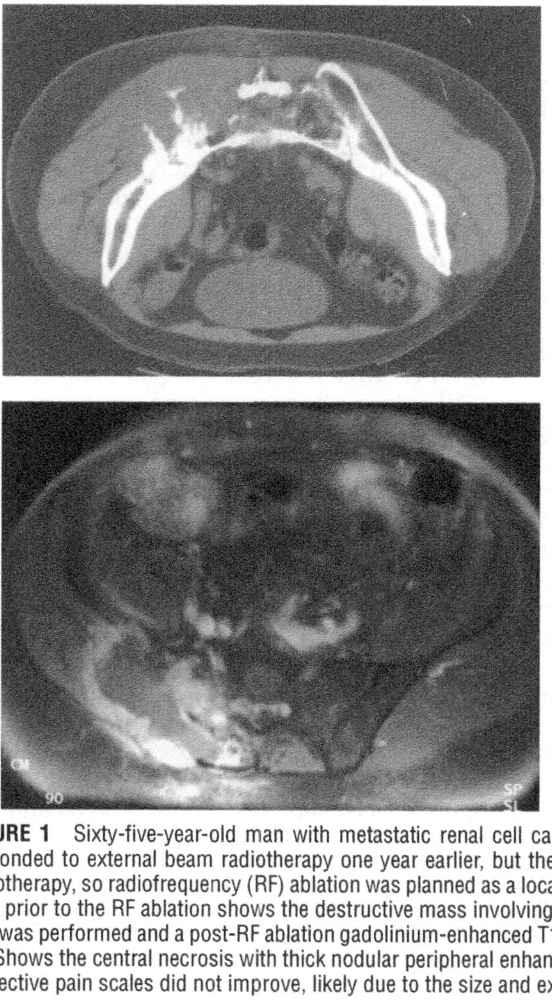

FIGURE 1 Sixty-five-year-old man with metastatic renal cell carcinoma to the right ilium. The patient had initially responded to external beam radiotherapy one year earlier, but then progressed. The patient could have no additional radiotherapy, so radiofrequency (RF) ablation was planned as a local treatment option. (**A**) Prone computed tomography (CT) prior to the RF ablation shows the destructive mass involving the posterior ilium and sacrum. CT-guided RF ablation was performed and a post-RF ablation gadolinium-enhanced T1-weighted axial magnetic resonance imaging image. (**B**) Shows the central necrosis with thick nodular peripheral enhancement consistent with residual tumor. The patient's subjective pain scales did not improve, likely due to the size and extensive involvement.

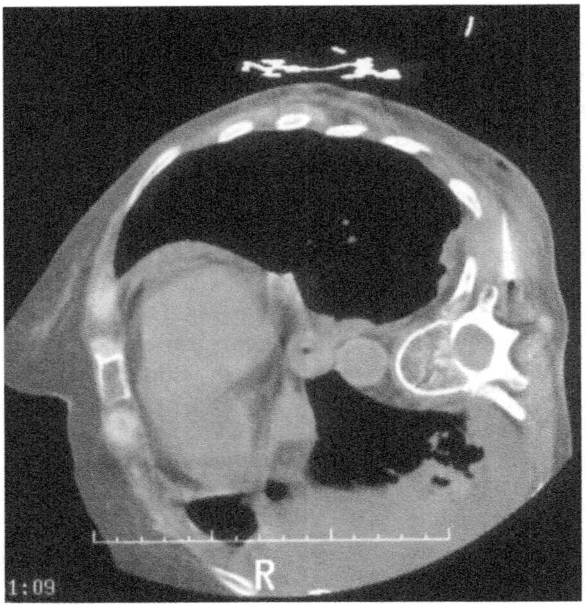

FIGURE 2 Seventy-three-year-old woman with metastatic esophageal carcinoma to the left posterior chest wall and adjacent ribs. With the patient in the right lateral decubitus position computed tomography–guided radiofrequency ablation was performed with placement of the electrode into the posterior soft tissues. The patient had symptomatic relief from the therapy.

a single-arm, paired comparison, observational study involving 12 patients with a single painful osteolytic metastasis; patients served as their own controls. Radiation therapy or chemotherapy had failed to provide pain relief. All treated lesions were osteolytic, with a combination of bone destruction and soft tissue mass. A single lesion was treated in all 12 patients. The size of the treated lesion ranged from 1 to 11 cm. One patient with a large lesion was treated in two separate sessions six weeks apart, and the remaining 11 patients were treated in a single RF session. Before RF ablation, the mean worst pain score in a 24-hour period in the 12 patients was 8.0 (range 6 to10). At four weeks post–RF treatment, the recorded mean worst pain had decreased to 3.1 ($P<0.001$). No major complications from RF ablation were observed in these 12 patients. Two large multi-center trials evaluating the efficacy of RF ablation in pain control for metastatic bone lesions have been completed. In a recently published article of one such multi-center study by Goetz et al. (28), a total of 41 of 43 patients (95%) experienced a decrease in pain following RF ablation with a 24-hour pre-procedural score of 7.9 followed by decreasing mean worst pain scores to 4.5 ($P<0.0001$) at week 4, 3.0 ($P<0.0001$) at week 12, and 1.4 ($P=0.0005$) at week 24 post–RF treatment. A similar decrease in the mean average pain was observed. This study reports highly significant and rapid reductions in pain scores as well as improvements in the quality of life for patients following RF ablation of painful bone metastases. Currently our institution led and helped complete a multi-center National Cancer Institute sponsored trial (ACRIN protocol #6661) evaluating the efficacy of CT-guided RF ablation for a solitary site of painful osseous metastatic disease. Fifty-six patients underwent RFA at nine centers. Metastatic sites included the pelvis in 24 patients, chest wall in 19, thoracolumbar spine in eight, and extremities in five. RFA showed significant effect in reducing pain at one and three month follow-up for all four pain assessments that were measured. The average increase in pain relief from pre-RFA to one month follow-up was 26.3 ($P<0.0001$) and the increase from pre-RFA to three month follow-up was 16.4 ($P=0.02$). The average increase in mood from pre-RFA to one month follow-up was 19.9 ($P<0.0001$) and the increase from pre-RFA to three month follow-up was 14.9 ($P=0.005$). The average decrease in pain intensity from pre-RFA to one month follow-up was 26.9 ($P<0.0001$) and the decrease from pre-RFA to three month follow-up was 14.2 ($P=0.02$). The odds of being in lower pain severity at one month follow-up was 14.0 ($P<.0001$) times higher than that at pre-RFA, and the odds at three month follow-up was 8.0 ($P<0.0001$) times higher than that at pre-RFA. There were no incidences of bleeding or infection. The only attributable adverse events of grade three or higher were nerve injury in two patients (29). Additional benefits in adding RFA to the palliation of patients with osseous metastases are the ability to perform concomitant chemotherapy and the avoidance of marrow suppression in more localized disease.

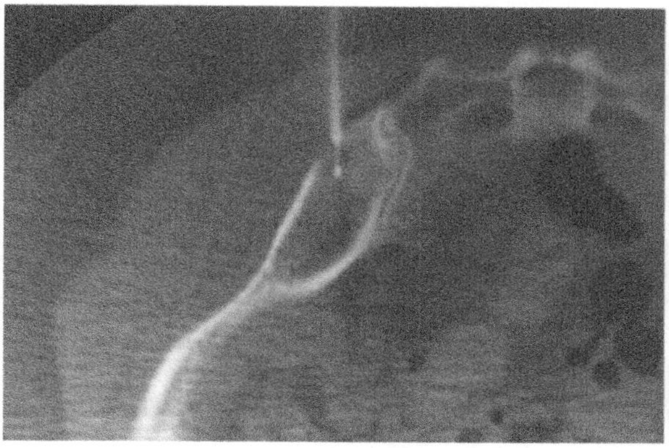

FIGURE 3 Fifty-three-year-old woman with lung cancer and a solitary bone metastasis detected by preoperative staging and biopsy. The patient was asymptomatic, and, therefore, radiation therapy was indicated. Given the size of the tumor, radiofrequency (RF) ablation was considered to be a good local treatment option to prevent future problems. Prone computed tomography image shows the RF electrode in the center of the metastasis. Because the cortex was intact, a trephine needle obtained access to the tumor and the outer sheath was used to place the electrode.

RFA may be used as an alternative to radiation therapy. In patients with very localized disease that can be safely treated with RFA, tumor control can be achieved (Fig. 3). In addition, in larger tumors that radiation therapy has a difficult time in controlling, a combination approach can be used (Fig. 4). In any event, RFA should always be considered a less invasive and less costly alternative when tumor burden is small and well confined, and an effective upfront cytoreduction treatment prior to conventional external beam radiotherapy.

RFA can also be applied to malignancies in the spinal axis, a common location for painful bone metastases. Spinal RF can be performed safely in the vertebral body without cytotoxic

(A) (B) (C)

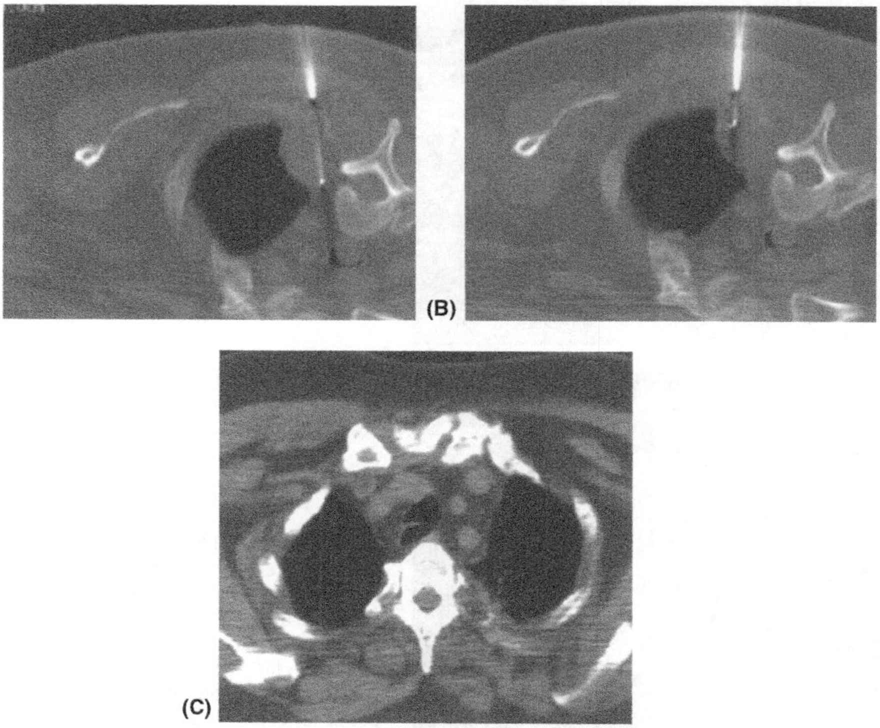

FIGURE 4 Fifty-one-year-old woman with metastatic hemangiopericytoma. A computed tomography (CT) showed an expansile rib metastasis adjacent to the neural foramen. Due to the size of the mass and known aggressive biology, combination upfront radiofrequency (RF) ablation followed by external beam radiation was planned. (**A, B**) Prone CT images show placement of the RF electrode into two separate regions of the mass to insure adequate thermocoagulation. (**C**) CT image one year later shows complete disappearance of the mass with some evidence of healing bone. The combination treatment prevented future neural compromise.

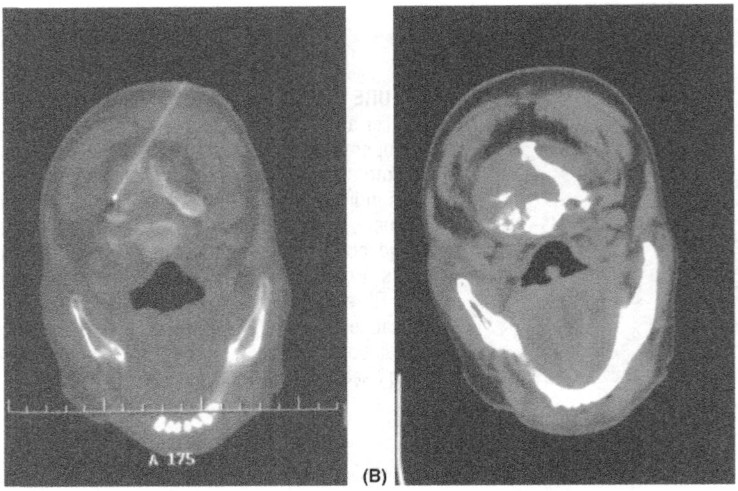

FIGURE 5 Seventy-eight-year-old man with metastatic lung cancer. The patient had a known posterior metastasis to the third cervical vertebrae that did not respond to radiotherapy. (**A**) Computed tomography scan prior to radiofrequency (RF) ablation shows osteolysis of the left lamina with an associated soft-tissue mass. (**B**) RF ablation was performed under conscious sedation without any neurological side effects due to the distance between the electrode and the cord and the presence of pulsating cerebrospinal fluid around the cord. The patient had pain relief from the treatment.

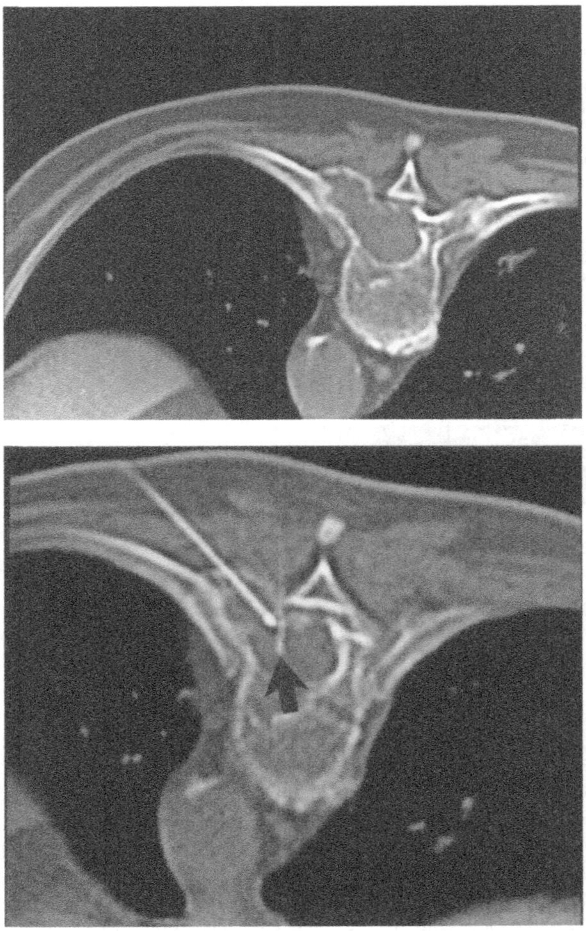

FIGURE 6 Seventy-five-year-old man with metastatic renal cell carcinoma. (**A**) Computed tomography (CT) image, with the patient prone, shows an osteolytic metastasis at the T-10 costovertebral junction with soft tissue encroaching upon the spinal canal. (**B**) Due to the soft-tissue component causing mass effect, the possibility of radiofrequency (RF)-induced nerve damage was a concern. Prone CT image shows the remote thermocouple adjacent to the thecal sac (*arrow*) to monitor the temperature effects from the RF electrode that was placed into the lateral-most portion of the mass.

temperature elevations observed within the spinal canal, if the vertebral body cortex is intact and a cerebrospinal fluid (CSF) space exists between the mass and spinal canal (26). Despite the use of internally cooled electrodes at maximum output, significant elevation of temperatures in the epidural space did not occur in this experimental study. Ex vivo studies confirmed decreased heat transmission in cancellous bone and an insulative effect of cortical bone. Additional factors that more than likely account for the differences in heat distribution observed are local heat sinks from the rich epidural venous plexus and CSF pulsations.

Perfusion-mediated tissue cooling has been demonstrated as negatively influencing the extent of coagulation, which can be produced in in-vivo liver, and decrease of blood flow by mechanical or pharmacologic means can increase the diameter of coagulation necrosis. The clinical application of RFA to the vertebral body may be important in the future. Osseous metastatic disease commonly involves the spine and often leads to debilitating pain, surgical decompression, and paraplegia. If a less invasive means of treating local tumor burden were available, then surgical decompression may be avoided in some cases. Clearly in cases where there is preserved cancellous or better yet cortical bone between the lesion and the spine, a margin of safety will be provided. In patients with extensive osteolysis with no intact cortex between tumor and spinal cord or nerve roots, RF may not be an option due to potential thermal injury to adjacent neural tissue. Theoretically if a CSF space was present between tumor and neural tissue, RF could be applied without unwanted neurotoxicity (Fig. 5). The use of a remote temperature sensor (Fig. 6) may offer a prudent margin of safety because the procedure could be terminated if deleterious temperature rises are observed adjacent to nervous tissue (i.e., greater than 45°C). In addition to destroying local tumor with heat, combination strengthening procedures such as vertebroplasty can be done at the same sitting (Fig. 7). Anecdotally we have performed RF and vertebroplasty in several patients, and the results have been clinically beneficial. Upfront RFA may contract the tumor, enabling improved distribution of cement.

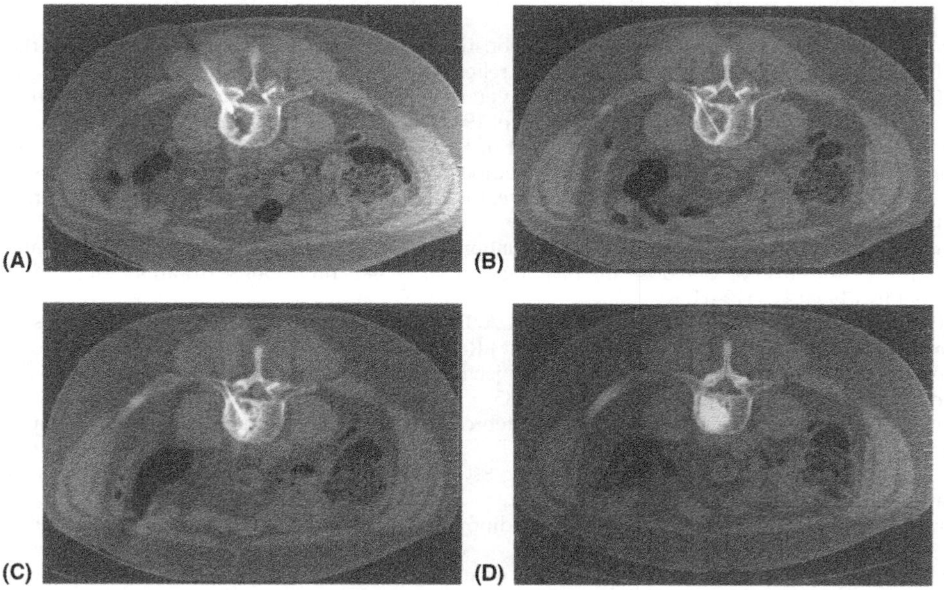

FIGURE 7 Fifty-seven-year-old man with metastatic renal cell carcinoma. The patient had an osteolytic metastasis in the fourth lumbar vertebral body that was irradiated one year earlier. The patient developed new back pain and a follow-up positron emission tomography scan and magnetic resonance imaging suggested tumor activity. (**A**) A prone computed tomography (CT) image shows the biopsy needle and central aspiration needle traversing the left L4 pedicle. On site cytopathology confirmed viable tumor. (**B**) A radiofrequency (RF) electrode was then placed through the outer sheath of the bone biopsy needle. (**C**) After the RF ablation cement was injected through a vertebroplasty needle under CT-fluoroscopy. (**D**) After 7 mL of cement, CT image confirms complete filling of the tumor cavity (*arrcw*).This case illustrates the ability to diagnose, treat, and strengthen with a single oupatient procedure.

In summary, RFA is an exciting new treatment modality that can be safely applied to metastatic bone disease. Its synergy with other therapies and outpatient nature make it an attractive treatment option for this fragile group of patients. The precise role of RF in this group of patients is not clearly defined, and continued work is underway in the hope that it will benefit those patients who currently have no treatment options and succumb to their disease without adequate analgesia.

REFERENCES

1. Stoll BA, Parbhoo S. Bone Metastasis: Monitoring and Treatment. New York: Raven Press, 1983.
2. Twycross RG. Management of pain in skeletal metastases. Clin Ortho 1995; 312: 187–196.
3. Janjan NA. Radiation for bone metastases: conventional techniques and the role of systemic pharmaceuticals. Cancer 1997; 80:1628–1645.
4. Quilty PM, Kirk D, Bolger JJ, et al. A comparison of the palliative effects of strontium- 89 and external beam radiotherapy in metastatic prostate cancer. Radiother Oncol 1994; 31:33–40.
5. Serafini AN, Houston SJ, Resche I, et al. Palliation of pain associated with metastatic bone cancer using samarium-153 lexidronam: a double blind placebo-controlled trial. J Clin Oncol 1998; 16: 1574–1581.
6. Hortobagyi GB, Theriault RL, Porter L, et al. Efficacy of pamidronate in reducing skeletal complications in patients with breast cancer and lytic bone metastases. Protocol 19 Aredia Breast Cancer Study Group.N Engl J Med 1999; 12:1785–1791.
7. Conte PF, Latreille J, Mauriac L, et al. Delay in progression of bone metastases in breast cancer patients treated with intravenous pamidronate: results from a multinational randomized controlled trial. The Aredia Multinational Cooperative Group. J Clin Oncol 1996; 14:2552–2559.
8. Tong D, Gillick L, Hendrickson FR. The palliation of symptomatic osseous metastases; final results of the study by the radiation therapy oncology group. Cancer 1982; 50:893–899.
9. Madsen EL. Painful bone metastasis: efficacy of radiotherapy assessed by the patients: a randomized trial comparing 4Gy versus 10Gy x 2. Int J Radiat Oncol Biol Phys 1983; 9:1775–1779.
10. Blitzer PH. Reanalysis of then RTOG study of the palliation of symptomatic osseous metastasis. Cancer 1984; 55:1468–1472.
11. Price P, Hoskin PJ, Easton D, Austin D, Palmer SG, Yarnold JR. Prospective randomized trial of single and multifraction radiotherapy schedules in the treatment of painful bony metastases. Radiother Oncol 1986; 6:247–255.
12. Arcangeli G, Micheli A, Arcangeli G, et al. The responsiveness of bone metastases to radiotherapy: the effect of site, histology and radiation dose on pain relief. Radiother Oncol 1989; 14:95–101.
13. Cole DJ. A randomized trial of a single treatment versus conventional fractionation in the palliative radiotherapy of painful bone metastases. Clin Oncol 1989; 1:59–62.
14. Poulter CA, Cosmatos D, Rubin P, et al. A report of RTOG 8206: a phase III study of whether the addition of single dose hemibody irradiation to standard fractionated local field irradiation is more effective than local field irradiation alone in the treatment of symptomatic osseous metastases. Int J Radiat Oncol Biol Phys 1992; 23:207–214.
15. Arcangeli G, Giovinazzo G, Saracino B, et al. Radiation therapy in the management of symptomatic bone metastases: the effect of total dose and histology on pain relief and response duration. Int J Radiat Oncol Biol Phys 1998; 42:1119–1126.
16. Ratanatharathorn V, Powers WE, Moss WT, Perez CA. Bone metastasis: review and critical analysis of random allocation trials of local field treatment. Int J Radiat Oncol Biol Phys 1999; 44:1–18.
17. Gangi A, Kastler BA, Klinkert A, Dietemann JL. Injection of alcohol into bone metastases under CT guidance. J Comput Assist Tomogr 1994; 18:932–935.
18. Murphy KJ, Deramond H. Percutaneous vertebroplasty in benign and malignant disease. Neuroimaging Clin N Am 2000; 10:535–545.
19. Dupuy DE, Safran H, Mayo-Smith WW, Goldberg SN. Radiofrequency ablation of painful osseous metastatic disease. Radiology 1998; 209(P):389.
20. De Giovanni JV. Treatment of arrhythmias by radiofrequency ablation. Arch Dis Childhood 1995; 73:385–387.
21. Sluijter ME. The use of radiofrequency lesions for pain relief in failed back patients. Int Disabil Stud 1988; 10:10–17.
22. Ogsbury JS, Simons H, Lehman RAW. Facet denervation in treatment of low back syndrome. Pain 1977; 2:257–263.
23. Rosenthal DI, Hornicek FJ, Wolf MW, Jennings LC, Gebhardt MC, Mankin HJ. Percutaneous radiofrequency coagulation of osteoid osteoma compared with operative treatment. J Bone Joint Surg—Am Vol. 1998; 80:815–821.
24. Goldberg SN, Gazelle GS, Solbiati L, Rittman WJ, Mueller PR. Radiofrequency tissue ablation: increased lesion diameter with a perfusion electrode. Acad Radiol 1996; 3:636–644.

25. Goldberg SN, Solbiati, Hahn PF, et al. Large-volume tissue ablation with radio frequency by using a clustered, internally cooled electrode technique: laboratory and clinical experience in liver metastases. Radiology 1998; 209:371–379.
26. Dupuy D, Ahmed M, Rodrigues B, et al. Percutaneous radiofrequency ablation of painful osseous metastases: a phase II trial. Proc Am Soc Clin Oncol 2001; 20:385A (abstract 1537).
27. Callstrom MR, Charboneau JW, Goetz MP, et al. Painful metastases involving bone: feasibility of percutaneous CT and US-guided radio-frequency ablation. Radiology 2002; 224:87–97.
28. Goetz MP, Callstrom MR, Charboneau JW, et al. Percutaneous image-guided radiofrequency ablation of painful metastases involving bone: a multicenter study. J Clin Oncol 2004; 22:300–306.
29. Dupuy DE, Liu D, Hartfeil D, et al. Percutaneous radiofrequency ablation of painful osseous metastases: A multi-center trial: American College of Radiology Imaging Network 6661. Proc Am Soc Clin Oncol 2007; In Press.

30 Cryoablation of Osseous Metastatic Disease

Leon Rybak
Department of Radiology, New York University/Hospital for Joint Diseases, New York, New York, U.S.A.

BACKGROUND AND HISTORY

Modern cryosurgery had its beginnings in the 1960s. Marcove is credited with introducing this technique to the field of orthopedic oncology in 1968 (1). Since then, it has gained widespread use as adjuvant therapy for ablation of the tumor margins in the treatment of both benign and malignant tumors with intralesional curettage (2–21). Most of this work is done with what is referred to as an open system, with direct application of the cryomaterial to the tumor surface. With the advent of closed systems, utilizing probes through which the cryogen is delivered, the applications of cryoablation have grown (22). Cryoablation has been long employed in the field of dermatology, and the past decade has seen its principles being applied to the treatment of urologic, hepatic, lung and breast tumors, and even in treatment of cardiac conditions. More recently, percutaneous cryoablation systems have been utilized in the treatment of selected primary bone tumors and metastases. Although early in the experience, preliminary reports have been encouraging (23,24).

MECHANISM OF ACTION

Cryoablation is thought to induce cell death through two sequential synergistic mechanisms. Rapid freezing results in the formation of intracellular ice, which can result in cell death by damage to intracellular organelles, as well as on a mechanical basis making the cell susceptible to shear injury. During the ensuing thaw, occlusion of the surrounding microvasculature results in ischemic necrosis. Research in the field of cryobiology has demonstrated that the critical temperature to achieve cell death is between −20°C to −40°C with temperatures higher than this resulting in supercooling of the tissues, but no intracellular ice formation. In addition, a cycle of treatment with an initial freeze followed by a thaw and a second freeze is found to be more effective. This is thought to be due to increased susceptibility of the tissues on repeat freezing to the freezing secondary to an increase in thermal conductivity and a decrease in the specific heat activity and vascularity of the tissues (25–29).

EQUIPMENT AND TECHNIQUE

Liquid nitrogen with a boiling point of −196°C is the coldest practical cryogenic agent. When placed in contact with a surface with a higher temperature, the liquid nitrogen will boil, extracting latent heat from its surroundings. In addition, if maintained in a pressurized environment, a gas, upon being allowed to expand, will lead to a drop in temperature. This is known as the Joule–Thomson effect. Most early applications involved the open application of liquid nitrogen into or onto the surface of a lesion. This involved pouring the liquid nitrogen into the tumor cavity, following intralesional excision or spraying along the margins with a probe (Fig. 1). Drawbacks of this approach include the formation of a thin layer of vapor, which may actually insulate the tissues, and an unpredictable delivery of the cryogen, resulting in temperature variability.

The initial closed systems also employed liquid nitrogen, which was continually circulated through the tip of the probes. However, liquid nitrogen is limited to probes of a diameter greater than 3 mm. Using pressurized argon gas (boiling point of −185.7°C), it is possible to create probes with a diameter as small as 1.4 mm, making directed treatment of small tumors possible (29).

FIGURE 1 Use of liquid nitrogen in an open system intraoperatively. *Source*: Photo courtesy of James Wittig, MD.

Systems are now available which utilize argon gas as a coolant and helium to aid in the thawing process. This makes it possible to reach temperatures as low as –100°C within a few seconds. An active thawing process is induced through the delivery of helium gas through the probe. With the current systems, it is possible to deploy up to eight probes at once.

What results at the tip of the probe is an "ice ball," which has a predictable geometry based on the length and diameter of the noninsulated tip of the probe. This ice ball can be visualized by various imaging technique including ultrasound, computed tomography (CT) and magnetic resonance imaging (MRI) (Fig. 2). Ultrasound, though very practical for certain applications, does not permit visualization deep into the most superficial portion of the ice ball. Most centers do not have the capability to employ MRI during the treatment process. This leaves CT imaging as the most practical and widely employed modality for this purpose.

It is important to remember that a temperature of 0°C should be assumed at the edge of the ice ball. Data published by the equipment manufacturers on the size and geometry of the different thermal zones surrounding the tip of the probe make it possible to predict the area of the ice ball at approximately –20°C or lower (Fig. 3). For complete necrosis of the tumor, it is important to extend the margins of the ice ball an appropriate distance beyond the tumor margins in order to ensure complete cell death. Although this is often the case with tumors in solid organs such as the liver and kidney, where the destruction of a small cuff of surrounding normal tissue will not result in undue morbidity to the patient, this is not the case in many musculoskeletal applications for several reasons. Much of the early data has been accumulated

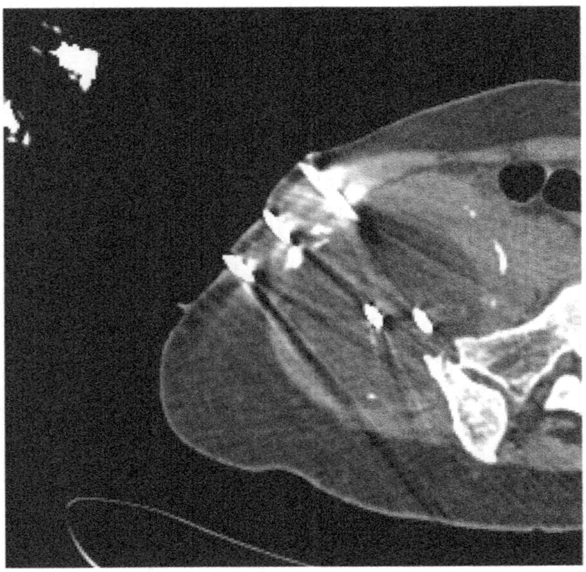

FIGURE 2 Axial computed tomographic image demonstrating the rounded low attenuation ice ball surrounding the cryoprobes.

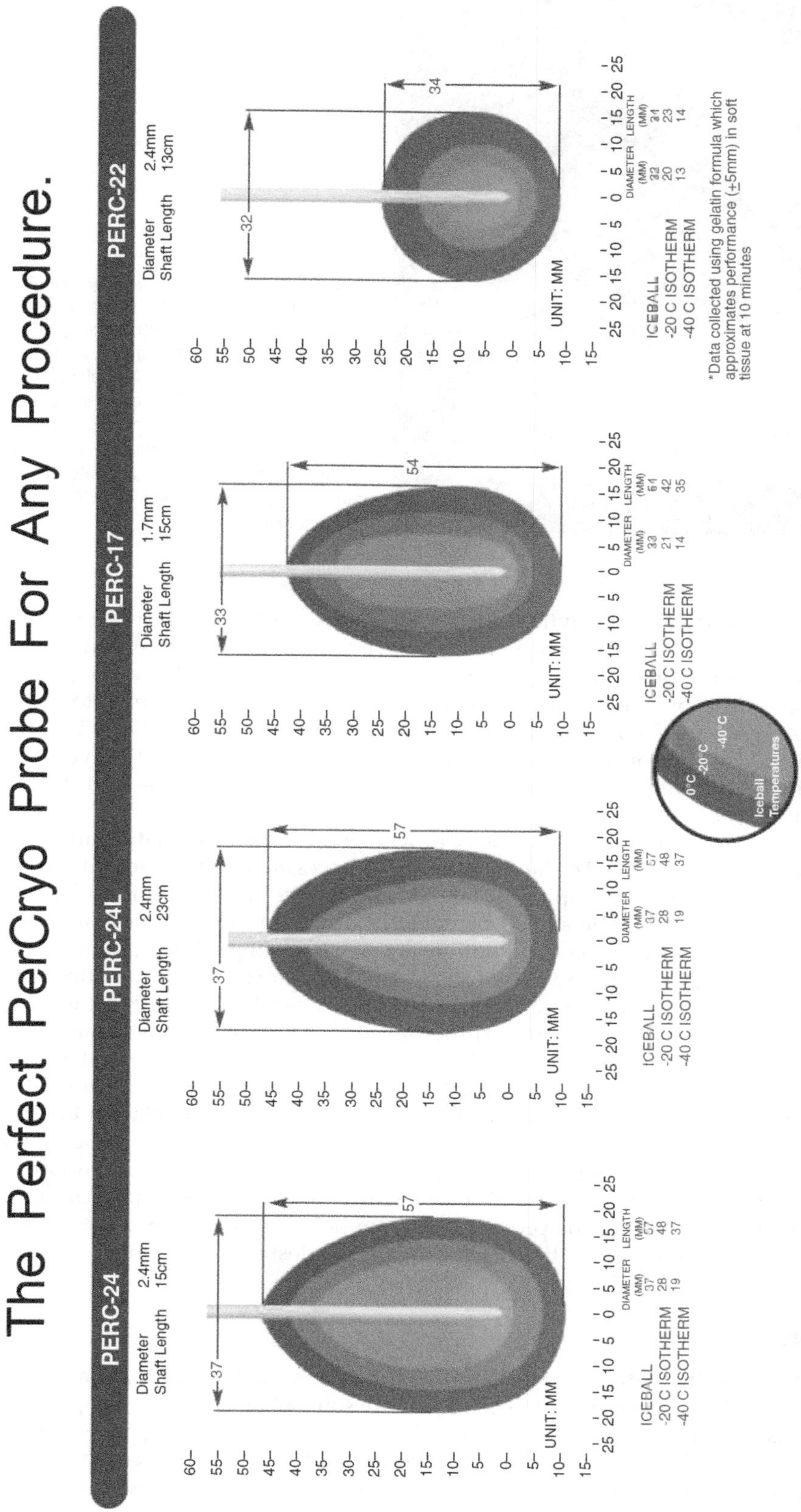

FIGURE 3 Illustration of the predicted ice formation achieved with several cryoprobes with the expected temperatures at various depths. *Source:* Photo courtesy of Endocare, Inc. of Irvine, California, U.S.A.

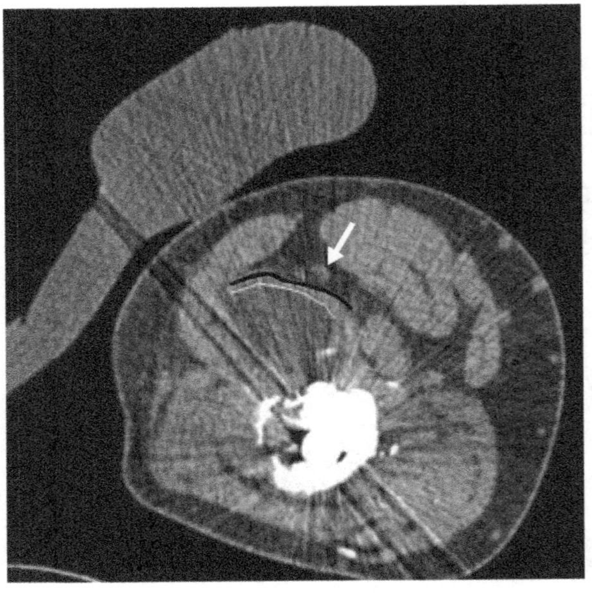

FIGURE 4 Axial image of the proximal tibia with the patient in the prone position. Note the low attenuation ice ball within the posterolaterally located tumor. The white arrow indicates adjacent neurovascular structures. A small cuff of untreated tumor has been left between the edge of the ice ball (*white line*) and the posterior margin of the mass (*black line*) in order to avoid neurologic complications. The structure lying along the skin surface is a sterile glove filled with warm water to prevent skin necrosis.

in treating large, predominantly lytic tumors which have destroyed the bone of origin and extended into the soft tissues. These lesions often have an intimate relationship with surrounding neurovascular structures, damage to which can have serious neurological sequelae for the patient. Preprocedure planning is very important for these lesions in order to avoid these complications, and a small cuff of untreated tumor may have to be left behind to spare these structures (Fig. 4). In addition, many of the patients with such lesions already have widespread metastatic disease and high tumor burden with the goal being palliation and debulking as opposed to cure.

Smaller tumor ablations can be carried out with good conscious sedation and the patient can be discharged the same day. The larger lesions may necessitate general anesthesia and overnight admission for control of postprocedural pain and monitoring for complications.

Certain points of technique should be noted here. The probes are fairly large and have very sharp tips, which may make penetration through pathologic bone possible. As of the writing of this article, there are no introducer sheaths made specifically for use with these probes. The possibility of skin freezing and necrosis should be taken into account when treating superficial lesions, where part of the active probe may traverse or approximate the skin surface. In these cases, it is important to warm the skin (we use warm sterile saline in a sterile glove) to prevent this complication (Fig. 4). It is important to test the probe before inserting it into the patient to ensure that it is functioning properly. This can be achieved by placing the probe in a small container of sterile saline and performing a test freeze. A small amount of ice at the tip of the probe is sufficient for this purpose. A good preprocedure MRI or CT is important to make planning possible. In this way, it is possible to estimate both the number of probes and tanks of gas which will be necessary for the procedure. If pain palliation is the goal, it is important to target any tumor–bone interface as this is felt to be the greatest source of pain.

INDICATIONS

As discussed earlier, the indications for cryotherapy in treating both primary and metastatic bone lesions is predicated in part on the goals that are somewhat unique as compared to treatment of lesions in solid viscera, such as the kidney or liver. Most of the patients have large lesions and advanced disease. The decision to employ cryotherapy should be made as part of a multidisciplinary team approach incorporating the opinions of the surgeon as well as the medical and radiation oncologist. Many of the lesions will have failed treatment with both

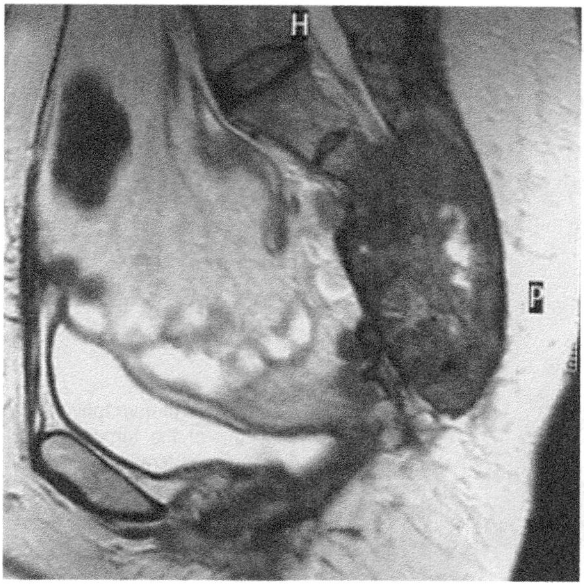

FIGURE 5 Metastatic rectal adenocarcinoma to the sacrum unresponsive to radiation and chemotherapy. The patient had severe pain uncontrolled by medication.

chemotherapy and radiation therapy. In cases where surgical resection would entail significant morbidity (i.e., hemipelvectomy for a large pelvic lesion), cryotherapy may be employed to debulk the tumor, thus making conservative treatment or a more limited surgical resection possible. In other cases, cryotherapy can be utilized for purposes of pain palliation in terminal patients with unresectable lesions (Figs. 5 and 6).

ADVANTAGES AND DISADVANTAGES

All forms of percutaneous ablation offer the advantage of decreased morbidity and mortality as compared to an open surgical procedure. Chemotherapy and radiation therapy carry the risks of systemic toxicity and the remote possibility of secondary sarcoma formation. With reference to other forms of percutaneous ablation, cryotherapy offers the additional advantages of direct visualization of the ice ball and decreased immediate postprocedural pain (23,24,30).

At present, the cryoprobes are still a bit large and unwieldy. It is necessary to use some forethought in placing multiple consecutive probes to ensure adequate clearance from the CT gantry and avoid placing undue torque on the needles due to awkward positioning of the attached tubing.

COMPLICATIONS

As mentioned previously, care should be taken to avoid freezing of any nearby neurovascular structures and the skin. Careful monitoring of the ice ball during the procedure should prevent these complications. Large vessels may act as a heat sink protecting both themselves and adjacent structures. If nerves are inadvertently incorporated into the periphery of the ice ball, where the temperature is greater than –20°C, a temporary neuropraxia may develop, which should resolve with time. If in the center of the ice ball, where temperatures of –40°C or lower predominate, permanent neurological damage may result (25). With large tumors that engulf adjacent nerve roots as can be the case with sacral tumors, this may be an acceptable risk. The patient should be made aware of the possible side effects, and a decision to proceed made after a careful evaluation of the risks and benefits.

Other possible complications include postprocedural pain, postablation syndrome and, in the case of intraosseous lesions, fracture. Although ice is known to have an anesthetic effect, temporary postprocedural pain exacerbation is still possible with the larger lesions. Postablation

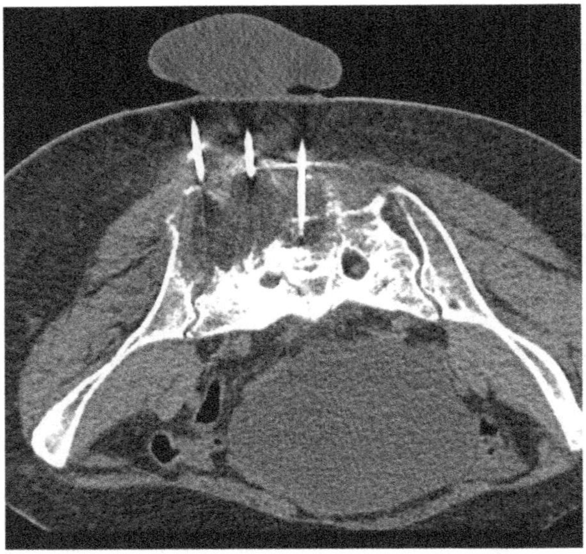

FIGURE 6 Axial computed tomography image taken during cryoablation of the tumor demonstrated in Figure 5. There was temporary relief of pain with no serious neurological sequalae, although the patient later succumbed to the disease.

syndrome is marked by transient symptoms including fever and generalized malaise. These symptoms are most likely mediated by systemic release of toxins from necrotic tumor, and there is an increased incidence of this syndrome in cases involving large ablation volumes (31). There is mention in the literature of fractures in the periprocedure period in ablations of intraosseous lesions. Many of these reports are in patients who underwent open intralesional curettage and cryoablation, although there are anecdotal reports of this complication in percutaneous treatments where probes were introduced into the medullary cavity. In such cases, it may be advisable to combine treatment with surgical instrumentation of the area for purposes of stabilization (32,33).

CONCLUSION

Great advances have been made in the area of percutaneous cryoablation in the past decade, which have made it possible to employ this modality in treatment of selected primary and metastatic bone lesions. As compared to radiofrequency ablation, it offers the advantages of direct visualization of the thermal ablation zone (ice ball) and decreased immediate periprocedure pain. Due to the extensive nature of the disease in many of these cases, the primary goals at this time remain palliation and debulking. Although scant published data exists, results to date have been promising.

REFERENCES

1. Marcove RC, Miller TR, Cahan WC. Treatment of primary and metastatic bone tumors by repetitive freezing. Bull NY Acad Med 1968; 44:532–544.
2. Marcove RC, Miller TR. Treatment of primary and metastatic bone tumors by cryosurgery. JAMA 1969; 207:1890–1894.
3. Marcove RC, Lyden JP, Huvos AG, Bullough PG. Giant cell tumors treated by cryosurgery. A report of twenty-five cases. J Bone Joint Surg (Am) 1973; 55:1633–1644.
4. Marcove RC, Stovell PB, Huvos AG, Bullough PG. The use of cryosurgery in the treatment of low and medium grade chondrosarcoma. A preliminary report. Clin Orthop 1977; 122:147–156.
5. Marcove RC, Weis LD, Vaghaiwalla MR, Pearson R. Cryosurgery in the treatment of giant cell tumors of bone. A report of 52 consecutive cases. Clin Orthop 1978; 134:275–289.
6. Marcove RC. A 17 year review of cryosurgery in the treatment of bone tumors. Clin Orthop 1982; 163:231–234.
7. Marcove RC, Zahr KA, Huvos AG, Ogihara W. Cryosurgery in osteogenic sarcoma: a report in three cases. Compr Ther 1984; 10:52–60.

8. Jacobs PA, Clemency RE Jr. The closed cryosurgical treatment of giant cell tumor. Clin Orthop 1985; 192:149–158.
9. Rock MG. Treatment of bone cysts and giant cell tumors. Curr Opin Orthopaed 1990; 1:423–434.
10. Malawer MM, Dunham W. Cryosurgery and acrylic cementation as surgical adjuncts in the treatment of aggressive (benign) bone tumors. Analysis of 25 patients below the age of 21. Clin Orthop 1991; 262:42–57.
11. Aboulafia AJ, Rosenbaum DH, Sicard-Rosenbaum L, Jelinek JS, Malawer MM. Treatment of large subchondral tumors of the knee with cryosurgery and composite reconstruction. Clin Orthop 1994; 307:189–199.
12. Schreuder HW, Conrad EU, Bruckner JD, Howlett AT, Sorensen LS. Treatment of simple bone cysts in children with curettage and cryosurgery. J Pediatr Orthop 1997; 17:814–820.
13. Schreuder HW, Veth RP, Pruszczynski M, Lemmens JA, Schraffordt Koops H, Molenaar WM. Aneurysmal bone cysts treated by curettage, cryotherapy and bone grafting. J Bone Joint Surg (Br) 1997; 79:20–25.
14. Schreuder HW, Pruszczynski M, Lemmens JA, Veth RP. Eosinophilic granuloma of bone: results of treatment with curettage, cryosurgery and bone grafting. J Pediatr Orthop B 1998; 7:253–256.
15. Schreuder HW, Pruszczynski M, Veth RP, Lemmens JA. Treatment of benign and low-grade malignant intramedullary chondroid tumors with curettage and cryosurgery. Eur J Surg Oncol 1998; 24: 120–126.
16. Bickels J, Meller I, Shmookler BM, Malawer MM. The role and biology of cryosurgery in the treatment of bone tumors. A review. Acta Orthop Scand 1999; 70:308–315.
17. Malawer MM, Bickels J, Meller I, Buch RG, Henshaw RM, Kollender Y. Cryosurgery in the treatment of giant cell tumor. Clin Orthop 1999; 2:176–188.
18. Wittig JC, Simpson BM, Bickels J, Kellar-Graney KL, Malawer MM. Giant cell tumor of the hand: superior results with curettage, cryosurgery, and cementation. J Hand Surg (Am) 2001; 26:546–555.
19. Kollender Y, Meller I, Bickels J, et al. Role of adjuvant cryosurgery in intralesional treatment of sacral tumors. Cancer 2003; 97:2830–2838.
20. Veth R, Schreuder B, van Beem H, Pruszczynski M, de Rooy J. Cryosurgery in aggressive, benign, and low-grade malignant bone tumours. Lancet Oncol 2005; 6:25–34.
21. Ahlmann ER, Menendez LR, Fedenko AN, Learch T. Influence of cryosurgery on treatment outcome of low-grade chondrosarcoma. Clin Orthop Relat Res 2006; 451:201–207.
22. Bickels J, Kollender Y, Merimsky O, Isaakov J, Petyan-Brand R, Meller I. Closed argon-based cryoablation of bone tumours. J Bone Joint Surg (Br) 2004; 86:714–718.
23. Beland M, Dupuy D, Mayo-Smith W. Percutaneous cryoablation of symptomatic extraabdominal metastatic disease: preliminary results. AJR 2005; 184:926–930.
24. Callstrom MR, Charboneau JM, Goetz MP, et al. Image-guided ablation of painful metastatic bone tumors: a new and effective approach to a difficult problem. Skeletal Radiol 2006; 35:1–15.
25. Korpan N. Basics of Cryosurgery. New York: Springer-Verlag/Wien 2001.
26. Robinson D, Halperin N, Nevo Z. Two freezing cycles ensure interface sterilization by cryosurgery during bone tumor resection. Cryobiology 2001; 43:4–10.
27. Hoffman NE, Bischof JC. The cryobiology of cryosurgical injury. Urology 2002; 60(suppl 2A):40–49.
28. Finelli A, Rewcastle JC, Jewett M. Cryotherapy and radiofrequency ablation: pathophysiologic basis and laboratory studies. Curr Opin Urol 2003; 13:187–191.
29. Baust JG, Gage AA. The molecular basis of cryosurgery. BJU International 2005; 95:1187–1191.
30. Allaf ME, Varkarakis IM, Bhayani SB, Inagaki T, Kavoussi LR, Solomon SB. Pain control requirements for percutaneous ablation of renal tumors: cryoablation versus radiofrequency ablation—initial observations. Radiology 2005; 237:366–370.
31. Goldberg SN, Grassi CJ, Cardella JF, et al. Image-guided tumor ablation: standardization of terminology and reporting criteria. Radiology 2005; 235:728–739.
32. Fisher AD, Williams DF, Bradley PF. The effect of cryosurgery on the strength of bone. Br J Oral Surg 1978; 15:215–222.
33. Pritsch T, Bickels J, Wu CC, Squires HH, Malawe MM. The risk for fractures after curettage and cryosurgery around the knee. Clin Orthop Relat Res 2007; [Epub ahead of print].

31 Intramuscular "Hemangioma" of Trunk and Extremities—Percutaneous Treatment

Odile Enjolras
Department of Neuroradiology, Lariboisière Hôpital, Assistance Publique–Hopîtaux de Paris, Paris, France

René Chapot
Department of Neuroradiology, Krupp Krankenhaus, Essen, Germany

Michel Wassef
Department of Pathology, Lariboisière Hôpital, Assistance Publique–Hopîtaux de Paris,Paris, France

INTRODUCTION

Vascular anomalies limited to skeletal muscles are uncommon. In the literature the word "hemangioma" or "intramuscular hemangioma" has been widely applied to a number of quite different histologic lesions, some being true tumors (when the name "hemangioma" is appropriate) and others comprising malformed vessels [when the name "vascular malformation" would be more justified than "hemangioma" (1,2)]. The use of a single name for lesions of different pathogenesis is confusing, and accordingly we will classify these intramuscular vascular lesions according to the International Society for the Study of Vascular Anomalies (ISSVA) as either vascular malformations or vascular tumors. Only two large series of vascular anomalies (diagnosed as "intramuscular hemangioma") have been reported, both reports coming from pathologists (3,4). Clinical data, radiological evaluation, and treatments outcome cannot clearly be drawn from a literature review; thorough reports are rare. It has been stated that intramuscular vascular lesions make up 0.8% of all vascular lesions ("angiomas") (5). However, in the 25-year experience of our interdisciplinary study group, involved in the management of patients of all ages, with the full range of vascular anomalies, the occurrence is probably less than 0.1%. No prevalence by gender is reported.

Clinically, intramuscular vascular anomalies of the extremities or trunk frequently manifest quite suddenly, in an adolescent or a young adult, occasionally in a younger very active child, sometimes following trauma. The most common intramuscular location is the lower limb (4). The majority of these lesions, if not all, are congenital in origin. Nevertheless, they usually do not produce symptoms until late childhood. Patients may complain of soft-tissue pain, localized swelling, and limitation of motion occurring during sports or even daily activities. Symptoms are intermittent and usually resolve with rest. Between the painful episodes, only a moderate swelling and tenderness may be noticed at clinical examination. Rarely the lesion pulsates and a bruit is heard [indicating the presence of arteriovenous fistulae (AVF)] (6). After some recurrent episodes, a tense swelling becomes permanent. Some patients have a growing palpable mass without pain, sometimes present for years (7).

With recent advances in imaging, these lesions are discovered earlier in life, ultrasound (US) and magnetic resonance imaging (MRI) being rapidly performed when pain appears (6). However, the workup of these lesions often raises a suspicion of malignancy (6–8). In many patients, if MR, angiographic, and US-Doppler findings (9,10) are not diagnostic, a biopsy is considered to reach a definitive diagnosis (2,11–13).

Frequently these lesions recur after excision (3–5). Consequently, complete excision is of paramount importance to avoid recurrence (6). However intraoperative hemorrhage may be dramatic and may lead to premature surgical procedure termination. For this reason, percutaneous sclerotherapy has been developed (*i*) in isolation, to try to control symptoms and growth or (*ii*) as a preoperative treatment, to facilitate surgical dissection and decrease morbidity (14).

PATHOLOGICAL CLASSIFICATION

As mentioned above, most authors have not precisely differentiated vascular malformations from vascular tumors. Allen and Enzinger, initially (3), and then Beham and Fletcher (4) pointed out that a predominant pattern of abnormal vessels is often evident in these lesions. They addressed the fact that many of these lesions are not true neoplasms; however, they kept the classical terminology of "intramuscular hemangioma."

Allen and Enzinger first (3) tried to classify their 89 lesions according to vessel size. Small-vessel lesions were prone to cause more diagnostic difficulties than large-vessel lesions. None of the small-vessel lesions were multiple, and this led to initial misdiagnosis (fear of malignancy). Adipose tissue intermingled with the malformed vessels, thrombi, hemosiderin, and fibrosis were more common in large-vessel lesions.

Beham and Fletcher (4) attempted to define a histological subclassification of their 74 intramuscular lesions (representing 4.4% of 1597 vascular lesions examined in their pathology department) according to vessel type and behavior. Subdividing them into capillary, venous, cavernous, arteriovenous, lymphatic, and complex lesions, they concluded that type and vessel size of intramuscular vascular anomalies have no influence on clinical symptoms and risk of recurrence when incompletely excised. A significant admixture of fat embedding the malformed vessels and dissecting the muscle is a common finding and these lesions are occasionally termed "angiolipoma" or "infiltrating angiolipoma" (5–15).

THE INTERNATIONAL SOCIETY FOR THE STUDY OF VASCULAR ANOMALIES CLASSIFICATION

In 1982, Mulliken and Glowacki proposed a "biologic classification of vascular birthmarks" (1). This classification distinguished hemangiomas from vascular malformations based upon clinical appearance, histopathologic features, and biologic behavior. Hemangiomas were described as tumors characterized by growth, hypercellularity, and endothelial proliferation. Malformations were considered congenital structural anomalies that are differentiated from hemangiomas based on their normal endothelial cell turnover and lack of excessive proliferation. They are composed of capillaries, veins, or lymphatics, or combinations of the above.

Takahashi et al., in 1994 (16), further characterized the biologic differences between hemangiomas and vascular malformations. In 2000, North et al. (17) described a phenotypic marker, glucose transporter 1 (GLUT 1), which is present in 100% of infantile hemangiomas and absent in other vascular tumors and in every type of vascular malformations. Differences between vascular tumors (hemangiomas expressing localized increase in angiogenic growth factors) and vascular malformations (developmental defects of the vascular tree) are supported by ongoing identification of mutated genes causing rare inherited forms of vascular malformations (18). The ISSVA classification (19) now refers to *vascular tumors* and *vascular malformations* (Table 1).

Thus, according to the ISSVA classification system, intramuscular vascular lesions of skeletal muscles are now classified as described in Table 2.

Slow-flow intramuscular vascular malformations, venous and lymphatic, represent the most frequent group of intramuscular vascular lesions. No pure capillary malformations exist within muscle (in contrast to what happens in skin where these are the most common type of vascular malformations, also known as "portwine stains"). However, a capillary component is

TABLE 1 International Society for the Study of Vascular Anomalies Classification of Vascular Anomalies

Vascular tumors (infantile tumors mainly)	Vascular malformations
Common: infantile capillary hemangioma	*Slow-flow*: capillary (CM), venous (VM), lymphatic (LM), or combined malformations
Rare: tufted angioma, kaposiform hemangioendothelioma, spindle cell hemangioendothelioma, rapidly involuting and noninvoluting congenital hemangiomas, etc.	*Fast-flow*: arteriovenous fistula and malformation (AVF, AVM)

TABLE 2 Classification of Intramuscular Vascular Lesions

Vascular tumors	Vascular malformations
Intramuscular hemangioma (in adolescent and young adults)	*Slow-flow*: venous (VM) and lymphatic (LM) malformations *Fast-flow*: arteriovenous malformations(AVM)

commonly associated. Fast-flow intramuscular arteriovenous malformations (AVMs) are quite rare. Intramuscular "hemangioma" is the accurate name for a rare intramuscular benign neoplasm, with distinctive radiological and pathologic features. The majority of infantile vascular tumors are not frequently found in skeletal muscle.

CLINICAL, RADIOLOGICAL, AND PATHOLOGIC CHARACTERISTICS

Venous malformations (*VMs*) are the most common intramuscular vascular anomalies: the patient is usually a child or an adolescent, sometimes a young adult, brought to consultation for localized pain, soft-tissue swelling, and intermittent functional impairment. Pain develops during physical activities (6) but also in the morning on waking up and during cold rainy weather. Physical signs include a tender mass under a normal skin [the situation is distinct clinically from an intramuscular VM of the extremities trunk with blue skin involvement (20)], no local inflammation, no pulsation, no bruit, and no general symptoms. If lower limb pain was present for several years, atrophy of the limb may develop as a consequence of the limitation of physical activities. Limitation of joint motion may also occur.

US-Doppler demonstrates sonolucent areas within one or several muscles. The calf and thigh musculature is the most common location. Some flow on color Doppler US may be found in these venous channels. Blood flow remains slow unlike in AVMs.

Plain films may reveal phleboliths, a helpful sign for the diagnosis of VM. Phleboliths may be visible as early as the first month of life, but usually appear in late childhood; they are less frequent in intramuscular VM than in superficial VMs.

MRI is the best imaging modality to localize and portray the lesions (21). Intermediate signal intensity on T1-weighted images and high signal intensity on T2- weighted images (lower than fat on T1-weighted images, and higher than fat on T2-weighted images) is most commonly found (10). Thrombosis and presence of adipose tissue likely explains the variations in signal intensity and patterns. Contrast between the vascular lesion and normal muscle is striking on T2-weighted sequences. Axial and, to a lesser degree, coronal sagittal images provide an anatomic view of the lesion extent (22). Varying morphologic patterns may be seen, from large venous pouches in compact lesions to extensive intramuscular infiltration in diffuse lesions. Severe muscle atrophy may be seen when a large skeletal muscle is filled with VMs. Extension of the intramuscular VM to the surrounding tissues, neighboring muscles, skin, bone, or joint is possible.

Arteriography provides little diagnostic information. It may show abnormal stagnation flow at the late venous phase.

When performed, pathological examination of a biopsy sample will show richly anastomosed, malformed, irregularly shaped veins, with focal lack of smooth muscle cells in their walls (13). The abnormal veins dissect the striated muscle fibers and are often associated with reactive adipose tissue. Organized thrombi and phleboliths may be seen.

Lymphatic malformation: Clinical signs and symptoms are quite similar to those of intramuscular VMs. A sudden swelling and localized tenderness may be seen. Otherwise, physical examination is normal.

US shows sonolucent fluid with cystic spaces. Color Doppler US excludes vascular flow. MRI with and without contrast best demonstrates the lymphatic cysts with hypo- or isointense signal intensity on T1-weighted sequences, and hyperintense signal intensity on T2-weighted sequences. In cystic lesions, a fluid–fluid level is commonly found due to layering of protein or blood. This is a helpful MR sign to differentiate lymphatic malformations from VM. Rim, septal, or no contrast enhancement may be observed after gadolinium injection in cystic lymphatic malformations (9). Angiography is not indicated.

Fluid aspiration and cytological analysis may be performed to rule out other intramuscular cystic tumors such as infantile fibrosarcoma, alveolar rhabdomyosarcoma, or papillary carcinoma of the thyroid, which, in our experience, may be confused with lymphatic malformations based on MR appearance. Pathologic samples of lymphatic malformations reveal large lymph-filled polyedric or irregularly dilated channels with thin walls and the presence of lymphocytes and macrophages aggregates, in the wall or lumen of the vessels.

Arteriovenous malformations: Symptoms are related to the deep, tense, and, sometimes, painful intramuscular mass. A single muscle is usually involved.

Doppler US evaluation shows high-velocity arterial flow and pulsatile venous flow. Angiography reveals AVF with enlarged arterial feeders and early venous return. MRI evidences flow voids (black tubular structures) corresponding to fast-flow vessels. Unlike intramuscular hemangiomas (see below) there is no parenchymal mass; however some increased signal intensity with a honeycomb appearance may be seen (2).

Histologically, there is irregularly thick-walled vessel, direct communications between vessels of different elastic components in their walls (AVF), and a capillary component dissecting the muscle.

True "intramuscular hemangioma": This is a poorly recognized and rare lesion, specifically affecting striated muscles. It develops in an adolescent or a young adult. To the best of our knowledge, spontaneous involution has never been reported. This is quite different from the course of an infantile hemangioma, which involutes during childhood. Growth is usually rapid and raises the suspicion of malignancy as a tense tender mass, distorting the body surface, develops inside a single muscle. The most common location in our experience is the neck and trunk, but any skeletal muscle may be affected.

The clinical workup may show contradictory or worrisome results: US-Doppler indicates the presence of fast-flow arterial feeders. MRI shows a well limited mass high signal intensity on both T1- and T2-weighted images, with strong enhancement after gadolinium injection, with flow-related internal signal voids. Angiography usually demonstrates intense hypervascularization from arteries and arterioles somewhat reminiscent of AVM, but without the early venous return. Because based on imaging characteristics a concern for malignancy is often present, a biopsy may be necessary.

Histologically, the venous component, formed of thick-walled large veins or thin-walled polyhedral cavities grouped in a nodular fashion, predominates in some cases. However, in most cases the lesion is quite similar to an AVM and is composed of thick-walled dilated vessels, mainly of venous type, associated with some arteries and capillaries aggregates. The capillaries may be grouped in lobules, containing isolated, sometimes atrophic, muscle cells, or may dissect the muscle fascicles. Direct communications can be seen between arteries and veins. Adipose tissue, fibrosis, lymphocytes, and eosinophils are present in varying amounts. In these mixed-type cases, the differential diagnosis between AVM and intramuscular hemangioma may be impossible without knowledge of the clinical and/or radiological data. Intramuscular hemangioma, however, is more frequent than purely intramuscular AVM. A number of lesions reported by pathologists (3–5) as small-vessel lesions or mixed type lesions (3–5) are probably such intramuscular hemangiomas. GLUT 1 (17) was negative in all of our own cases of intramuscular hemangioma—a finding indicating that these lesions are not late-growing infantile hemangiomas (the course of an infantile hemangioma consists of rapid growth during months after birth, and slow subsequent involution over 3 to 10 years). Interestingly, infantile hemangiomas never grow in muscle. This is a distinct type of vascular tumor, of specifically intramuscular location.

TREATMENT: EMBOLIC AGENTS, THERAPEUTIC SCHEMES

Percutaneous treatment varies according to the type of vascular malformation. Within each type of vascular malformation, the treatment may also differ with location and size.

VARIOUS EMBOLIC AGENTS MAY BE USED, WITH VARYING EFFICIENCIES AND ADVERSE EFFECTS

Particles: Calibrated microparticles (*Embospheres*®or *PVA, Contour*®) are injected in the arteries through a catheter. They progress with the arterial flow until reaching a blockage. Their efficiency is transient because there is a progressive recanalization of the embolized vessels.

They are only indicated preoperatively to reduce blood loss, the delay between embolization and surgery being less than two days. *Cyanoacrylate* (*NBCA, Histoacryl*®) is a liquid glue that polymerizes and hardens when it comes in contact with blood. It induces arterial occlusion and vessel wall destruction at the site of injection. It may be used in AVMs.

Distal navigation in the arterial tree to reach the shunt or direct puncture with a needle is required to occlude the AV shunts. Proximal injection of cyanoacrylate is inefficient and dangerous. Histoacryl may also be injected preoperatively in large VMs to better define the limits of the VM and reduce the surgical blood loss. Ethibloc® is a fluid sclerosing agent based on zein, a protein extracted from corn; its high viscosity limits the risk of local diffusion in the normal surrounding vascular network. Lipiodol1® is added to the solution to make it radio-opaque. It is injected through a needle after direct puncture of venous and lymphatic malformations. Absolute ethanol is used as a sclerosing agent with a high efficacy. Adjunction of opacifying agents is recommended to assess the extent of diffusion space. Ethanol may be injected in arterial feeders of an AVM through a catheter or directly percutaneous in the AV shunts through a needle to occlude and destroy the vessels. It is also used in venous or lymphatic malformations through direct needle puncture. Its high efficiency is counterbalanced by a high rate of complications, due to the distal occlusion of all arteries when used intra-arterially and to the diffusion into the surrounding tissues when injected in venous and lymphatic malformations. The volume of ethanol injected during ethanol embolization or sclerotherapy of vascular malformations is directly correlated with serum ethanol levels with a risk of acute ethanol intoxication when the patients receive up to 1.0 mL/kg-body weight ethanol during one procedure; this puts the patient at risk for respiratory depression, cardiac arrhythmia, seizure, rhabdomyolysis, and hypoglycemia (23).

Treatment may have two different goals. It may be used as a preoperative embolization to reduce blood loss during surgery, or as an exclusive treatment. Both treatments can be used in each type of vascular malformation, except for the true intramuscular hemangioma in which there is no indication of exclusive embolization. In case of exclusive embolization, the aim of treatment is usually a palliative rather than a definite cure. It may require multiple sessions. For slow-flow intramuscular lesions (venous or lymphatic), percutaneous sclerosing treatment is the best method of approach, while transarterial embolization and percutaneous approach may both be considered for fast-flow intramuscular AVMs.

THERAPEUTIC SCHEMES FOR EACH TYPE OF VASCULAR MALFORMATION

(a) *Venous malformation*: Exclusive sclerosis aims at reducing the size of the malformation. Definite cure is rarely obtained. VMs may be filled with the sclerosing agent through a needle after direct puncture because VMs cannot be easily reached by arterial or venous endovascular navigation. US guidance may help localize the venous channels. Care must be taken to assess correct initial positioning of the needle and to avoid displacement of the needle during the injection. Diffusion of the sclerosing agent in the normal surrounding vascular network connected to the VM must also be avoided. Injection of a radio-opaque sclerosing agent under fluoroscopy enables detection of extravasation and diffusion in the normal veins (24). Different sclerosing agents may be used. Ethanol has the highest efficiency and may be used alone (25). However, it may diffuse rapidly from the VM to the normal surrounding veins due to its low viscosity. This is reduced by compression of the draining veins using a tourniquet for limb VMs. Definite nerve damage may occur when nerves are located close to or within the VM. To reduce the diffusion of ethanol and lower its toxicity, a mixture of ethanol, Ethibloc®, and Lipiodol® may be used. The concentration of ethanol used ranges from 20% to 70% to take advantage of the sclerosing ability of ethanol, the viscosity of Ethibloc®, and the radio-opacity of Lipiodol®. The amount of sclerosing agent injected depends on the size of the dysplasic veins. Thrombosis occurs shortly after the procedure and is followed by immediate swelling, lasting for one to four weeks. Efficacy of sclerosis is variable, and best appreciated as clinical resolution of functional impairment.

Successive sclerosing sessions may be necessary to achieve significant malformation size reduction. Localization of the VM inside muscle leads to post-procedure stiffness and pain, which responds well to nonsteroidal anti-inflammatory drugs. Efficacy of sclerosis is variable, and is again best appreciated as resolution of functional impairment. Even after significant clinical volume reduction, there is usually partial persistence of the VM as shown on follow-up MR examination.

Absolute ethanol sclerosing procedures have a higher rate of local and systemic complications. Lee et al. who used absolute ethanol in 30 patients with VMs of various locations obtained an immediate good result in 92% and an overall improvement rate of 96% with the follow-up (26). Complications occurred in 16% of the 98 therapeutic sessions (26). Complications included bullae, tissue necrosis, deep venous thrombosis, pulmonary embolism, pulmonary hypertension, and peripheral nerve palsy (26). Extensive VMs of the extremities may be the site of a localized intravascular chronic coagulopathy (LIC), with low fibrinogen and high D dimer levels (27). Sclerosing procedures may worsen the LIC, or even create clotting disturbances in patients with no LIC detectable before the procedure, with a risk of bleeding or thrombosis and acute pain in their lesions (28). Treatment of such LIC is based on low-molecular-weight heparin (27). In cases of an infiltrative VM without large lakes, surgical removal of the involved muscle may be indicated. Embolization can be proposed as an adjunct prior to surgery. Treatment consists in particle embolization of the feeding arteries of the involved muscle, thus reducing the venous filling. In huge VMs, preoperative injection of histoacryl glue in the venous pouches helps delineate the margins of the VM that should be excised.

(b) *Lymphatic malformation*: Sclerosis is more effective and has a lower recurrence rate in lymphatic malformations than in VMs. It is performed in cases of macrocystic lymphatic malformations but usually not in microcystic lesions due to a risk of adjacent soft-tissue necrosis and scarring. Also, sclerosis cannot be readily applied to diffuse lesions involving a complete limb. Procedures are quite similar to that used for VMs, using direct puncture of the lesions, sometimes under US guidance. Embolic agents used are either pure ethanol or a mixture of ethanol, Ethibloc, and Lipiodol.

(c) *Arteriovenous malformations*: The aim of treatment is to occlude the arteriovenous shunts with a permanent occlusion material. Particles are therefore not indicated. Occlusion of the shunts may be obtained with histoacryl or ethanol after navigation with a microcatheter or direct needle puncture. The site of injection of the embolic agent must be as close as possible to the AV shunts. Otherwise, there is a proximal occlusion with immediate refilling of the shunts through an anastomotic network. Coils are therefore not optimal. Proximal injection of a liquid occlusive agent is more effective, but it carries a risk of occlusion of the normal arteries, with subsequent necrosis. This is particularly relevant to ethanol, whose ability for distal vascular penetration is high. The best indications for embolization are small compact AVMs where a definite cure can be obtained. Large AVMs involving several muscles (a rare situation for fast-flow intramuscular anomalies), AVMs with a diffuse nidus, and AVMs located at the distal part of the extremities are less adapted for embolization, except as a palliative means in expanding lesions inducing cardiac failure.

(d) *Intramuscular hemangioma*: This hypervascular tumor should be treated surgically because the aim is complete resection without functional sequelae. Occlusion of the arterial feeders by embolization is indicated preoperatively to prevent excess intraoperative bleeding.

The feeders may be selectively catheterized and occluded by various means, including particles or NBCA. The procedure is performed on the same day as surgical excision or one day before surgery.

CONCLUSION

The so-called "intramuscular hemangiomas" constitute a heterogeneous group of vascular lesions. They represent a diagnostic challenge due to their rarity, their variable imaging patterns, and even sometimes difficult precise histologic classification. Most intramuscular vascular lesions are not "hemangiomas" because the majority are not true neoplasms but rather are malformations. Percutaneous sclerotherapy may be a useful palliative treatment for vascular malformations of intramuscular location. It often does not truly ablate the lesion, but is effective in reducing symptoms (pain and size of the lesion). This treatment is more effective in intramuscular venous or lymphatic malformation than in AVMs. On the other hand, the rare intramuscular tumor (the "intramuscular hemangioma") is not improved by percutaneous sclerotherapy. Pre-operative arterial particle embolization associated with direct puncture embolization with glue helps the surgeon dissect and excise the lesion without excessive bleeding.

REFERENCES

1. Mulliken JB, Glowacki J. Hemangiomas and vascular malformations in infants and children: a classification based on endothelial characteristics. Plast Reconstr Surg 1982; 69:412–420.
2. Kim EY, Ahn JM, Yoon HK, et al. Intramuscular vascular malformations of an extremity: findings on MR imaging and pathologic correlation. Skeletal Radiol 1999; 28:515–521.
3. Allen PW, Enzinger FM. Hemangioma of skeletal muscle. An analysis of 89 cases. Cancer 1972; 29:8–22.
4. Beham A, Fletcher CDM. Intramuscular hemangioma: a clinicopathologic analysis of 74 cases. Histopathology 1991; 18:53–59.
5. Weiss SW, Goldblum JR. Benign tumors and tumor-like lesions of blood vessels: intramuscular hemangioma. In: Enzinger and Weiss's, Soft Tissue Tumors. 4th ed. Mosby, 2001:869–872.
6. Quinn PS, Sieunarine K, Lawrence-Brown M, Tan P. Intramuscular hemangioma: hookwire localization prior to surgical excision: report of four cases. ANZ J Surg 2001; 71:62–66.
7. Wild AT, Raab P, Krauspe R. Hemangioma of skeletal muscle. Arch Orthop Trauma Surg 2000; 120:139–143.
8. Teo ELHJ, Strouse PJ, Hernandez RJ. MR imaging differentiation of soft-tissue hemangiomas from malignant soft-tissue masses. AJR 2000; 174:1623–1628.
9. Burrows PE, Laor T, Paltiel H, Robertson RL. Diagnostic imaging in the evaluation of vascular birthmarks. Dermatol Clinics 1998; 16:455–488.
10. Dubois J, Garel L, Grignon A, et al. Imaging of hemangiomas and vascular malformations in children. Acad Radiol 1998; 5:390–400.
11. Jenner G, Soderlund V, Bauer HF, Brosijo O. MR imaging of skeletal muscle hemangiomas. A report of 16 cases. Acta Radiol 1996; 37:140–144.
12. May DA, Disler DG, Jones EA, Balkissoon AA, Manaster BJ. Abnormal signal intensity in skeletal muscle at MR imaging: patterns, pearls and pitfalls. Radiographics 2000; 20: S295–S315.
13. Wassef M, Enjolras O. Les malformations vasculaires superficielles: classification et histopathologie. Ann Pathol 1999; 19:253–263.
14. Amram CPT, Cohen MC, Youkey JR, et al. Intramuscular hemangioma. JAMA 1983; 249:2680–2682.
15. Pribyl C, Burke SW, Roberts JM, Mackenzie F, Johnston CE. Infiltrating angiolipoma or intramuscular hemangioma? A report of five cases. J Pediatr Orthop 1986; 6:172–176.
16. Takahashi K, Mulliken JB, Kozakewich H. Cellular markers that distinguish the phases of hemangioma during infancy and childhood. J Clin Invest 1994; 93:2357–2364.
17. North PE, Waner M, Mizeracki A, Mihm MC Jr. G1UT1: a newly discovered immunohistochemical marker for juvenile hemangiomas. Hum Pathol 2000; 31:11–22.
18. Vikkula M, Boon LM, Mulliken JB. Molecular genetics of vascular malformations. Matrix Biol 2001; 20:327–335.
19. Enjolras O, Mulliken JB. Vascular tumors and vascular malformations (new issues). Adv Dermatol 1998; 13:375–423.
20. Enjolras O, Ciabrini D, Mazoyer E, Laurian C, Herbreteau D. Extensive pure venous malformations in the upper or lower limb: a review of 27 cases. J Am Acad Dermatol 1997; 36(Pt 1):219–225.
21. Yuh WTC, Kathol MH, Sein MA, Ehara S, Chiu L. Hemangiomas of skeletal muscle: MR findings in five patients. AJR 1987; 149:765–768.
22. Nack J, Gustafson L. Intramuscular hemangioma. Report of a case and literature review. J Am Podiatr Med Assoc 1990; 80:441–443.
23. Mason KP, Michna E, Zurakowski D, Koka BV, Burrows PE. Serum ethanol levels in children and adults after ethanol embolization or sclerotherapy for vascular anomalies. Radiology 2000; 217:127–132.
24. Herbreteau D, Aymard A, Jhaveri HS, et al. Current management of cervicofacial superficial vascular malformations and hemangiomas. In: Connors JJ, Wojack JC, eds. Interventional Neuroradiology. WB Saunders, 1999:317–327 (Chapter 26).
25. Yakes WFJ. Management of extracranial head and neck and paraspinal vascular malformations. In: Connors JJ, Wojack JC, eds. Interventional Neuroradiology. WB Saunders, 1999:327–337 (Chapter 27)
26. Lee BB, Kim DI, Huh S, et al. New experiences with absolute ethanol sclerotherapy in the management of a complex form of congenital venous malformation. J Vasc Surg 2001; 33:764–772.
27. Mazoyer E, Enjolras O, Laurian C, Houdart E, Drouet L. Coagulation abnormalities associated with extensive venous malformations of the limbs: differentiation from Kasabach-Merritt Syndrome. Clin Lab Haematol 2002 Aug; 24:243–251.
28. Mason KP, Neufeld EJ, Karian VE, Zurakowski D, Koka BV, Burrows PE. Coagulation abnormalities in pediatric and adult patients after sclerotherapy or embolization of vascular anomalies. Am J Roentgenol 2001; 177:1359–1363.

32 Specific Uses of Interventional Musculoskeletal Magnetic Resonance Imaging

Gerhard Adam
Department of Diagnostic and Interventional Radiology, University Hospital Hamburg Eppendorf, Hamburg, Germany

Arno Bücker, Jörg Neuerburg, and Rolf W. Günther
Department of Diagnostic Radiology, University of Technology, Aachen, Germany

INTRODUCTION

Magnetic resonance (MR) imaging has become a routine imaging procedure. Due to the continued developments of MR scanners, and the evolution of fast imaging technologies, real-time or near–real-time imaging can now be performed. This, combined with the introduction of more open MR scanners and the clinical use of short-bore whole-body magnets, has led to an increase in research and clinical focus on interventional MR.

Earlier authors have shown the feasibility of MR guidance for open and stereotactically guided biopsies in the head (1,2) and neck region (3,4), tumor ablation procedures using laser (5), cryo- (6–9) and radio frequency (10,11) probes as well as intravascular interventions (12–18). This chapter will review the current status of MR guidance for musculoskeletal interventions, addressing the system requirements, the different materials for biopsies and other procedures, and the current clinical indications.

SYSTEM REQUIREMENTS

MR imaging offers advantages when compared with other imaging modalities utilized for musculoskeletal interventions. It has a superior soft-tissue contrast, the possibility to choose nonstandard imaging planes, and the lack of ionizing radiation. MR also has certain disadvantages—limited workspace and patient access (in a closed bore magnet system) and limited temporal or spatial resolution (in an open MR system). Principally, there are three basic types of MR systems commercially available which may be used for interventional procedures.

Vertical, open, low-field scanners (open C-arm configuration) (Fig. 1A) are operated between 0.064 and 0.7 T and allow a lateral approach to the patient (19). Because the signal-to-noise ratio (SNR) depends directly on the field strength, acquisition time is longer for many diagnostic imaging sequences as compared to higher field strength magnets. However, specific fast imaging techniques such as steady-state free precision, for example, True FISP, have been implemented on these scanners.

Open midfield scanners, such as that manufactured by General Electric (Fig. 1B), consist of two vertically positioned super-conducting magnets, allowing a workspace of about 60 cm between the magnets (20). Still, the SNR is inferior as compared to closed-bore, super-conducting, high-field systems. In addition, the homogeneity of the magnetic field is also inferior as compared to closed systems (e.g., ±7.5 ppm in a 30 cm sphere for the Signa SP). Moreover, these large scan rooms are difficult to shield, and because the magnets weigh eight tons, sitting is quite difficult.

Short-bore, high-field system combined with X-ray fluoroscopy (hybrid system) is also utilized. These types of magnets are those typically used for "conventional" MR imaging. Modern systems offer a gantry space of about 60 cm. If the construction of the patient table is modified so that it can be moved and repositioned exactly during a procedure, these scanners

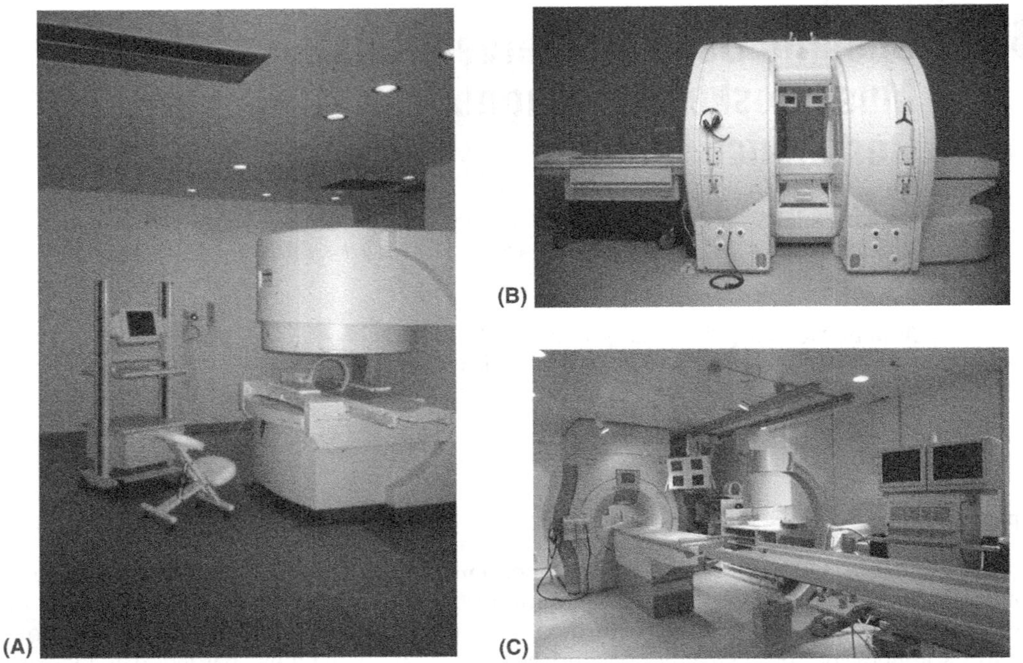

FIGURE 1 (**A**) Open low-field system, with in-room console, monitors, and in-room chair (Siemens Magnetom Open). (**B**) Open mid-field system (gradient echo Signa SP). (**C**) Hybrid system, with installation of a 1.5 T system, together with a fluoroscopy unit in one room (Philips ACS NT and BC 212 fluoroscopy unit).

may be used for interventional procedures (21–23). The only restriction for interventions is the limited space within the gantry, which may make an intervention in obese patients impossible. These magnets offer advantages of high-field MR, including various contrast preparation sequence techniques and ultrafast (real-time) imaging capabilities. The fluoroscopy system needs dedicated shielding to prevent image distortion by the residual magnetic field when the patient is moved to the fluoroscopy system. The center of the fluoroscopy system has to be outside the 5 Gauss line. Usually the electronics of the fluoroscopy system are switched off when the patient is moved via the floating table top into the magnet.

More recently, a "true hybrid" interventional system has been released (24). In this system a flat panel radiographic detector has been integrated in the patient table of an open Signa SP system, allowing MR acquisition and generation of X-ray images without patient repositioning.

The appropriate choice of the magnet site depends primarily on the frequency in which interventional procedures are performed. A closed-bore, high-field whole-body scanner combined with a fluoroscopy system may be a good overall compromise, because it can be used not only for interventions, but also for high quality whole-body applications. In many circumstances, a conventional short-bore, high–field strength scanner without fluoroscopy can be utilized. For sites with larger volumes of procedures, an open, lower field strength scanner is adequate. For the highest volume sites, where the scanner is combined with an operating room, the double doughnut gradient echo (GE) is a good system.

Some modifications of the MR suite are required for interventional procedures. Firstly, a communication system between the magnet room and the main console of the scanners, which is usually outside the scan room, is required. In addition, it is helpful if some of the basic functions are controllable from the scan room, either using a foot pedal or a button on the magnet casing, in order to view a previously performed sequence. Furthermore, in-room monitors are essential to allow direct viewing after acquiring an image set or a single image. These are often ceiling mounted. As mentioned above, the MR table may be modified so that smooth and exact positioning of the patient is possible. The exact table position should be indicated on both the room console and directly on the magnet bore casing. For interventional purposes, either the body coil or dedicated surface coils may be used.

Because the majority of musculoskeletal interventions can be performed under local anesthesia, general anesthesia is rarely required. However, because it may be necessary in some patients (e.g., in pediatric interventions), outlets through the shielding for medical gases are required. In addition, MR-compatible anesthesia monitoring equipment has to be available for anesthetized patients.

MATERIALS FOR MAGNETIC RESONANCE–GUIDED INTERVENTIONS

Localization

The biopsy pathway from the entrance point on the skin of the patient to the target (lesion) can be determined in several ways. In open systems, the entrance point may be indicated by the finger of the radiologist. In closed systems, the easiest way is the use of an MR-compatible grid, allowing biopsy planning on T1- or T2-weighted sequences. For planning using T1-weighted or T2-weighted sequences, one may use home made grids consisting of catheter fragments filled with either a diluted gadolinium solution (normally 1:200) or water. An alternative is to mark the entrance point on the skin with a nitrocapsule. After demarcation on the patient's skin, the distance and the angle of the biopsy pathway can be decided using a protocol with the standard software of most MR scanners. Other, more sophisticated tools are also available. They consist of instrument holders equipped with laser diodes referring to a fixed laser source within the MR room, which allow on-the-fly the determination of the pathway by overlying the data in x, y, and z directions on a previously acquired MR image (25).

Biopsy Instruments

The appropriate choice of the biopsy instrument depends on the clinical requirements. For corticosteroid injection or MR arthrography, MR compatible needles ranging from 18 to 22 g in size are obtainable. For biopsies, fine needles, cutting needles (Fig. 2), or MR-compatible drilling systems need to be used.

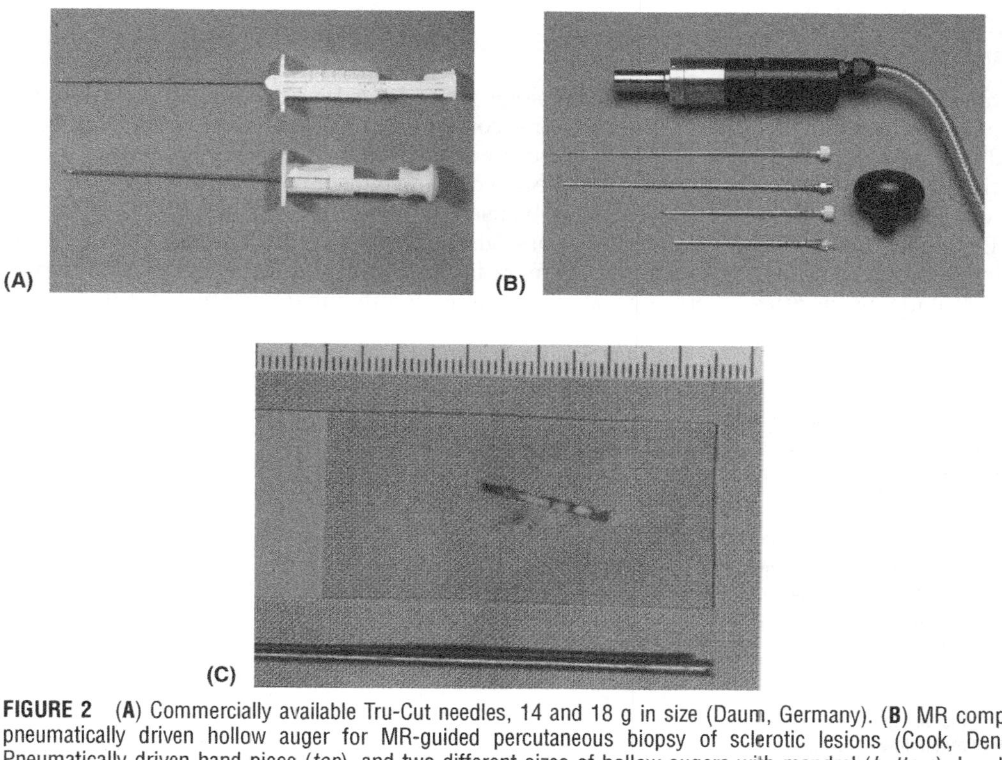

FIGURE 2 (**A**) Commercially available Tru-Cut needles, 14 and 18 g in size (Daum, Germany). (**B**) MR compatible, pneumatically driven hollow auger for MR-guided percutaneous biopsy of sclerotic lesions (Cook, Denmark). Pneumatically driven hand piece (*top*), and two different sizes of hollow augers with mandrel (*bottom*). In addition, a manual tool holder (*right*) is available. (**C**) Biopsy specimen acquired with the MR compatible biopsy system.

Needle selection rules similar to those followed in fluoroscopy- or computed tomography (CT)-guided biopsies are utilized. If the lesion is purely cystic with a thin cortical rim, a fine needle is sufficient, because it allows the penetration of the cortical bone and the aspiration of intralesional fluid. If the lesion is covered by a thick layer of cortical bone, or if the lesion is osteosclerotic, a drill is necessary.

MR-guided core biopsy needles are available in sizes ranging from 14 to 18 g. Drilling systems are either manually or mechanically driven (pneumatically, to ensure MR compatibility) (Fig. 2B and C) (26,27).

A problem specific to MR is difficulty secondary to susceptibility artifacts (28,29). Most of the MR compatible instruments are manufactured from nitinol alloys. Although these alloys are nonferromagnetic, they cause variable amounts of artifact. The artifact size depends on the size of the needle, the exact composition of the needle, whether it is used with a mandrel, the field strength of the MR scanner, the type of sequence used [GRE worst, fast spin echo (FSE) best], and the needle position with respect to the main magnetic field and to the phase- and frequency-encoding gradients (Table 1). In addition, the artifact size depends on the echo time (TE) and the bandwidth of the frequency-encoding gradient (Fig. 3).

Sequences for Magnetic Resonance–Guided Musculoskeletal Interventions

The appropriate choice of the localizing sequence depends on the type of lesion. In general, the sequence displaying the target with the highest spatial and contrast resolution should be applied. However, if the lesion can only be targeted with difficulty, or via a potentially dangerous pathway, a compromise between temporal and spatial resolution should be used. Although multiple types of fast imaging techniques are available, they predominantly utilize FSE or fast gradient echo techniques. Also, using radial or spiral *k* space–sampling strategies in combination with back projection for image reconstruction, the acquisition time can be reduced further (30–32). Typical sequence parameters utilized for fast gradient and FSE techniques are listed in Table 2.

CLINICAL INDICATIONS

Biopsies

Percutaneous bone biopsies are common radiologic procedures for bone biopsy, including the histologic confirmation of suspected primary or secondary malignant and benign bone tumors. In addition, percutaneous bone biopsy may be used to confirm an infectious process and to determine the causative microorganism in suspected osteomyelitis (27,33,34). For these purposes, either core biopsy needles or fine needles may be used.

Until now, CT or X-ray fluoroscopy has been used mostly to guide biopsies. MR imaging offers superior soft-tissue contrast and lacks any radiation burden and can be also used to perform percutaneous biopsies. In our own series, we demonstrated that bone biopsies under MR

TABLE 1 Parameters Influencing the Artifact Size

Spin echo technique	Low
Gradient echo technique	High
TE: long	High
TE: short	Low
Frequency-encoding gradient alongside the instrument	Low
Frequency-encoding gradient perpendicular to the instrument	High
Bandwidth: broad	High
Bandwidth: small	Low
B0: high	High
B0: low	Low
Amount of metal: high	High
Amount of metal: low	Low

Abbreviation: TE, echo time.

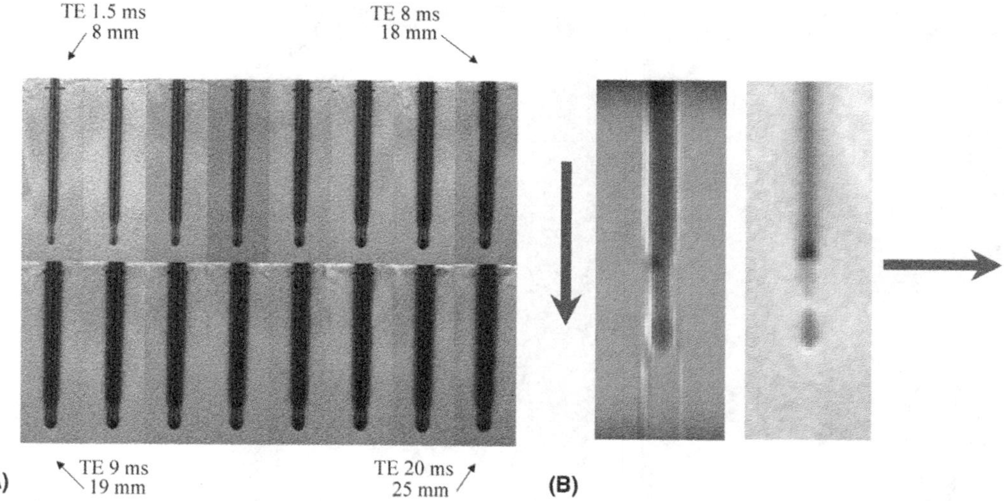

FIGURE 3 (**A**) Influence of the increasing echo time on the magnetic resonance (MR) appearance of a 14 g (equal to 2 mm) core biopsy needle (Tru-Cut). At 1.5 msec, the needle is displayed a 8 mm thick structure and increases subsequently with increasing echo time up to 25 mm at an echo time of 20 msec. (**B**) Influence on the MR appearance of the an MR compatible 14 g needle, which is aligned to the phase-encoding gradient (*arrow*) (*left*) as well as perpendicular to the gradient (*right*). *Abbreviation*: TE, echo time.

imaging guidance provides results similar to those of biopsies performed under CT guidance (26,27). MR offers the unique feature of allowing histologic evaluation of nonspecific bone marrow edema (Fig. 4).

MR can be also used for percutaneous core decompression in patients with avascular necrosis of the femoral or humeral head (Fig. 5). However, the experience in this field is still limited and more clinical experience is required. The complications of MR-guided percutaneous biopsies of bone lesions are similar to those of any percutaneous biopsy. These include infection, bleeding, nerve injury, and breakage of the biopsy device. None of these complications has been observed in our own series.

MR guidance cannot be currently recommended for spinal lesions, which often require a transpedicular biopsy pathway. A safe access to the lesion depends here on a meticulous biopsy planning. In such cases, we still prefer CT as a guidance tool, because it offers the best spatial resolution. In addition, for the percutaneous ablation of the nidus in osteoid osteomas, CT is still the method of choice, because it allows more exact targeting of the nidus.

TABLE 2 Sequence Parameters for Fast Interventional MRI

	GE	GE+seg. EPI	FSE	LoLo
FOV	450 × 450	375 × 275	375 × 275	250 × 250
Matrix	256 × 256	128 × 128	256 × 187	256 × 256
Slice thickn. (mm)	10	7	10	10
TR	8 msec	246 msec	576 msec minimum	592 msec minimum
TE	3.6 msec	4.6 msec	100 msec effective	104 msec effective
Flip angle	25°	80°	90°	90°
AC	2	2	2	1
Time	4.1 sec	15 sec for 16 slices EPI factor 3	Turbo factor 79	Turbo factor 77

Note: Sequence parameters for fast imaging techniques used for interventional magnetic resonance imaging. LoLo applies the 90° excitation pulse, orthogonally rotated to the following 180° refocusing pulses. Only the spins first excited by the 90° pulse and then refocused by the 180° pulse will give signal to the image. No fold over artifacts can occur in this type of fast small field of view imaging.
Abbreviations: GE, gradient echo; GE + EPI, gradient echo planar; FSE, fast spin echo imaging; LoLo, Local look; TE, echo time.

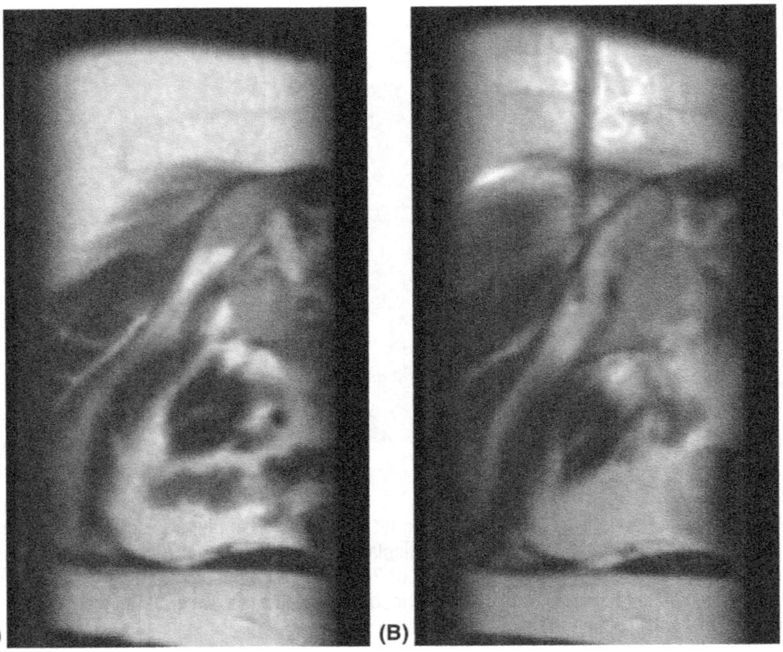

(A) (B)

FIGURE 4 Bone-marrow edema of unknown origin in the right iliac bone. The patient underwent magnetic resonance–guided bone biopsy in prone position using a hollow auger under local anesthesia. On pathohistology, osteomyelitis was confirmed. Fast spin echo technique.

Percutaneous biopsies of the appendicular skeleton and pelvis, however, can be considered as safe, and based upon the still limited experience, accurate procedures (Fig. 6). For many lesions, it may be used as an alternative targeting modality to CT.

Magnetic Resonance Arthrography

MR arthrography offers superior detailed information about the normal and pathologic anatomy of joints, especially of the shoulder (35,36). Since the introduction of MR arthrography, several groups demonstrated that direct arthrography leads to a better delineation of labral tears and rotator cuff lesions in acute and chronic injuries. Normally, the diluted contrast agent is injected under fluoroscopic guidance outside the MR scanner before the patient is transferred to the MR unit. However, successful direct puncture of the shoulder joint under MR guidance has been reported in a series of three patients in an open 0.5 T scanner using a three-dimensional digitizer system (37,38). No complications were reported and all punctures were diagnostic,

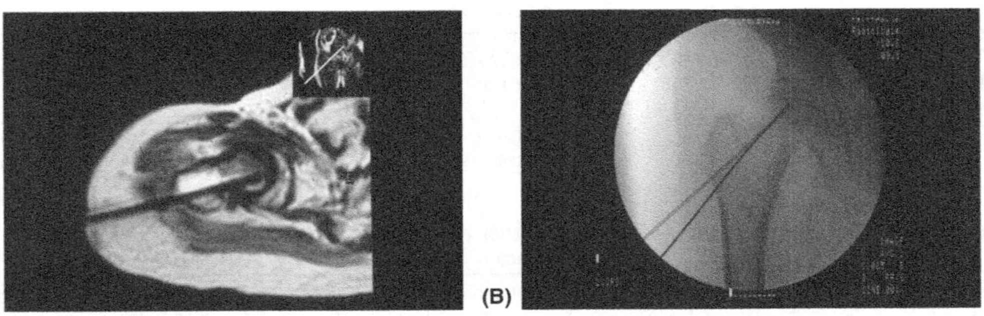

(A) (B)

FIGURE 5 (**A**) Magnetic resonance (MR)–guided core decompression of an avascular necrosis of the femoral head using a MR compatible hollow auger. For the procedure, a paracoronal approach has been chosen. For control, a T1-weighted sequence has been applied. (**B**) Fluoroscopical control of the MR-guided procedure.

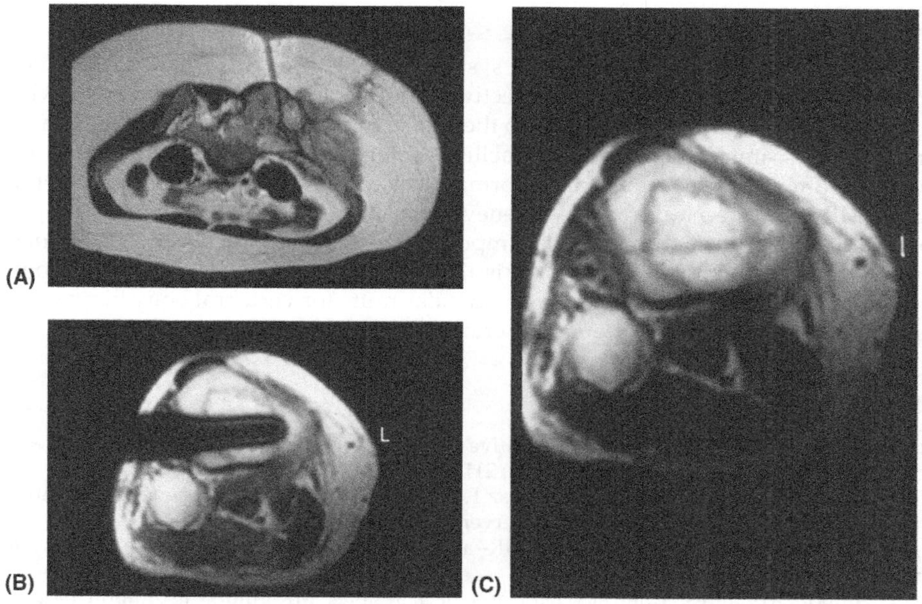

FIGURE 6 (**A**) Magnetic resonance (MR)–guided biopsy of a large chondrogenous tumor, proved to be a low-grade chondrosarcoma on pathohistology. Patient in prone position. For biopsy, a Tru-Cut needle has been used. Biopsy was performed using a T1-w fast gradient echo sequence. (**B**) Biopsy of a metaphyseal lesion of the proximal femur, verified as bone infarction on histology. Note the large artifact of the hollow auger. The needle is perpendicular to the frequency-encoding gradient in the axial slice orientation on this T1-weighted gradient echo image. (**C**) After the biopsy probe has been taken, the biopsy channel is well visualized on MR.

with the pathology confirmed on arthroscopy. Petersilge et al. (39) reported successful joint puncture on a 0.2 T system using an anterior or a modified anterosuperior approach. In these series no adverse events were reported, except inadvertent injection of the contrast agent into the subacromial or subdeltoidal bursa.

In our institution, we perform the puncture of the shoulder under fluoroscopic control inside the magnet room, with the patient moved directly to the MR scanner via the floating tabletop. For interventional-arthrographic purposes, the gadolinium chelate solution (e.g., gadpentetate dimeglumine, Magnevist, Schering, Berlin, Germany) is diluted 1:100 or 1:200 in saline solution. None of the commercially available gadolinium chelates have been approved for intra-articular injection.

Local Drug Therapy

Injection of corticosteroids or local anesthetics is a well-known procedure in musculoskeletal radiology. Until now, fluoroscopy or CT guidance has been used. Because both methods are using ionizing radiation and the patient population is usually younger, MR is an attractive alternative. The technique used is similar to the procedures described under CT guidance. Pereira et al. (40,41) described their first experiences in 12 patients with 24 steroid injections into the sacroiliac joints. All procedures were successfully carried out on an open 0.2 T open system, and clinical improvement was observed in 10 of 12 patients suffering from persistent buttock pain.

Celiac plexus block has also been carried out under MR guidance. Hol et al. (42) performed it in a patient population suffering from pain caused by pancreatitis or tumors of the pancreas. They reported good-to-moderate pain relief in 13 of 14 patients. No complication was reported, and the needle pathway was always achieved appropriately under MR guidance.

Other neurolytic procedures such as sympathicolysis in patients suffering from peripheral artery occlusive diseases may also be carried out under MR guidance. The intramuscular injection of botulinus toxin under MR guidance has also been performed in our institution.

CONCLUSION

Interventional procedures of the musculoskeletal system may be performed under MR guidance. These procedures appear to be safe and effective. After acquiring some experience with this guidance tool, the procedure time is similar to those of CT guided procedures. They can be performed on most MR scanners. Open systems facilitate interventional procedures. However, they typically do not offer the same technical performance as closed high-field machines. In the majority of cases, MR is an alternative to CT. Whenever possible, MR may be used in younger patients, where radiation protection may be an important issue. Also, in selected cases such as bone marrow edema of unknown origin, MR is the method of choice as a guidance tool. In anatomically difficult regions, such as the transpedicular route for vertebral body biopsy, CT is still the method of choice because it offers a superior spatial resolution.

REFERENCES

1. Schwartz RB, Hsu L, Wong TZ, et al. Intraoperative MR guidance for intracranial neurosurgery: experience with the first 200 cases. Radiology 1999; 211:477–478.
2. Wong TZ, SchwartzRB, Black P, Alexander E, Loesz F. Intraoperative magnetic resonance guidance for intracranial neurosurgical procedures. Sem Intervent Radiol 1999; 16:23–30.
3. Lufkin R, Layfiled L. Coaxial needle system of MR- and CT-guided aspiration cytology. J Comput Assist Tomogr 1989; 13:1105–1107.
4. Duckwiler G, Lufkin RB, Teresi L, et al. Head and neck lesions: MR-guided aspiration biopsy. Radiology 1989; 170:519–522.
5. Vogl TJ, Mack MG, Müller PK, Straub R, Engelmann K, Eichler K. Interventional MR: interstitial therapy. Eur Radiol 1999; 9:1479–1487.
6. Tacke J, Adam G, Speetzen R, et al. MR-guided interstitial cryotherapy of the liver with a novel, nitrogen-cooled cryoprobe. Magn Reson Med 1998; 39:354–360.
7. Tacke J, Speetzen, Adam G, et al. Experimental MR imaging-guided interstitial cryotherapy of the brain. Am J Neuroradiol 2001; 22:431–440.
8. Haage P, Tacke J. MRI-controlled percutaneous cryotherapy of liver metastasis. Radiologe 2001; 41:77–83.
9. Tacke J, Adam G, Haage P, Sellhaus B, Grosskortenhaus S, Günther RW. MR-guided percutaneous cryotherapy of the liver: in vivo evaluation with histologic correlation in an animal model. J Magn Reson Imaging 2001; 13:50–56.
10. Merkle EM, Shonk JR, Zheng L, Duerk JL, Lewin JS. MR imaging-guided radio frequency thermal ablation in the porcine brain at 0.2 T. Eur Radiol 2001; 11:884–892.
11. Merkle EM, Haaga JR, Duerk JL, Jacobs GH, Brambs HJ, Lewin JS. MR imaging guided radio-frequency thermal ablation in the pancreas in a porcine model with modified clinical C-arm system. Radiology 1999; 213:461–467.
12. Glowinski A, Kürsch KJ, Adam G, Bücker A, Noll TG, Günther RW. Device visualization for interventional MRI using local magnetic fields: basic theory and its application to catheter visualization. IEEE Trans Med Imaging 1998; 17:786–793.
13. Adam G, Glowinski A, Neuerburg J, Büker A, van Vaals JJ, Güther W. Visualization of MR compatible catheters by electrically induced field inhomogeneities: evaluation in vivo. J Magn Reson Imaging 1998; 8:209–213.
14. Büker A, Neuerburg JM, Adam G, et al. Stentplazierung unter Echtzeit-MR-Kontrolle: erste tierexperimentelle Erfahrungen. Fortschr Röntgenstr 1998; 169:655–657.
15. Buecker A, Neuerburg JM, Adam GB, et al. Real-time MR fluoroscopy for MR-guided iliac artery stent placement. J Magn Reson Imaging 2000; 12:616–622.
16. Leung DA, Debatin JF, Wildermuth S, et al. Active visualization of intravascular catheters with MRI: in vitro and in vivo evaluation. AJR 1995; 164:1265–1270.
17. Dumoulin CL, Souza SP, Darrow RD. Real time position monitoring of invasive devices using magnetic resonance. Magn Reson Med 1993; 29:411–415.
18. Manke C, Nitz WR, Djavidani B, et al. MR imaging-guided stent placement in iliac arterial stenoses: a feasibility study. Radiology 2001; 219:527–534.
19. Koskinnen SK, Parkkola RK, Karhu J, Komu MES, Kormano MJ. Orthopedic and interventional applications at low field MRI with horizontally open configuration. Radiologe 1997; 37:819–824.
20. Schenck JF, Jolesz FA, Roemer PB, et al. Superconducting open configuration MRI system for image guided therapy. Radiology 1995; 195:805–814.
21. Adam G, Neuerburg J, Bücker A, et al. Interventional MR: first clinical experience on a 1.5 T system combined with C-arm fluoroscopy. Invest Radiol 1997; 32:19–97.
22. Adam G, Büker A, Glowinski A, Nolte-Ernsting C, Neuerburg C, Güther RW. Interventionelle MR-Tomographie: Geräekonzepte. Radiologe 1998; 38:168–172.

23. van Vaals JJ, Mens G, Martin AJ, et al. Interactive real time MR on a standard scanner: requirements and clinical applications. In: Naruse S, Watari H, eds. Ultra Fast Magnetic Resonance Imaging in Medicine. Amsterdam: Elsevier Science, 1999:111–117.
24. Fahrig R, Butts K, Rowlands JA, et al. A truly hybrid interventional MR/X-ray system: feasibility demonstration. J Magn Reson Imaging 2001; 13:294–300.
25. Newman RW, Penner EA. External referencing systems. In: Debatin JF, Adam G, eds. Interventional Magnetic Resonance Imaging. Berlin: Springer, 1998.
26. Neuerburg J, Adam G, Bücker A, et al. Neues MR- (und CT-) kompatibles Knochenbiopsiesystem: erste klinische Ergebnisse. Fortschr Röntgenstr 1998; 169:515–520.
27. Neuerburg JM, Adam G, Buecker A, et al. MRI-guided biopsy of bone in a hybrid system. J Magn Reson Imaging 1998; 8:85–90.
28. Frahm C, Gehl HB, Melchert U, Weiss HD. Visualization of magnetic resonance compatible needles at 1.5 and 0.2 T. Cardiovasc Intervent Radiol 1996; 19:335–340.
29. Gehl HB, Frahm C. MR gesteuerte Biopsien. Radiologe 1998; 38:194–199.
30. Bücker A, Adam GB, Neuerburg JM, Glowinski A, van Vaals JJ, Günther RW. MR-gesteuerte Biopsien mit einer hochauflösenden schnellen T2-gewichteten Turbospinechosequenz "LoLo": erste klinische Erfahrungen. Fortschr Röntgenstr 1997; 167: 491–495.
31. Buecker A, Adam G, Neuerburg JM, Glowinski A, van Vaals JJ, Guenther RW. MR-guided biopsy using a T2-weighted single-shot zoom-imaging sequence (local look technique). J Magn Reson Imag 1998; 8:955–959.
32. Bücker A, Adam G, Neuerburg JM, et al. Echtzeit-MR mit radialer k-Raumabtastung zur Überwachung angiographischer Interventionen. Fortschr Röntgenstr 1998; 169: 542–546.
33. Koenig CW, Duda SH, Truebenbach J, et al. MR-guided biopsy of musculoskeletal lesions in a low field system. J Magn Reson Imaging 2001; 13:761–768.
34. Salomonovitz E. MR imaging-guided biopsy and therapeutic intervention in closedconfiguration magnet: single center series of 361 punctures. AJR 2001; 177:159–163.
35. Palmer WE, Caslowitz PL, Chew FS. MR arthrography of the shoulder: normal intraarticular structures and common abnormalities. AJR 1995; 164:141–146.
36. Kopka L, Funke M, Fischer U, Keating D, Oestmann J, Grabbe E. MR arthrography of the shoulder with gadopentetate dimeglumine: influence of concentration, iodinated contrast material and time on signal intensity. AJR 1994; 163:621–623.
37. Trattnig S, Breitenseher M, Pretterklieber M, et al. MR-guided joint puncture and realtime MR assisted contrast media application. Acta Radiol 1997; 38:1047–1049.
38. Hilfiker PR, Weishaupt D, Schmid M, Dubno B, Hodler J, Debatin JF. Real-time MR-guided joint puncture and arthrography: preliminary results. Eur Radiol 1999; 9: 201–204.
39. Petersilge CA, Lewin JS, Duerk JL, Hatem SF. MR arthrography of the shoulder: rethinking traditional imaging procedures to meet the technical requirements of MR imaging guidance. AJR 1997; 169:1453–1457.
40. Pereira PL, Gunaydin I, Duda SH, et al. Corticoisteroid injections of the sacroiliac joint during magnetic resonance: preliminary results. J Radiol 2000; 81:223–226.
41. Pereira PL, Gunaydin I, Trubenbach J, et al. Interventional MR imaging for injection of sacroiliac joints in patients with sacroiliitis. AJR 2000; 175:265–266.
42. Hol PK, Kvarstein G, Viken O, Smedby O, Tonnessen TL. MRI-guided celiac plexus block. J Magn Reson Imaging 2000; 12:562–564.

33 Catheterization of the Long Bones

Nicolas Sans, Erik Estivalezes, Denise Galy, Pascal Swider, Hélène Chiavassa, and Jean-Jacques Railhac
Department of Radiology, Purpan University Hospital, Place du Docteur Baylac, Toulouse, France

INTRODUCTION

In interventional radiology and orthopedics, technological innovations have opened new perspectives for minimally invasive approaches avoiding open surgery. We developed a technique to permit access to the entire contents of long bones, using a minimally invasive approach based on vascular catheterization.

TECHNIQUE

Most interventional techniques in the skeleton include direct percutaneous biopsy and therapeutic procedures such as percutaneous resection of osteoid osteoma, injections of ethibloc or calcitonin in bone cysts, and acrylic cement injection. Precise localization under fluoroscopic or computed tomography (CT) guidance is necessary for such procedures. In the future, magnetic resonance imaging (MRI) will undoubtedly be useful for localization and treatment of focal bone-marrow anomalies. Moreover, since the introduction of open magnets or short-bore closed magnets, MRI has become an interesting tool for guidance in interventional procedures (1).

The first step of the technique (penetration of the shaft using a drill) has been commonly used for more than 10 years for percutaneous resection of bone tumors such as osteoid osteoma. It is a safe procedure, but fluoroscopic or CT guidance is necessary to prevent sideways slipping and to make holes in the correct position with regard to the shaft. Thus, an oblique angle of 40° seems to be the best compromise to allow satisfactory progression of the catheterization instruments. The soft tissues must also be protected during the deep-bone approach because of the high-speed rotation of the motor-driven drill (2).

During catheterization, progression within the medullary canal is easy. However, progression within the cancellous bone of the metaphysis and epiphysis is more difficult and requires a more rigid device. We therefore developed a prototype device, which we tested in vitro on various animal and human long bones.

The method of catheterization (Fig. 1) can be divided into three steps: (*i*) penetration of the cortex from the shaft, (*ii*) progression of the catheter within the shaft, and (*iii*) drilling of a targeted lesion in the metaphysis or extremity.

First, cortical drilling is carried out with existing orthopedic instrumentation (Aesculap AG, Tuttlingen, Germany) (Fig. 2) used for CT-guided percutaneous resection of osteoid osteoma (3,4). The cortex is drilled with 6 or 9 mm drills used alone or pushed over 2 mm Kirschner wires using a coaxial Seldinger technique. The diameter of the drilling machine must be selected carefully. Biomechanical studies have shown that a hole made in cortical bone leads to a loss of mechanical strength, which does not exceed 30% as long as the size of the hole does not exceed 30% of the shaft diameter (5). Care should be taken in selecting the size of the drill, and relief from weight bearing of the involved limb should be recommended to minimize the risk of secondary fracture on the approach. As in Seldinger's technique, the use of guidewires allows catheters to advance without difficulty in the marrow cavity and they can be changed if necessary, as in vascular interventional procedures. The main requirement for the new device we developed was that it should be sufficiently rigid to progress within the medullary cavity but flexible enough to bend within the epiphysis of the bone toward a precise location. It should also be as small as possible to minimize bone weakening. A modified endoscope thus seems to be the best alternative to reach the epiphysis of a long bone by a percutaneous approach.

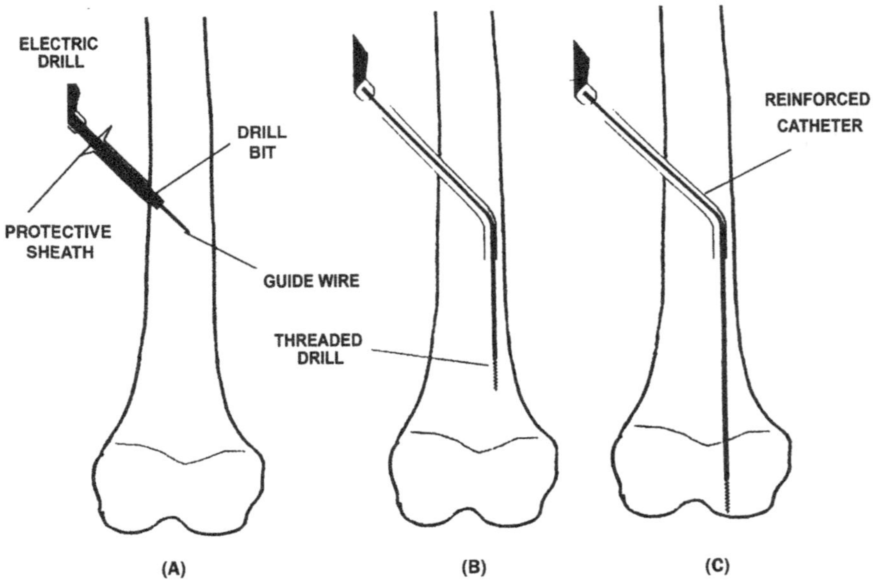

FIGURE 1 The prototype and detail of the drive cable and the burr at its extremity.

The position and the movement of the drill in relation to the main axis of the shaft are monitored under fluoroscopic guidance. The adjacent soft tissues are protected from the drill by a sheath in order to prevent thermal and mechanical damage (2). Oblique penetration at an angle of about 40° to the cortex toward the epiphyseal region appeared to be the best compromise to allow satisfactory motion within the medullary cavity and to prevent deviation during the percutaneous approach.

Second, progress within the medullary cavity with various available vascular catheters, for example, guidewires, dilators, or drainage catheters is performed. To allow catheterization of the bone extremity, a flexible, reinforced catheter is introduced into the shaft. Finally, the extremity is catheterized and the cancellous bone of the metaphysis and epiphysis penetrated. For drilling of spongy bone, a prototype catheterization device was developed, and the principle of the endoscope appeared to be very suitable for this purpose.

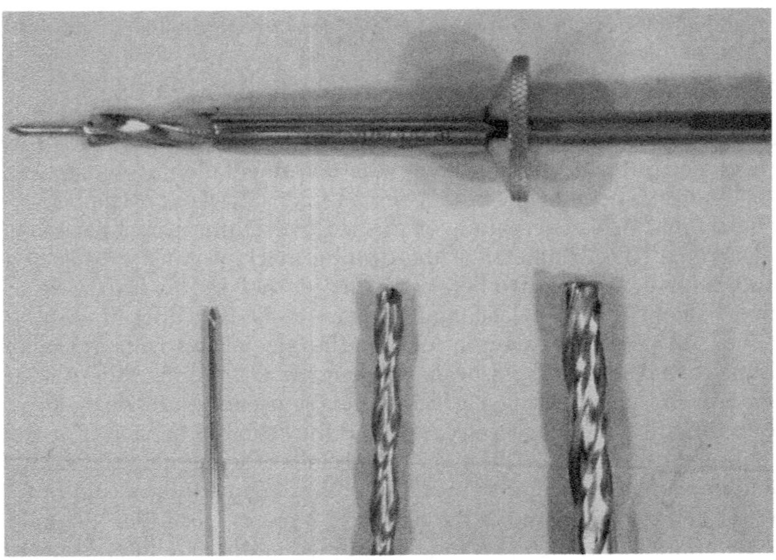

FIGURE 2 Drill-resection system: (*left* to *right*) 9-mm-diameter toothed drill, 9-mm-diameter drill, guidewire and coaxial system assembled for cortical drilling (*above*).

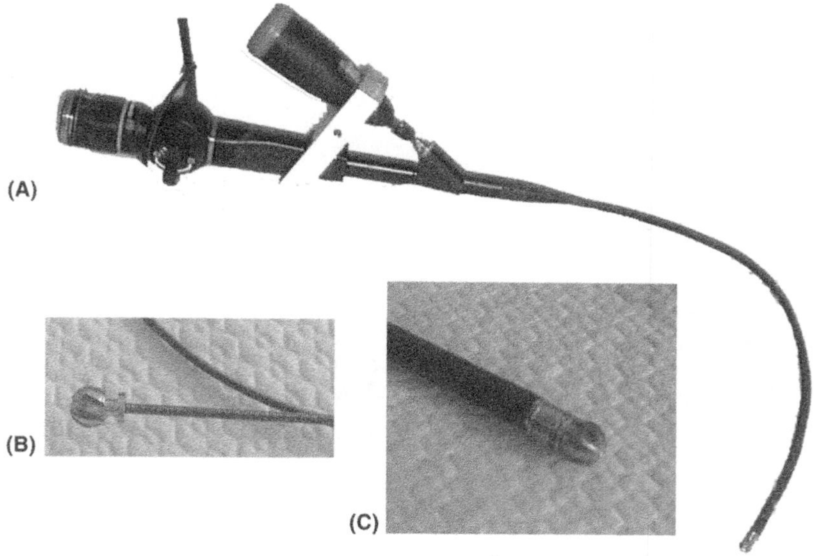

FIGURE 3 Different steps of the procedure: (**A**) Placement with an electric drill of the drill bit, using a Kirschner wire as a guide. Protection of soft tissues with a sheath. (**B**) Insertion of a dedicated sheath guiding a threaded drill through the long axis of the bone. (**C**) Replacement of the threaded drill by the modified endoscope, allowing drilling through the epiphysis of the bone.

The new device consists of an ureteroscope, a drive cable, and a burr. The burr is fixed at the end of the ureteroscope and is driven by the drive cable passing through the instrumental canal. The prototype is illustrated in Figure 3. Like a conventional endoscope, the mobility of this prototype allows the operator to drill and reach the extremity of the bone (Fig. 4).

Figure 5 shows the gross anatomy of a sheep's bone, which is cut into two parts after catheterization. Penetration of the burr in the spongy bone is clearly seen.

Clinical studies must be awaited, but already a certain number of applications seem feasible. The indications for such a technique may increase with the development of specific catheterization instruments: for aspiration of neoplastic cells to decrease tumor volume (especially in

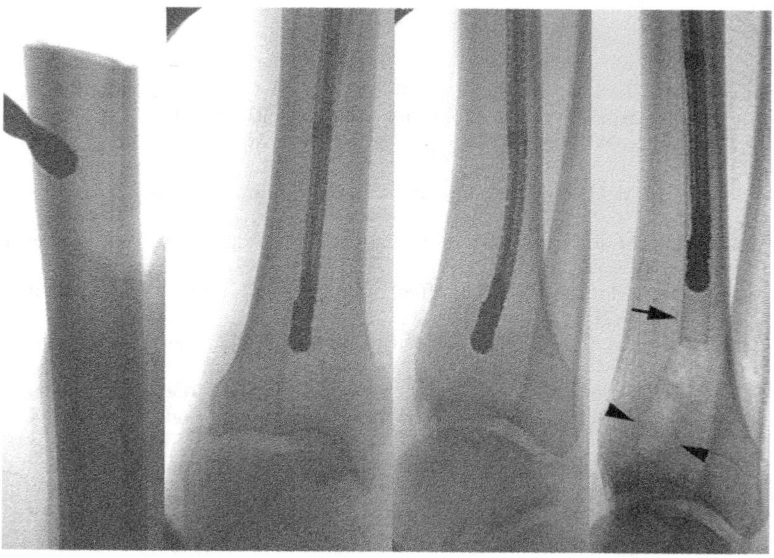

FIGURE 4 Tibial catheterization. The mobility of the prototype allows the operator to drill as far as the epiphysis. Note the reinforced catheter (*arrow*) placed within the marrow cavity and the hole made within the spongy epiphyseal bone (*arrowhead*).

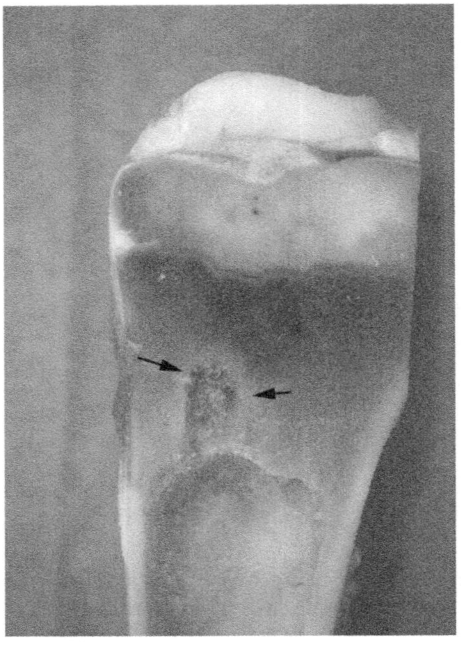

FIGURE 5 Gross anatomy of a sheep bone after catheterization. Note the cavity made by the burr within the spongy bone (*arrow*).

hematology) (6), abscess drainage (7–9), bone-marrow sampling, in situ injections of antimitotic drug or bone-growth stimulation factors (10), or preventive cement injections in weakened bones. The latter is a particularly promising perspective with the development of liquid cements that can be injected into tumoral or osteoporotic lesions.

CONCLUSIONS

This technique allows the bone to be approached through a safe pathway, avoiding vascular or nervous structures as well as joints and regions where bone stress is greatest. In the future, this technique should be of interest for diagnostic or therapeutic purposes, and subsequent clinical studies are required to prove its long-term value.

REFERENCES

1. Adam G, Bücker A, Nolte-Ernsting C, Tacke J, Günther RW. Interventional MR imaging: percutaneous abdominal and skeletal biopsies and drainages of the abdomen. Eur Radiol 1999; 9:1471.
2. Kattapuram SV, Rosenthal DI, Phillips WC. Trephine biopsy of the skeleton with the aid of a hand drill. Radiology 1984; 152:231.
3. Assoun J, Railhac JJ, Bonnevialle P, et al. Osteoid osteoma: percutaneous resection with CT guidance. Radiology 1993; 188:541.
4. Sans N, Galy-Fourcade D, Assoun J, et al. Osteoid osteoma: CT-guided percutaneous resection and follow-up in 38 patients. Radiology 1999; 212:687.
5. Clark CR, Morgan C, Sonstegard DA, Matthews LS. The effect of biopsy-hole shape and size on bone strength. J Bone Joint Surg Am 1977; 59:213.
6. Dollahite HA, Tatum L, Moinuddin SM, Carnesale PG. Aspiration biopsy of primary neoplasms of bone. J Bone Joint Surg Am 1989; 71:1166.
7. Compere EL, Metzger WI, Mitra RN. The treatment of pyogenic bone and joint infections by closed irrigation (circulation) with a non-toxic detergent and one or more antibiotics. J Bone Joint Surg Am 1967; 49:614.
8. Hoffer FA, Emans J. Percutaneous drainage of subperiosteal abscess: a potential treatment for osteomyelitis. Pediatr Radiol 1996; 26:879.
9. Pfister A, Ochsner PE. Experiences with closed irrigation-suction drainage and simultaneous administration of an antiseptic. Unfallchirurg 1993; 96:332.
10. Cook SD, Baffes GC, Wolfe MW, et al. The effect of recombinant human osteogenic protein-1 on healing of large segmental bone defects. J Bone Joint Surg Am 1994; 76:827.

34 Percutaneous Therapy of Cystic Bony Lesions

Francesco Aparisi
University Hospital La Fe and Nueve de Octubre Valencia, Valencia, Spain

INTRODUCTION

A focal region of radiolucency, or "cystic" lesion occurs when there is disruption of the steady state between bone formation and bone resorption. Matrix deposition, abnormal metabolic activity, or increased blood flow all may lead to these focal areas of osteolysis.

There is a spectrum of lytic lesions ranging from primary malignant tumors and metastases to subarticular geodes and unicameral bone cysts (Fig. 1). Biologically aggressive lesions show an irregular appearance, with poorly defined margins and a broad interface. Less aggressive lesions are termed bony cysts, with homogeneous lysis, well-defined sclerotic or nonsclerotic margins, cortical bone thinning without destruction, and, in some cases, fluid–fluid levels on computed tomography (CT) or magnetic resonance imaging (MRI).

Occasionally, conventional imaging may yield a specific diagnosis. In many cases, however, more advanced imaging or even percutaneous biopsy is required.

I recommend treating many types of lytic bone lesions to avoid potential pathological fractures.

Histologically there are three types of bone cysts, the simple bone cyst containing serofibrinous fluid, the ganglion or synovial cyst with mucinous material, and the aneurysmal bone cyst (Fig. 2), with central blood products. Because the central fluid is produced by the outer capsule, this fluid, to some degree, reflects the biologic behavior—serofibrinous in areas of low activity, mucinous in lesions with medium activity, and hemorrhagic in active processes.

Cysts require treatment following fracture when thought to be at risk for pathologic fracture, or when symptomatic. The classic therapy has been surgical resection of the membrane with internal bone grafting. Occasionally, the adjacent bone is inadequate to support the load without metallic fixation. Also, recurrence can occur when the capsule is incompletely resected.

An alternative treatment, after aspirating to confirm the presence of fluid, is to lyse the capsule and stimulate new bone production. These objectives can be achieved via controlled ischemia or foreign body–induced inflammation.

This ischemia is created iatrogenically by obstructing the blood supply or via the use of irritating chemical products.

Arterial coils can be placed to interrupt the blood flow to the lesion. This technique historically has had limited value due to the scarcity of large feeding vessels in these lesions and is currently considered obsolete. The notable exception is aneurysmal bone cysts. In this situation very selective embolization may be performed. To obtain effective and maintained ischemia, I recommend the use of metallic coils rather than particles. This method is technically difficult in younger patients.

An alternative procedure is to directly inject particles into the lesion, also with the intention of provoking ischemia. The site of puncture is selected utilizing CT guidance. Similar to feeding-vessel embolization, this technique has had limited success.

Method: Angiography with selective examination.
Materials: 18-gauge needle, 5 F catheter, and others for selective examination. Particles or metallic coils.
Remarks: Most useful in very highly vascularized lesions such as anuerysmal bone cysts. Follow-up shows progressive thickening of the wall.

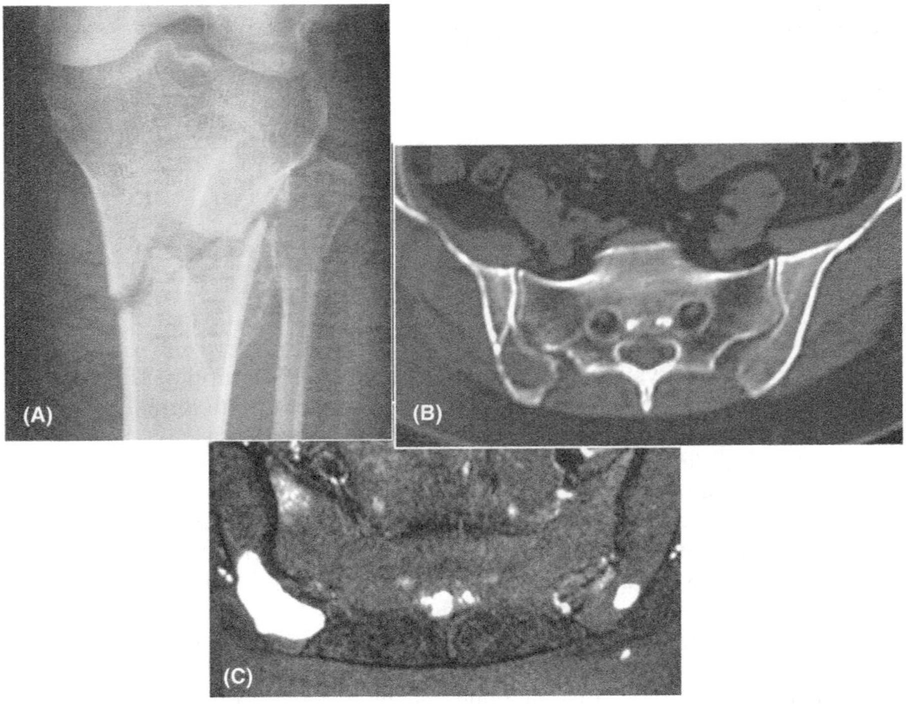

FIGURE 1 (**A**) Anteroposterior X-ray view with pathological fracture through a cystic lesion. (**B**) Axial computed tomography of the iliac bone. Proved simple cyst involving the posterior iliac spine. (**C**) Axial gradient echo T2W image of the iliac and sacrum. A high-signal cyst is indentured in the ilium. The additional focal area of marrow edema in the anterior sacrum (**C**) may be the result of altered mechanics and bone weakening. Cyst with cortical expansion.

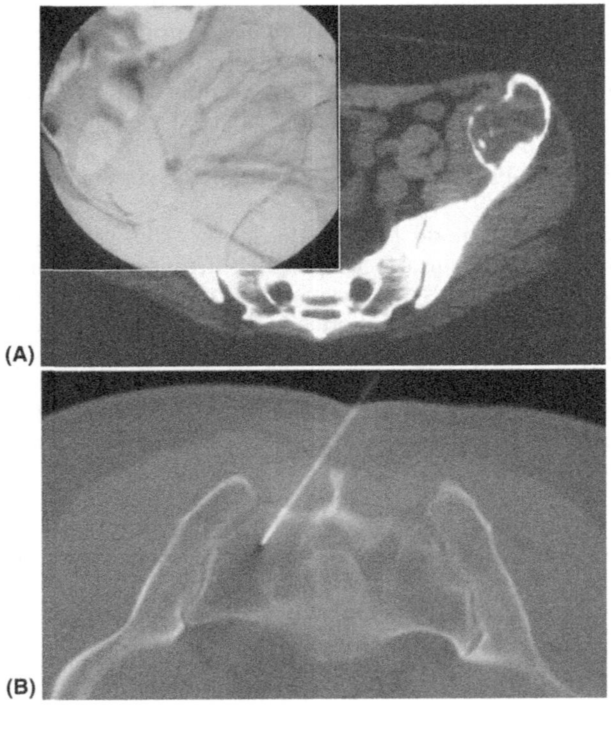

FIGURE 2 (**A**) Axial computed tomography scanner of an aneurysmal bone cyst in the left iliac wing with fluid–fluid levels. Inset shows angiography and coil embolization. (**B**) Direct puncture of a cyst-like lesion involving the sacrum.

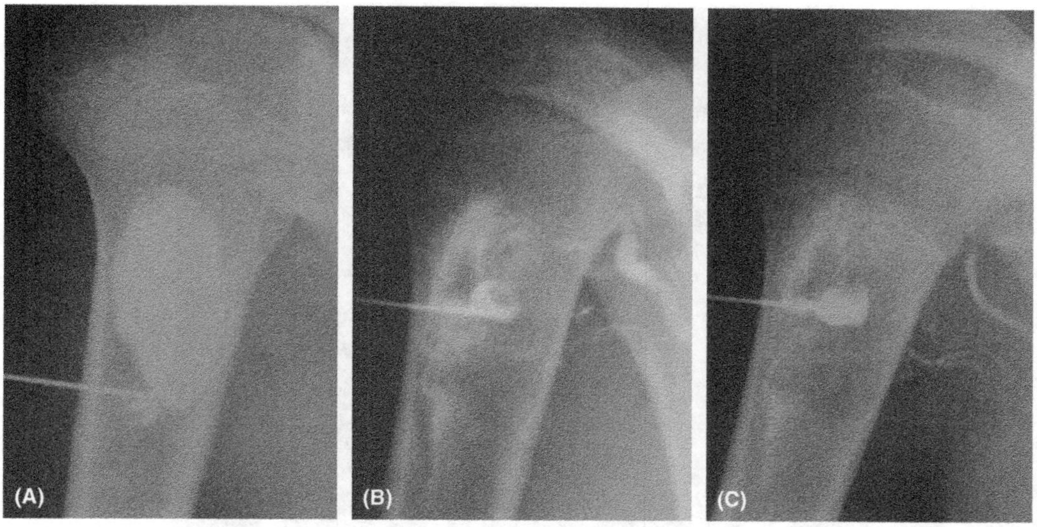

FIGURE 3 (**A**) Anteroposterior view of the proximal humeral shaft showing a typical single bone cyst filled with iodinated contrast after the puncture with a 20-gauge needle. (**B**) Same patient one month later, after one dose of 160 mg methyl methacrylate and prednisolone injection. Note the sclerotic margin and visible blood vessels. (**C**) End result of the treatment three months later.

Indirect chemical techniques are more effective than direct vascular ones. These products destroy the capsule and stimulate new bone formation. This occurs secondary to a combination of ischemia, inflammation, and direct toxicity.

The first products employed for the direct treatment of the cysts were various types of corticosteroids. We currently use methylprednisolone acetate (Fig. 3). This treatment was originally described as an operative procedure but may be performed percutaneously in a fluoroscopy suite.

The direct puncture should be made with a fine needle (18- to 20-gauge). First the periosteum is anesthetized, and then the superior aspect of the lesion is localized. The needle is then advanced into the cavity, utilizing a twisting maneuver. An additional needle may be placed to ensure ease of aspiration. In either situation, the cyst fluid is initially aspirated and evaluated visually and via cytology. This technique can be used in long-bone cysts, juxtarticular cysts, and in the spine.

During or before the procedure, iodinated contrast should be injected to determine the size and number of cavities within the cyst. It is important to compare these images with available cross-sectional studies to be sure that all parts of the lesion are accessible from the initial puncture. It is mandatory to fill all cyst loculations. If areas are not visualized on the initial contrast injection, performance of a second or third puncture may be necessary. Failure of the steroid to completely fill the lesion is an important cause of recurrence.

The approximate dose is 160 mg for each injection and three of them are carried out at one-month intervals. The steroid injection produces capsular thickening with hyperemia as seen on contrast-enhanced MRI. It is believed that the hyperemia leads to an uncoupling of the osteoclast–osteoblast balance, with the result being new bone formation.

Normally for patient follow-up, we utilize conventional radiography (Fig. 4). When planning a second steroid dose, iodinated contrast is injected into the lesion. When effective, the treatment produces capsular thickening and visualization of collateral circulation.

The corticosteroid dose rarely leads to side effects or adverse reactions (Fig. 5). Recurrences do occur, often related to incomplete filling. These recurrences are somewhat more frequent in the calcaneous and with very large cysts. The incidence of recurrence is decreased when opacification of all components of the lesion is verified prior to treatment. The best results are obtained in younger patients.

Method: Local anesthesia is provided and a direct puncture is made in the superior region of the lesion. Usually a handheld drill is sufficient. All loculations must be filled with iodinated

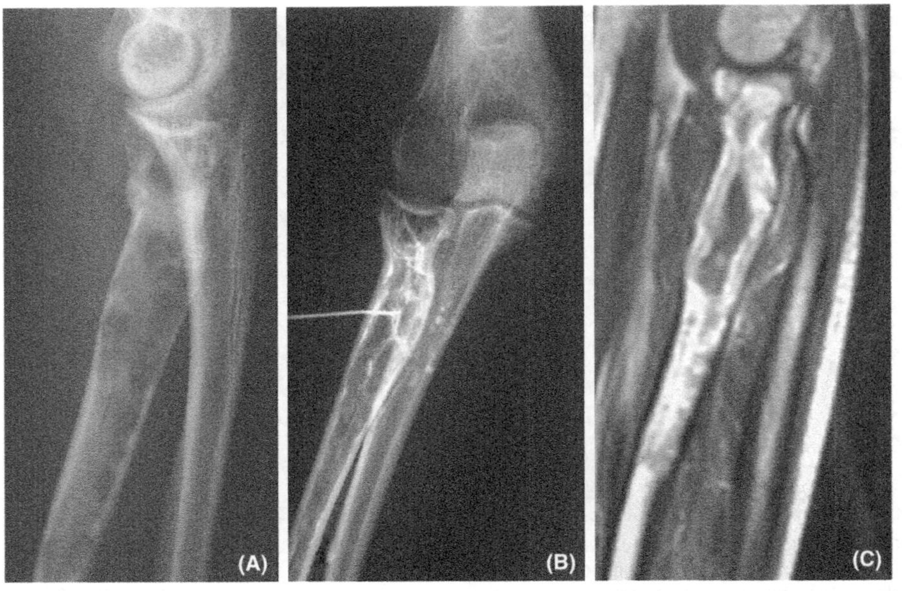

FIGURE 4 (**A**) Anteroposterior (AP) X-ray view of the forearm showing a typical single bone cyst. The cortical bone shows areas of endosteal scalloping corresponding with the cyst loculations. (**B**) AP X-ray after treatment, confirming the cyst size with iodinated contrast injection around the lesion. (**C**) Coronal SE TIW image after gadolinium injection. A thick capsule is noted.

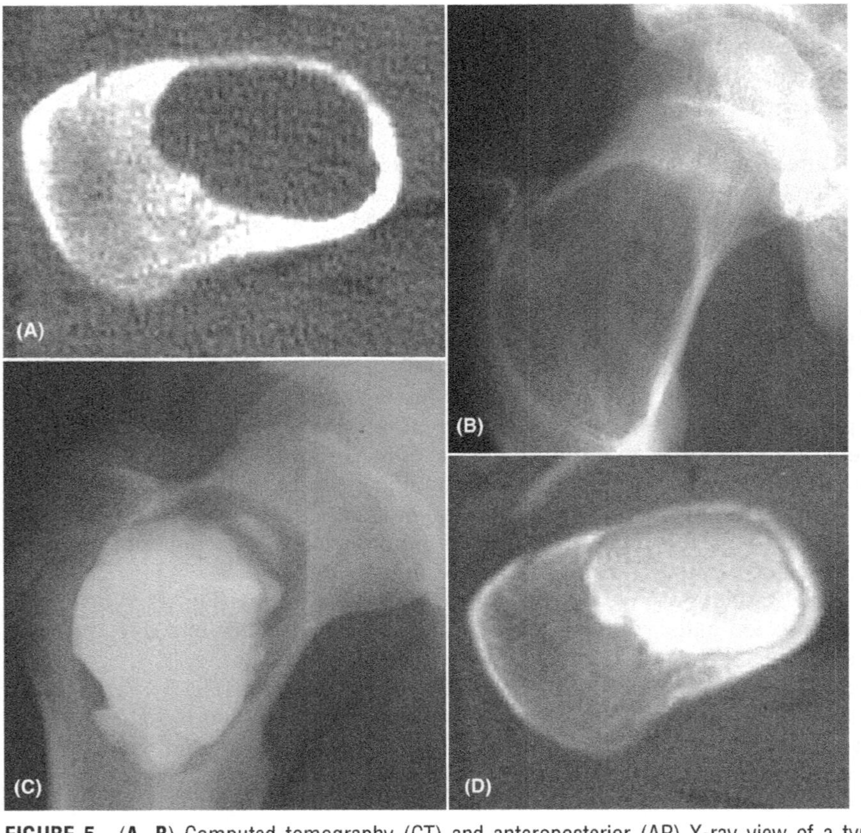

FIGURE 5 (**A, B**) Computed tomography (CT) and anteroposterior (AP) X-ray view of a typical single bone cyst involving the femoral neck. (**C, D**) AP X-ray view and CT-after to assess the cyst capacity following steroid treatment. Absence of bone answer or capsular thickening.

contrast. If necessary, each loculation might require a separate puncture. To avoid pain, the author recommends placing two needles. After treatment, the patient should rest for one to two hours.

Materials: 18 to 20-gauge needles, local anesthesia, iodinated contrast, and methylprednisolone acetate, 160 mg for each puncture.

Remarks: Steroid injection is a very safe procedure with few side effects and is useful in intermediate-sized lesions. This treatment is recommended for simple bone cysts. Recurrences can be minimized with attention to technique.

An alternative to steroid injection is bone marrow injection into the cyst. The marrow can be obtained from the patient's iliac crest and directly injected into the lesion. This treatment produces an osteoblastic reaction through a nonspecific mechanism. This technique can be used as a second-line procedure in the event of recurrence (Fig. 6).

In cases of recurrence, instead of marrow injection, or in very active lesions, alcohol may be injected. Using this method, scarring and chemical irritation to the capsule is produced with resultant thickening.

Absolute alcohol can be used in adults (Fig. 7). Or as an alternative, Ethibloc® composed of corn protein and ethanol, can be injected. The Ethibloc causes a foreign body reaction inflammatory response with the result also being capsular thickening.

These are the preferred treatments in areas where the steroids are not effective, such as the calcaneous and in very active or large simple or aneurysmal bone cysts. The author prefers the use of Ethibloc in pediatric patients as well. Theoretically, this is less dangerous than the absolute alcohol because of its increased viscosity.

No precise Ethibloc dose is accepted, but an excessive dose can produce serious inflammatory side effects. Because the therapeutic effect is based on hyperemia, there is a fairly narrow window between positive increase in blood flow and pathologic hyperemia (Fig. 8). We recommend follow-up MRI examinations to assess the extent of the inflammatory reaction. An aseptic abscess can also appear following Ethibloc injection.

Method: Direct puncture, avoiding areas of cortical destruction. If Ethibloc extravasates, a soft-tissue granuloma can form. A baseline contrast injection should be performed to determine the number of lobules. Careful dose titration and only partial cyst filling should be attempted.

Materials: Local anesthesia, iodinated contrast, 14-gauge needle, and Ethibloc.

Remarks: The Ethibloc is a quick and useful treatment in many cases and leads to rapid wall thickening. Failures, however, can occur.

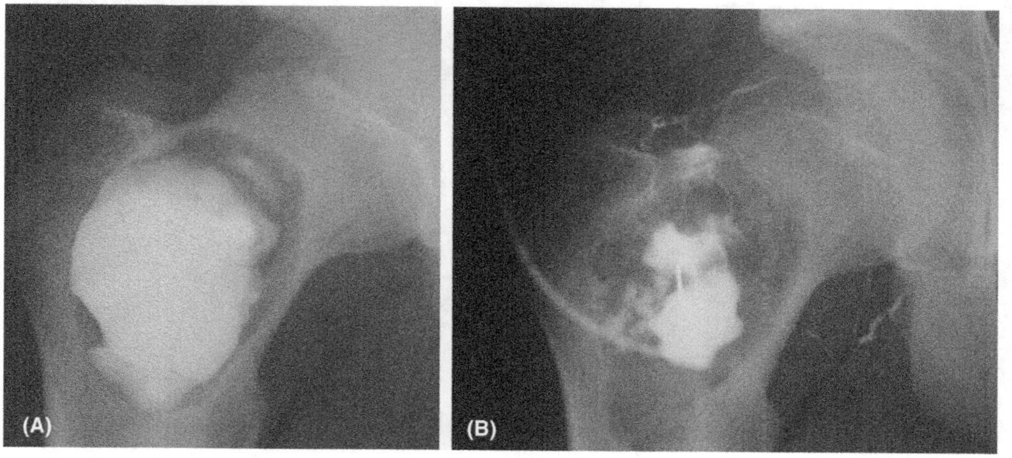

FIGURE 6 (**A**) Anteroposterior (AP) X-ray view of the femoral neck after a steroid treatment, previously seen in Figure 5. Note the absence of the thick wall. (**B**) AP X-ray view, one month after repeat bone marrow injection. Note the development of a thick capsule and new vessels.

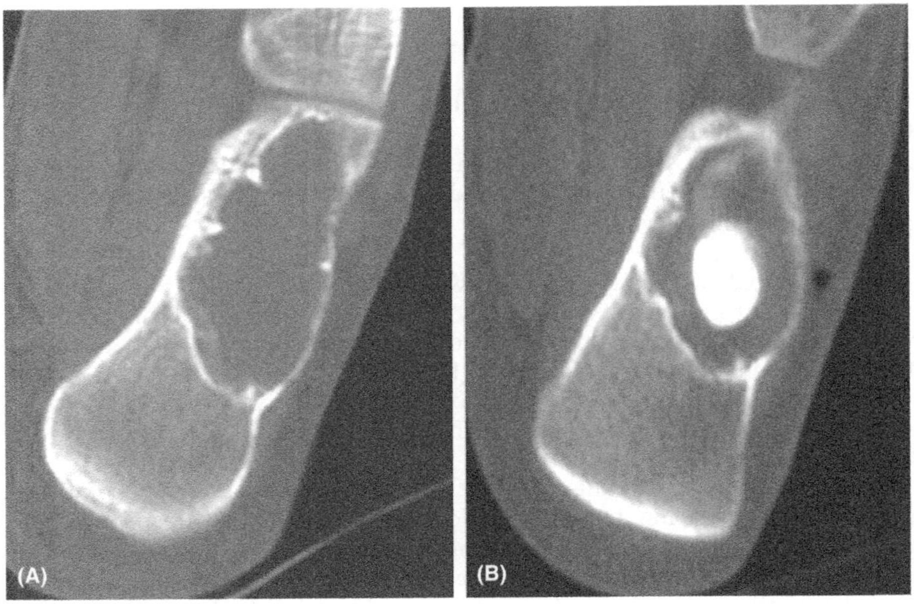

FIGURE 7 (**A**) Computed tomography of calcaneous bone showing an expansile, low-grade single bone cyst. (**B**) One month after 4-cc alcohol injection. Single bone cyst with thick walls due to a granulomatous reaction. Magnetic resonance imaging confirmed the thickening of the cyst wall.

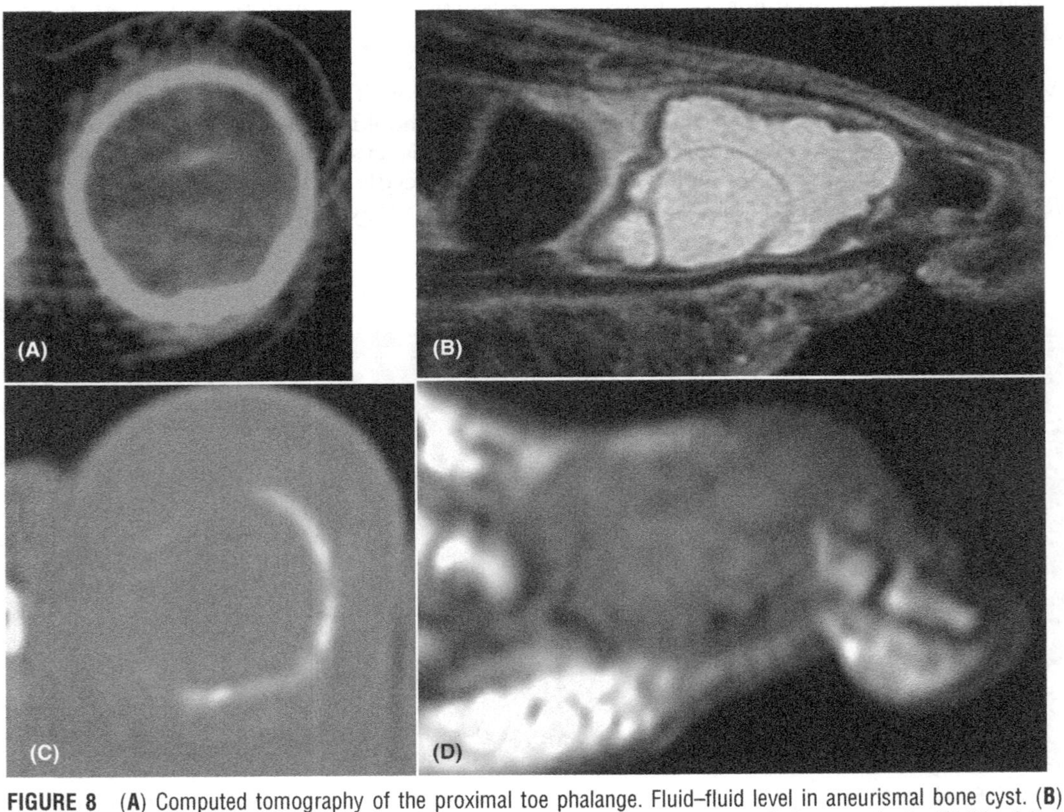

FIGURE 8 (**A**) Computed tomography of the proximal toe phalange. Fluid–fluid level in aneurismal bone cyst. (**B**) Sagittal T2W magnetic resonance imaging (MRI) shows a lobulated expansile lesion. (**C**) One month after Ethibloc injection, showing prominent wall thinning. (**D**) T1W MRI image also shows cortical disruption with a soft-tissue mass.

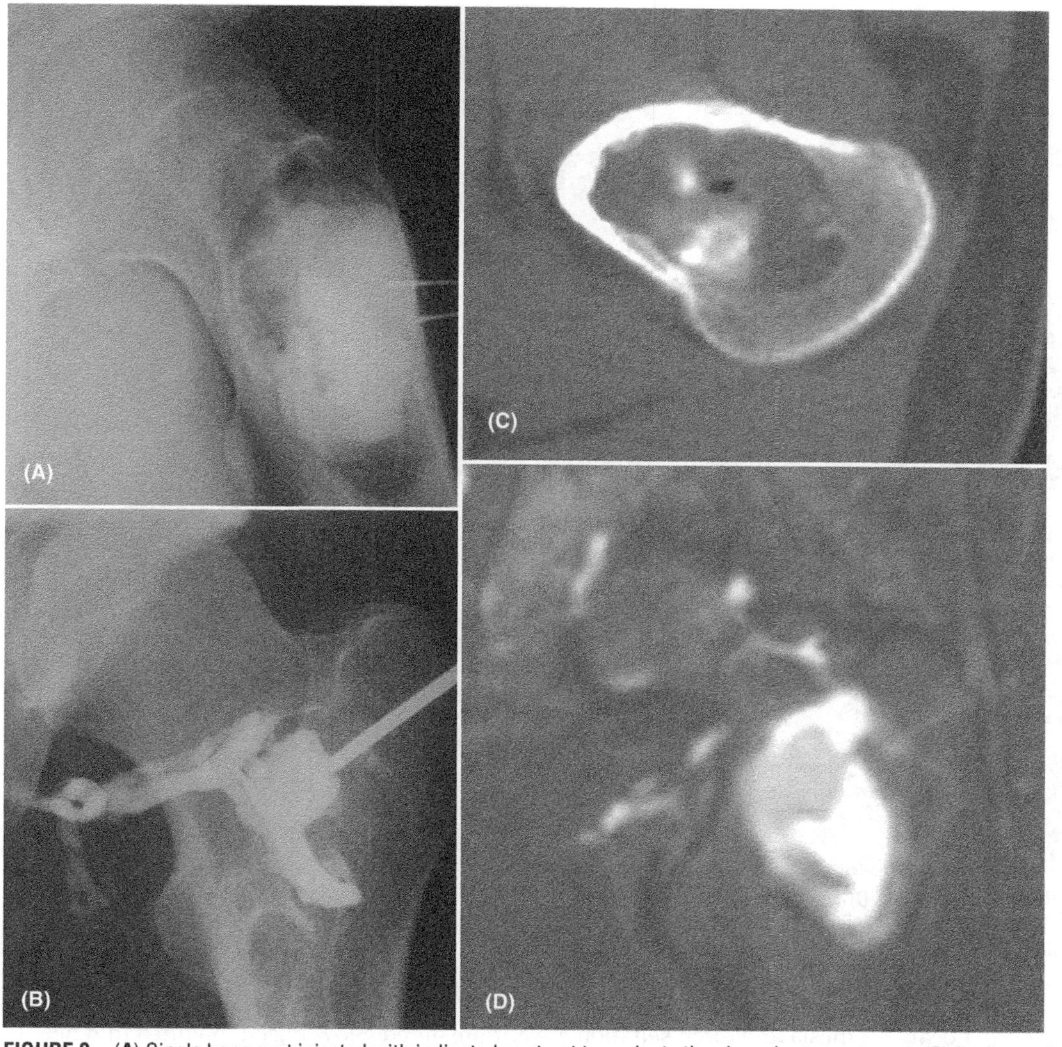

FIGURE 9 (**A**) Single bone cyst injected with iodinated contrast to evaluate the size prior to treatment. (**B**) Radiograph one month after ethanol treatment shows a marked decrease in cyst size, with visualization of large draining veins. (**C**) Computed tomography shows wall thickening. (**D**) Magnetic resonance imaging, GRE image. Shows central reactive tissue.

Mucinous cysts, juxta-articular ganglions, and large anuerysmal and simple bone cysts respond well to alcohol installation (Fig. 9). The ethanol provokes an inflammatory reaction, obliteration of blood vessels, and tissue necrosis.

Extreme care must be used when the lesion is close to the articular surface, because intra-articular alcohol can destroy cartilage and, potentially, the synovium. A baseline contrast injection is performed again in this situation to confirm that the cyst does not communicate with the joint surface.

In cases of ganglion cysts, we initially aspirate, and then inject a local anesthetic with a 14-gauge needle to decrease the viscosity of the lesional matrix. The alcohol is then administered by means of puncture with a 14- to 20-gauge needle. The recommended dose is 3 to 5 cc injected slowly to avoid pain, under fluoroscopic guidance.

MRI performed after alcohol injection will show perilesional bony edema and extensive capsular thickening (Fig. 10). Over time, the cavity will fill in with fibrous tissue and enhancement will decrease.

Usually the results of alcohol injection are excellent.

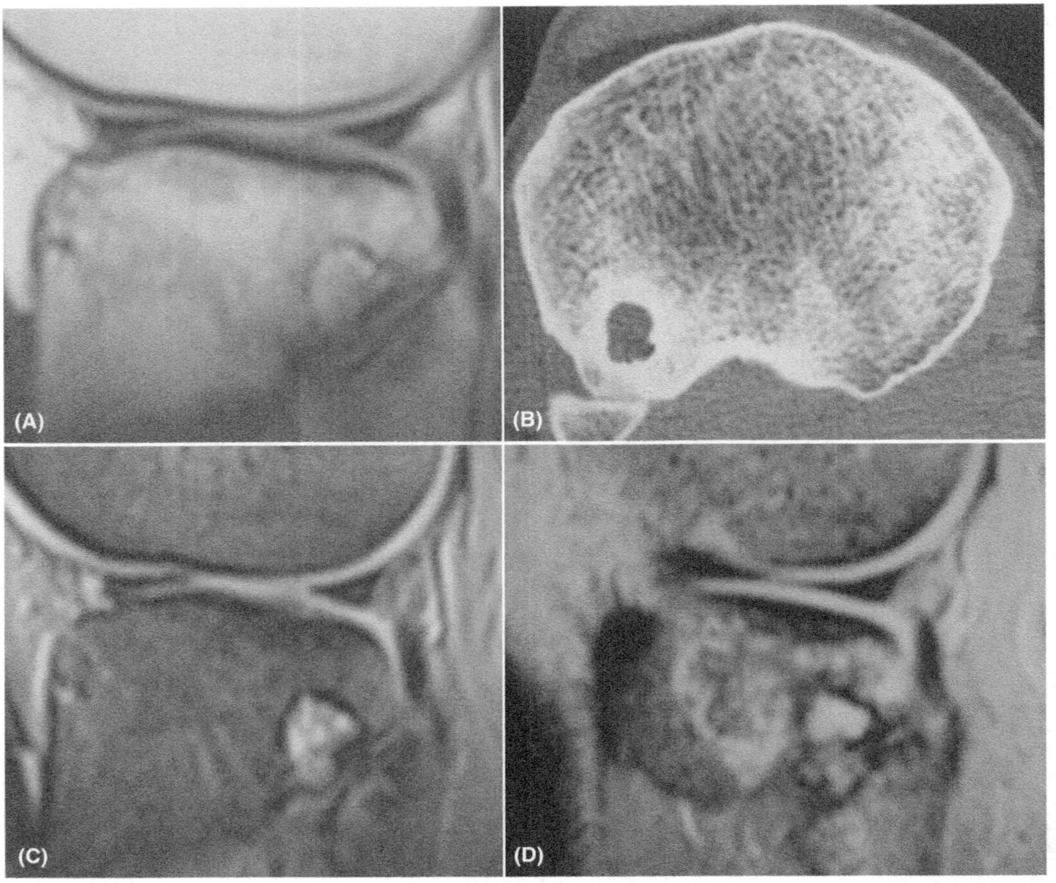

FIGURE 10 (**A** and **C**) Magnetic resonance imaging (MRI) Sagittal view TIW and T2W images, juxta-articular cyst without reactive perilesional edema. (**B**) Computed tomography shows the margins of the lesion. (**D**) MRI following alcohol treatment shows the new perilesional edema and thickening of the lesion capsule.

Method: Local anesthesia and often sedation are administered. Slow injection with a thin needle under fluoroscopic guidance to verify that all lobules are filled.
Materials: 14- to 20-gauge needle, local anesthesia, sterile absolute ethanol, and fluoroscopy.
Remarks: Alcohol injections yield rapid results with moderate risk. This method is best used for synovial and aneurysmal bone cysts.

As an alternative treatment for the juxta-articular ganglion, cement can be injected (Fig. 11). The increasing use of vertebroplasty has allowed the diffusion of cement handling skills and the development of kits for its injection. This is a very attractive treatment in those cases that merit rapid increase in the adjacent bone's resistance to stress and trauma. Similar to alcohol, methyl methacrylate is chondrotoxic. Therefore when lesions are juxta-articular we have utilized synthetic hydroxyapatite as an alternative. This injection is similar to bone cement and we use the same equipment as that used in vertebroplasty, including the needle and injector device. The advantage of hydroxyapatite, in contrast to the acrylic products, is the absence of exothermic reaction; however one loses the immediate strength available via cementoplasty, although both products are able to decrease the articular pain promptly.

To fill the cavity we must prepare the cement after evaluating the capacity of the cyst and observing for possible local leakage of iodinated contrast.

The use of a nonexothermic material allows us treat safely spinal lesions that have broken the posterior wall of the vertebral body (Fig. 12). The preliminary results are suggestive of success, and with this method we may treat more degenerative cysts.

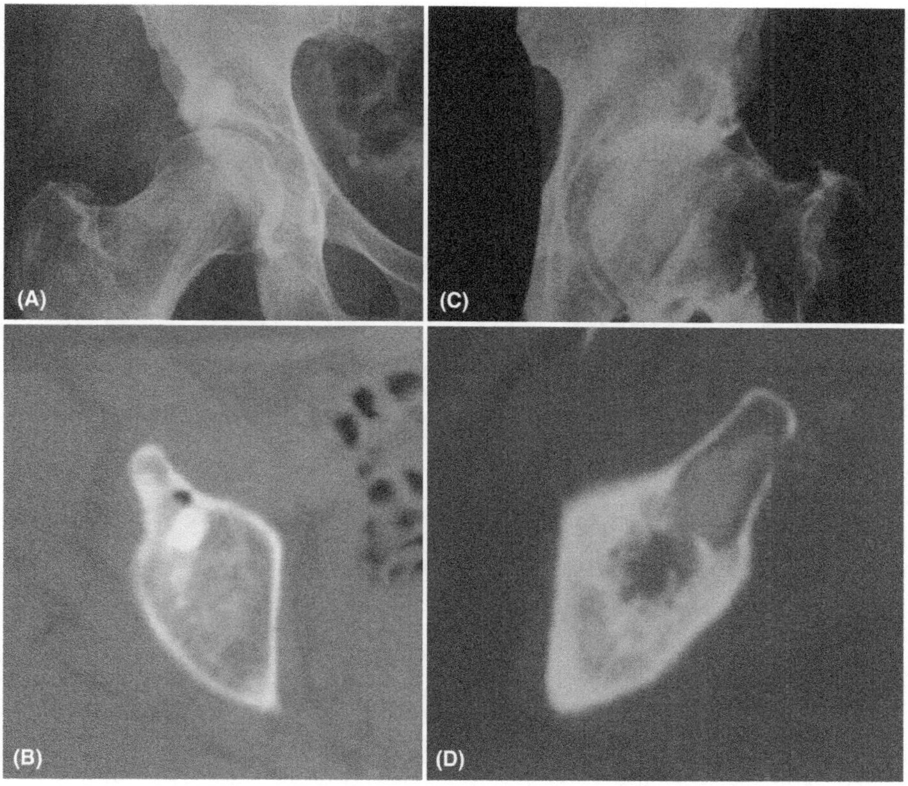

FIGURE 11 (**A**, **B**) Juxta-articular ganglion, filled with acrylic cement. (**C**, **D**) Juxta-articular ganglion, filled with synthetic hydroxyapatite.

Method: The paste must be injected using needles and vertebroplasty instruments. The puncture should avoid nerve roots and important vessels, deep analgesia is necessary.

Materials: Surgical acrylic cement or synthetic hydroxyapatite. 14-gauge needles.

Remarks: Sometimes it is better to fill the cavity with a solid product, especially in lesions close to the articular surface. We can use acryclic bone cement, but this material produces an increase in local tissue temperature that can be dangerous to the articular surface. To avoid chondrolysis we can use synthetic hydroxyapatite, a biologically well-accepted material without chondral toxity. The selection of the product to be injected can be determined after the extent and location of contrast extravasation is noted.

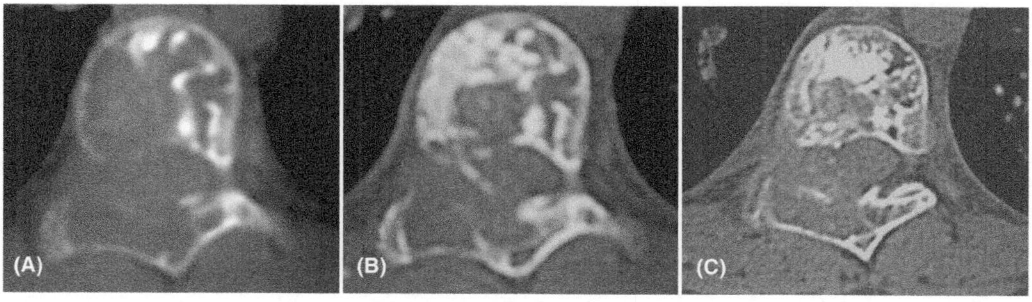

FIGURE 12 (**A**) Computed tomography of ABC. Vertebral body and posterior arch are involved. (**B**) One month later, the vertebral body becomes partially ossified. (**C**) Two months later, the soft-tissue component and posterior arch begin to be ossified.

FURTHER READING

Adamsbaum C, Leclet H, Kalifa G. Intralesional Ethibloc injections in bone cysts. Semin Musculoskelet Radiol 1997; 1(2):301–304.

Carrata A, Garbagna P, Mapelli S, Zucchi V. The treatment of simple bone cysts by topical infiltrations of methylprednisolone acetate: technique and results. Eur J Radiol 1983; 3(1):3–8.

Delloye C, Docquier PL, Comu O, et al. Simple bone cysts treated with aspiration and a single bone marrow injection. A preliminary report. Int Orthop 1998; 22(2):134–138.

Garg NK, Carty H, Walsh HP, Dorgan JC, Bruce CE. Percutaneous Ethibloc injection in aneurysmal bone cysts. Skeletal Radiol 2000; 29(4):211–216.

Leclair A, Gangi A, Lacaze F, et al. Rapid chondrolysis after an intra-articular leak of bone cement in treatment of a benign acetabular subchondral cyst: an unusual complication of percutaneous injection of acrylic cement. Skeletal Radiol 2000; 29(5):275–278.

Leclet H, Adamsbaum C. Intraosseous cyst injection. Radiol Clin North Am 1998; 36(3):581–587. Review.

Lokiec F, Wientroub S. Simple bone cyst: etiology, classification, pathology, and treatment modalities. J Pediatr Orthop B 1998; 7(4):262–273. Review.

Kose N, Gokturk E, Turgut A, Gunal I, Seber S. Percutaneous autologous bone marrow grafting for simple bone cysts. Bull Hosp J Dis 1999; 58(2):105–110.

Rossi C, Ricci S, Boriani S, et al. Percutaneous transcatheter arterial embolization of bone and soft tissue tumors. Skeletal Radiol 1990; 19(8):555–560.

Oppenheim WL, Galleno H. Operative treatment versus steroid injection in the management of unicameral bone cysts. J Pediatr Orthop 1984; 4(1):1–7.

Wilkins RM. Unicameral bone cysts. J Am Acad Orthop Surg 2000; 8(4):217–224. Review.

35 Ethanol Injection in Vertebral Hemangiomas

Francesco Aparisi
University Hospital La Fe and Nueve de Octubre Valencia, Valencia, Spain

INTRODUCTION

The vertebral hemangioma (VH) is a benign condition, asymptomatic in most cases. However, neurologic symptoms and local pain are sometimes present in relation to morphological changes, fracture, bone expansion, or presence of a soft-tissue mass. Radiologically, VHs are characterized by a thickening of the vertical bone trabeculae, giving a typical striated appearance, and the rarefaction of other trabeculae with vascular proliferation or fat deposition in the medullary spaces. There is no general agreement concerning the possibility of predicting the clinical behavior from radiographs or magnetic resonance imaging (MRI).

VH is more a vascular malformation than a true neoplasm. It is found in 12% of autopsic spines. Most patients are asymptomatic and the lesion is discovered accidentally during radiologic evaluation. Three categories of VHs have been described: asymptomatic, symptomatic (painful), and compressive hemangiomas (Figs. 1 and 2). Asymptomatic VHs are most frequent. Symptomatic VHs are seen in patients evaluated radiographically for back pain of uncertain origin. Compressive VHs that behave as an aggressive tumor may have the following radiologic signs: (*i*) location between T3 and T9, (*ii*) involvement of the entire vertebral body, (*iii*) involvement of the neural arch, (*iv*) irregular trabeculation, (*v*) expanded and poorly defined cortex, and (*vi*) presence of soft-tissue mass. Three or more of the above-mentioned signs may indicate a potentially symptomatic VH in patients with back pain of uncertain origin. The pain is the main reason for treatment, although the presence of a complication such as bone fracture, soft-tissue mass, or bone expansion with neural canal stenosis may also require treatment.

Symptomatic and compressive VHs can be treated in several ways. These techniques include radiotherapy, surgery, arterial embolization, and various types of vertebroplasty. Ethanol injections were first used successfully in treatment of VH in two patients with spinal cord compression. However, since this first publication, a case of VH with spinal cord compression in which alcohol injections caused severe but transient neurologic complications has been reported. Thus, there is controversy about the use of ethanol injection in the treatment of VHs. The purpose of this chapter is to assess the efficacy of percutaneous embolization with an alcohol solution in the treatment of hemangiomas. This treatment aims at obtaining a sclerosis of the vascular channels and a decrease of the soft-tissue component.

In our experience, the criteria for ethanol treatment are limited to the presence of pain and the absence of major risk factors. In any case, the complication rate associated with ethanol injection should be lower than that of the surgical treatment.

TECHNIQUE

Both fluoroscopic and computed tomography (CT) guidance can be used. After preparing an aseptic field, local anesthesia to the skin and the targeted vertebral arch is performed using a spinal needle. Three types of approach can be used to reach the vertebral body—the transpedicular, posterolateral oblique, and the costovertebral route at the thoracic level. To avoid risks, I recommend the transpedicular as the safest approach (Fig. 3). A 14-gauge bone-biopsy needle [Angiomed (Karlsruhe, Germany)] is inserted through the skin and pedicle into the vertebral body. In some cases, this needle is replaced by an 18-gauge needle, mainly in cases of VH with a soft-tissue involvement. In cases of hard bone, a hammer can be used, but use of a manual drill device is preferable.

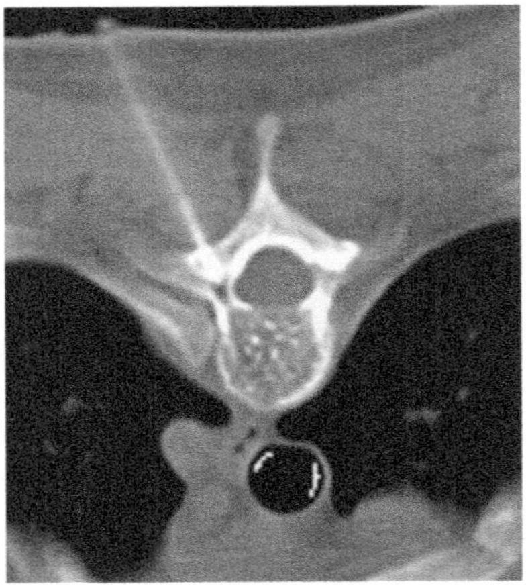

FIGURE 1 Symptomatic noncomplicated hemangioma, without involvement of the neural arch. Typical jail bar picture.

A preliminary contrast media opacification test should be performed prior to ethanol injection to check for vascular safety. The procedure is safe when the contrast medium is retained by the VH, whereas the procedure should not be performed in case of fast washout. Therefore, 2 mL of contrast medium are manually injected into the VH and a fast angiographic sequence of radiographs or a short helical CT-scan acquisition is carried out to analyze the washout of the VH (Fig. 4). In safe cases, the injected contrast remains in the medullary area. The presence of thin epidural veins is not a contraindication to the ethanol injection. Contraindications are the presence of a significant venous reflux with early opacification of paravertebral veins or suspicion of leakage into the spinal canal. Then 5 mL of absolute ethanol is injected slowly. Ethanol may be mixed with iodine contrast to check the distribution of the alcohol (Figs. 5 and 6). The presence of an anesthesiologist is not required but optimal. An anesthesiologist was present in half of the procedures we have performed.

AUTHOR'S EXPERIENCE

We have treated 28 cases of VH with percutaneous ethanol injections and have evaluated the results at a mean follow-up of two years (range one to seven years) (1). During the same period

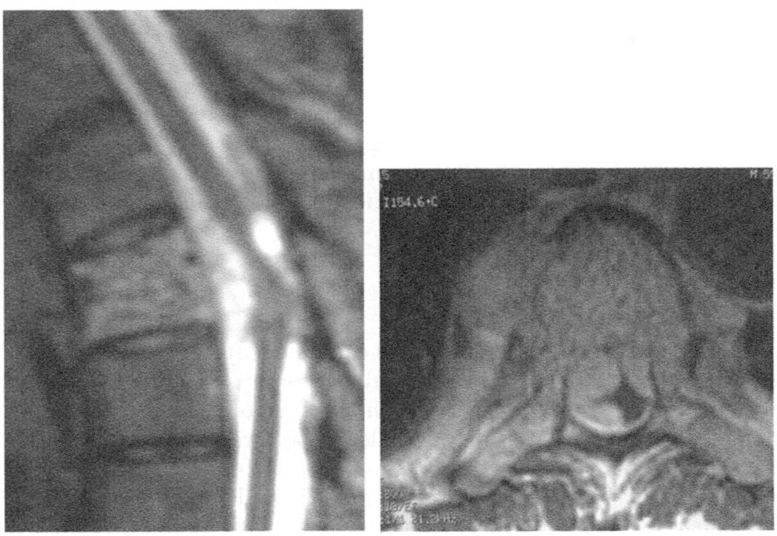

FIGURE 2 Symptomatic and complicated vertebral hemangioma, widening of pedicles, and presence of a soft-tissue mass into the spinal canal.

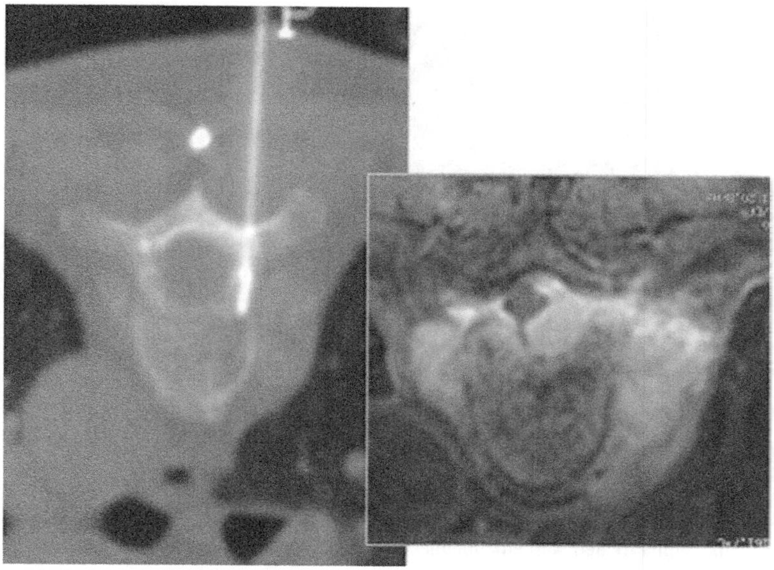

FIGURE 3 Control thoracic hemangioma. Transpedicular approach under computed tomography scan.

of time, ethanol injection was contraindicated in 10 cases because contrast medium was not retained by the VH during the preliminary opacification test.

Clinical Presentation and Radiologic Features

The most common presenting clinical feature was local pain (all patients) and neurologic symptoms resulting from spinal cord or nerve root compression (four patients). The clinical symptoms of the four patients with neurologic symptoms are described in Table 1. Twenty-four patients did not receive any kind of treatment before the ethanol injection, but one patient had

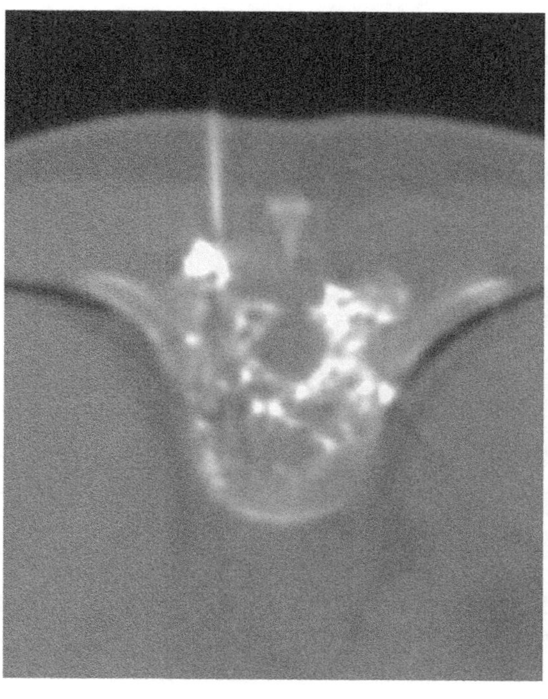

FIGURE 4 Computed tomography scan obtained after manual injection of contrast medium. The contrast is retained in the cancellous bone. There is no venous reflux.

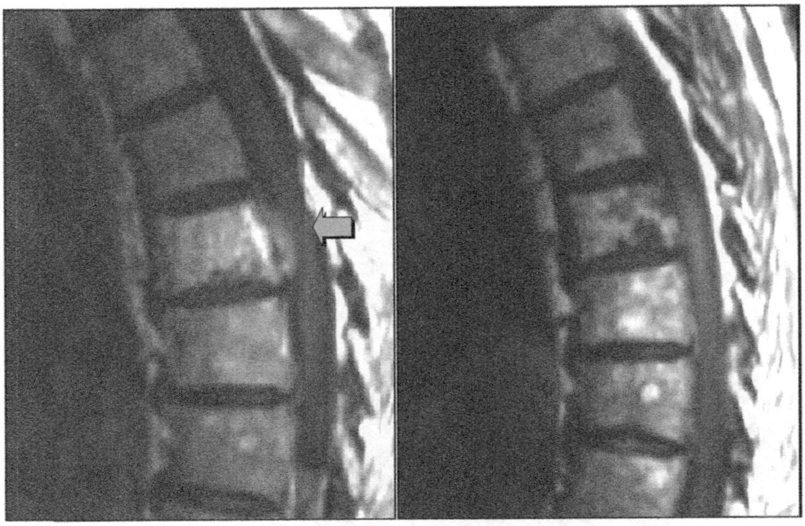

FIGURE 5 Complicated thoracic hemangioma treated with ethanol injection. Two injections were necessary to obtain sclerosis and decrease of the soft-tissue mass.

undergone surgery previously with an unsuccessful result. Three cases exhibited a vertebral compression fracture, and seven cases were complicated by a spinal cord compression before ethanol injection. In four of these seven cases, the spinal cord compression was due to a soft-tissue mass formed in the epidural space.

Complications

There were no neurologic complications. In two cases, an inadvertent puncture of the thecal sac occurred, with leakage of cerebrospinal fluid, requiring needle repositioning.

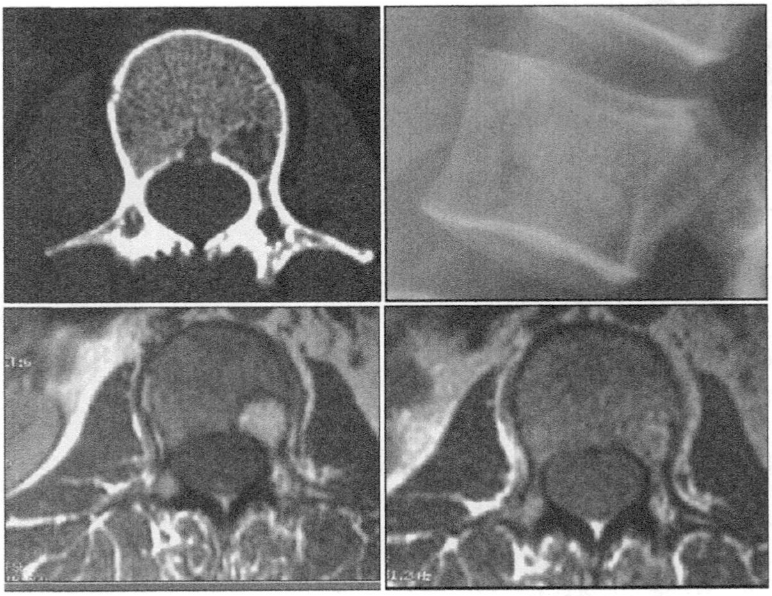

...initial... ...and three months post-ethanol

FIGURE 6 Noncomplicated hemangioma involving the pedicle. After treatment, the bone becomes nearly normal.

Results and Follow-Up

Only three cases had a residual active hemangioma after the first injection. In two cases, a second injection was performed, and in one case, a third injection was needed to obtain a complete obliteration. The final result was excellent in 26 out of the 28 patients. In one patient, pain relief was only 50%. The remaining patient, who had a vertebral fracture, experienced 75% pain relief. In this case, surgical stabilization of the fracture was done. After one ethanol injection, early relief of local pain was obtained in all patients; this effect was permanent in 20 of them. The clinical follow-up of the four patients with neurologic symptoms is described in Table 1.

At MRI, the intralesional signal-intensity on T1-weighted images decreased in 15 cases and increased in one case. No signal change was noted in 12 cases on T1-weighted and T2-weighted images. There was a decrease of the soft tissue mass in three cases. Thickening of the wall of the lesion was noted in one case. In three cases, a ring surrounding the lesion was observed, and in six cases a heterogeneous pattern was seen. We found no relationship between follow-up MRI findings and the clinical results.

DISCUSSION

Our experience in 28 VH cases treated with ethanol injection indicates that it is a safe and attractive method. It preserves the trabecular support and relieves pain, although the pattern of pain relief is variable. Ethanol injection can be repeated if a complete obliteration is not achieved after the first injection (Fig. 7).

There is one report of neurologic complications after ethanol injection into a VH, which is probably due to thrombosis of the epidural veins at the level of the injection, although leakage of alcohol into the subarachnoid space cannot be ruled out (14). Authors suggested that the use of a metrizamide powder to visualize the ethanol distribution on radiographic images may prevent this type of complication. To avoid neurologic complications, both the alcohol infusion rate and the VH washout should be taken into account. It is interesting to point out that in some cases, one injection is not enough to obtain complete obliteration of the VH. In our study, only three cases had a residual hemangioma after the first injection. Two cases had a second injection, and in one case a third injection was needed to obtain a complete obliteration of the hemangioma.

TABLE 1 Clinical Findings and Results of Ethanol Injection in Four Vertebral Hemangiomas with Neurologic Symptoms

Age, sex, and vertebral level	Preoperative neurologic status	Number of ethanol injections performed	Clinical follow-up after ethanol injection
49-year-old man, T12	Back pain	2	Back pain improved 2 mo after first injection
	Ataxic gait, bilateral hyper-reflexy of the lower limbs, and normal sphincters tone, motor function, and sensitivity		Improvement of ataxic gait 8 mo after the second injection
66-year-old man, T8	Mild proximal weakness of the lower limbs	1 (bipedicular)	Motor deficit resolved 6 wk after the injection
22-year-old man, T8	Right leg pain Bilateral T2 nerve root pain worsened by coughing and activity	1	Disappearance of the right leg level
77-year-old man, T7	Back pain 2	2	Improvement one wk after the first injection
	Paresthesias and weakness and slight hyper-reflexy of the lower limbs		Normal neurologic status 3 mo after the second injection

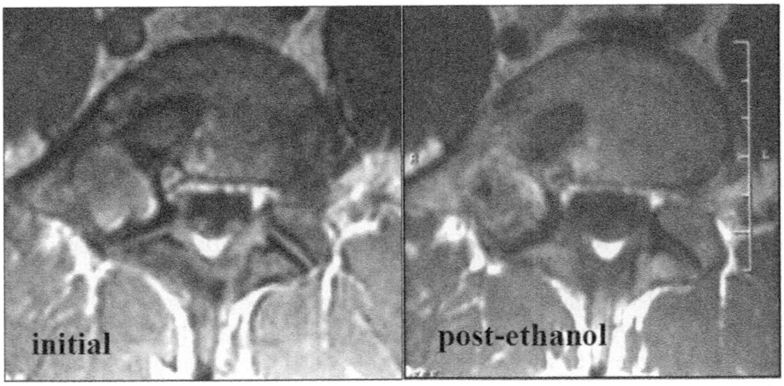

FIGURE 7 Lumbar hemangioma treated with alcohol injection. Seven years later, the lesion was completely repaired.

KEY POINTS

- Both symptomatic and asymptomatic VH can be treated with alcohol injection.
- A fast washout of the VH during a preliminary opacification test is a contraindication to the method.
- The procedure can be repeated, if needed.

REFERENCES

1. Bas T, Aparisi F, Bas JL. Efficacy and safety of ethanol injections in 18 cases of vertebral hemangioma: a mean follow-up of 2 years. Spine 2001 15; 26:1577–1582.
2. Cotten A, Boutry N, Cortet B, et al. Percutaneous vertebroplasty: state of the art. Radiographics 1998; 18:311–320.
3. Deramond H, Damasson R, Galibert P. Percutaneous vertebroplasty with acrylic cement in the treatment of aggressive spinal angiomas. Rachis 1989; 1:143–153.
4. Ellman B, Parkhill B, Curry T, et al. Ablation of renal tumors with absolute ethanol: a new technique. Radiology 1981; 141:619–626.
5. Fox MW, Onofrio BM. The natural history and management of symptomatic and asymptomatic hemangiomas. J Neurosurg 1993; 78:36–45.
6. Galibert P, Deramond H. Percutaneous acrylic vertebroplasty as a treatment of vertebral angioma as well as painful and debilitating diseases. Chirurgie 1990; 116:326–334.
7. Guedea F, Majo J, Guardia E, et al. The role of radiation therapy in vertebral hemangiomas without neurologic signs. Int Orthop 1994; 118:77–79.
8. Heiss J, Doppman JL, Oldfield EN. Brief report: relief of spinal cord compression from vertebral hemangioma by intralesional injection of absolute ethanol. N Engl J Med 1994; 25:508–511.
9. Ide C, Gangi A, Rimelin A, et al. Vertebral hemangiomas with spinal cord compression: the place of preoperative percutaneous vertebroplasty with methyl methacrylate. Neuroradiology 1996; 38:585–589.
10. Lapras C, Mottolese C, Deruty R, et al. Injection percutaneé de méthyl-methacrylate dans le traitement de l'ostéoporose et de l'ostéolyse vertébrale grave (Technique de P. Galibert). Ann Chir 1989; 43:371–376.
11. Laredo JD, Assouline E, Gelbert F, et al. Vertebral hemangiomas: fat content as a sign of aggressiveness. Radiology 1990; 177:467–472.
12. Laredo JD, Reizine D, Bard M, Merland J. Vertebral hemangiomas: radiologic evaluation. Radiology 1986; 161:183–189.
13. Nicolini U, Zuliani G, Caravelli E, et al. Alcohol injection: a new method for treating placental chorioangiomas. Lancet 1999; 353:1674–1675.
14. Niemeyer T, McClellan J, Webb J, Jaspan T. Brown-Sequard syndrome after management of vertebral hemangioma with intralesional alcohol: a case report. Spine 1999; 24:1845–1847.
15. Smith TP, Koci T, Mehringer CM. Transarterial embolization of vertebral hemangioma. J Vasc Interv Radiol 1993; 4:681–685.
16. Yakes WF, Parker SH, Gibson MD, et al. Symptomatic vascular malformations: ethanol embolotherapy. Radiology 1989; 170:1059–1066.

Index